Economics: A Mathematical Introduct

Economics:

A Mathematical

Introduction

C. J. McKenna and Ray Rees

OXFORD UNIVERSITY PRESS

Oxford University Press, Walton Street, Oxford OX2 6DP

Oxford New York Toronto
Delhi Bombay Calcutta Madras Karachi
Kuala Lumpur Singapore Hong Kong Tokyo
Nairobi Dar es Salaam Cape Town
Melbourne Auckland Madrid
and associated companies in
Berlin Ibadan

Oxford is a trade mark of Oxford University Press

Published in the United States
by Oxford University Press Inc., New York

British Library Cataloguing in Publication Data
Data available
ISBN 0–19–877292–0
ISBN 0–19–877291–2 (pbk)

Library of Congress Cataloging in Publication Data
Data available
ISBN 0–19–877292–0
ISBN 0–19–877291–2 (pbk)

Printed in Great Britain by
Biddles Ltd, Guildford, King's Lynn

For James and Kate
Zac and Dan

Preface

Many students embarking upon courses in introductory or intermediate economics have already completed calculus courses at school, college, or university, but then have to sit through verbal and diagrammatic expositions of marginal concepts and solutions to optimization problems which are (not always adequate) translations from the calculus. This may not only bore them and waste their time: it often leaves them confused, when they would find an explicitly mathematical treatment enlightening. The first aim we had in writing this book therefore was to provide a reference text for a course in microeconomic and macroeconomic principles designed for such students.

The book is not intended to be a text on 'mathematics for economists'. Throughout, we use the relevant mathematics in a straightforward, matter of fact way without explanation of it *as such*. Appendix I, however, briefly reviews all the mathematics we use in the book, and serves to define its mathematical prerequisites. At the same time we do see the book as suitable for use as a supplementary text on courses in mathematics for economists. Because of space constraints, the several excellent textbooks in this area are limited in the economic applications they can cover. We hope that teachers of these courses will find that many chapters in both microeconomic and macroeconomic sections of this book are good applications of the mathematics as well as being interesting and challenging ways of deepening the student's understanding of economics. The summaries at the ends of the chapters will give the instructor an idea of the book's level and coverage of economics.

We would like to thank Venk Sadanand for his comments, Lois Lamble for typing the book so admirably, and John Sheriff for preparing answers to the problems.

C. J. McKenna
R. Rees

February 1991

Contents

List of Figures

List of Figures

List of Tables

1   Introduction

It is now established beyond doubt that economics is a 'mathematical' subject. Though it is perfectly possible to learn economics in a non-mathematical way, one is always then reading a translation rather than the original and that is usually second best. The reason economics is mathematical is not that economists want to seem 'scientific', and lust after the intellectual respectability of subjects like physics, but because of its innate nature. It is concerned with quite long chains of reasoning about the interactions among numerical variables in complex systems, and mathematics is the most powerful and effective language we have for doing this.

At the same time, economics is not *only* mathematics. Although it may not seem so to the student, the purely mathematical part of an economic analysis may often be the easiest (or assumptions can be made to make it so!). The hard parts come before and after the mathematics: the formulation of the model to be analysed, and the interpretation of the results. The criterion to be applied to any piece of mathematical economics is: how important are the economic issues it tries to deal with and by how much does the analysis in the end contribute to our understanding of them? The use of mathematics in economics has greatly expanded the set of issues we are able to analyse systematically, has led to much clearer and more sharply defined conclusions, and has greatly improved the quality of reasoning in economics.

The preceding paragraph introduced the term 'model'. In fact economics can be thought of as a collection of models. The process of *doing*, as opposed to talking about, economics, is entirely one of formulating, analysing, and interpreting models. Accordingly, the remainder of this introductory chapter is concerned with a discussion of what models are and how they are made.

1.1 Economic models

The process of economic modelling consists of the following steps:
1. defining the variables;
2. specifying functions, identities, and equations which show how the variables are related;
3. choosing a solution principle;
4. solving the model;
5. interpreting the solution;
6. analysing perturbations of the solution.

You will see this process being repeated time after time in this book, although we will not usually draw attention to the particular step we are taking as we take it. However, if you want to begin to set up your own models, as we very much hope you do, as well as to understand fully those which already exist, you should try to identify these steps explicitly as you read through the book. We now say a little more about each of them.

1.1.1 The variables

Variables are thought of as numerical quantities. For example prices, incomes, the quantities of goods people buy, the number of unemployed workers in an economy are all represented as real numbers. For purposes of mathematical analysis their numerical values are represented by letters p, y, q, u, just as in algebra.

We distinguish between *endogenous* and *exogenous* variables. Endogenous variables are those the model is trying to explain, their values are to be determined within the model. Exogenous variables are those whose values are determined outside the model, and so are taken as given when analysing the model. The possibility that these values might change, in some way which is not explained within the model, is what gives rise to step 6, the analysis of perturbations in the model's solution.

The question of what variables to include in a model, which to take as endogenous, which as exogenous, is entirely determined by the issues to be analysed, and much of the art of modelling lies in giving a good answer to this question. For example, if we were trying to explain what had been happening in the housing market in a small town, we would regard house prices, the size of the housing stock, the numbers of houses bought and sold, rents in the private and public rental sectors, and vacancy rates as the key endogenous variables; consumers' incomes, population size, interest rates, and mortgage availability as key exogenous variables. But some at least of these exogenous variables, for example interest rates and mortgage availability, would become endogenous in a model of the national housing market. Also, one could argue about population size: in the short run it could be taken as exogenous, but in the longer run, the population of a town may well be influenced by its house prices.

There is always an element of relativism in model-building: explanation of some things will have to wait until later – what seems the best way to proceed to get a good answer to the problem immediately at hand? Trying to explain *everything* at the very outset will usually lead precisely nowhere. From the point of view of the economic system as a whole, there are very few truly exogenous variables, particularly if we take a long-term perspective – even the weather, via the effect of industrial emissions and timber production (deforestation), may be endogenous. But most progress seems to be made by setting up models which take only small groups of variables as endogenous at any one time.

1.1.2 Functions, identities, and equations

A function is a statement of a relationship among variables. In setting up a model, we have to specify how the value of one variable depends on the values taken by one or more others, and the standard mathematical notation is used. Often a function will be denoted by a letter which suggests the name of a relationship. For example, a demand function will be denoted by $D(.)$, a supply function $S(.)$, a consumption function $C(.)$, and so on.

In economic models, functions are usually specified in a *qualitative* way. They are usually assumed to be differentiable, and we specify only the signs of the first and perhaps the second derivatives, that is, the general shapes of the functions. This contrasts to, say, physics, where experimentation will have suggested specific numerical functions or formulae. The economic world is not as regular and determinate as the physical one. In economics, specific functions are often chosen simply to illustrate more general relationships and more for their mathematical convenience than for their conformity with measurements in the 'real world'. For example, the 'Cobb-Douglas function' $y = x_1^a x_2^b$, when used as a utility function in the theory of consumer demand (see Chapters 6, 7), has the unfortunate implication that the amount of a good a consumer wishes to buy does not depend on the price of any other good, just its own. This is highly counter-factual, but nevertheless this function is much used because it is convenient as an illustration of some important points.

An identity is a definitional statement that necessarily holds true for all values of the variables it contains. In economics, identities are usually *accounting definitions*: they describe how one value magnitude is made up of others. For example the famous $Y \equiv C + I$ says that the value of the entire output of goods and services in the economy, Y, is made up of the value of those goods which are used for current consumption, C, and the value of those used for investment, I. As we have done here, such an identity is usually written with a three-bar equality sign, to emphasize that it is true by definition, for all possible values of the variables.

Equations are statements which hold only for some values of the variables, and not for others. Thus, the statement $5x = 35 - 2x$ is true only if $x = 5$ and not otherwise. Unlike an identity, x can be thought of as taking values for which the equation is not satisfied.

An important difference between an equation and an identity, which is sometimes blurred in economics, is that it is always legitimate to say that the differentials or derivatives of both sides of an identity are equal, but it is not legitimate to say that the differentials or derivatives of both sides of an equation are equal, unless we have some way of proving that they are. For example, suppose that Y, C, and I are all functions of time, t. Then, since $Y(t) \equiv C(t) + I(t)$ is true for all t by definition, we must have $dY/dt \equiv dC/dt + dI/dt$. On the other hand, we might have a value of x such that $f(x) = g(x)$ as an equation, but it will not be true in general that $f'(x) = g'(x)$. For example $f'(x)$ might be positive, $g'(x)$ negative, as in the above example, where $f'(x) = 5$, $g'(x) = -2$. Only if we have some reason for asserting that x changes in such a way as to maintain the equality of the functions, can we equate the derivatives.

The functions, identities, and equations of a model constitute its structure. They determine how the variables interact. Our specification of them essentially represents our perception of how the particular part of the economy we are modelling works, of what determines or influences what, and how.

1.1.3 Solution principle

By this we mean a rule we apply to the functions, equations, and identities in a model to produce a solution, which is a statement about the values of the endogenous variables determined by the model. There are two types of solution principle in economics, both of which are widely used in this book:

(i) *the equilibrium principle*: solution values of the endogenous variables are those which satisfy a particular set of simultaneous equations. These equations define a state of rest or balance in the model: if they are not satisfied values of the variables will tend to change, while the values of the variables which do satisfy the equations will not change as long as the equations themselves stay the same.

The key aspect of the modelling process here is to define the equilibrium conditions – the set of equations that have to be satisfied – and also to make sure that the model has been specified in such a way that a solution *does* exist – that there *are* values of the endogenous variables which satisfy the equations. Usually, for example, it will be necessary to ensure that there are as many independent, consistent equilibrium conditions (equations) as there are endogenous variables (unknowns) but this alone need not be sufficient to guarantee solution. It can be quite a difficult mathematical problem to confirm that a particular model has an equilibrium solution. At the level of this book, however, it is usual not to worry about this problem and to assume the models we set up do have solutions. We shall certainly make this assumption.

(ii) *the optimization principle*: many models are solved by maximizing or minimizing some function with respect to the endogenous variables, often subject to some side conditions on the variables known as constraints. The result of the optimization is a set of equations which can then be solved (at least in principle) for the values of the endogenous variables. The key modelling step here is to formulate the optimization problem appropriately and to ensure that a solution to it does exist and is correctly characterized by the optimality conditions. Again, we shall usually assume that this is the case.

Choice of which solution principle to use will usually be obvious from the nature of the problem we are analysing. When it is a problem of individual decision-taking – a consumer, a firm, or a policy maker has to make some kind of choice – then the solution principle will almost invariably be optimization. On the other hand, when we are dealing with impersonal interacting forces, as in a market, or in the economy as a whole, then we shall need the equilibrium principle. This explains, incidentally, why there is virtually no application of the optimization principle in the macroeconomics section of the book.

Debate has raged over the years as to whether people and organizations 'really optimize' when they take decisions. Certainly, you will not see people in a supermarket writing down the constrained maximization problems that we solve when we analyse their choices (although you do see people using pocket calculators).

At the same time, if you were to ask someone at the check-out counter, if the basket of goods they were buying was the one they preferred out of all those they could afford, the answer, if printable, would surely be yes. But that is all the optimization principle is essentially saying. We do not require people to go through the same *processes* of analysis as we do when we analyse models (though it is noteworthy that many large firms use methods of operations research, which directly embody optimization principles, in taking their decisions), it is enough if they are systematically trying to do the best they can – that is all optimization really means. If so, we should then be making reasonably accurate statements about the *outcomes* of their decisions, whatever processes they use to make them. It should also be pointed out that no solution principle with the effectiveness, power, and simplicity of the optimization principle has yet been proposed, which can capture the notion that people do not optimize, but simply 'behave'.

1.1.4 The solution

Applying a solution principle to the model will, if it has been properly formulated, result in a solution. We can, however, mean one of two things by the term 'solution'. If a model has been set up in terms of specific functions and equations with numerical parameters and if numerical values of the exogenous variables have been assigned, then a solution of the model will be a set of numbers, one numerical value for each endogenous variable.

However, this kind of solution is typically only found in examples and exercises designed to reinforce understanding of the theory. If the model has been set up in terms of general functions, with only qualitative restrictions on their shapes, then we cannot actually *obtain* a solution, we can only *characterize* it. By this is meant that we can only make statements about the conditions the solution has to satisfy. For example, in a general model of a market for a good, we cannot find the numerical value of the equilibrium price, but we can characterize or describe it as the price which equates demand and supply. Similarly, in a general model of a firm's profit-maximizing output choice, we cannot find a numerical value for output, but we can characterize it by the condition that it must equate marginal revenue to marginal cost. So, we usually feel we have solved a model when we have arrived at a set of conditions which the actual solution values of the variables, whatever they may be, must satisfy.

The reason for this approach is of course the desire for generality. We do not want to draw conclusions that apply to just one market, or just one firm, but markets in general or firms in general. The general solution of a model is something which, in principle, could be applied to any one in the set of economic entities – markets, firms, consumers, entire economies – it is designed to analyse, to give a correct solution once the functions and numerical parameters specific to that entity are plugged in.

1.1.5 Interpretation

Once the model's solution has been derived, or characterized, we need to see what it implies for the economic problem with which we started. Of course, in a logical sense, the formulation of the model, discussed here as steps 1–3, completely determines its outcome, and it is well known in many types of model that particular choices of formulation will have specific implications for the solution – thus we can have 'monetarist' and 'Keynesian' macroeconomic models, for example, in which choice of the shapes of some functions can have profoundly different implications about the importance of fiscal policy in an economy.

It is also interesting to observe that economists can sometimes be surprised by what 'emerges out of the mathematics' of their models, and this happens more frequently, the more truly innovative the models are. In any case, the analysis of a model is incomplete until the mathematical conditions characterizing the solution have been translated into economic terms, and the implications of the results for the economic problem at hand drawn out as fully as possible.

1.1.6 Perturbation of the solution

The solution of a model will generally depend on the values of those variables which have at the outset been assumed exogenous. For example, as we shall see in the next chapter, in analysing the determinants of price and quantity traded in a market we assume consumers' incomes are held fixed, and the solution values for price and quantity can be expected to depend on the exogenous values of incomes. It is then often of interest to examine how the solution values of the endogenous variables change when the exogenous variables change, and indeed this may have been one of the main aims of setting up the model in the first place. This stage of the analysis is usually called 'comparative static equilibrium analysis', or just 'comparative statics', because we usually evaluate the effects of the changes in exogenous variables by comparing the different static equilibrium positions to which they give rise.

This brief outline of the steps involved in economic modelling is intended to help you see the framework which underlies all the models in the rest of this book. We hope it will also help you to start constructing models of your own.

Exercise 1.1

1. A farmer grows two crops on a fixed amount of land. He has to decide what acreage of each crop to grow. Construct a model which will predict his decision, specifying carefully: the variables, which are endogenous, which exogenous; the functions and equations you need; the solution principle. Note: your model does not *have* to conform to any standard economic model you may know about. Discuss what your model tells you about the decisions that farmers take.

1.2 Summary

Economics is essentially a mathematical subject, and the subject has gained a great deal from the application of mathematics to economic analysis.

Economics is a collection of models. 'Doing economics' consists of formulating, analysing, and interpreting models.

Modelling involves six steps:

- specifying variables and deciding which are endogenous and which exogenous;
- specifying functions, identities, and equations among the variables, which capture the essence of the problem you want to analyse;
- choosing and applying a solution principle;
- characterizing the solution;
- interpreting the solution and what it implies for the problem you started with;
- carrying out a comparative statics analysis.

Look out for these steps in the models we set out in the rest of the book, and try as soon as possible to start building your own.

PART I: MICROECONOMICS

Any economic system has to solve the problem of allocating scarce resources among alternative uses. Microeconomics is primarily concerned with the study of how a market economy solves this problem. As consumers, individuals in the economy decide how much of the various available goods and services to buy; and as owners of the inputs, labour, land, and capital, they decide how much of these to supply to firms which want to use them in production. Firms decide how much of the inputs to hire or buy, and how much output to produce. These separate decisions have to be coordinated and made consistent, and the main aim of microeconomics is to achieve an understanding of how, and how efficiently, markets do this.

In Chapters 2 to 5 we examine the central concepts and questions concerning the functioning of a single competitive market or a small system of interrelated markets. The central concept is that of the equilibrium of demand and supply. Chapters 6 and 7 then examine the foundations of the theory of market demand by analysing in some depth a model of consumer's buying decisions. Chapters 8 and 9 develop the theory of market supply from a model of the firm's profit maximizing production and selling decisions. It must however be recognized that many markets are not in fact perfectly competitive, and Chapters 10 and 11 examine models of markets which are dominated by a single seller (monopoly) or a small number of sellers (oligopoly). The particular features of the markets for labour and capital are analysed in Chapters 12 and 13. Chapter 14, on general equilibrium, then returns to the central issue: an economy consists of a system of interrelated markets whose simultaneous equilibrium determines a resource allocation and we obtain an understanding of how a competitive market economy solves the problems of resource allocation by means of a formal model of this equilibrium. Finally, Chapter 15 on Welfare Economics, considers the question of the efficiency of the market mechanism, and then applies the concepts of microeconomics to analyse a number of important problems of economic policy such as pollution, taxation, import tariffs, and the use of scarce natural resources.

2   Market Analysis

Markets exist in a wide variety of forms. First, they may differ in geographical extent. Contrast, for example, the market in foreign currencies, which is virtually world-wide, to a street market dealing in fruit and vegetables. Secondly, the goods and services traded on them can be of widely differing types, from forms of labour services, through durable goods services such as housing, to tangible commodities like steel and coal. Finally, they may be highly organized, with sophisticated mechanisms for bringing buyers and sellers together, as in the markets for stocks and shares, houses, and package holidays, or they may be quite disorganized, leaving buyers to find sellers as best they can.

From the great diversity of market forms that we observe in reality, we want to abstract and model the essential elements. These are, on the one side, a set of buyers wishing to exchange money for a good or service, and on the other side, a set of sellers wishing to exchange the good or service for money. The interaction between buyers and sellers results in a market price, or rate of exchange between money and the good in question, and a total quantity traded. We want to construct a model to explain this market outcome, and to predict how it will change in response to changes in underlying market conditions.

2.1 Demand, supply, and equilibrium

The view of the market as an arena in which buyers and sellers trade suggests the method we should use to construct a model. We group together all the variables that determine buyers' demands, on the one hand, and all the variables which determine sellers' supply, on the other. The market outcome — a price and a quantity traded — is then determined by the interaction of these two sets of variables.

A key distinction we must make is between *desired* or *ex ante*, and *realised* or *ex post*, quantities. A buyer may go into a market with a quantity in mind that he or she wishes to buy, and this may well differ from the quantity he or she ends up in fact buying. Likewise a seller may actually sell an amount that differs from the quantity he or she wants to sell. Clearly, actual sales must equal actual purchases, but desired sales and purchases may well differ. It seems reasonable to assert that what drives the activity in a market are the desires and wishes of buyers and sellers, while actual quantities traded are the outcome. So, in constructing a market model, we have first to specify what determines *desired* purchases or sales, and then explain how these interact to determine a quantity actually bought and sold, and a market price.

The central aspect of this explanation is the *equilibrium method*. An equilibrium

is a state of rest or balance: interacting forces are such that there is no tendency for
the variables they determine to change. We proceed by setting out a condition for
the market to be in an equilibrium and then find the values of price and quantity
traded, if they exist, that satisfy this condition. This we take to be the market
outcome we are seeking to explain.

We denote demand, the amount buyers wish to buy, as q_D. This should be
thought of as a non-negative number of units of a particular good or service *per
unit of time*, e.g., so many pounds of bananas per hour, or so many taxicab rides per
day. The time dimension of quantities is important. For example, the statement:
'one million taxicab rides were taken in Central London' is meaningless unless
the period of time to which this relates is specified.

In a similar way we denote supply, the amount sellers wish to sell, by q_S.
Again, it is usually a non-negative number, and must be measured in the same
units, including the time dimension, as demand.

Finally, we denote the amount actually traded by q (how should it be measured?).

We express the idea that quantity demanded is determined by a set of other
variables by using the mathematical idea of a function: q_D is a function of some
set of variables. We have to specify this function. For virtually all goods, the
important determining variables are:

1. the good's own price, p;
2. prices of other goods, $p_1, \ldots, p_n$, complements or substitutes to the good in
   question;
3. aggregate income of the good's consumers, y.

However, for specific goods other variables may well also be important. For
example, the demand for houses will also depend on the mortgage interest rate
and the availability of mortgages; the demand for ice-cream will depend on the
weather, and so on.

We write

$$q_D = D(p, p_1, \ldots, p_n, y) \tag{2.1}$$

as the *demand function* for the good. We do not wish to exclude the possibility of
other variables being relevant by this notation, but here we simply concentrate on
those variables relevant to goods in general. Note also that factors such as needs,
tastes, and preferences are not treated as variables in this function but, as we shall
see in Chapter 6, determine the form of the function itself.

Likewise, we have to specify the variables which enter into the *supply function*.
The most important of these are:

1. the good's own price, p;
2. prices of inputs used in producing the good, $(w_1, \ldots, w_m)$;

so that we write the supply function as

$$q_S = S(p, w_1, \ldots, w_m). \tag{2.2}$$

Again, this is not to exclude the possibility that for specific goods other variables
might be important in determining supply. For example, agricultural supply is very

much affected by weather conditions. The general form of the supply function is determined by the technological characteristics of production of the good, as we shall see in Chapter 8.

In analysing how demand and supply interact in a market, we first assume that all the variables that are not determined on that market, i.e. that are *exogenous* to the market, take fixed values — only the *endogenous* variables p, q_D, q_S, and q are free to vary. The equilibrium of the market is then given by the condition

$$q_S = q = q_D. \tag{2.3}$$

Using (2.1) and (2.2) we can re-write this condition as

$$D(p) = S(p) \tag{2.4}$$

where, because the exogenous variables are assumed fixed, they are suppressed.

We can regard (2.4) as determining the market equilibrium price and hence demand, supply, and quantity traded. Mathematically, solving for this price looks like a trivial problem, with just one equation and one unknown. To explore this, let us assume we have the specific case of linear D and S functions, with

$$D(p) = a + bp; \qquad S(p) = \alpha + \beta p \tag{2.5}$$

(we can think of the effect of the exogenous variables as being captured in the constant terms a and α respectively). Then, using (2.4), we have the equilibrium price and quantities in the market defined by these demand and supply functions as:

$$p^* = \frac{a - \alpha}{\beta - b}; \quad q^* = q_D^* = q_S^* = \alpha + \beta \left(\frac{a - \alpha}{\beta - b} \right) = \frac{a\beta - \alpha b}{\beta - b}. \tag{2.6}$$

These equations give us the usual solution for a market equilibrium illustrated in Figure 2.1. Or do they?

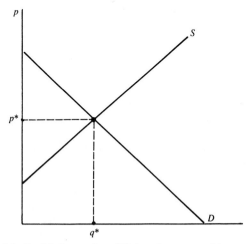

Fig. 2.1: Straightforward equilibrium in a competitive market

Recall that we cannot have negative prices and quantities, and so p^* and q^* cannot be negative. Now look at (2.6) simply as equations, without any preconceptions created by diagrammatic economics you might know. Then we immediately note some possibilities not shown in Fig. 2.1.

(i)   Suppose $b = \beta$. Then p^* and q^* in (2.6) are not defined, since we cannot divide by zero. So, no market equilibrium exists in this case.

(ii)   Suppose $b < 0$ and $\beta > 0$ (as is implicit in Fig. 2.1), so the denominator in (2.6) is certainly positive, but that $\alpha > a > 0$. Then, from (2.6) $p^* < 0$ which is not permissible. On the other hand $q^* > 0$.

(iii) Suppose again that $b < 0$ and $\beta > 0$, and that $a > 0 > \alpha$, so that $p^* > 0$, but that $\alpha b > a \beta$. Then $q^* < 0$, which is also not permissible.

Do we learn anything from these 'awkward cases'? Fig. 2.2 suggests we can.

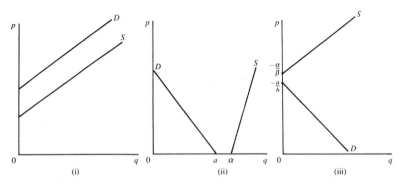

Fig. 2.2: Other cases

In (i) we have parallel demand and supply curves, i.e. $b = \beta$, so that no equilibrium exists. The fact that supply and demand curves are normally assumed to have the slopes shown in Fig. 2.1, with $b < 0$ and $\beta > 0$, suggests we should not lose too much sleep about this case, though we shall find later perfectly good theoretical reasons for the reverse, i.e. for positively sloped demand curves and negatively sloped supply curves. The general point, however, is that if we wish to use a mathematical model to analyse equilibrium, we must be sure to provide for the existence of an equilibrium.

In (ii) we have $\alpha > a$: at every positive price supply exceeds demand, and so the only possible 'equilibrium' in the positive quadrant is at $p = 0$. In that case $q = q_D < q_S$: the good is in excess supply even at a zero price. This is an example of a *free good* — sand in the desert, the air we breathe, birdsong. The interest in the example is that it suggests that the 'freeness' of goods is entirely a matter of demand and supply. If demand should grow and supply shrink, then price could become positive — the good would become *relatively scarce*.

In (iii) we have a case in which at every positive quantity the supply curve is above the demand curve. We could therefore regard the equilibrium as at $q^* = 0$: the market is inactive, and so no price is observable. This is a good for which buyers are not prepared to pay a price which sellers would require to make it available. This again is clearly a matter of demand and supply — for example if demand should grow sufficiently the market could become active.

Exercise 2.1

1. Suppose that in a market at a particular point in time the amount buyers want to buy differs from the amount sellers want to sell. How much of the good and service do you think would actually be traded?

2. Which of the following conditions do you think would hold at a market equilibrium, and which would hold out of equilibrium? Could some do both? Give reasons for your answer.

   (i) The amount buyers want to buy exceeds the amount traded.
   (ii) The amount sellers want to sell equals the amount traded.
   (iii) The amount buyers want to buy is less than the amount sellers want to sell and equals the amount traded.
   (iv) The amount sellers want to sell and the amount buyers want to buy both equal the amount traded.

3. The definition of a market given here is very abstract. Actual markets exist in a great variety of types and scales. Compare what you think would be the features of the following markets that economists would regard as being most important:

   (i) the stock market;
   (ii) a street market dealing in fruit and vegetables;
   (iii) the market in foreign currencies;
   (iv) the market for accountancy services;
   (v) the market for coal;
   (vi) the market for unskilled manual labour in a particular town.

4. What problems might arise in assuming that the market outcome we are seeking to explain is an equilibrium outcome?

5. Given the demand and supply functions

   (a) $q_D = 200 - 2p$; $q_S = 100 + p$,
   (b) $q_D = 3p^{-1/2}$; $q_S = p$,
   (c) $q_D = 10 + p$; $q_S = 20 + p/2$,
   (d) $q_D = 20 - p^2$; $q_S = 12 + p$,

   find equilibrium price and quantity in each case. Sketch the supply and demand curves.

6. Given a demand function $q_D = 15 - 2p$, and a supply function

$$q_S = 0 \quad \text{for} \quad p < 5,$$

$$q_S = 10 + p \quad \text{for} \quad p \geq 5,$$

analyse the problem of finding an equilibrium (hint: draw a graph). Can you give a plausible economic explanation for this supply function?

7. Show, by drawing diagrams analogous to Fig. 2.2, that the 'awkward cases' discussed in this section are not simply a consequence of the linearity of the supply and demand functions.

8. Can you give examples of
   (i)   goods which may once have been free goods but which now have positive prices;
   (ii) goods which may once have had inactive markets but which now have active ones?

2.2 Changes in equilibrium

In the linear example of the previous section it is clear that the equilibrium solution depends on the parameters α and a, which we suggested could be taken as summarizing the influence of the exogenous variables. If therefore the value of at least one of these were to change, so would the equilibrium position. More generally, we could write the equilibrium condition as

$$D(p, p_1, \ldots, p_n, y) = S(p, w_1, \ldots, w_m). \tag{2.7}$$

Then, if the values $p_1, \ldots, p_n, y$, and $w_1, \ldots, w_m$ are all fixed and known, we can solve for equilibrium p^*, q^*, q_D^*, and q_S^* just as before (ignoring awkward cases). The question we now want to explore is this: what is the effect on the equilibrium solution, of a change in value of an exogenous variable? In particular, we want to see how equilibrium price and quantity are affected by

(i)   a change in consumers' income y;
(ii) a change in the price of a complement, say p_1;
(iii) a change in the price of a substitute, say p_2;
(iv) a change in the price of an input, say w_1.

Mathematically, we interpret this question as the problem of evaluating the signs of the derivatives dp^*/dp_1, dp^*/dp_2, dp^*/dy, dp^*/dw_1 and the corresponding quantity derivatives dq^*/dp_1, ... etc.

The idea underlying the procedure is the following. We can rewrite the equilibrium condition (2.7) as

$$D(p^*, p_1, p_2, y) - S(p^*, w_1) = 0 \tag{2.8}$$

(where we have again suppressed exogenous variables that remain constant). The left-hand side of (2.8) can be regarded as an implicit function in the variables p^*, p_1, p_2, y, and w_1. When we change one variable, say y, we allow p^* to change, while requiring the equilibrium condition still to hold. Thus given the differential dy, we find the differential dp^* which maintains the equality in (2.8). The ratio dp^*/dy can then be taken as giving us the derivative we seek. We will have found the change in p^* required to maintain market equilibrium given the change in y, and thus we will have identified the change in equilibrium price caused by a change in y. Let us see how the procedure works out mathematically.

Totally differentiating through (2.8) gives:

$$\frac{\partial D}{\partial p}dp^* + \frac{\partial D}{\partial p_1}dp_1 + \frac{\partial D}{\partial p_2}dp_2 + \frac{\partial D}{\partial y}dy - \frac{\partial S}{\partial p}dp^* - \frac{\partial S}{\partial w_1}dw_1 = 0. \qquad (2.9)$$

Note that the partial derivatives in (2.9), $\partial D/\partial p$, $\partial D/\partial p_1$, etc., are *specific numbers*, *not* general expressions — they are the derivatives evaluated at a specific point, (p^*, q_D^*) or (p^*, q_S^*).

Now we set $dp_1 = dp_2 = dw_1 = 0$, since we wish to concentrate on the effect of a change in y. In that case (2.9) becomes

$$\left(\frac{\partial D}{\partial p} - \frac{\partial S}{\partial p}\right)dp^* + \frac{\partial D}{\partial y}dy = 0 \qquad (2.10)$$

and so

$$\frac{dp^*}{dy} = \frac{\partial D/\partial y}{(\partial S/\partial p - \partial D/\partial p)}. \qquad (2.11)$$

We have to evaluate the sign of this expression. Assume first that we have the well-behaved case in which $\partial D/\partial p < 0$ and $\partial S/\partial p > 0$, i.e. demand curves slope downward and supply curves upward. Then the denominator of (2.11) is certainly positive. Hence we can say

(i) $dp^*/dy > 0$, an increase in income increases equilibrium price, if $\partial D/\partial y > 0$, i.e. the good is a normal good.

(ii) $dp^*/dy < 0$, an increase reduces equilibrium price, if $\partial D/\partial y < 0$, i.e. the good is an inferior good.

In the same way, we can obtain

$$\frac{dp^*}{dp_1} = \frac{\partial D/\partial p_1}{(\partial S/\partial p - \partial D/\partial p)}, \qquad (2.12)$$

$$\frac{dp^*}{dp_2} = \frac{\partial D/\partial p_2}{(\partial S/\partial p - \partial D/\partial p)}, \qquad (2.13)$$

$$\frac{dp^*}{dw_1} = \frac{-\partial S/\partial w_1}{(\partial S/\partial p - \partial D/\partial p)}. \qquad (2.14)$$

In each case the required sign of the derivative is determined by the sign of the numerator on the right-hand side. If p_1 is the price of a complement to the good in question then $\partial D/\partial p_1 < 0$. Therefore (2.12) tells us that an increase in the price of a complement would reduce the equilibrium price of the good in question. If p_2 is the price of a substitute then $\partial D/\partial p_2 > 0$ (explain) and (2.13) tells us that a fall in the price of a substitute would also reduce the equilibrium price of a good. Finally, for reasons which will be made precise in Chapter 9, a rise in an input price reduces supply of a good, or $\partial S/\partial w_1 < 0$, so (2.14) tells us that an increase in an input price would raise the equilibrium price of the good.

The effects on equilibrium quantities can be easily derived by noting that

$$q_D^* = D(p^*, p_1, p_2, y), \qquad q_S^* = S(p^*, w_1). \qquad (2.15)$$

It follows that when $dw_1 = 0$, since $q_D^* = q_S^*$ at equilibrium, we can use the 'function of a function' rule of differentiation to obtain

$$\frac{dq_D^*}{dy} = \frac{dq_S^*}{dy} = \frac{dq_S^*}{dp^*} \frac{dp^*}{dy} \qquad (2.16)$$

and similarly

$$\frac{dq_D^*}{dp_1} = \frac{dq_S^*}{dp^*} \frac{dp^*}{dp_1}; \qquad \frac{dq_D^*}{dp_2} = \frac{dq_S^*}{dp^*} \frac{dp^*}{dp_2}. \qquad (2.17)$$

While, with $dp_1 = dp_2 = dy = 0$, by using the same approach we have

$$\frac{dq_S^*}{dw_1} = \frac{dq_D^*}{dw_1} = \frac{dq_D^*}{dp^*} \frac{dp^*}{dw_1}. \qquad (2.18)$$

Then, since $dq_S^*/dp^* = \partial S(p^*, w_1)/\partial p^* > 0$, (2.15)–(2.17) tell us that the change in equilibrium quantity is, in each case, in the same direction as the change in equilibrium price. While, since $dq_D^*/dp^* = \partial D(p^*, p_1, p_2, y)/\partial p^* < 0$, (2.18) tells us that the change in equilibrium quantity is in the opposite direction to that in the equilibrium price.

This section has set out, in a general form, the kind of 'comparative static equilibrium analysis' which is usually conducted diagrammatically by shifting around supply and demand curves. When we are dealing with a single market, the diagrammatic approach is probably simpler and clearer. A good understanding of the method of this section is, however, essential when we progress to systems of two or more interrelated markets, for which diagrammatic methods are very cumbersome and inadequate.

Exercise 2.2

1. Given the demand and supply functions

$$q_D = a - bp - cp_1 + ep_2 + fy,$$
$$q_S = \alpha + \beta p - \gamma w_1,$$

derive expressions for the derivatives in (2.11)–(2.14) and (2.16)–(2.18). Use supply and demand diagrams to illustrate the changes that are being evaluated in each case.

2. Given the demand and supply functions

$$q_D = a - bp + cy \qquad a, b, c > 0,$$
$$q_S = \alpha + \beta p \qquad \alpha, \beta > 0,$$

derive expressions for dp^*/dy and dq^*/dy. Formulate conditions under which these derivatives will and will not have the same signs as those in equations (2.11) and (2.16). Illustrate your answers with diagrams.

3. Show how the methods of this section can be used to answer the following questions:

  (i)   What would be the effect on the price and output of coal of a fall in the price of oil?

  (ii) What would be the effect on the price and sales of convenience foods of increasing consumer incomes?

  (iii) How would falling prices of home computers affect the price and sales of computer games?

  (iv) How would better than average weather conditions affect the market for wine?

2.3 Stability of equilibrium

We have taken as the market outcome in which we are interested the equilibrium price and quantity traded. Our predictions about the workings of market forces are essentially based on analysis of equilibrium outcomes. To be valid, we must be sure that the market will in fact reach this equilibrium position, if it is ever out of it. We need to specify how the market operates out of equilibrium, and ask whether what happens leads the market toward or away from equilibrium. This is the question of *stability*. A market is said to be stable if, out of equilibrium, changes take place which push it to an equilibrium. It is *unstable* in the converse case. The analysis of the conditions under which each of these cases will occur is called stability analysis.

We now have to introduce time into the analysis explicitly. We denote time by the real number $t \geq 0$, with $t = 0$ as the initial point in time. Price, demand, supply, and quantity traded are all now functions of time, $p(t), q_D(t), q_S(t)$, and $q(t)$. We again suppress the exogenous variables and take linear demand and supply functions: $q_D(t) = a - bp(t)$, and $q_S(t) = \alpha + \beta p(t)$. Note that we take the parameters of these demand functions to be independent of time or *time invariant*. Among other things, this involves the assumption that a state of disequilibrium (when buyers and sellers are not able to realize their plans) at one point in time, *does not* affect supplies or demands at other points in time — not a terribly realistic assumption. However, this gives us the simplest model, which is also a mathematical formulation of the kind of story usually told. It is argued in diagrammatic treatments that if supply exceeds demand price will be bid down, while if demand exceeds supply price will be bid up. Somehow, it is asserted, this results in equilibrium. Let us now look at this more rigorously.

We can express this idea of how the market works out of equilibrium by the relationship

$$\frac{dp}{dt} = \gamma[q_D(t) - q_S(t)] \qquad \gamma > 0. \tag{2.19}$$

In words, the speed with which price changes out of equilibrium is proportional to the gap between demand and supply. The parameter γ is called the 'speed of adjustment'. At $t = 0$, if we have $p(0) \neq p^* = (a - \alpha)/(\beta + b)$, then the price will change, upward if $q_D(0) > q_S(0)$ and downward if $q_D(0) < q_S(0)$. The question we want to answer is: under what conditions will price converge to the equilibrium?

If we substitute the demand and supply functions into (2.19) we obtain a first-order linear differential equation

$$\frac{dp}{dt} = \gamma(a - \alpha) - \gamma(b + \beta)p(t). \tag{2.20}$$

Defining, for convenience, $A \equiv \gamma(a - \alpha)$ and $B \equiv \gamma(b + \beta)$, the general solution to this equation is

$$p(t) = Ce^{-Bt} + \frac{A}{B} \tag{2.21}$$

where C is a constant. Note that $A/B = p^*$. Hence whether $p(t)$ tends to p^* as $t \to \infty$ depends on whether $Ce^{-Bt} \to 0$ as $t \to \infty$, and hence we have the condition

$$B = \gamma(b + \beta) > 0 \tag{2.22}$$

as necessary and sufficient for the market to be stable.

First, note that since $\gamma > 0$ it can be divided out without changing this condition, i.e. the speed of adjustment does not determine whether, but only how rapidly, convergence to equilibrium takes place.

Next, we note that if b and $\beta > 0$ — the usual case — then the condition is certainly satisfied. Thus the postulated adjustment process is sufficient for stability when the demand curve slopes downward and the supply curve upward. Equation (2.21) tells us that the time path of price will show steady convergence to the equilibrium value.

However, if $b < 0$ and $\beta > 0$, then we require $|b| < |\beta|$ for (2.22) to hold, while if $b > 0$ and $\beta < 0$ we require $|b| > |\beta|$. If these conditions are not satisfied, then the adjustment process leads the market away from, and not toward, the equilibrium.

It may seem a little improbable that, in a real market, price would diverge steadily away from equilibrium in one direction or the other. This may give one confidence that the stability conditions are commonly satisfied, perhaps because demand curves slope downward and supply curves upward. However, we do also tend to come across assertions that, in particular markets, disequilibrium adjustments are *destabilizing*, for example in markets for foreign exchange, stocks and shares, and some agricultural products. The analysis of this section lends some support for such assertions. It shows that markets are not *necessarily* stable. However, the model analysed here is much too simple to provide a suitable treatment of such markets. It really says very little about how individuals in the market behave when price is not at its equilibrium value. In particular, it says nothing about what *expectations* they might form, and how these expectations could influence their current behaviour and hence the time path of price. We shall therefore return to this subject in Chapter 5.

Finally, note that the question of *whether* the market converges to equilibrium, on which this section has concentrated, may not be as important a question as *how quickly* it moves toward equilibrium. If the wage rate is above the equilibrium wage, then there is excess supply of labour, i.e. unemployment. It may be small consolation to know that, ultimately, the wage rate will converge to equilibrium, thus restoring full employment, if the speed of adjustment is very slow so that unemployment persists for a long time.

Exercise 2.3

1. Describe the process by which price would respond to an excess of demand over supply in each of the following markets:

    (i) a street market for fruit and vegetables;
    (ii) the market for sterling (or dollars, or Deutschmarks);
    (iii) the market for houses.

2. Draw diagrams to illustrate the various cases in which the stability condition (2.22) is or is not satisfied. Describe how the market moves from a position in which initial price is above equilibrium, in each case.

3. Suppose we replace the 'adjustment rule' in (2.19) by the following:

$$\frac{dq}{dt} = \gamma[p_D(t) - p_S(t)] \qquad \gamma > 0$$

where $p_D(t) = (a - q_D(t))/b$, $p_S(t) = (-\alpha + q_S(t))/\beta$. Interpret this rule, then carry out the stability analysis and compare results to those set out in this section.

2.4 Summary

In economics, a market is defined in an abstract way, as a system of interacting forces. These are the forces of supply and demand. We distinguish between desired, planned, or *ex ante* demands and supplies, and actual, realized, or *ex post* quantities bought and sold.

Ex post quantities are necessarily equal, but *ex ante* demand and supply are equal if and only if an equilibrium price prevails, in which case the market is in a state of balance.

By specifying the variables that determine demand and defining a demand function, and by specifying the variables that determine supply and defining a supply function, we can solve for equilibrium price and quantity traded. We do this by solving the equation that follows from the equilibrium condition: demand equals supply.

This equilibrium depends on the values of the exogenous variables, such as consumers' income, prices of complements and substitutes, and input prices. If one of these changes, the market equilibrium will in general change. The purpose of comparative statics analysis is to identify the directions of change in the endogenous variables — price and quantity traded — following from specified changes in the exogenous variables.

This is done first by totally differentiating the equilibrium condition and solving for the ratio of the differential of the endogenous variable to the differential of the exogenous variable. We then have to work out the sign of the resulting expression, given the assumptions that have been made about the signs of the partial derivatives of the demand and supply functions.

For the equilibrium point to be of interest we require the market to be stable, which means that if the market is out of equilibrium changes will be taking place that will cause market price and quantity traded to converge to their equilibrium values.

We made in this chapter the simplest possible assumption about the market adjustment process: out of equilibrium, market price changes over time at a speed which is proportional to the excess of demand over supply. The factor of proportionality (γ) determines how quickly equilibrium is reached but not *whether* it is reached.

Whether or not the market is stable depends on the signs and relative magnitudes of the slopes of the supply and demand curves. In the 'standard case' the market is always stable. However, it is certainly a possibility that a market could be unstable.

3 Elasticities

In Chapter 2 we used demand and supply functions to analyse market equilibrium but said very little about their nature in mathematical terms. A full analysis of the underlying determinants of these functions will be carried out in Chapters 6 and 7. Here we explore the question of the measurement of the responsiveness of demand or supply to its determining variables, as a way of expressing the underlying shapes of these functions. We then apply the concepts of this and the preceding chapters to the analysis of indirect taxation.

3.1 Definition of elasticities

Take first the demand function. Consider the main determining variables of demand: the good's own price, p; the price of complements, as typified by p_1; the price of substitutes, as typified by p_2; and total consumers' income y. We want to examine the general shape of the demand function $D(p, p_1, p_2, y)$.

By holding all but one of the independent variables constant, we define a *ceteris paribus* relationship between demand and the remaining independent variable. For example, if we plot the curve defined by $D(p, \bar{p}_1, \bar{p}_2, \bar{y})$ (where a bar denotes a fixed value of a variable) we obtain the usual demand curve, and $\partial D / \partial p$ is the slope of this. The position of this curve will change if we change the fixed value of one of the other variables, as we have already seen. If we plot the curve $D(\bar{p}, \bar{p}_1, \bar{p}_2, y)$, we obtain a relationship between demand and income known as the Engel curve, and $\partial D / \partial y$ is the slope of this. And so on for $D(\bar{p}, p_1, \bar{p}_2, \bar{y})$, and $D(\bar{p}, \bar{p}_1, p_2, \bar{y})$ (although these curves do not appear to have been given specific names).

When we consider any one of these *ceteris paribus* relationships, a question that often arises is: how responsive is demand to a change in the variable concerned? One measure of responsiveness that could be used to answer this question is the partial derivative, but this does not really give a satisfactory answer. To be told that the demand for mangoes changes by 5 tonnes when the price changes by 1¢ does not really tell us whether mango demand responsiveness to price is high or low. It certainly does not help us to compare mangoes to, say, apples, where the absolute levels of price and demand may be very different. Obviously, what we need is some sort of *relative* measure, and this is given by the concept of elasticity.

The elasticity of a function is the *proportionate* change in the value of the dependent variable, divided by the *proportionate* change in the value of the independent variable that caused the change in dependent variable. So, to calculate the *own price elasticity of demand*, for example, we take the proportionate change in quantity demanded divided by the proportionate change in the good's own price. This is a pure number, independent of units, and so can be compared across different goods and services.

There are, however, some ambiguities in this definition. Consider the own price elasticity and, suppressing the other variables, take demand $D(p)$ at two price levels, p^0 and p^1. We define

$$\Delta D = D(p^1) - D(p^0); \qquad \Delta p = p^1 - p^0. \qquad (3.1)$$

Then we have the first ambiguity: for any demand function except the linear one, the value of ΔD depends on the value of Δp, and different elasticity measures may be derived from the same point $(p^0, D(p^0))$ on the demand curve depending on the p^1 we take.

The second ambiguity arises when we wish to derive *proportionate* changes in demand and price. We have to divide the changes ΔD and Δp by, respectively a level of D and a level of p to arrive at these, but we have a choice of initial values, p^0 and $D(p^0)$, or final values, p^1 and $D(p^1)$, or any average of initial and final values. Each choice will usually result (even in the linear demand case) in a different elasticity measure.

A natural way to avoid these difficulties is to define elasticity at a point by using the concept of a derivative. Thus $\partial D/\partial p = \lim_{\Delta p \to 0} \Delta D/\Delta p$ is the slope of the demand curve at $(p^0, D(p^0))$, and is uniquely defined (given that $D(p)$ is differentiable at that point). We then define the own price elasticity as

$$e_p \equiv (-)\frac{\partial D}{\partial p}\frac{p^0}{D(p^0)}. \qquad (3.2)$$

The $(-)$ is included in the definition because normally $\partial D/\partial p < 0$ whereas it is usually convenient to have e_p as a positive number.

In a similar way, we define the elasticities

$$e_y \equiv \frac{\partial D}{\partial y}\frac{y^0}{D(y^0)}; \quad e_{p_1} \equiv \frac{\partial D}{\partial p_1}\frac{p_1^0}{D(p_1^0)}; \quad e_{p_2} \equiv \frac{\partial D}{\partial p_2}\frac{p_2^0}{D(p_2^0)}. \qquad (3.3)$$

The first of these is the income elasticity of demand, the other two are referred to as cross-price elasticities. As it happens, it is not usual to define e_{p_1} to be positive, even though $\partial D/\partial p_1 < 0$, and so cross-price elasticities of complements are negative and of substitutes are positive.

The definition of elasticities does not add anything to our knowledge of the demand function, but elasticities themselves may be useful ways of describing particular properties of demand functions. For example, to be told that the income elasticity of demand for electricity is about 2 allows us to predict (*ceteris paribus*) that expected growth of 3% in consumers income over the coming year should increase electricity demand by 6%.

The most important application of own price elasticity is in its relation to total expenditure on a good, which (in the absence of taxation — see the following section) is also the total revenue received by suppliers. Assume the quantity traded is equal to demand, q_D. Then total expenditure, denoted by E, at some price, p^0, and quantity, $D(p^0)$, is given by

$$E = p^0 q_D^0 = p^0 D(p^0).$$ (3.4)

Using the rule for differentiating products of functions, we then obtain

$$\frac{\partial E}{\partial p} = D(p^0) + p^0 \frac{\partial D}{\partial p} = D(p^0)(1 - e_p).$$ (3.5)

Thus the effect on expenditure of a change in price is proportional to the elasticity of demand.

Note further that, using the function of a function rule,

$$\frac{\partial E}{\partial q_D} = \frac{\partial E}{\partial p} \div \frac{\partial D}{\partial p} = p^0 + D(p^0)\frac{\partial p}{\partial D} = p^0(1 - 1/e_p).$$ (3.6)

Equation (3.6) gives essentially the same information as (3.5), but relates the change in expenditure to change in quantity (it is in this form that the relationship is used, for example, in the theory of monopoly — see Chapter 10).

We then have, using (3.5) and (3.6),

(*a*) $e_p = 1 \Rightarrow \partial E/\partial p = \partial E/\partial q = 0$. Changes in price and quantity leave expenditure unaffected.

(*b*) $e_p > 1 \Rightarrow \partial E/\partial p < 0$, $\partial E/\partial q > 0$. Reduction in price, implying an increase in quantity, increases total expenditure.

(*c*) $e_p < 1 \Rightarrow \partial E/\partial p > 0$, $\partial E/\partial q < 0$. Reduction in price, implying an increase in quantity, reduces total expenditure.

Elasticities of the supply function are defined in an analogous way. Thus, if we take supply as a function of own price and a representative input price, we write $S(p, w_1)$ and define

$$s_p \equiv \frac{\partial S}{\partial p} \frac{p^0}{S(p^0)}; \qquad s_{w_1} \equiv \frac{\partial S}{\partial w_1} \frac{w_1^0}{S(w_1^0)},$$ (3.7)

where the elasticities are evaluated at the points $(p^0, S(p^0))$ and $(w_1^0, S(w_1^0))$ respectively. Again, note that these elasticities are simply measures of a property of the given function. We still have to understand why the function takes the form it does, and therefore why the elasticities take the values they do; and this is the subject of Chapter 8.

Finally, we can produce a useful mathematical fact. Recall that if we have a function $y = \log x$, then $dy/dx = 1/x$. It follows that if we have a (*ceteris paribus*) demand function $q_D = D(p)$, we have

$$(-)\frac{\partial \log D(p)}{\partial \log p} = (-)\frac{p}{D(p)}\frac{\partial D(p)}{\partial p} = e_p,$$ (3.8)

which tells us that an elasticity can be derived by logarithmic differentiation. This apparently arcane fact turns out to have some useful applications, as you will see in question 3 of the following exercise.

Exercise 3.1

1. Show that along a linear demand curve e_p varies from zero to infinity.
2. Draw Engel curves for a normal and inferior good respectively. What will be the sign of the income elasticity of demand in each case?
3. Given the demand function

$$q_D = ap^{-b_1}y^{b_2}p_1^{-b_3}p_2^{b_4} \qquad b_1, b_2, b_3, b_4 > 0$$

   what are the four elasticities of demand? [Hint: show what $\log q_D$ equals and recall the definition of elasticity as a logarithmic derivative.] Under what conditions will total expenditure be constant along the (*ceteris paribus*) demand curve derived from this function? Sketch the demand curve.
4. Given a linear demand function, $q_D = a - bp$, show that $dE/dp = a - 2bp$. Given the *inverse* demand function $p = \alpha - \beta q_D$, where $\alpha = a/b$ and $\beta = 1/b$, show that $dE/dq_D = \alpha - 2\beta q_D$. Use either of these to show that total expenditure is at a maximum when $e_p = 1$.
5. Show how the effects of an increase in price or reduction in quantity on total expenditure are related to the own price elasticity of demand.

3.2 Taxation: an application

We can now put the market model of Chapter 2 and the elasticity concept of section 3.1 to work in analysing the effects of taxes and subsidies on a good's price and quantity traded. We take the case of a specific tax, i.e., a fixed sum of money levied per unit of the good. Denote this amount by t. The effect of the imposition of a tax is to drive a wedge between the price consumers pay, which we continue to denote by p, and the price producers receive, which we now denote by x.

There are two equivalent ways of thinking about the relation between p and x. We could think of consumers paying p, the government subtracting its tax of t, and leaving producers with what is left, $p - t = x$. Alternatively, we could think of the government adding the tax t onto the price, x, producers wish to receive for a given quantity supplied, to obtain the price consumers would have to pay for that quantity, $p = x + t$. Obviously, these two expressions are algebraically equivalent and lead to identical results for the model we study here. However, in some contexts it may turn out that one way of viewing the tax is more convenient than the other (for example see the analysis of discriminating monopoly in Chapter 10 below).

Consumers' demand depends on the price they pay, and so we write the (*ceteris paribus*) demand function as $D(p)$. Sellers' supply depends on the price they receive, and so we write the supply function as $S(x)$. Then we can write the market equilibrium condition in either of two equivalent ways, depending on whether in the first instance we want to solve for p or x:

$$D(p) = S(p - t), \tag{3.9}$$

$$D(x + t) = S(x). \tag{3.10}$$

Since t is given exogenously, (3.9) determines p immediately, and x then follows by subtracting t; while (3.10) determines x directly and p then follows by adding t.

Taking the linear case for example, where $D(p) = a - bp$, $S(p-t) = \alpha + \beta(p-t)$, we have the equilibrium condition:

$$a - bp = \alpha + \beta(p - t) \tag{3.11}$$

implying that the equilibrium market price after the tax is imposed, $\hat{p}$, is given by

$$\hat{p} = \frac{a - \alpha}{b + \beta} + \frac{\beta t}{b + \beta}. \tag{3.12}$$

So the equilibrium producer price is

$$\hat{x} = \hat{p} - t = \frac{a - \alpha}{b + \beta} + \frac{\beta t}{b + \beta} - t = \frac{a - \alpha}{b + \beta} - \frac{bt}{b + \beta} \tag{3.13}$$

(using $(b + \beta)/(b + \beta) = 1$).

Recall that the equilibrium price in the absence of tax is $p^* = (a - \alpha)/(b + \beta)$, and so (3.12) and (3.13) tell us that the effect of imposing a tax is as follows:

(a) Market price p rises unless $\beta = 0$, i.e. the supply curve is vertical. Assume from now on that $\beta > 0$.

(b) The producer price falls unless $b = 0$, i.e. the demand curve is vertical. Assume from now on that $b > 0$.

(c) The *incidence* of the tax is divided between buyers and sellers, with the proportion of the tax $\beta/(b + \beta)$ being borne by buyers, and the remaining proportion $b/(b + \beta)$ being borne by sellers. That is, buyers end up paying $\beta t/(b+\beta)$ *more* for each unit of the good, and sellers end up receiving $bt/(b+\beta)$ *less* for each unit of the good. Thus, the relative incidence of the tax depends on the slopes of the demand and supply curves.

(d) As long as $b > 0$ and $\beta > 0$, market price rises and producer price falls by less than the amount of the tax.

(e) Quantity traded falls, to

$$\hat{q} = \frac{a\beta + \alpha b}{b + \beta} - \frac{b\beta t}{b + \beta} = q^* - \frac{b\beta t}{b + \beta}$$

(since $q^* = (a\beta + \alpha b)/(b + \beta)$).

(*f*) Hence tax revenue is

$$t\hat{q} = tq^* - \frac{b\beta t^2}{b + \beta}.$$

(*g*) Revenues accruing to sellers fall from p^*q^* to $\hat{x}\hat{q}$.
(*h*) Expenditure of buyers changes from p^*q^* to $0\hat{p}\hat{q}$, but whether this is an increase
       or decrease depends upon the elasticity of demand at (p^*, q^*).

Fig. 3.1 illustrates all these results.

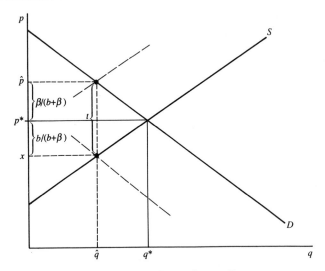

Fig. 3.1: The effects of a specific tax

Now let us return to the general conditions in (3.9) and (3.10). Totally differentiating each and rearranging we have

$$\frac{dp}{dt} = \frac{\partial S/\partial p}{(\partial S/\partial p - \partial D/\partial p)}, \tag{3.14}$$

$$\frac{dx}{dt} = \frac{\partial D/\partial p}{(\partial S/\partial p - \partial D/\partial p)}. \tag{3.15}$$

Thus, as long as $\partial D/\partial p < 0$ and $\partial S/\partial p > 0$, we have the results just obtained for
the linear case confirmed for the general case: consumers pay more ($dp/dt > 0$),
producers receive less ($dx/dt < 0$), but the change in each case is less than the
amount of the tax and the changes sum to the total of the tax ($dp/dt + |dx/dt| = 1$).

We conclude this section by considering the following question: what level of
tax would *maximize* the government's tax revenue? We define the tax revenue as
$T = tq$, where q of course is the equilibrium quantity traded given the tax. Then
T will be maximized at the point where $dT/dt = 0$ (assuming also at that point that
$d^2T/dt^2 < 0$), and so, using the rule for differentiating a product, this implies the
condition

$$\frac{dT}{dt} = q + \frac{tdq}{dt} = 0, \quad \text{or} \quad q(1 - \tau) = 0 \tag{3.16}$$

where $\tau \equiv -(t/q)(dq/dt)$ is the elasticity of equilibrium quantity traded with respect to the tax rate. Thus, if a 1% increase in the tax rate is expected to cause about a 1% reduction in quantity traded, the tax rate is around the level that maximizes tax revenue. If the fall in quantity traded would be less than this, then $\tau < 1, dT/dt > 0$ and so the government could, if it wished, increase tax revenue by increasing the tax. If the fall in quantity traded would be greater than 1%, $\tau > 1, dT/dt < 0$, and the government could actually increase its tax revenue, as well as its popularity, by reducing the rate of tax.

Exercise 3.2

1. Show that in the case of linear demand and supply functions, the tax rate which max-imizes tax revenue is given by $t^* = (a\beta + \alpha b)/2b\beta$. Sketch the function relating tax revenue T to tax rate t.
2. An *ad valorem* tax is levied as a proportion of the producer price, i.e. $p = (1 + v)x$ where v is the rate of tax. Derive the results (*a*)–(*g*) of this section for this type of tax.
3. Suppose that a specific subsidy, s units of money per unit of good, is paid, either to consumers or to producers. Show how it affects equilibrium consumer and producer prices and how the benefits of it are shared between them. [Hint: a subsidy is a negative tax].
4. Assuming linear supply and demand functions, under what conditions would the impo-sition of a specific tax reduce market price and increase quantity?
5. Suppose initially $t = 0$ in the linear demand and supply case. Then a tax dt is imposed. Show that the elasticity of demand at (p^*, q^*) determines the effect on expenditure.

3.3 Summary

Elasticities are numbers which give us an idea of the shape of a function, in terms of the sensitivity or responsiveness of the dependent variable to change in an independent variable. One advantage they have as measures of responsiveness is that they are pure numbers, independent of the units in which quantities and prices are measured, and so they can readily be compared for different goods.

If we take finite changes in independent and dependent variables, the elasticity measures are not uniquely defined. So, we define *point elasticities*: we take proportional rates of change at a point on the function.

The most important and widely used elasticities are the own price elasticity e_p defined in (3.2), the income elasticity e_y, and the cross-price elasticities defined in (3.3).

The main reason for interest in e_p is that its value tells us how consumers' expenditure (= producers' revenue) will change as price and quantity change.

An elasticity can also be conveniently expressed as a logarithmic derivative.

We can apply market analysis and the concept of elasticity to the analysis of the effects of taxation on a good's price, quantity traded, consumers' expenditure, and producers' revenue. We find that the price consumers pay rises and the price producers receive falls, in each case by less than the amount of the tax. The sum of these two price effects equals the amount of the tax and their relative magnitudes depend on the slopes of the supply and demand curves. Quantity traded and producers' revenue fall. Consumers' expenditure may increase or decrease, depending on the own price elasticity of demand.

It is of interest to consider the tax rate which maximizes tax revenue. This is the value of the tax rate at which the elasticity of tax revenue with respect to the tax rate is just equal to unity. This suggests that if this elasticity were at a value greater than unity, a government could increase tax revenue by reducing the tax.

4 Multi-market Equilibrium

In Chapter 2 we considered the effect on the market equilibrium for one good of a change in the price of a related good — a complement or a substitute. It is, however, unsatisfactory to treat the price of the latter as exogenous. We would expect the change in equilibrium in one market to feed back on to the equilibrium in the other market, causing a further change in price, and hence further changes in the first market, and so on. For example, in Fig. 4.1, in the market for coffee, we assume that adverse weather conditions at harvest time cause the supply curve to shift from S_1^0 to S_1^1. This causes a rise in the equilibrium price of coffee to p_1^1. This in turn causes an increase in demand for tea to D_2^1, and so a rise in the price of tea to p_2^1. But this increase in price of tea will cause a rise in demand for coffee to D_1^1, and a *further* rise in the coffee price. And so on.

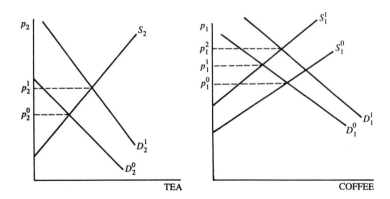

Fig. 4.1: Effect of a shift in supply of coffee on the market for tea

There are two questions raised by this process of mutual interaction between the two markets, which the diagrammatic analysis is not really well designed to answer. One is whether the process will in fact settle down to a new equilibrium — from the diagram it is not at all obvious that it will. For example, suppose the outward shifts in a demand curve caused by changes in the *other* good's price become larger and larger. The second question is, where is the new equilibrium, assuming one *is* reached, in relation to the initial position?

The first question has an answer, but the methods required to reach it are beyond the scope of this book. We shall therefore assume a new equilibrium *is* ultimately reached, and consider only the second question, that of the relation between the initial and final equilibrium positions.

4.1 Multi-market equilibrium and comparative statics

We analyse a model of two interdependent markets, in which both prices are endogenous. We take as the exogenous variables: consumers' income, y, which appears in both demand functions; and a weather variable, θ, which appears in the supply function of just one of the goods, and is so measured that an increase in θ increases supply of the good at every price.

Let q_{D_i} and q_{S_i} denote demand for and supply of good $i = 1, 2$, with p_i the corresponding price. For equilibrium to exist on both markets, we require simultaneous satisfaction of the two conditions

$$D_1(p_1, p_2, y) - S_1(p_1, \theta) = 0, \tag{4.1}$$

$$D_2(p_1, p_2, y) - S_2(p_2) = 0. \tag{4.2}$$

Note that we do not assume that the goods are related in supply as well as demand— sellers do not switch from supply of one good to supply of another in response to relative price changes. This is somewhat special and you are asked in an exercise to generalize this.

Given the values of θ and y, we assume that these two equations are sufficient to determine equilibrium values p_i^*, q_i^* in the two markets. We are interested in the effects on this equilibrium of changes in the exogenous variables. Hence we differentiate totally through (4.1) and (4.2) to obtain

$$\left(\frac{\partial D_1}{\partial p_1} - \frac{\partial S_1}{\partial p_1} \right) dp_1^* + \frac{\partial D_1}{\partial p_2} dp_2^* + \frac{\partial D_1}{\partial y} dy - \frac{\partial S_1}{\partial \theta} d\theta = 0, \tag{4.3}$$

$$\frac{\partial D_2}{\partial p_1} dp_1^* + \left(\frac{\partial D_2}{\partial p_2} - \frac{\partial S_2}{\partial p_2} \right) dp_2^* + \frac{\partial D_2}{\partial y} dy = 0. \tag{4.4}$$

The essence of the mathematical approach is to regard (4.3) and (4.4) as a pair of *linear* simultaneous equations, with the *unknowns* being dp_1^* and dp_2^*, since we wish to solve for the changes in equilibrium prices. The *parameters* are, on the one hand, the coefficients of these unknowns, and on the other hand, the terms in the exogenous variables θ and y. The justification for taking the derivatives $\partial D_1/\partial p_1, \partial D_1/\partial p_2, \partial D_2/\partial p_1, \partial D_2/\partial p_2, \partial S_1/\partial p_1, \partial S_2/\partial p_2$ as parameters is that these derivatives are evaluated at the equilibrium point, so they are fixed numbers rather than general expressions. The same applies to the derivatives $\partial D_1/\partial y, \partial D_2/\partial y$, and $\partial S_1/\partial \theta$. The differentials dy and $d\theta$ are also taken as (very small) fixed numbers. Hence, in matrix notation, we have the system

$$\begin{bmatrix} \dfrac{\partial D_1}{\partial p_1} - \dfrac{\partial S_1}{\partial p_1} & \dfrac{\partial D_1}{\partial p_2} \\[2ex] \dfrac{\partial D_2}{\partial p_1} & \dfrac{\partial D_2}{\partial p_2} - \dfrac{\partial S_2}{\partial p_2} \end{bmatrix} \begin{bmatrix} dp_1^* \\[2ex] dp_2^* \end{bmatrix} = \begin{bmatrix} \dfrac{\partial S_1}{\partial \theta} d\theta - \dfrac{\partial D_1}{\partial y} dy \\[2ex] -\dfrac{\partial D_2}{\partial y} dy \end{bmatrix} \tag{4.5}$$

which we can solve in the same way as any other pair of simultaneous equations. Let $|A|$ denote the determinant of the matrix on the left-hand side of (4.5). Then

$$|A| = \left(\frac{\partial D_1}{\partial p_1} - \frac{\partial S_1}{\partial p_1}\right) \cdot \left(\frac{\partial D_2}{\partial p_2} - \frac{\partial S_2}{\partial p_2}\right) - \frac{\partial D_1}{\partial p_2}\frac{\partial D_2}{\partial p_1}. \tag{4.6}$$

The sign of $|A|$ is, of course, going to be important. Now, if the *ceteris paribus* demand curves in both markets slope downward, and the supply curves upward, then the first product in (4.6) is certainly positive, since it is the product of two negative numbers. We expect $\partial D_1/\partial p_2$ and $\partial D_2/\partial p_1$ to have the same signs, in which case the second product in (4.6) is negative, and so the sign of $|A|$ is undetermined. However, it seems not too implausible to make a *dominant diagonal assumption* (so called because the first product in (4.6) involves the terms along the diagonal of $|A|$), that the sum of the effects on its own demand and supply of a change in a price is larger than its effect on the demand of the other good. This then implies that $|A| > 0$. (In fact it can be shown that this dominant diagonal assumption is sufficient for equilibrium in this two-market model to be stable.)

We want to consider the effect of a change in just one exogenous variable at a time. So, set $dy = 0$, and use Cramer's Rule in (4.5) to obtain

$$dp_1^* = \frac{1}{|A|}\left(\frac{\partial D_2}{\partial p_2} - \frac{\partial S_2}{\partial p_2}\right)\frac{\partial S_1}{\partial \theta}d\theta, \tag{4.7}$$

$$dp_2^* = \frac{-1}{|A|}\left(\frac{\partial D_2}{\partial p_1}\frac{\partial S_1}{\partial \theta}\right)d\theta. \tag{4.8}$$

To evaluate the sign of dp_1^*, suppose that $d\theta > 0$, and recall we assumed that $\partial S_1/\partial \theta > 0$, i.e., an increase in θ increases supply in market 1. If demand and supply curves have the normal slopes, as we have presumed, then $dp_1^* < 0$. Thus, a 'favourable supply shock' in market 1 reduces the price of good 1 after the interactions and feedback effects in market 2 are taken into account. The sign of dp_2^* will then depend on the sign of $\partial D_2/\partial p_1$. If the goods are substitutes, $\partial D_2/\partial p_1 > 0$ and so $dp_2^* < 0$; if complements, $dp_2^* > 0$. So, to summarize: given the dominant diagonal assumption, a favourable supply shock in market 1 will reduce price in that market, and will also reduce price in market 2 if the goods are substitutes, and increase price in market 2 if they are complements.

Note that this result would also have been obtained if we had carried out the diagrammatic analysis ignoring any feedback effects. It might then be wondered why it should be necessary to go through the algebra as we have done. The answer is of course that this result will not be obtained under all circumstances: for example, if the dominant diagonal assumption were not made, and it is assumed instead that $|A| < 0$, then the signs of dp_1^* and dp_2^* are reversed. The advantage of the present approach is that it makes clear what has to be assumed to obtain one particular result. It also makes it much easier to carry out a systematic analysis of alternative possibilities. This is an important general property of the mathematical analysis. In many models, we find that the effects of some change on the equilibrium position are not simple and unambiguous; they depend on the relative magnitudes of terms with opposite signs. The usefulness of the mathematical analysis in this case is that it enables us to work systematically through the logical possibilities, clarifying the conditions under which the various possible results may hold.

Now let us consider the consequences of an increase in consumers' incomes, y. Setting $d\theta = 0$ in (4.5) and using Cramer's Rule again, we have

$$dp_1^* = \frac{1}{|A|}\left(\frac{\partial D_1}{\partial p_2}\frac{\partial D_2}{\partial y} - \frac{\partial D_1}{\partial y}\left(\frac{\partial D_2}{\partial p_2} - \frac{\partial S_2}{\partial p_2}\right)\right)dy, \tag{4.9}$$

$$dp_2^* = \frac{1}{|A|}\left(\frac{\partial D_2}{\partial p_1}\frac{\partial D_1}{\partial y} - \frac{\partial D_2}{\partial y}\left(\frac{\partial D_1}{\partial p_1} - \frac{\partial S_1}{\partial p_1}\right)\right)dy. \tag{4.10}$$

These results tell us that if both goods are normal and substitutes ($\partial D_1/\partial p_2$, $\partial D_2/\partial p_1 > 0$), and their demand and supply curves have the usual slopes ($\partial D_i/\partial p_i - \partial S_i/\partial p_i < 0$, $i = 1, 2$), then both prices certainly increase with an increase in consumers' income, given again the dominant diagonal assumption that implies $|A| > 0$.

However, if they are both normal but complements, then the first terms in the main brackets in (4.9) and (4.10) are negative, while the second terms are positive. Thus the answer is not clearcut, and it is possible that at least one of the changes in equilibrium price could be negative. Thus we have the rather counter-intuitive result: an increase in consumers' income could lead to a fall in price of one of the two complementary goods, even though they are both normal.

Let us investigate this further. Suppose $dp_1^* < 0$. Then, with $dy > 0$ and $|A| > 0$, from (4.9) this implies

$$\frac{\partial D_1}{\partial p_2}\frac{\partial D_2}{\partial y} < \frac{\partial D_1}{\partial y}\left(\frac{\partial D_2}{\partial p_2} - \frac{\partial S_2}{\partial p_2}\right). \tag{4.11}$$

Note that both sides of this inequality are negative, and so for it to be true algebraically we must have

$$\left|\frac{\partial D_1}{\partial p_2}\frac{\partial D_2}{\partial y}\right| > \left|\frac{\partial D_1}{\partial y}\left(\frac{\partial D_2}{\partial p_2} - \frac{\partial S_2}{\partial p_2}\right)\right| \tag{4.12}$$

(where $|\ldots|$ now denotes 'absolute value') and so

$$\frac{\dfrac{\partial D_2}{\partial y}}{\dfrac{\partial D_1}{\partial y}} > \frac{\left|\left(\dfrac{\partial D_2}{\partial p_2} - \dfrac{\partial S_2}{\partial p_2}\right)\right|}{\left|\dfrac{\partial D_1}{\partial p_2}\right|}. \tag{4.13}$$

Now from the dominant diagonal assumption we know that the right-hand side of
(4.13) is greater than 1 — own price effects are stronger than cross-price effects.
Thus the inequality tells us that, for p_1 to fall, the effect of the increase in income
on demand for good 2 must be larger than that on good 1. Moreover, the ratio of
the two income derivatives must be larger, the larger the sum of absolute values
of demand and supply derivatives, and the smaller the cross-price effect, i.e., the
weaker the complementarity relation.

Fig. 4.2 illustrates this case. Initial equilibrium is at prices p_1^0, p_2^0. The increase
in income shifts the demand curves initially to D_1^1 and D_2^1 respectively, where the
effect is much larger for good 2 than for good 1. The effect of the rise in the price
of good 2 is to shift down the demand curve for good 1 to D_1^2. The effect of the
initial rise in the price of good 1 is to shift down the demand curve for good 2 to
D_2^2. Such are these cross-effects, that D_1^2 is below the initial demand curve D_1^0, so
price has fallen, while D_2^2 is still above D_2^0, so price has risen. An exercise asks
you to examine the precise role each of the terms in (4.13) plays in this example.

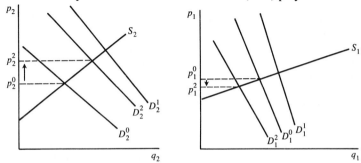

Fig. 4.2: Feedback effects of changes in related markets

The question then arises: could we have a *really* counter-intuitive result, and
have *both* prices falling in the complementary goods cases? The answer is no, if
the dominant diagonal assumption holds. To see this, note that the inequality in
(4.12) must hold for both goods if both prices are to fall, i.e., we must have

$$\left| \frac{\partial D_i}{\partial p_j} \frac{\partial D_j}{\partial y} \right| > \left| \frac{\partial D_i}{\partial y} \left(\frac{\partial D_j}{\partial p_i} - \frac{\partial S_j}{\partial p_j} \right) \right| \qquad i, j = 1, 2, \ i \neq j. \qquad (4.14)$$

It follows, by multiplying together the two left-hand sides of the inequalities, and
the two right-hand sides, that

$$\left| \frac{\partial D_1}{\partial y} \frac{\partial D_2}{\partial y} \right| \cdot \left| \frac{\partial D_1}{\partial p_2} \frac{\partial D_2}{\partial p_1} \right| > \left| \frac{\partial D_1}{\partial y} \frac{\partial D_2}{\partial y} \right| \cdot \left| \left(\frac{\partial D_1}{\partial p_1} - \frac{\partial S_1}{\partial p_1} \right) \left(\frac{\partial D_2}{\partial p_2} - \frac{\partial S_2}{\partial p_2} \right) \right|. \qquad (4.15)$$

But after dividing out the common term $| (\partial D_1/\partial y)(\partial D_2/\partial y) |$ we see that the result
violates the dominant diagonal assumption. Thus, given this assumption, at most
one of the equilibrium prices can fall.

A number of other cases could be analysed, for example where one or both
goods may be inferior. We leave consideration of these to an exercise.

Exercise 4.1

1. Set out as fully as you can, for the following pairs of markets, the way in which a supply change in one is likely to affect the price in the other, and cause a further feedback effect on itself:
   (a) the markets for houses and mortgages;
   (b) the markets for automobiles and gasoline;
   (c) the markets for coal and coalminers;
   (d) the markets for package holidays in Spain and Greece;
   (e) the markets for beef and corn.
2. Suppose that in the two-market example, the supply of, as well as the demand for, each good is a function of both prices. Give some examples of goods for which this might be true. Then carry out the analysis of the effects of a change in consumers' incomes (appearing in both demand functions) and an exogenous supply shock (in only one supply function) on equilibrium prices. Explain carefully the economic meaning of the extension you have to make to the dominant diagonal assumption.
3. Extend the analysis of this section to the changes in equilibrium *quantities*. [Hints: in the case of the change in θ, note that one of the supply functions does not change, and for the other good the results will be ambiguous; state the conditions for various cases. In the case of a change in y, note that neither supply function changes.]
4. For the analysis of the effects of a supply shock ($d\theta > 0$) in this section, consider fully the implications of assuming
   (a) all demand and one or more supply curves have negative slopes
   (b) one or more demand and all supply curves have positive slopes
   paying particular attention to the dominant diagonal assumption in each case. Illustrate your answers with diagrams.
5. For the analysis of the effects of an increase in consumers' income, consider the implications of assuming that one good is normal and one is inferior, in the case where the goods are substitutes. Illustrate your answer on a diagram.
6. Using a diagram, explain exactly why each of the terms in the condition (4.13) has the effect it has in determining that $dp_1^* < 0$.
7. Using the methods of this section, set up models and derive answers to the following questions:
   (a) How would an abnormally cold winter affect prices of oil and coal?
   (b) How would cost-reducing innovations in personal computer design affect the prices and quantities sold of personal computers and software?

4.2 An example: price controls in a multi-market setting

Economic policy-makers often appear to believe that their goals can be achieved by imposing non-equilibrium prices in particular markets. For example, the poor are believed to be helped by setting maximum rentals below the market equilibrium level for apartments or houses; borrowers are thought to benefit from ceilings on interest rates; farmers are supposed to be helped by having prices of their outputs set above equilibrium levels; low-paid workers may be thought to benefit from a minimum wage set above equilibrium.

In the usual diagrammatic treatment of such policies, it is shown that the immediate consequence is: an excess demand, if a maximum price below equilibrium is imposed; or an excess supply, if a minimum price above equilibrium is set. In the former case, buyers will have to be rationed in some way. This may be done by some form of queuing, or by a formal system of rationing by use of 'ration coupons'. In the latter case, sellers will have to be rationed, for example by imposing output quotas on producers, as in agricultural markets, or by queuing, favouritism, or even by gender or ethnic discrimination. In general, when market price is prevented from carrying out its *rationing function*, that of allocating available supply among buyers, or available demand among sellers, then some non-price mechanism is required, since rationing of *some* kind is always necessary.

Indeed, it should be noted that price-based market forces will often, at least to some extent, reassert themselves. When goods are rationed, a black market may develop on which they can be bought, without ration coupons, at prices above the prescribed maximum (and quite probably above the equilibrium price as well). Bribes may be paid to those administering the rationing system. In the market for apartment rentals subject to rent control, 'key money', or the sale of 'fixtures and fittings' at inflated prices, are devices by which a higher price can in effect be charged.

It is straightforward to show the effects of price controls mathematically, though little is added to the diagrammatic treatment if we confine attention to just one market. However, the main point of this chapter is: markets cannot usually be considered in isolation. Goods and services subject to price control usually have substitutes that are not necessarily similarly controlled. How does price control in one market affect the equilibrium in a related market, and what are the feedback effects of this on the controlled market? We now go on to set up and analyse a model which can answer this question.

We consider a market, of which European agricultural markets are an example, where price is fixed above the equilibrium, but the price-fixing authority buys up and stores the resulting excess supply. We can simplify the analysis without losing anything essential if we assume linear supply and demand functions. Thus the initial equilibrium is given by the counterparts of equations (4.1) and (4.2):

$$D_1(p_1, p_2) \equiv a_1 - b_{11}p_1 + b_{12}p_2 = \alpha_1 + \beta_{11}p_1 - \beta_{12}p_2 \equiv S_1(p_1, p_2), \qquad (4.16)$$

$$D_2(p_1, p_2) \equiv a_2 + b_{21}p_1 - b_{22}p_2 = \alpha_2 - \beta_{21}p_1 + \beta_{22}p_2 \equiv S_2(p_1, p_2), \qquad (4.17)$$

where the parameters $a_i, \alpha_i, b_{ij}, \beta_{ij}$ are all positive, $i, j = 1, 2$. Note that the signs on the parameters $b_{ij}, \beta_{ij}, i, j = 1, 2, i \neq j$, imply that the goods are substitutes in both demand and supply (explain fully why this is so).

We can rewrite this system as

$$(b_{11} + \beta_{11})p_1 - (b_{12} + \beta_{12})p_2 = a_1 - \alpha_1, \qquad (4.18)$$

$$-(b_{21} + \beta_{21})p_1 + (b_{22} + \beta_{22})p_2 = a_2 - \alpha_2. \qquad (4.19)$$

Using Cramer's Rule we have as solutions for equilibrium prices

$$p_1^* = \frac{(a_1 - \alpha_1)(b_{22} + \beta_{22}) + (a_2 - \alpha_2)(b_{12} + \beta_{12})}{(b_{11} + \beta_{11})(b_{22} + \beta_{22}) - (b_{12} + \beta_{12})(b_{21} + \beta_{21})}, \qquad (4.20)$$

$$p_2^* = \frac{(a_2 - \alpha_2)(b_{11} + \beta_{11}) + (a_1 - \alpha_1)(b_{21} + \beta_{21})}{(b_{11} + \beta_{11})(b_{22} + \beta_{22}) - (b_{12} + \beta_{12})(b_{21} + \beta_{21})}. \qquad (4.21)$$

We would naturally assume $a_i > \alpha_i$, $i = 1, 2$; that is, at zero prices demand exceeds supply. However, the equations show that this is not alone sufficient for $p_i^* > 0$. We again require a 'dominant diagonal assumption' to ensure that the denominators in (4.20) and (4.21) are positive. We again make that assumption.

Suppose now that in market 1 a price $p_1^0 > p_1^*$ is imposed. Market 2, however, is left to find its own equilibrium. Since p_1 is now exogenously given we have to drop condition (4.16), and the equilibrium price p_2^0 in market 2 is determined by

$$-(b_{21} + \beta_{21})p_1^0 + (b_{22} + \beta_{22})p_2 = a_2 - \alpha_2, \qquad (4.22)$$

implying that the new equilibrium price in market 2 is

$$p_2^0 = \frac{(a_2 - \alpha_2) + (b_{21} + \beta_{21})p_1^0}{(b_{22} + \beta_{22})}. \qquad (4.23)$$

What is the relationship between p_2^0 and the previous equilibrium value p_2^*? We can show that $p_2^0 > p_2^*$, and that the difference $p_2^0 - p_2^*$ is smaller (given the dominant diagonal assumption) than the difference $p_1^0 - p_1^*$. To see this, note first that we must have

$$p_2^* = \frac{(a_2 - \alpha_2) + (b_{21} + \beta_{21})p_1^*}{(b_{22} + \beta_{22})} \qquad (4.24)$$

(make sure you can explain why). Then, define:

$$p_1^0 = p_1^* + \Delta p_1, \quad p_2^0 = p_2^* + \Delta p_2.$$

Substituting into (4.23) we have

$$p_2^* + \Delta p_2 = \frac{(a_2 - \alpha_2) + (b_{21} + \beta_{21})(p_1^* + \Delta p_1)}{(b_{22} + \beta_{22})} = p_2^* + \frac{(b_{21} + \beta_{21})}{(b_{22} + \beta_{22})}\Delta p_1. \quad (4.25)$$

Thus $\Delta p_2 = k\Delta p_1$, where $k \equiv (b_{21} + \beta_{21})/(b_{22} + \beta_{22}) < 1$ given the dominant diagonal assumption.

Thus, increasing the price above equilibrium in one market raises the equilibrium price in an uncontrolled market for a substitute good, but by less than the price increase in the original market. (Use this result to explain why companies which brew beer in Europe must have benefited from the Common Agricultural Policy's effect of increasing the price of wine.)

We now have to consider the feedback effect of the price rise in market 2 on market 1. There can be no effect on p_1, since this is fixed at p_1^0. Rather, there will be an effect on the excess supply of good 1, which we can write as

$$E_{S_1}(p_1, p_2) = S_1(p_1, p_2) - D_1(p_1, p_2). \tag{4.26}$$

We know that at the initial equilibrium,

$$E_{S_1}(p_1^*, p_2^*) = 0 \tag{4.27}$$

which is simply a way of writing the equilibrium condition. When p_1 is raised to p_1^0, but with no change in p_2, we must have

$$E_{S_1}(p_1^0, p_2^*) > 0 \tag{4.28}$$

or supply exceeds demand. We can now show that the effect of the rise in price in market 2 is to reduce excess supply in market 1. Thus we note that, from (4.16):

$$E_{S_1}(p_1^0, p_2) = (\alpha_1 - a_1) + (\beta_{11} + b_{11})p_1^0 - (\beta_{12} + b_{12})p_2. \tag{4.29}$$

It then follows that

$$E_{S_1}(p_1^0, p_2^0) - E_{S_1}(p_1^0, p_2^*) = -(\beta_{12} + b_{12})(p_2^0 - p_2^*) < 0. \tag{4.30}$$

Could the excess supply be eliminated entirely? You should now show, by writing out the expression for

$$E_{S_1}(p_1^0, p_2^0) - E_{S_1}(p_1^*, p_2^*),$$

that, given the dominant diagonal assumption, there must always in the end be *some* excess supply.

The fact that excess supply will end up lower, as a result of the feedback effect from market 2, than would have been estimated from a consideration of market 1 alone, is good news for those administering the price support policy. This is because the cost of buying up the excess supply is given by $p_1^0 E_{S_1}(p_1^0, p_2^0)$, which is less than $p_1^0 E_{S_1}(p_1^0, p_2^*)$. Thus, one way of looking at this multi-market analysis is as a better way of estimating the cost of the price support policy, in terms of the expenditure (funded usually out of general taxation) required to finance it. It is also important to note that the economic effects of the policy are not confined to the one good whose price is being fixed above equilibrium.

A very important feature of this model was that the excess supply of the good at the regulated price was bought up by the price control authority. Therefore, in a sense, there was no disequilibrium: sellers actually were selling what they wanted to sell at the fixed price. We would expect things to be different if excess supply were *not* bought up. Then, sellers would find themselves with some unsold output, and we would expect this to cause them to switch to market 2 despite the higher-than-equilibrium price in market 1. Analysis of this truly disequilibrium case, and of its counterpart when price is fixed below equilibrium, is very interesting but has to be conducted at a more advanced level than that of this book.

Exercise 4.2

1. Work through the analysis of this section on the assumption that goods 1 and 2 are complements rather than substitutes. Explain in particular the nature of the feedback effects on excess supply in market 1 in this case.
2. Suppose government fixes market price below equilibrium but meets excess demand by buying in foreign supplies at the previous equilibrium price. Analyse the effects of this on the market for a substitute.

4.3 Summary

In general, changes in one market affect the equilibrium position in related markets for complements and substitutes. These induced effects then cause feedback effects on the initial market. The purpose of multi-market analysis is to take all these effects into account.

Because of the complexity of the cross effects it will usually be necessary to make some assumption about relative size of various effects if we are going to be able to sign the outcome. The most usual is the dominant diagonal assumption: effects of a change in a good's price on its own supply and demand are stronger than the effects on another good's supply and demand.

We often find that the kinds of effects we expect to take place from a single-market analysis do not in fact occur when market interactions are taken into account. For example, two goods may be normal, but if they are close complements then an increase in income may cause one price to rise and the other *to fall*.

Price controls in a market create excess demand or excess supply. If the price control authority takes steps to neutralize this, e.g., by buying up excess supply when it fixes price above equilibrium, then no true disequilibrium results.

However, even if this is the case, there will be repercussions on related markets and feedback effects on the initial market. For example, the price of a substitute will rise and excess supply of the controlled good will be less than would have been estimated from a consideration of this market alone, ignoring feedback effects.

Where excess demand or supply at the controlled price is *not* neutralized it will be necessary to ration buyers or sellers. Black markets and other practices may develop — all a tribute to the strength of market forces.

5 Dynamic Market Models

The word 'dynamic' denotes change over time. In dynamic economic models we analyze the process of change in the variables the model is designed to explain. The models are of two types. In dynamic *equilibrium* models, we assume that at each point in time the conditions for equilibrium are satisfied, but that a steady process of change in the variables or parameters of the model induces the equilibrium solutions to change. We then seek to analyse the resulting equilibrium time paths. In dynamic *disequilibrium* models, we are interested in the process of adjustment from one static equilibrium to another. A major question of concern is to find the conditions under which a new equilibrium will in fact be reached, that is, the conditions under which equilibrium in the model is *stable*. In the next two sections we examine examples of both these types.

The mathematical methods we use in analysing the more sophisticated dynamic models are usually those of differential and difference equations. Problems involving these rapidly become quite complex and so the trade-off between 'realism' on the one hand, and simplicity on the other, is even more acute in this area than in other parts of economics. So, we will not hesitate to make simplifying assumptions in order to make the problem tractable at the level of mathematics assumed in this book. The model of the next section shows that some quite interesting results can be derived without even having to solve difference or differential equations.

5.1 Equilibrium paths in a market

Over time, we expect consumers' incomes to change, there will be changes in the technology for producing goods, consumers' preferences may well change, and these and many other factors will tend to cause changes in the equilibrium prices of goods and quantities traded. We shall now examine in a simple model how such factors as these may interact to determine the *long-run trends* in price and quantity of a good. We assume that a steady rate of growth in consumers' income will affect demand, and a steady rate of technological change will affect supply. Although actual prices and quantities may fluctuate about the resulting trends because of the influence of 'random shocks' or short-run changes, we confine ourselves to analysis of these long-term trends.

The market demand function is given by:

$$q_D(t) = D(p(t), y(t)) = ap^{-\epsilon}y^{\alpha} \tag{5.1}$$

where p is market price, y is consumers' income, $\epsilon > 0$ is the price elasticity of demand and α is income elasticity of demand (since $\epsilon = -\partial \log D/\partial \log p$, $\alpha = \partial \log D/\partial \log y$). The notation on the left-hand side of (5.1) emphasizes that the variables in the model are now to be thought of as functions of time, t. We may have $\alpha \gtrless 0$, i.e., the good may be normal or inferior. Given some initial date $t = 0$, we assume

$$y(t) = e^{gt} y_0 \tag{5.2}$$

where $y_0 = y(0)$ is an initial income level and g is the (trend or long-run) growth rate of income, assumed exogenous (confirm that $g = (dy/dt)(1/y) = d \log y/dt$). Thus, letting $\gamma \equiv a y_0^{\alpha} = $ constant, we can write the market demand function as

$$D(p(t), t) = \gamma p^{-\epsilon} e^{\alpha g t}. \tag{5.3}$$

The effect of technological change on supply is, we shall assume, to increase supply, at any given price, steadily through time. Thus we have

$$q_s(t) = S(p(t), t) = e^{rt} \beta p^{\eta} \tag{5.4}$$

where $\eta > 0$ is the elasticity of supply, and r is the rate of technological change. We require that at each t, the market must be in equilibrium, and so we have:

$$D(p(t), t) = S(p(t), t) \implies \gamma p^{-\epsilon} e^{\alpha g t} = e^{rt} \beta p^{\eta}, \tag{5.5}$$

implying the equilibrium solutions for price and quantity:

$$p^*(t) = \left(\frac{\gamma e^{(\alpha g - r)t}}{\beta} \right)^{\frac{1}{\epsilon + \eta}} = k_1 e^{\delta_1 t}, \tag{5.6}$$

$$q^*(t) = e^{rt} \beta p^{*\eta} = k_2 e^{\delta_2 t}, \tag{5.7}$$

where

$$k_1 \equiv (\gamma/\beta)^{1/(\epsilon + \eta)} > 0, \tag{5.8}$$

$$k_2 \equiv \beta k_1^{\eta} > 0, \tag{5.9}$$

$$\delta_1 \equiv (\alpha g - r)/(\epsilon + \eta) \gtrless 0, \tag{5.10}$$

$$\delta_2 \equiv (\epsilon r + \eta \alpha g)/(\epsilon + \eta) \gtrless 0. \tag{5.11}$$

Since growth rates of equilibrium price and quantity are given respectively by

$$\hat{p}^* = \frac{d \log p^*(t)}{dt} = \delta_1, \tag{5.12}$$

$$\hat{q}^* = \frac{d \log q^*(t)}{dt} = \delta_2, \qquad (5.13)$$

these growth rates are clearly constant in the present model and their determining factors are relatively easy to analyse. This simplicity is a direct result of the multiplicative forms of the supply and demand functions in (5.3) and (5.4), as well as of the assumption of constant rates of change of income and 'technology'. In problems involving growth, it is often *linearity in logarithms*, rather than linearity in the variables themselves, that gives rise to the simplest cases.

First consider the growth path of the equilibrium price. The trend in price will be upward, downward, or perfectly flat according as $\delta_1 > 0$, $\delta_1 < 0$, or $\delta_1 = 0$. Since $\epsilon + \eta > 0$, from (5.10) we see that this depends entirely on the sign of $(\alpha g - r)$. Given $g, r > 0$ (consumers' income is growing, technology is 'improving'), we then have the possibilities (see also question 1 of Exercise 5.1):

1. $\alpha < 0$, that is the good is an inferior good. In that case $\hat{p}^* = \delta_1 < 0$ and the trend of market price is certainly downward. Falling demand and increasing supply must cause price to fall.
2. $\alpha > 0$, that is the good is a normal good. In that case we have three sub-cases:
   (a) $\alpha g > r$, so that $\hat{p}^* = \delta_1 > 0$. In this case, the growth in demand, which depends both on the rate of growth of income, g, and the income elasticity of demand, α, is sufficiently rapid that it overcomes the price-reducing effect of technological change, and causes a rising trend in price.
   (b) $\alpha g = r$, so $\hat{p}^* = \delta_1 = 0$. The two conflicting pressures on price just offset each other.
   (c) $\alpha g < r$, so $\hat{p}^* = \delta_1 < 0$. The price-reducing effect of technological improvement overcomes the price-increasing effect of demand growth.

Now consider the trend in quantity. Again we can ignore $\epsilon + \eta$ and consider only the sign of $(\epsilon r + \eta \alpha g)$. There are again two main cases:
1. $\alpha > 0$, the good is normal. Then we have that $\hat{q}^* = \delta_2 > 0$. Quantity of the good traded *must* grow over time.
2. $\alpha < 0$, the good is inferior and so demand is falling. There are three sub-cases:
   (a) $\epsilon r > -\eta \alpha g$. The quantity-expanding effect of technological change outweighs the contracting effect of declining demand.
   (b) $\epsilon r = -\eta \alpha g$. The conflicting pressures are just in balance.
   (c) $\epsilon r < -\eta \alpha g$. The contracting effect of declining demand outweighs the expansionary effect of technological change.

Table 5.1 summarizes these results. Each cell of the table shows the condition on the parameters which gives the corresponding pair of price and quantity trends, and an empty cell implies that that case cannot arise (recall that we assume throughout that ϵ, η, r, and g are all positive). Thus on the given assumptions we could never observe a rising price *and* falling quantity, or constant price *and* constant quantity.

Table 5.1 Effects of income growth and technological progress on equilibrium price and quantity trends

	$\hat{p}^* > 0$	$\hat{p}^* = 0$	$\hat{p}^* < 0$
$\hat{q}^* > 0$	$\alpha g > r > 0$	$\alpha g = r > 0$	$0 < \alpha g < r$ or $\epsilon r > -\eta \alpha g > 0$
$\hat{q}^* = 0$			$\epsilon r = -\eta \alpha g > 0$
$\hat{q}^* < 0$			$-\eta \alpha g > \epsilon r > 0$

The analysis we have carried out in this section is somewhat mechanical. All behaviour of buyers and sellers is summarized in the supply and demand functions and there is no explicit 'dynamic decision-taking' as such. For example, we have not considered the question: how might buyers' and sellers' perception of the price trends we have analysed feed back on to the decisions they take, and hence influence the process itself? Nevertheless, though it is quite limited, the analysis can be quite useful, as we hope some of the following exercises will show.

Exercise 5.1

1. Use supply and demand diagrams to illustrate the various cases and results analysed in this section, and in particular use them to confirm Table 5.1.
2. If there is inflation, i.e., a process of increasing levels of prices in general, then we need to distinguish between the nominal and real price of a good. The former is its money price, the latter its price in terms of real purchasing power. If p is nominal price, and P is a measure of the general price level, then real price $p_R = p/P$. In general, the price that is relevant for microeconomic analysis is the real price. That is, we assume that demand and supply functions are defined on real price rather than nominal price. Take the model of this section and show the consequences of replacing p by $p_R = p/P$ in (5.3) and (5.4). [Hint: assume some initial price level P_0 and assume a constant inflation rate i so that $P(t) = e^{it}P_0$.] In (5.2) how should income be defined?
3. Use Table 5.1 to point out the problems involved in taking pairs of values of price and quantity at successive points in time and using these to construct 'the market demand curve'. Under what conditions (state them both diagrammatically *and* algebraically) *would* such a procedure yield a reasonably close — but systematically biased — representation of this demand curve?
4. In the light of the analysis of this section, discuss the following observations:

(*a*) although normal goods and highly income elastic, the prices of TV sets and international air travel have fallen considerably in real terms over the past decade:

(*b*) the real price of houses has increased phenomenally over the past twenty-five years;

(*c*) under the Common Market's agricultural policy, the result of fixing prices above equilibrium has been rapidly growing surpluses of agricultural products.

5. Suppose that instead of technological change *increasing* supply at every price, a process of increase in real wages causes decreasing supply at every price. Show how the model of this section can be adapted to analyse this case, and derive the counterpart of Table 5.1 for it. Can you think of specific markets whose price and quantity trends seem well-described by this model?

5.2 Market disequilibrium

In this section we examine some models concerned with the way in which expectations about future prices influence how markets work. Expectations can be defined as beliefs about future values of variables or states of the world. They are important because they are likely to influence current decisions in many contexts. This is of central concern in situations of *disequilibrium*. In static equilibrium, price is constant and all buyers and sellers are actually making the trades that they wanted or planned. So it seems reasonable to assume that, in the absence of change in underlying determinants of the equilibrium, people expect to go on doing what they are doing, and do so. Out of equilibrium, on the other hand, prices will be changing. We should take account of the fact that expectations of changes will influence what people do now, which in turn determines the nature of the changes themselves. Clearly, this kind of process must be modelled by a dynamic analysis, i.e., one in which time and changes in variables over time enter explicitly.

We begin with the *cobweb model* which, while not very satisfactory in itself, is a useful starting point. The core of the model is the idea of a *supply lag*, which can perhaps most clearly be explained in terms of agricultural products. At some point in time a crop is sown, and one period later — let us call it a year, though this may well differ in reality — it is harvested. Ignoring the uncertainties due to weather, pests, etc., the size of the harvest is determined by how much seed is sown. Assume also that the entire harvest is put on to the market to fetch whatever price it can. Hence the supply at time t, q_S^t, is determined by a decision, the planting decision, made at time $t - 1$. We would hypothesize that the decision on what acreage to plant is made in light of the expectations the farmers have at $t - 1$ of what the market price will be at time t. Denote this by p_{t-1}^e: the superscript e denotes it as an expectation of a price one year ahead and the subscript $t - 1$ indicates when the expectation was held (and the decision taken). Thus we write the (linear) supply function

$$q_S^t = \alpha + \beta p_{t-1}^e. \tag{5.14}$$

Supply at t is completely determined by the price expected at $t - 1$. Note that the parameters α and β are time invariant.

We assume that there are no lags on the demand side: buyers can adjust their purchases instantaneously to *current* price, p_t. Hence we have the usual kind of (linear) demand function

$$q_D^t = a - bp_t. \tag{5.15}$$

In order to be able to solve this model we need to specify precisely how farmers' price expectations are formed. The assumption that characterizes the cobweb model, and the source of most dissatisfaction with it, is that of *naive expectations*, i.e.

$$p_{t-1}^e = p_{t-1}. \tag{5.16}$$

This says that farmers always expect that the current price will still be the price in one year's time, and they take their planting decision accordingly. Naive indeed: it is as if they believe they are always in equilibrium.

Substituting from (5.16) into (5.14), and using the equilibrium condition $q_D^t = q_S^t$, then gives the basic first-order difference equation of the system,

$$p_t = \frac{a - \alpha}{b} - \frac{\beta}{b} p_{t-1} \qquad t = 1, 2, ..., \tag{5.17}$$

and we assume we have a given initial value of price, p_0. Note that the equilibrium price in this market is again $p^* = (a - \alpha)/(b + \beta)$. One way of seeing this is to note that an equilibrium must persist over two consecutive periods, i.e., $p_{t-1} = p_t = p^*$, and substituting p^* for p_{t-1} *and* p_t in (5.17) and solving for it then gives the equilibrium value (do it). It follows from this that if $p_0 = p^*$ then (5.17) has the constant solution $p_t = p^*$. We assume $p_0 \neq p^*$, i.e., the market is initially out of equilibrium. Then the solution to the first-order linear difference equation in (5.17) is

$$p_t = \left(\frac{-\beta}{b}\right)^t (p_0 - p^*) + p^* \qquad t = 1, 2, ... \tag{5.18}$$

from which we can determine the behaviour of market price over time. There are three cases to consider and these are illustrated in Fig. 5.1.

First, if $\beta/b < 1$, then $(-\beta/b)^t$ goes to zero, alternately through negative (t odd) and positive (t even) values. Thus p_t will converge to p^* through damped oscillations, as Fig. 5.1 illustrates.

Next, if $\beta/b > 1$, then $(-\beta/b)^t$ goes to infinity, and so p_t diverges from p^* through increasing oscillations, as Fig. 5.1 illustrates.

Finally, if $\beta/b = 1$, then $(-\beta/b)^t$ oscillates from -1 (t odd) to $+1$ (t even), and so p_t oscillates permanently between the values p^0 and $p^* + (p^* - p^0)$, never converging, as the figure shows.

Thus we have, as the stability condition in this model, that $\beta < b$, i.e., that the slope of the supply curve to the p-axis is less than that of the demand curve (in the conventional supply–demand diagram the supply curve must be *steeper* than the demand curve, referred to the q-axis).

Fig. 5.2 illustrates the stable and unstable cases in a supply–demand diagram, which also shows why the model is called 'the cobweb'.

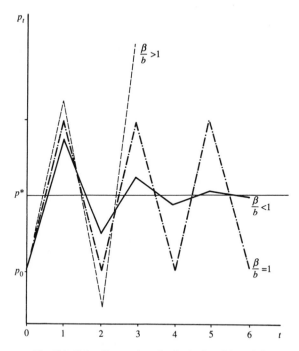

Fig. 5.1: Price fluctuations in the 'cobweb' model

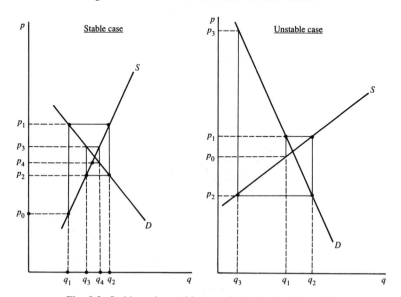

Fig. 5.2: Stable and unstable cases in the 'cobweb' model

If we compare the stability condition in the cobweb model to that in the market model of Chapter 2 we see that it is no longer sufficient for stability that the supply and demand curves have the usual slopes — the introduction of supply lags has changed that. It should also be noted that the importance of supply lags extends well beyond agricultural markets. For example, in any manufacturing industry, firms may be able to adjust current supply along a short-run supply curve that is not vertical, but nevertheless their future supplies (the position of their future short-run supply curves) will depend on current decisions about the scale of future capital stock, i.e., on current investment decisions. These in turn will depend on current expectations of future prices. So we have for manufacturing industries the same kind of supply lag as in the cobweb model, with the (slight) added complication that the short-run curve in any period is positively sloped instead of vertical. Thus the cobweb model is concerned with a very common feature of markets in general.

The main source of dissatisfaction with the cobweb model is the naive expectations assumption. Out of equilibrium, period after period sellers' expectations are consistently falsified — price next period is not the same as it was this period — and yet they go on believing that it will be. There is something very irrational in this, both in the everyday sense, and in the sense that it is not very profitable behaviour. Consider the possibility of a clever farmer who realizes that, because price is low today, everyone else will be cutting their acreage of the crop. Then he calculates that next period the price will be high so he could expand *his* acreage now. But then why should not *more* farmers reason like this, and form their expectations *rationally*, in the sense that they try to predict what the actual market outcome will be? Pursuing this question leads to a way of modelling expectations known as the *rational expectations* approach, and this has radical implications for the analysis of stability.

The idea underlying the theory of rational expectations is this: the expectations that rational economic agents form are the predictions of the economic model itself. After all, if *we* know the structure of the market model, and it is costly to make mistaken price predictions, then there are profits to be made in selling our market model to the market participants!

To see the implications of this, suppose that all N sellers in the market have identical supply functions of the linear form (with zero intercept)

$$q_i^t = \gamma p_{t-1}^i \quad i = 1, 2, ..., N \tag{5.19}$$

where p_{t-1}^i is the price that seller i, at $t - 1$, *expects* to prevail at t. Suppose $n \geq 0$ of the N sellers are naive, in that they form their expectations according to $p_{t-1}^i = p_{t-1}$, as before. However, suppose that $r = N - n \geq 0$ of the sellers are rational, in that they:

(a) know the market demand function, $a - bp_t$, and the supply parameter, γ, as well as the number of sellers who are naive and rational respectively;

(b) form their expectations of next period's price by solving for that period's equilibrium, as we are now going to do.

The total supply forthcoming from the naive sellers will be n times the supply from any one of them, since they are identical. From (5.19) this is

$$q_{(n)}^t = nq_i^t = n\gamma p_{t-1}^i = n\gamma p_{t-1}. \tag{5.20}$$

The total supply from the rational sellers will be r times the supply from any one of them:

$$q_{(r)}^t = rq_i^t = r\gamma p_{t-1}^i = r\gamma p_t. \tag{5.21}$$

The key point is the last equality in (5.21). The price expectation of the rational seller is the actual price p_t at t. How do they find this out? By solving the equilibrium condition $q_D^t = q_{(n)}^t + q_{(r)}^t$, i.e., by solving for p_t in

$$a - bp_t = n\gamma p_{t-1} + r\gamma p_t = \gamma(np_{t-1} + rp_t). \tag{5.22}$$

To facilitate comparison with the cobweb model, note that the market supply curve slope β in that model is equal to $N\gamma$, since in that model all N sellers are naive. Hence we can rewrite (5.22) as

$$a - bp_t = \beta(sp_t + (1 - s)p_{t-1}) \tag{5.23}$$

where $s = r/N$ is the proportion of sellers in the market who are rational. Then, by solving (5.23) for p_t, we obtain the linear first-order difference equation

$$p_t = \frac{a}{b + s\beta} - \frac{\beta(1 - s)}{b + s\beta}p_{t-1} \qquad t = 1, 2, \dots \tag{5.24}$$

Note that the rational sellers can be thought of as computing this difference equation, so that at each point in time they can use it to forecast *correctly* next period's actual price.

The market equilibrium price is again $p^* = a/(b + \beta)$. To see this, note again that for equilibrium we require $p_t = p_{t-1} = p^*$, and so substituting this into (5.24) and taking the right-hand term in p^* to the left-hand side gives

$$\left[1 + \frac{\beta(1 - s)}{b + s\beta}\right]p^* = \left[\frac{b + s\beta + \beta(1 - s)}{b + s\beta}\right]p^* = \frac{a}{b + s\beta} \tag{5.25}$$

so that

$$p^* = \frac{a}{b + s\beta + \beta(1 - s)} = \frac{a}{b + \beta} \tag{5.26}$$

as required. Note therefore that the *equilibrium* price is independent of the values of n and r, being in effect determined by the condition

$$a - bp^* = N\gamma p^* = \beta p^*. \tag{5.27}$$

The proportion of rational sellers in the market determines only the stability of equilibrium and, if stable, the rate of convergence, rather than the equilibrium itself.

The solution of the difference equation in (5.24) is given by:

$$p_t = \left(\frac{-(1-s)\beta}{b+s\beta} \right)^t (p_0 - p^*) + p^* \qquad t = 1, 2, \dots \qquad (5.28)$$

This can be compared with (5.18), reproduced here for convenience:

$$p_t = \left(\frac{-\beta}{b} \right)^t (p_0 - p^*) + p^*. \qquad (5.29)$$

We see immediately that the stability of the system depends crucially on s, the proportion of sellers who are rational. If $s = 0$ then we have of course the original cobweb model. As s increases, the absolute value of $-\beta(1-s)/(b+s\beta)$ clearly declines, thus making it more likely that the system will converge. In other words, in the present model, we may have values of b and β which would have caused instability in the cobweb model but, for suitable s, are consistent with stability here. (See question 6 of Exercise 5.2.)

Finally, note the most radical result: if $s = 1$, then, from (5.28), $p_t = p^*, t = 1, 2, \dots$ In other words, if there is a disequilibrium at $t = 0$, this is immediately corrected next period: the system moves directly to equilibrium and stays there. Thus, if *all* sellers are rational, not only is the system *always* stable, but equilibrium is immediately restored after a change which creates disequilibrium.

We should clarify the economic reasoning underlying rational expectations. Essentially, it is costly to forecast the wrong price: if a seller supplies $\hat{q}_S$ at t when the market price is actually p^*, and $\hat{q}_S \neq q_S^* = \gamma p^*$, he will have lost profit as a result, since q_S^* is profit-maximizing supply at price p^*. Given that we assume that sellers take decisions rationally, it seems a reasonable extension to assume that they also try to forecast prices rationally. But if they all do this, then they will forecast the equilibrium price and so plan to supply the equilibrium quantity, and so equilibrium will be achieved! They will forecast the equilibrium price because if all are rational ($s = 1$ in (5.28)) each will seek the solution to (5.27), which then gives the equilibrium price.

It may be felt that the rationality postulate in this model is *too* strong, and its implication, that the market will always be in equilibrium except when hit by unforeseen shocks, not plausible. This latter point, of course, is a matter for empirical testing. On the first point, perhaps the real criticism is not that of rationality as such, but of the informational requirements of the model. *If* sellers did know the underlying model then it is hard to see why they would not use it in the way described. But that they do know it may be an excessively strong assumption to make. Certainly, many firms and industry associations employ economists and econometricians precisely for the purpose of estimating market relationships and forecasting prices — interestingly, agriculture was one of the earliest areas of application of modern econometric methods — and so the real world may well be much closer to the model of rational expectations than that of naive expectations. Precisely how close, however, is still a matter of debate.

Exercise 5.2

1. Suppose that supply instantaneously responds to price, but that demand takes one period to adjust to changes in price. Give examples of goods for which this may be true. Set up a model of such a market and analyse its stability, on the assumptions both of naive and rational expectations.
   Can you solve second-order linear difference equations? If not, ignore the following question; if so, answer it.
2. Suppose that sellers form their expectations not quite naively, but according to a weighted average of current and past price, i.e.

$$p^e_{t-1} = \lambda p_{t-1} + (1 - \lambda)p_{t-2}, \quad 0 < \lambda < 1.$$

   Formulate the cobweb model for this case and analyse its stability. Compare this model of expectations with that of rational expectations.
3. If the demand function in a market is $q^t_D = 10 - 2p_t$, and each of the 100 identical sellers in the market has the supply function $q^t_i = 0.03\,p^i_{t-1}$, find the minimum proportion of rational sellers in the market necessary for equilibrium to be stable. Then generalize to give this minimum stable value of s for general parameters b and β.
4. In both cobweb and rational expectations models, what determines how *quickly* price converges to equilibrium, assuming it does so? Analyse the effect of having rational sellers in the market on the speed of convergence.
5. Take a pair of axes, label one b, the other β, and show the areas of the positive quadrant corresponding to (b, β) pairs for which the simple cobweb model is respectively stable and unstable. Then show what effect introducing the proportion of rational sellers s, where $0 < s \leq 1$, has on these areas.

5.3 Summary

Dynamic models of markets analyse the process of change through time in the variables the models seek to explain, and may be based either on the assumption that at each point in time the system is in equilibrium, or on the assumption that the system is moving between static equilibrium positions.

In an equilibrium model in which demand is affected by growth in consumers' incomes and supply by technological change, we are able to show how the time paths of price and quantity traded depend on the relative values of a number of key parameters: price elasticities of demand and supply; income elasticity of demand; the growth rate of income and the rate of technological change. The simplest model which allows us to see clearly how these factors interact involves multiplicative (i.e., log-linear) demand and supply functions and constant growth rates of exogenous variables.

In a model of the out-of-equilibrium behaviour of a market, based on the existence of a supply lag, we see that a fundamental issue is the assumption we make about how market participants form expectations.

Given the assumption of naive expectations, we are able to provide simple conditions under which a market will or will not be stable. However, naive expectations seem highly irrational, in that sellers do not appear to learn from experience and take consistently unprofitable decisions.

If we introduce the assumption that some proportion of sellers are rational, in that they are fully informed about all relevant market parameters (including this proportion of rational sellers!) and form their expectations of future price by solving the market model, we then find

    (*a*)   the class of parameter values for which equilibrium in the market is stable is wider than under naive expectations;

    (*b*)   the speed of convergence to a new equilibrium increases.

In the limit, when all sellers are rational, we find that equilibrium is always stable and the market takes only one period to restore itself to equilibrium.

Actual markets probably lie somewhere between the two extremes of naive and (fully) rational expectations. To ascertain how close to one or the other a particular market may lie is ultimately a matter for empirical research.

6 Theory of Consumer Demand

As we have seen in the previous chapters, demand and supply functions are the basic tools used to analyse the way markets work. Given their central importance, it seems worth while to try to understand how they can be derived from the decisions of individual consumers and firms. We therefore wish to construct models of consumer decision-taking and of firm decision-taking which will give us a deeper understanding of the nature of the aggregate market relationships. In this chapter and the next we analyse the determinants of demand.

A consumer typically takes three types of decisions:

(i)   how much of each good or service to buy;
(ii) how much time to spend working, and how much to spend at leisure;
(iii) whether to spend more than current income on buying goods and services, implying the need to borrow; or to spend less than current income, implying saving.

The decisions will tend to be closely interrelated, but as a first step we will consider them separately.

The first kind of decision determines the consumer's demands for goods and services: aggregating over consumers' demands for any one of them gives us the market demand function.

The second decision determines the individual's *short-run* labour supply, and hence, by aggregation, market labour supply. We call this 'short-run' because it presupposes the individual has already taken the fundamental decisions as to what skills to acquire and where to live, so that the type and location of job he or she can do are already determined and all that matters is how much time is spent working. In a sense, the latter is the less interesting problem, especially as, in many jobs, there may be constraints on hours of work, for example a fixed working week with limited possibilities of 'overtime'. Nevertheless, in this book we concentrate on this question, and consider it further in Chapter 12, where we analyse some aspects of the working of labour markets.

Finally, the third type of decision determines whether the consumer is a supplier (lender) or demander (borrower) on the *capital market*, the market for loans. Aggregating over individuals in this market gives us the demand for and supply of loans, and enables us to analyse the determination of their price, the market *interest rate*. We consider this aspect of consumer choice further in Chapter 13, where we analyse how capital markets work.

A central concept in the analysis of each of these decisions is the *utility function*. It is important to be clear about the nature of this function. The word 'utility' is simply a catch-all term for whatever it is that makes us want to consume goods and services. It indicates pleasure, satisfaction, wish fulfilment, physical comfort and well-being, snobbery, ostentation, and so on. Now the term 'utility function' may give the impression that we are somehow able to define some numerically measurable quantity of utility and describe how it varies as a function of other things. This is however emphatically not the case – we do *not* assume it is possible to measure numerically the subjective sensations consumers experience when they consume goods and services. We adopt a different approach entirely. We now go on to consider this approach.

6.1 Ordinal utility theory

First of all we define a *bundle of goods* as an n-tuple of numbers, $(x_1, x_2, \ldots, x_n)$, where there are n goods in total, and x_i measures the quantity of the ith good in the bundle, $i = 1, 2, \ldots, n$. Then we assume simply that the individual can always rank two alternative bundles of goods in terms of preference or indifference: he or she can always tell us that one is preferable to the other or that they are equally good. Underlying this statement of preference or indifference may well be some judgement about the pleasure, etc. each bundle of goods may yield, but we do not need to enquire into this and we certainly do not need to measure it. We then assume that, by means of a series of pairwise comparisons of the given set of bundles of goods, the consumer can arrange them all in an order of preference which, by assumption, has a particular kind of structure. Each bundle belongs to one and only one *indifference set*, i.e., a set of bundles between which the consumer is indifferent. The preference ordering can be described in terms of these indifference sets: the indifference sets give us a partition of the entire set of bundles.

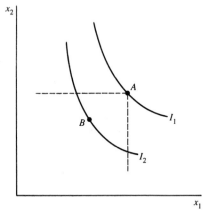

Fig. 6.1: Typical indifference sets

Some typical indifference sets, for bundles consisting just of two goods, are graphed in Fig. 6.1. A 'bundle of goods' is represented by a point in the figure, and the set of bundles or points lying along a curve are all indifferent to each other, and so form an indifference set. Not unreasonably the curves are therefore known as indifference curves. Since in general we assume that the consumer will always prefer more of a good to less of it (in fact that could be taken as a definition of what we mean by a good), a bundle such as A on I_1 must be preferred to a bundle such as B on I_2, and so all the bundles on I_1 must be preferred to all those on any lower indifference curve (see question 1 of Exercise 6.1). Thus, the higher the indifference curve a bundle is on, the higher it is in the consumer's preference ordering. The indifference curves we have drawn embody the important assumption about consumer preferences known as *convexity of preferences*, i.e., they are drawn convex to the origin. For more on this see question 6 of Exercise 6.1.

The basic information on tastes and preferences that we have for the consumer is the preference ordering, as described by the system of indifference sets or, graphically in two dimensions, the indifference curves. If we were content to carry out the analysis entirely in diagrammatic terms, this is all we would need. However, this is too limited a method: we would like to have a *numerical representation* of the consumer's preference ordering in the form of a function that we can work with analytically. We define this in the following way. The consumer's preference ordering arranges the bundles $(x_1, x_2, \ldots, x_n)$ into indifference sets. We adopt the convention: to each bundle $(x_1, x_2, \ldots, x_n)$ we assign a number u which indicates its place in the ordering, according to the rule, if one bundle is higher in the ordering than (is preferred to) another, then it must be given a higher number; if two bundles are in the same indifference set, they must be given the same number. We then write the function $u = u(x_1, x_2, \ldots, x_n)$ to indicate how the numbers we assign depend on the bundles we take.

The function $u(.)$ is usually referred to as an *ordinal utility function*. The use of the word utility is perhaps a little unfortunate, since the function does not measure utility in any sense. The term 'preference function' would be much better, but economists have simply been unable to purge their language of the old term utility. The word ordinal, usually dropped once the introduction to the function has been made, is included to try to avoid confusion: the function merely indicates the *order* of the bundles in the preference ranking — the 'utility numbers' u are simply place-markers.

Note that this implies that we are not very restrictive about the precise numerical function we use; the function simply has to increase with preferred bundles and stay constant over bundles indifferent to each other. It follows that if a particular function $u(x_1, x_2, \ldots, x_n)$ is a correct representation of a particular consumer's preference ordering, then so must be any other function $v(x_1, x_2, \ldots, x_n)$, as long as $v(.)$ increases when $u(.)$ increases and stays constant when $u(.)$ stays constant.

For example, suppose that we found that we could represent a consumer's preference ordering by the function

$$u = 2x_1^{1/2}x_2x_3 \tag{6.1}$$

in the 3-good case. This tells us for example that

(*a*) the bundle $(10, 4, 64)$ is preferred to the bundle $(8, 9, 27)$ (check it out);
(*b*) the bundle $(4, 8, 2)$ is indifferent to the bundle $(64, 1, 4)$ (again, check it out);

(c) any bundles satisfying the relationship $x_1 = 25(x_2 x_3)^{-2}$ are indifferent to each other (explain why; what is their utility number?).

Then, an *equally good* representation of the preference ordering is the function

$$v = 10x_1^3 x_2^6 x_3^6 = 5[2x_1^{1/2} x_2 x_3]^6 = 5u^6. \tag{6.2}$$

This is because any bundles which have the same u-number will have the same v-number, and if a bundle has a higher u-number than another it must also have a higher v-number than that other.

In going from $u(.)$ in (6.1) to $v(.)$ in (6.2) we made a transformation of a particular kind, called a *positive monotonic transformation*, because it makes v a strictly increasing function of u. This illustrates the general property of ordinal utility functions: if some function $u(.)$ is a representation of a preference ordering, then so is the function $v(.)$ provided that there is some positive monotonic transformation $T[.]$ such that

$$v = T[u] \tag{6.3}$$

where $dv/du = T' > 0$. Another way of putting this is to say that an ordinal utility function is unique only up to a positive monotonic transformation.

We can note a final concept concerning an ordinal utility function. We first assume that the utility function is differentiable. Consider a given indifference set by fixing a value of u, say u^0, and putting

$$u(x_1, x_2, \ldots, x_n) = u^0. \tag{6.4}$$

We consider differentials $dx_1, dx_2, \ldots, dx_n$ which are such as to leave the value of u unchanged at u^0, implying

$$du = u_1 dx_1 + u_2 dx_2 + \ldots + u_n dx_n = 0. \tag{6.5}$$

Set all except two of these differentials, say dx_1 and dx_2, to zero. Then we have

$$du = u_1 dx_1 + u_2 dx_2 = 0 \Rightarrow -\frac{dx_2}{dx_1} = \frac{u_1}{u_2}. \tag{6.6}$$

We define the term $-dx_2/dx_1$ as the *marginal rate of substitution* between goods 1 and 2. It can be interpreted as the rate at which one good must be substituted for the other to stay within the same indifference set. Diagrammatically, it is the (absolute value of the) slope of an indifference curve drawn in the (x_1, x_2)-space. Note that in general its value depends on the values of *all* n variables $x_1, x_2, \ldots, x_n$. We call $u_1, u_2, \ldots, u_n$ the *marginal utilities* of the respective goods, and thus the marginal rate of substitution at a point is given by the ratio of the goods' marginal utilities at that point.

From now on in this chapter we shall be using utility functions without further explanation or justification. So, it is useful to summarize here what we have assumed about them:

(i)   they do not measure utility in any sense; they simply indicate the place of a bundle in the consumer's preference ordering over the set of bundles. This ordering is the fundamental information we take as given;

(ii) they are differentiable to any order we might require;

(iii) they are unique up to a positive monotonic transformation. Among other things, this implies that all of the conclusions we derive from them (for which see the rest of this chapter) must continue to hold if we apply such a transformation to the particular utility function we have been dealing with, otherwise they are not valid conclusions;

(iv) when graphed in two dimensions (taking any section through an n-dimensional graph) the indifference curves are negatively sloped and convex to the origin;

(v)   they are increasing in the x_i (this, and the negative slope of the indifference curves, is due to the assumption that all the x_i are *goods*. We relax this assumption later).

Exercise 6.1

1. A consumer's preferences are said to be *transitive* if, when bundle A is preferred to bundle B, and bundle B to bundle C, then A is preferred to C, and likewise for indifference. Show that transitivity of preferences implies that indifference sets or curves cannot intersect. Show also that it implies that in Fig. 6.1, all the bundles on I_1 must be preferred to all those on any lower indifference curve.

2. Given the ordinal utility function $u = ax_1^{b_1} x_2^{b_2} \ldots x_n^{b_n}$, derive an expression for the marginal rate of substitution between some pair of the goods. Then apply the transformation $v = \alpha u^\beta$ and again derive the marginal rate of substitution between the same two goods. What do you notice about the results in each case? From this, generalize to show that given any function $u(x_1, x_2, \ldots, x_n)$, applying a positive monotonic transformation $T[u]$, with $T' > 0$, leaves the marginal rate of substitution at a point unchanged. [Hint: use the function of a function rule of differentiation.]

3. Explain why:
   (*a*) getting on to the highest possible indifference curve;
   (*b*) choosing an element in the highest possible indifference set;
   (*c*) maximizing the value of the ordinal utility function;
   (*d*) doing the best you can,
   are all *equivalent* statements about consumer behaviour.

4. If x_i is a good, $i = 1, \ldots, n$, what sign will u_i have? What sign will u_{ii} have? (Careful!)

5. In the two-good case, a consumer's marginal rate of substitution is given by $-dx_2/dx_1 = u_1(x_1, x_2)/u_2(x_1, x_2)$. Derive an expression for d^2x_2/dx_1^2 (a good test of your skills at differentiation). Relate this derivative to the curvature of the indifference curves in Fig. 6.1 and say what sign the derivative ought to have, to be consistent with that curvature.

6. *Convexity of indifference curves*: The shape of the indifference curves in Fig. 6.1 is said to reflect 'diminishing marginal rate of substitution', i.e., $-dx_2/dx_1$ falls as x_1 increases. Give a common-sense interpretation of this property. Sketch indifference curves for which the marginal rate of substitution is (*a*) constant, (*b*) increasing.

6.2 Demand functions

We assume from now on that a consumer's preference ordering over bundles of goods $(x_1, x_2, \ldots, x_n)$ can be represented by an ordinal utility function $u(x_1, x_2, \ldots, x_n)$ (we also drop the word ordinal for brevity). The central assumption we make about the consumer's behaviour is that he or she *optimizes*: given any available set of bundles of goods, the consumer chooses that bundle which maximizes the utility function (recall question 3 of Exercise 6.1). The consumer's demand for a good is then the quantity chosen as a result of this maximization, and this will clearly depend on precisely what set of bundles of goods is available. We now have to see what determines this.

In this chapter, we assume that the consumer neither borrows nor lends, and that the labour supply decision has also been taken. This means that the income available for expenditure on goods, which we denote by y, is fixed. The basic constraint that faces the consumer in the absence of borrowing or lending is that expenditure must equal income, y. Now total expenditure on any bundle of goods $(x_1, x_2, \ldots, x_n)$ is found by multiplying the quantity bought of each good by its price and then summing these expenditures, and so we can write the consumer's *budget constraint* as

$$p_1 x_1 + p_2 x_2 + \ldots + p_n x_n = y. \tag{6.7}$$

This equation defines the set of bundles $(x_1, x_2, \ldots, x_n)$ the consumer can just afford to buy, *given* the prices $p_1, p_2, \ldots, p_n$ and income y. Out of the set, the consumer wants to choose the bundle that maximizes the utility function. Mathematically, that means we want to solve the constrained maximization problem:

$$\max \; u(x_1, x_2, \ldots, x_n) \quad \text{s.t.} \; p_1 x_1 + p_2 x_2 + \ldots + p_n x_n = y. \tag{6.8}$$

So, using the Lagrange method, we have as first order conditions

$$
\begin{aligned}
u_1 - \lambda p_1 &= 0, \\
u_2 - \lambda p_2 &= 0, \\
&\cdots\cdots\cdots \\
u_n - \lambda p_n &= 0, \\
\sum_{i=1}^{n} p_i x_i - y &= 0,
\end{aligned}
\tag{6.9}
$$

which gives us $n + 1$ conditions in the $n + 1$ unknowns, $x_1, \ldots, x_n$, and λ. The assumptions we have made about the utility function imply that we can solve for these unknowns. The key point to note is that these solution values will in general depend on all the parameters of the problem, and in particular the values of the prices $p_1, \ldots, p_n$, and income y. We express this by writing the solution quantities as functions of the parameters, i.e.

$$x_1 = D_1(p_1, p_2, \ldots, p_n, y),$$
$$x_2 = D_2(p_1, p_2, \ldots, p_n, y),$$
$$\cdots\cdots\cdots\cdots\cdots\cdots\cdots$$
$$x_n = D_n(p_1, p_2, \ldots, p_n, y). \tag{6.10}$$

As the notation is meant to suggest, these functions are precisely the demand functions we are looking for: they express the quantity of each good the consumer wants to buy – the solution to the optimization problem – as a function of price and income. We can use them therefore to consider for the individual consumer the various *ceteris paribus* relationships, such as the demand and Engel curves, that we defined for the market as a whole in Chapter 3. Before doing this, however, we look more closely at what is involved in going from the statement of the problem to the demand functions in (6.10).

Taking any two of the conditions in (6.10), say the first two, we can rearrange and then divide them to obtain

$$\frac{u_1}{u_2} = \frac{p_1}{p_2}. \tag{6.11}$$

This says that at the optimum, the marginal rate of substitution between the two goods is equal to the ratio of their prices. In diagrammatic terms this corresponds to a tangency between an indifference curve and the budget constraint (specialized to two dimensions – draw the diagram).

Now we consider a specific example of the problem. Suppose that there are only two goods and that the consumer's preferences can be represented by the utility function

$$u = x_1^{b_1} x_2^{b_2}. \tag{6.12}$$

Then the Lagrange function for the problem is $ax_1^{b_1} x_2^{b_2} + \lambda(y - p_1 x_1 - p_2 x_2)$, and the first-order conditions are

$$b_1 x_1^{(b_1 - 1)} x_2^{b_2} - \lambda p_1 = 0, \tag{6.13}$$
$$b_2 x_1^{b_1} x_2^{(b_2 - 1)} - \lambda p_2 = 0, \tag{6.14}$$
$$p_1 x_1 + p_2 x_2 - y = 0. \tag{6.15}$$

Rearranging the first two conditions and dividing the first by the second gives

$$\frac{b_1 x_1^{(b_1 - 1)} x_2^{b_2}}{b_2 x_1^{b_1} x_2^{(b_2 - 1)}} = \frac{b_1}{b_2} \frac{x_2}{x_1} = \frac{p_1}{p_2}. \tag{6.16}$$

We can therefore solve for x_1 in terms of x_2 to obtain

$$x_1 = \frac{p_2}{p_1} \frac{b_1}{b_2} x_2. \tag{6.17}$$

Note that, since p_2, p_1, b_1, and b_2 are all given parameters, (6.17) is a linear equation – it defines a straight line through the origin, as shown in Fig. 6.2. It is a general expression for *all* the points of tangency between budget lines with slope $-p_1/p_2$, and the consumer's indifference curves, since it is derived from the condition $u_1/u_2 = p_1/p_2$. The curve or line tracing out these points is usually called the income consumption curve. (6.17) is by itself insufficient to solve the consumer's problem since it is just one equation in two unknowns. It is simply a statement of the general condition that the solution be at *some* point of tangency. However, when we put (6.17) together with the budget constraint (6.15), we can substitute for x_1 to obtain

$$p_1\left(\frac{p_2}{p_1}\frac{b_1}{b_2}x_2\right) + p_2x_2 = y \implies x_2 = \frac{1}{(1 + b_1/b_2)}\frac{y}{p_2} = \frac{s_2y}{p_2} \tag{6.18}$$

where $s_2 = 1/(1 + b_1/b_2) = b_2/(b_1 + b_2)$.

By substituting back into (6.17) we can solve for x_1 (do it!) to obtain

$$x_1 = \frac{s_1y}{p_1} \tag{6.19}$$

where $s_1 = 1/(1 + b_2/b_1) = b_1/(b_1 + b_2) = 1 - s_2$.

Then, (6.18) and (6.19) give us the consumer's demand functions for x_2 and x_1 respectively. Diagrammatically, what we have done is to find the point of intersection of the budget constraint with the income consumption curve (refer to Fig. 6.2), or, equivalently, the point of tangency between the budget constraint and an indifference curve.

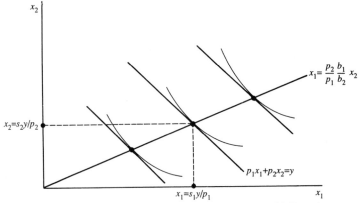

Fig. 6.2: Solution of the consumer's problem in the Cobb-Douglas case

It is always of interest to examine the properties of a demand function, by which we mean here the nature of the *ceteris paribus* relationships we derive from it. In the present case, we note:

(i)   Both goods are normal, with Engel curves which are straight lines through the origin, with slopes $\partial x_i/\partial y = s_i/p_i$, $i = 1, 2$.

(ii) Both demand curves have negative slopes, since $\partial x_i/\partial p_i = -s_i y/p_i^2$, $i = 1, 2$, though neither is linear.

(iii) Multiplying each x_i by p_i, we have $p_i x_i = s_i y$, $i = 1, 2$. In words, expenditure on each good is a fixed share of income regardless of price, and so the demand curves are rectangular hyperbolae: expenditure is constant at all prices.

(iv) A corollary of (iii), the own price elasticities along the demand curves are constant and equal to 1. This is becauase $\log x_i = \log(s_i y) - \log p_i$ and so $e_p = -\partial \log x_i/\partial \log p_i = 1$.

(v) Finally, a rather special and unattractive property of the demand curve derived from the particular type of utility function used in this example, we have $\partial x_1/\partial p_2 = \partial x_2/\partial p_1 = 0$, since only the good's own price appears in its demand function.

This example should have helped understanding of the general solution given earlier. We now would like to say something about the properties of the general demand functions in (6.10), in particular about the nature of responses of demands to own prices, $\partial x_i/\partial p_i$, and to income $\partial x_i/\partial y$, $i = 1, \ldots, n$. However, we do not have a lot to go on: essentially, apart from technical properties like differentiability of the utility function, all we have is the assumption that the consumer's preferences are described (in any 2 dimensions) by a set of non-intersecting, convex-to-the-origin indifference curves. The first point to make is that we cannot in general infer from that assumption the sign of the derivative $\partial x_i/\partial y$, i.e., the slope of the consumer's Engel curve. The simplest way to see this is by drawing diagrams, and question 2 of Exercise 6.2 asks you to do this. All we can do is to *classify* goods as either *normal* ($\partial x_i/\partial y \geq 0$) or *inferior* ($\partial x_i/\partial y < 0$). It turns out that as a result of this we cannot in general place a definite sign on the own price derivative $\partial x_i/\partial p_i$ of the demand function D_i, $i = 1, \ldots, n$. That is, we cannot say that individual demand curves definitely do slope downward. It really does not seem very helpful if all we can say is that in response to a price change a consumer's demand for the good may go up, down, or indeed stay the same. Clearly this calls for further investigation, and we shall conduct that in the next chapter. In the mean time, in the remaining section of this chapter, we shall *assume* that $\partial x_i/\partial p_i < 0$, and consider the question of what this analysis has told us about market demand.

Exercise 6.2

1. Derive the demand functions for a consumer with the utility function $u = a(x_1 - c_1)^{b_1}(x_2 - c_2)^{b_2}$ where $x_i \geq c_i$, $i = 1, 2$, and comment on their properties. How would you interpret the c_i?

2. By drawing a suitable diagram, for a two-good case, show that the assumption that the consumer's preferences can be represented by a set of non-intersecting, convex-to-the-origin indifference curves does not rule out the possibility that a good is inferior. By drawing a second diagram, show that the assumption does not rule out the possibility that when the price of a good *falls* the consumer may *reduce* the amount she wants to buy. Finally, show also that *both* cases $\partial x_2/\partial p_1 > 0$ and $\partial x_2/\partial p_1 < 0$ are possible.

3. Apply the transformation $v = \alpha u^\beta$ to the utility function in (6.12). Then show that the consumer demand functions are unaffected. Explain why this is the case (recall question 2 of Exercise 6.1). Generalize to show that applying a positive monotonic transformation to any utility function leaves the demand functions unchanged.

4. Suppose that in the general model of (6.8), all prices and the consumer's income are multiplied by a factor $k > 1$. What happens to the consumer's demand? Explain your result. Explain why this implies that we can always *normalize* prices and income by multiplying them all by some factor k.

6.3 Individual and market demand functions

We motivated the analysis of individual consumer demand in this chapter by saying that we wanted to provide some theoretical foundation for that valuable analytical tool, the market demand function. This chapter has shown how prices, the consumer's income, and preferences (i.e., tastes, habits, desires, and drives) interact to determine the individual demand function. Suppose, then, that we have individual demand functions for a good. We introduce a superscript h to denote any one of the H individuals in the market for a good and, for simplicity, we drop the subscript on the good, denote its price by p, and ignore the prices of other goods. Then individual h's demand function can be written

$$x^h = D^h(p, y^h) \qquad h = 1, \ldots, H. \tag{6.20}$$

The notation expresses the idea that demands and incomes may well vary across consumers, but all consumers face the same price p. The market demand x is then the sum of the individual demands

$$x = \sum_{h=1}^{H} x^h = \sum_{h=1}^{H} D^h(p, y^h) = D(p, y^1, y^2, \ldots, y^H). \tag{6.21}$$

With all consumers' incomes held constant, we have

$$\frac{\partial x}{\partial p} = \frac{\partial D}{\partial p} = \sum_{h=1}^{H} \frac{\partial D^h}{\partial p}, \tag{6.22}$$

which tells us that the slope of the market demand curve is simply the sum of the slopes of the individual demand curves. In the next chapter we shall show that the slope of the individual demand curve depends on two things:

(a) *the substitution effect*: the extent to which a change in price of a good, with all other prices and real income held constant, causes it to be substituted for other goods;

(b) *the income effect*: the extent to which a change in money income, with all prices held constant, changes the demand for the good.

Thus, the slope of the market demand curve depends on the aggregate of these effects across consumers. The more readily substitutable for other goods a particular good is, and the greater the increase in consumption which would follow from an increase in consumers' incomes, the greater the slope of its market demand curve. Inferior goods, or goods with few possibilities of substitution, will tend to have demand curves with smaller slopes.

However, *composition effects* are also important. That is, the precise pattern of demand responsiveness to price across consumers also plays a role. To see this it is simplest to work in terms of elasticities. By analogy with the definition of elasticity in Chapter 3, define the *h*th consumer's own price elasticity of demand as

$$e_p^h = -\frac{p}{x^h}\frac{\partial D^h}{\partial p}. \tag{6.23}$$

Also, define the *h*th consumer's *consumption weight* as $w^h = x^h/x$. Recall that market price elasticity is $e_p = -(p/x)(\partial D/\partial p) = -(p/x)(\sum_{h=1}^{H} \partial D^h/\partial p)$. Then we have

$$e_p = -\frac{p}{x}\sum_{h=1}^{H}\frac{\partial D^h}{\partial p} = -\frac{p}{x}\sum_{h=1}^{H}\left(\frac{x^h}{x^h}\right)\frac{\partial D^h}{\partial p} = -\sum_{h=1}^{H}\left(\frac{x^h}{x}\right)\frac{p}{x^h}\frac{\partial D^h}{\partial p} = \sum_{h=1}^{H}w^h e_p^h. \tag{6.24}$$

That is, the market price elasticity is a weighted sum of the individual consumers' price elasticities, where the weights are the shares consumers have in total market consumption. This implies for example that if relatively large consumers of a good have high price elasticities, then the market demand for the good will tend to be highly elastic. Question 1 of Exercise 6.3 asks you to explore this further.

Turning now to the relation between demand and income, we notice an important difference between the market demand function in (6.21) and that introduced in Chapter 2. Here, the market demand is a function of the entire set of consumer incomes $\{y^1, \dots, y^H\}$, whereas earlier we wrote market demand as a function of aggregate consumers' income $y = \sum_{h=1}^{H} y^h$. Now in general, summing a given set of individual demand functions will *not* give us a function of the *sum* of individual incomes. For example, if each consumer has the type of demand function given in (6.18), with s^h the share of the *h*th consumer's income spent on this good, we have

$$x = \sum_{h=1}^{H}x^h = \sum_{h=1}^{H}\frac{s^h y^h}{p} = \frac{1}{p}\sum_{h=1}^{H}s^h y^h. \tag{6.25}$$

Clearly, we can factor out the s^h in (6.25) *only if* they are all equal, say to $\hat{s}$, in which case we have

$$x = \frac{\hat{s}}{p}\sum_{h=1}^{H}y^h = \frac{\hat{s}y}{p} \tag{6.26}$$

as required.

We can generalize from this example. The necessary and sufficient conditions under which the sum of individual demand functions is a function of the sum of incomes is that every consumer's demand function take the form

$$x^h = a^h(p) + b(p)y^h. \tag{6.27}$$

That is, there may be an element specific to consumer h, such as a^h, but if this is not a constant, then it must depend *only* on price, not income; the coefficient of income must be the same for all consumers (b does not carry an h superscript), although it may also depend on price. Clearly, (6.26) is a special case of (6.27), with $a^h(p) = 0$, all h, and $b(p) = \hat{s}/p$.

Condition (6.27), which is known as an *aggregation condition*, is clearly very strong and unlikely to be satisfied in reality. It requires that every consumer's Engel curve be linear with the same slope: an increase of \$1 in income leads to the same change in demand for all consumers and at all income levels.

If the aggregation condition (6.27) is not satisfied, then we have to recognize that market demand may be affected by a change in the distribution of income as well as by a change in the total. Also, the effect on market demand of a given change in total income may depend on how it is distributed among consumers: clearly, if relatively more of the income increase flows to consumers with relatively high income elasticities of demand for the good then demand will increase by more than if these consumers received relatively less of the income increase.

To make this more precise, recall that $e_y \equiv (\partial D/\partial y)(y/x)$ is the income elasticity of market demand, and let $e_y^h \equiv (\partial D^h/\partial y^h)(y^h/x^h)$ be consumer h's income elasticity of demand. Let dy^h/dy represent the effect on h's income of a change in total consumer's income (with of course $\sum_{h=1}^{H} dy^h/dy = 1$). Then we have, using the function of a function rule,

$$\frac{\partial D}{\partial y} = \sum_{h=1}^{H} \frac{\partial D^h}{\partial y^h} \frac{dy^h}{dy} = \sum_{h=1}^{H} \frac{\partial D^h}{\partial y^h} \left(\frac{y^h}{y^h}\right) \left(\frac{x^h}{x^h}\right) \frac{dy^h}{dy}$$

$$= \sum_{h=1}^{H} e_y^h \left(\frac{x^h}{y^h}\right) \frac{dy^h}{dy}. \tag{6.28}$$

Now, define $\delta^h \equiv (dy^h/dy)(y/y^h)$ as the elasticity of h's income with respect to a change in total income, and $w^h \equiv x^h/x$ is again h's share in total demand. Then using (6.28) we have

$$e_y = \frac{\partial D}{\partial y} \frac{y}{x} = \sum_{h=1}^{H} e_y^h \frac{x^h}{x} \frac{y}{y^h} \frac{dy^h}{dy} = \sum_{h=1}^{H} e_y^h w^h \delta^h. \tag{6.29}$$

Thus income elasticity of market demand depends on the sum of individual income elasticities weighted by demand shares, and further weighted by the proportionate effects on individual incomes of changes in aggregate income. Aggregate income increases with significantly different distributions of δ^h could then imply quite different market income elasticities of demand.

In spite of all this it is useful, both in theoretical and empirical work, to write market demand as a function of aggregate consumers' income. It seems excessively pedantic to write market demand functions in terms of the set of all consumer incomes, and in any case any one consumer's income change would have a negligible effect on market demand. To express demand as a function of the incomes of major sub-groups of consumers may be more worth while but, provided we are aware that distributional effects *could* be important, the type of demand function used in Chapter 2 should be adequate for many purposes. The main implication of the discussion in this section is that a market demand function defined on aggregate consumers' income cannot in general be thought of as resulting from the straightforward summation of individual demand functions. However, if consumers' money incomes are fixed, the market demand curve, relating total demand to price, *can* be thought of as the aggregate of individual demand curves because of the relationship shown in (6.22) and (6.24).

Exercise 6.3

1. There are two consumers of a good. The first has the demand function $x_1 = 100p^{-1}$. The second has the demand function $x_2 = 50p^{-2}$. What is the elasticity of market demand at prices of $5, $10, and $15 respectively?
2. Rich consumers buy 2/3 of the current output of a good, poor consumers buy 1/3. Rich consumers' income elasticity for the good is 2, that of poor consumers is 1. Currently rich consumers have 3/4 of total income, and poor consumers have 1/4. Calculate the income elasticity of demand for the good given an increase in total income
   (*a*) goes entirely to rich consumers;
   (*b*) goes entirely to poor consumers;
   (*c*) is split equally between them.
   [Hint: insert the values of e_y^h, w^h, and δ^h into (6.29), for $h = 1, 2$.]
3. Prove that it is sufficient for the sum of individual demand functions to be a function of the sum of incomes, that the individual demand functions take the form in (6.27).

6.4 Summary

The central concept in the theory of consumer choice is the *preference ordering*. This shows how the consumer ranks alternatives (bundles of goods) in terms of preference and indifference, and is regarded as a given datum.

An *ordinal utility function* is any function defined on bundles of goods, which gives a numerical representation of the preference ordering, in that it increases as we move higher in the preference ordering, and stays constant as we move through an indifference set (a set of bundles among which the consumer is indifferent, or which are ranked equally in the preference ordering).

Though not essential, it is analytically very convenient to work with a utility function. The idea that the consumer will always want to choose the bundle of goods highest in the preference ordering out of those that are available, is translated into the principle of maximizing the utility function subject to the budget constraint. This gives us a straightforward mathematical procedure for deriving the consumer's demand functions. These show how the chosen value of each good depends on prices and income. The utility function determines the form and the parameters of these demand functions.

We would then like to specify the general properties of these demand functions, in terms of the signs of their partial derivatives. However, a simple two-dimensional diagrammatic analysis tells us that the assumptions we have made about the nature of the consumer's preference ordering are too general to allow specific restrictions to be placed on these signs. We can show that an increase in income may cause an increase or decrease in demand for a good; a fall in price of a good may cause its own demand to increase or decrease, and likewise for the demand of another good.

If we regard the purpose of the theory of the consumer to be to 'explain why demand curves have negative slopes' then we would appear to have failed. However, this conclusion would be premature. The deeper analysis carried out in the next chapter shows we can be quite precise about the circumstances in which demand curves *will* have negative slopes, and gives us a good explanation of what determines the slope of an individual's demand curve.

When we sum the demand functions of individual consumers we find that in general market demand is a function of price and the entire set of consumers' incomes. Only under quite strong assumptions on the form of the individual demand function would it be the case that aggregate market demand is a function of aggregate consumers' income.

This tells us essentially that market demand will be a function both of the *total* and of the *distribution* of consumers' incomes. We could observe a change in demand with total income constant if distribution changed significantly. The effect of a given change in total income may depend on how it is distributed. Provided we remember this, it is a useful notational simplification to define market demand functions on prices and aggregate income.

On the other hand, because all consumers face the same price, the market demand curve is the sum of all individual demand curves (with incomes held constant).

7 Duality Theory and the Slutsky Equation

Our main purpose in this chapter is to investigate further the properties of the demand functions derived in section 3 of the previous chapter, and, in particular, to say something more illuminating about the demand derivatives $\partial x_i/\partial p_i$, $i = 1, \ldots, n$, than that they can be positive, negative, or indeed zero. In order to do this we shall develop some aspects of what is known as *duality theory*, a subject of considerable interest in its own right. We use this theory to discuss the subject of consumer surplus, which, as we shall see in Chapter 15 below, is an important concept in applied microeconomic analysis.

7.1 Expenditure minimization

In the previous chapter we took the problem: maximize the consumer's utility subject to a budget constraint. We now consider the *dual* to this problem: instead of setting income constant, we set the value of the utility function constant. We then minimize the expenditure required to achieve this value of utility. The contrast between the two problems (the original utility maximization problem is called the *primal*) is shown in Fig. 7.1.

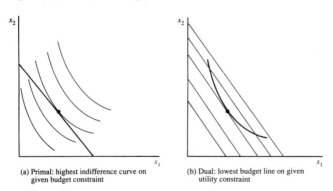

(a) Primal: highest indifference curve on given budget constraint

(b) Dual: lowest budget line on given utility constraint

Fig. 7.1: Primal and dual problems

In (a) of the figure, the primal problem is to find the consumption bundle on the given budget constraint that is also on the highest possible indifference curve. In (b) of the figure, the dual problem is to find the consumption bundle on the given indifference curve that is also on the lowest possible budget line. The figure also makes one important point very clear: if in the primal problem we fix as the constraint the budget line which obtains at the solution to the dual problem; or equivalently, if in the dual problem we fix as the constraint the indifference curve which obtains at the solution to the primal problem, then we have as a solution exactly the same consumption bundle in each case.

It is also worth noting that the dual problem is not simply an artificial construct, but has an economic interpretation in its own right. We can think of a given indifference curve, and associated value of the utility function, as defining a particular 'standard of living' or level of *real income* for the consumer. Then the dual problem is asking: at the given prices, what is the lowest possible cost of achieving a particular standard of living? It also permits us to analyse the effects of price changes on the cost of achieving a particular standard of living, i.e., 'the cost of living'. These have long been questions of considerable interest to economists, statisticians, and policy makers.

Let us now examine the dual problem mathematically. We can write it as

$$\min y = p_1 x_1 + p_2 x_2 + \ldots + p_n x_n \quad \text{s.t. } u(x_1, x_2, \ldots, x_n) = u,$$

where y can now be thought of as 'required expenditure' rather than income – it is free to vary, but utility, which is fixed at some value u, is not.

Forming the Lagrange function $p_1 x_1 + \ldots + p_n x_n - \mu(u(x_1, \ldots, x_n) - u)$, we have as first-order conditions

$$p_1 - \mu u_1 = 0,$$

$$p_2 - \mu u_2 = 0,$$

$$\cdots\cdots\cdots\cdots \tag{7.1}$$

$$p_n - \mu u_n = 0,$$

$$u(x_1, \ldots, x_n) - u = 0.$$

Again, we have $n + 1$ conditions to determine the $n + 1$ unknowns, $x_1, \ldots, x_n$, and μ. Taking any pair of the first n conditions in (7.1), say the first two, we have

$$\frac{u_1}{u_2} = \frac{p_1}{p_2}. \tag{7.2}$$

Thus, we have exactly the same kind of tangency condition as we have in the utility maximization problem. However, we may not in general obtain exactly the same consumption bundle as a solution, because of course the last condition, the constraint, is different.

We can, however, solve these conditions, as before, to obtain values of $x_1, x_2, \ldots, x_n$, and again these values depend on the parameters of the problem, in this case the prices $p_1, p_2, \ldots, p_n$ and u. Thus we can express these solution values as functions of the parameters

$$x_1 = H_1(p_1, \ldots, p_n, u),$$

$$x_2 = H_2(p_1, \ldots, p_n, u),$$

$$\cdots\cdots\cdots\cdots\cdots\cdots \tag{7.3}$$

$$x_n = H_n(p_1, \ldots, p_n, u).$$

It seems reasonable to interpret these functions as demand functions. We call them *Hicksian* demand functions. The obvious way in which they differ from the demand functions in (6.10), which we now refer to as the *Marshallian* demand functions, is that they have u, and not y, as an argument. This leads to an important difference in interpretation of the demand curves derived from the demand functions. When we consider the demand curve we derive from any $D_i(.)$, we hold income, y, and all other prices constant and observe how x_i varies with p_i, $i = 1, \ldots, n$. When we consider deriving a demand curve from any H_i, we hold utility, u, and all other prices constant and observe how x_i varies with p_i, $i = 1, \ldots, n$. Thus the first could be called a 'constant money income demand curve' (Marshallian), and the second a 'constant utility demand curve' (Hicksian). In general, we would not expect them to be the same, and Fig. 7.2 illustrates one case of the relation between them.

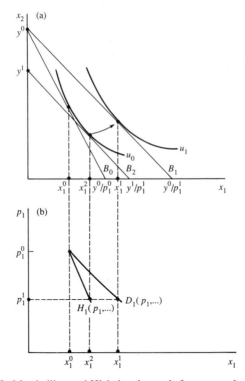

Fig. 7.2: Marshallian and Hicksian demands for a normal good

We assume that the price of x_2, $p_2 = 1$, while the price of x_1 is initially p_1^0 and then it falls to p_1^1. In (a) of the figure, *with money income constant*, at y^0, this causes the budget constraint to change from B_0 to B_1 and the consumer's demand for x_1 changes from x_1^0 to x_1^1. The value of utility has increased from u^0 to u^1. In (b) of the figure, we plot the consumer's demand for x_1 against its price p_1, and the points (x_1^0, p_1^0) and (x_1^1, p_1^1) are then two points on the consumer's Marshallian demand curve $D_1(p_1, \ldots)$. In (a), to hold utility constant at u^0, the fall in price must result in the new budget constraint B_2, corresponding to a money income y^1. We could think of achieving this budget constraint by taking away from the consumer an amount of money income equal to $(y^0 - y^1)$. The result is a change in demand for x_1 from x_1^0 to x_1^2. So, in (b) of the figure we have the two points (x_1^0, p_1^0) and (x_1^2, p_1^1) on the consumer's Hicksian demand curve $H_1(p_1, \ldots)$, which shows how x_1 varies with p_1 when u is held constant at u^0 and y is adjusted accordingly.

The increase in demand from x_1^0 to x_1^2, caused by the fall in price with u held constant, is usually called the *substitution effect*. Clearly then, the slope of the Hicksian demand curve shows the substitution effect. It is obvious from the curvature of the indifference curve that the Hicksian demand curve *must* have a negative slope: moving a sequence of budget lines with flatter and flatter slopes around a given indifference curve must result in a sequence of tangency points with increasing values of x_1 (try it).

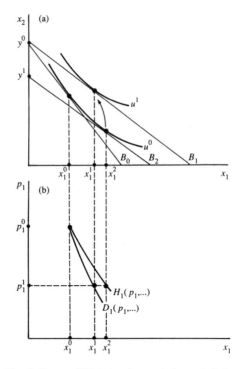

Fig. 7.3: Marshallian and Hicksian demands for an inferior good

The change in demand $x_1^1 - x_1^2$ is usually called the *income effect*. This is because, in (a) of Fig. 7.2, this change in demand comes about because of a parallel shift in the budget constraint from B_2 to B_1 which, at prices p_1^1 and $p_2 = 1$, would be achieved by increasing the consumer's money income from y^1 to y^0. Thus, the *overall* demand change $x_1^1 - x_1^0$, which determines (in conjunction with the price change $p_1^0 - p_1^1$) the slope of the Marshallian demand curve, is decomposed into: a substitution effect, the move along the Hicksian demand curve; and an income effect, the horizontal distance between Hicksian and Marshallian demand curves.

We can now suggest intuitively why studying the relation between the Hicksian and Marshallian demand curves helps us resolve the ambiguity about the slope of the Marshallian demand curve. In the case shown in Fig. 7.2(b), this slope is clearly negative, because the change in x_1 due to the income effect reinforces the change in x_1 due to the substitution effect (which itself *must* be an increase, for a reduction in price). In other words here we have a normal good, for which $\partial x_1/\partial y \geq 0$. Suppose instead x_1 was inferior. Then the move from x_1^2 to x_1^1 would involve a *reduction* in x_1, as shown in Fig. 7.3(a), so that $x_1^2 > x_1^1$. This means that the Marshallian demand curve in Fig. 7.3(b), though still negatively sloped, lies inside the Hicksian. However, in Fig. 7.4 the income effect more than offsets the substitution effect and so the Marshallian demand curve has a positive slope. In this case x_1 is called a Giffen good.

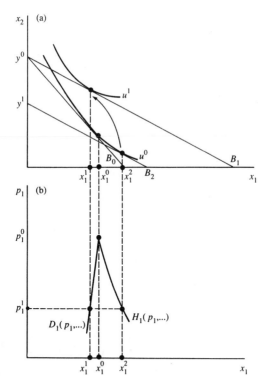

Fig. 7.4: Marshallian and Hicksian demands for a Giffen good

This analysis allows us to sharpen up considerably what the theory allows us to say about the slope of the Marshallian demand curve. We can say:

(a) for a normal good, with $\partial x_i/\partial y \geq 0$, we will always have $\partial x_i/\partial p_i < 0$.

(b) for an inferior good, with $\partial x_i/\partial y < 0$, we may have $\partial x_i/\partial p_i \gtrless 0$.

We now want to put this diagrammatic argument on a more rigorous footing. To do this we first need the *expenditure function*, which, as well as helping us establish the relation between Marshallian and Hicksian demand functions, has a number of important applications, some of which we will study in the next section.

Go back to the statement of the expenditure minimization problem, and note that the conditions (7.1) determine the solutions (7.3). Hence, at the optimal solution, the value of the *minimal* expenditure is

$$y = p_1 x_1 + \ldots + p_n x_n = p_1 H_1(p_1, \ldots, p_n, u) + \ldots + p_n H_n(p_1, \ldots, p_n, u)$$
$$= y(p_1, \ldots, p_n, u). \tag{7.4}$$

This says that the minimum expenditure required to achieve a particular level of utility, u, depends, via the minimization process, upon the level of utility and also the levels of prices. For example, in Fig. 7.2(a) we saw that when p_1 fell from p_1^0 to p_1^1, y had to fall from y^0 to y^1 to keep utility constant at u^0. Thus we would have

$$y^0 = y(p_1^0, p_2, u^0),$$
$$y^1 = y(p_1^1, p_2, u^0).$$

Then (7.4) gives minimum expenditures as a general function of prices and utility. Consider now the derivative $\partial y/\partial p_i$, $i = 1 \ldots n$. From (7.4) total differentiation with every price except p_i constant gives

$$dy = \left\{ H_i(p_1, \ldots, p_n, u) + p_1 \frac{\partial H_1}{\partial p_i} + \ldots + p_n \frac{\partial H_n}{\partial p_i} \right\} dp_i. \tag{7.5}$$

Note that we can write the utility constraint, using (7.3), as

$$u\left(H_1(p_1, \ldots, p_n, u), \ldots, H_n(p_1, \ldots, p_n, u) \right) = u \tag{7.6}$$

where u is of course fixed. Differentiating through (7.6) gives

$$du = \left\{ u_1 \frac{\partial H_1}{\partial p_i} + \ldots + u_n \frac{\partial H_n}{\partial p_i} \right\} dp_i = 0. \tag{7.7}$$

We can clearly multiply through (7.7) by μ without changing the equality. Moreover, we can subtract the result, which still equals zero, from the right-hand side of (7.5) without changing the equality, to obtain

$$dy = \left\{ H_i(p_1, \ldots, p_n, u) + p_1 \frac{\partial H_1}{\partial p_i} + \ldots + p_n \frac{\partial H_n}{\partial p_i} \right\} dp_i - \mu \left\{ u_1 \frac{\partial H_1}{\partial p_i} + \ldots + u_n \frac{\partial H_n}{\partial p_i} \right\} dp_i$$

$$= H_i(p_1, \ldots, p_n, u) dp_i + \left\{ (p_1 - \mu u_1) \frac{\partial H_1}{\partial p_i} + \ldots + (p_n - \mu u_n) \frac{\partial H_n}{\partial p_i} \right\} dp_i. \tag{7.8}$$

But from the condition in (7.1), each term in the braces in (7.8) is zero, and so we have the result:

$$\frac{\partial y}{\partial p_i} = H_i(p_1, \ldots, p_n, u) = x_i. \tag{7.9}$$

This result, known as *Shephard's lemma*, has a very simple interpretation. Suppose you are currently spending y per week on buying your preferred bundle of goods, and in particular are buying three bottles of wine per week. Then, (7.9) is saying that if the price of a bottle of wine goes up by, say, a cent, to a first approximation you will need 3 cents a week more in expenditure to maintain the same standard of living. Note, however, that this 'intuitively obvious' result should be treated with care — the key words are 'to a first approximation'. If the price of wine rose by \$1 a bottle, it would be a gross over-statement to say that you would need \$3 a week more to maintain your standard of living. This is because you can do this (stay on the same indifference curve) by substituting other goods, whose prices have not changed, for wine, thus reducing the need for increased income. Shephard's lemma shows that in the neighbourhood of the initial equilibrium, for very small price changes, these substitution effects can be ignored, but this becomes a worse approximation the bigger the price change.

We can now give a precise statement of the relationship between Marshallian and Hicksian demand functions. Suppose that in the expenditure minimization problem, the value of the utility function chosen as the constraint is that value achieved at the solution to the utility maximization problem (implying that the value of expenditure at the solution to the expenditure minimization problem is equal to the income constraint in the utility maximization problem). Then the solutions $x_1, x_2, \ldots, x_n$, to the two problems are identical, and so using (6.10) and (7.3) we can write

$$D_i(p_1, \ldots, p_n, y) = H_i(p_1, \ldots, p_n, u). \tag{7.10}$$

Since $y = y(p_1, \ldots, p_n, u)$, we can substitute in D_i to obtain

$$D_i\big(p_1, \ldots, p_n, y(p_1, \ldots, p_n, u)\big) = H_i(p_1, \ldots, p_n, u) \tag{7.11}$$

Consider now taking some change dp_i, holding other prices and u constant but allowing y to change in whatever way is required to maintain the equality in (7.11). Then we obtain, using the 'function of a function' rule of differentiation,

$$dD_i = \left\{\frac{\partial D_i}{\partial p_i} + \frac{\partial D_i}{\partial y}\frac{\partial y}{\partial p_i}\right\}dp_i = dH_i = \frac{\partial H_i}{\partial p_i}dp_i. \tag{7.12}$$

Using Shephard's lemma ($\partial y/\partial p_i = x_i$) and rearranging (7.12) then gives the *Slutsky equation*

$$\frac{\partial D_i}{\partial p_i} = \frac{\partial H_i}{\partial p_i} - x_i\frac{\partial D_i}{\partial y}. \tag{7.13}$$

This says that the slope of the Marshallian demand function can be decomposed into the slope of the Hicksian demand function (the substitution effect) and an income effect, which depends partly on how much of the good is bought and partly on how demand for it responds to income changes. Since $\partial H_i / \partial p_i < 0$, (7.13) confirms our earlier conclusions about the conditions under which the Marshallian demand curve will or will not have a negative slope.

We can illustrate the whole of the theory of this chapter with an example. We again take the utility function $u = x_1^{b_1} x_2^{b_2}$. Then the expenditure minimization problem becomes

$$\min \; y = p_1 x_1 + p_2 x_2 \quad \text{s.t.} \; x_1^{b_1} x_2^{b_2} = u.$$

The Lagrange function is $p_1 x_1 + p_2 x_2 - \mu(x_1^{b_1} x_2^{b_2} - u)$, and the first-order conditions are

$$p_1 - \mu b_1 x_1^{b_1 - 1} x_2^{b_2} = 0,$$

$$p_2 - \mu b_2 x_1^{b_1} x_2^{b_2 - 1} = 0, \qquad (7.14)$$

$$x_1^{b_1} x_2^{b_2} - u = 0.$$

The first two conditions again give, on eliminating μ,

$$\frac{b_1 \, x_2}{b_2 \, x_1} = \frac{p_1}{p_2} \;\Rightarrow\; x_1 = \frac{p_2 \, b_1}{p_1 \, b_2} x_2. \qquad (7.15)$$

We could call (7.15) a 'utility consumption curve' since it shows the relation between x_1 and x_2 as we take successively higher indifference curves and find their points of tangency with the budget lines. Diagrammatically, it is a straight line through the origin, and, as long as the price ratio is the same, is indistinguishable from the income consumption curve we found in the previous chapter.

To solve for x_2, we substitute (7.15) into the utility constraint in (7.14) to obtain

$$\left(\frac{p_2 \, b_1}{p_1 \, b_2} x_2\right)^{b_1} x_2^{b_2} - u = 0 \;\Rightarrow\; \left(\frac{p_2}{p_1}\right)^{b_1} \left(\frac{b_1}{b_2}\right)^{b_1} x_2^{b_1 + b_2} = u \;\Rightarrow\; x_2 = \alpha_2 \left(\frac{p_1}{p_2}\right)^{\beta_2} u^{\gamma}$$

$$(7.16)$$

where $\alpha_2 \equiv (b_2/b_1)^{\beta_2}$; $\beta_2 \equiv b_1/(b_1 + b_2)$; $\gamma \equiv 1/(b_1 + b_2)$.
Substituting (7.16) back into (7.15) (do it!) we find

$$x_1 = \alpha_1 \left(\frac{p_2}{p_1}\right)^{\beta_1} u^{\gamma} \qquad (7.17)$$

where $\alpha_1 = (b_1/b_2)^{\beta_1}$; $\beta_1 = b_2/(b_1 + b_2) = 1 - \beta_2$.

Then (7.16) and (7.17) give us the Hicksian demand functions. Clearly, they are different functions from the Marshallian demand functions in (6.18) and (6.19) of the previous chapter. Not only do they include u instead of y, but now demands are functions of *both* prices, i.e., of the slope p_1/p_2 of the budget constraint. We can obtain the expenditure function from

$$y = p_1 x_1 + p_2 x_2 = p_1 \alpha_1 \left(\frac{p_2}{p_1}\right)^{\beta_1} u^{\gamma} + p_2 \alpha_2 \left(\frac{p_1}{p_2}\right)^{\beta_2} u^{\gamma} = (\alpha_1 + \alpha_2) p_1^{\beta_2} p_2^{\beta_1} u^{\gamma} \quad (7.18)$$

(recall that $1 - \beta_1 = \beta_2$). Question 3 of Exercise 7.1 asks you to confirm both Shephard's lemma and the Slutsky equation from (7.18).

We can note one final point from this example, which is an important general proposition. The parameters of the Hicksian demand functions depend entirely on the values of b_1 and b_2, as do the parameters of the Marshallian demand function. Now in general we would expect to be able to observe and estimate (at best) only Marshallian demand functions and not Hicksian demand functions. The former involve observable demand responses to observable changes in prices and income, while the latter involve the *intrinsically* non-observable u, and are based on a hypothetical variation in income — to maintain u constant — which would be very difficult to make in practice. On the other hand, in many applications of demand theory, as we shall see in the next section, it is the Hicksian rather than the Marshallian demand function we would really like to have. What this example shows is that if we can estimate the parameters of the consumer's Marshallian demand function, we can use them to determine the parameters of the Hicksian demand function, since all involve the same underlying information, the values b_1 and b_2. This extends beyond the precise examples given here to any set of estimated demand functions (provided they are based on some underlying utility function or expenditure function with the appropriate properties).

Exercise 7.1

1. What are the effects of:
   (a) applying a positive monotonic transformation to the utility function;
   (b) increasing all prices by the same factor k,
   on the Hicksian demand functions and the expenditure function?
2. A consumer has the utility function $u = x_1 x_2$, an income of \$100, and faces prices $p_1 = \$1, p_2 = \$2$. Graph her Hicksian and Marshallian demand functions for x_1 for $p_1 \leq \$1$.
3. By differentiating the expenditure function in (7.18), confirm Shephard's lemma. By taking derivatives of the demand functions in (6.18), (6.19), (7.16), and (7.17), confirm the Slutsky equation.
4. Econometric estimation results in the following demand functions for a consumer:

$$x_1 = 0.4 \, y/p_1; \quad x_2 = 0.6 \, y/p_2.$$

Assuming the special case of the utility function used in this section applies, use this information to find the consumer's Hicksian demand functions and expenditure function. [Hint: use the definition of s_1 in (6.19) to find values for b_1 and b_2. Recall that b_1 and b_2 cannot be uniquely defined, so set $b_1 + b_2 = 1$.]

7.2 Consumer surplus

In many areas of applied economics, we often want to know the answer to the question: what is the effect of a change in price on consumer welfare? For example, we might be interested in the questions:

What is the effect on consumer welfare of fixing prices of foodstuffs above their world market equilibrium levels, as in the Common Agricultural Policy of the EEC?

How much do consumers benefit from an investment which reduces the price of a good they consume, or makes available a good which previously was not available?

Instead of covering costs by charging a price per unit bought, could consumers be made better off by charging them a lump sum (entrance fee, standing charge, annual subscription) for the right to be able to buy the good, and then a lower price per unit?

By how much does taxing a good reduce consumers' welfare?

Each of these questions, and many more like them arising in applied economics, involve the concept of *consumer surplus*. This gives a money measure of the benefit (utility) consumers derive from consuming a good. The original definition of the concept, as set out by Dupuit and Marshall, was quite straightforward, though rather non-rigorous. Consider the consumer's ordinary or Marshallian demand function for a good, $x = D(p)$, where for simplicity we suppress all independent variables other than the good's own price. We can invert this function to give the *inverse demand function* $p = D^{-1}(x) = p(x)$. We then make an important *normative* interpretation of this demand curve: we assert that the price, p, can be taken as a money measure of the benefit the consumer derives from the marginal bit of her consumption of the good. That is, if $B(x)$ is a function giving the *total* money value of the benefit a consumer derives from consumption x, then

$$\frac{dB(x)}{dx} = p(x). \tag{7.19}$$

But then it follows immediately that at any consumption level x_0

$$B(x_0) = \int_0^{x_0} \frac{dB(x)}{dx} dx = \int_0^{x_0} p(x) dx. \tag{7.20}$$

Now if the consumer is paying the price $p(x_0)$ for *all* the x_0 units she consumes, it follows that she is enjoying a surplus of total benefit over expenditure, given by

$$S(x_0) = B(x_0) - p(x_0) x_0, \tag{7.21}$$

and this is therefore her 'consumer surplus'. Fig. 7.5 illustrates this. Total benefit $B(x_0)$ is the area under the demand curve, total expenditure is the rectangle, and so consumer surplus is the triangular area indicated.

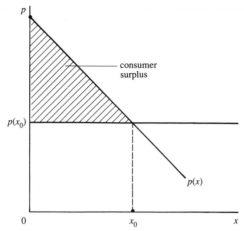

Fig. 7.5: Consumer surplus

For example, suppose that the consumer's Marshallian demand function is $x = a - bp$ (where again all other variables are subsumed into a). Then the inverse demand function is $p = (a - x)/b = \alpha - \beta x$, where $\alpha = a/b$, $\beta = 1/b$. Then

$$B(x_0) = \int_0^{x_0} (\alpha - \beta x)dx = \alpha x_0 - \beta x_0^2/2$$

and her consumer surplus is

$$S(x_0) = (\alpha x_0 - \beta x_0^2/2) - (\alpha x_0 - \beta x_0^2) = \beta x_0^2/2.$$

From Fig. 7.5, recalling that the area of a triangle is 1/2 base height, you see that $\beta x_0^2/2$ is precisely the area of the consumer surplus triangle.

This exposition is useful in illustrating the idea that, if market price measures the value of the *marginal* unit the consumer buys, there may in general be a surplus of total benefit over expenditure accruing to the consumer. However, the discussion is rather *ad hoc*, and we would now like to put it on a sounder, more rigorous footing by relating it to the analysis of consumer choice given in the previous section.

First we redefine the question. Suppose that initially the consumer has income y^0, faces prices $p_1^0, p_2^0, \ldots, p_n^0$, and chooses a bundle of goods $(x_1^0, x_2^0, \ldots, x_n^0)$. She will then have a utility level $u^0 = u(x_1^0, \ldots, x_n^0)$. We know then from the analysis of section 7.1 that the value of her expenditure function is

$$y^0 = y(p_1^0, \ldots, p_n^0, u^0). \tag{7.22}$$

For consistency with the previous exposition, suppose that p_1^0 is just high enough to make the consumer choose $x_1^0 = 0$. Now suppose the price of x_1 falls to p_1^1. The consumer, with her income of y^0, will now choose a bundle of goods $(x_1^1, x_2^1, \ldots, x_n^1)$, and will have a new (higher) utility level $u^1 = u(x_1^1, x_2^1, \ldots, x_n^1)$. It also follows that $y^0 = y(p_1^0, p_2^0, \ldots, p_n^0, u^1)$. We now ask the question: what is a money measure of the benefit the consumer has derived from the fall in price of x_1 from p_1^0 to p_1^1? To this it seems reasonable to give either of two answers:

(a) a money measure of the benefit the consumer derives from the price fall is the maximum amount of income *she is prepared to pay* in order to have the lower price p_1^1 rather than the higher price p_1^0;

(b) a money measure of the benefit the consumer derives from the price fall is the minimum amount of income *she would have to be given* to make the initial situation with the higher price p_1^0 just as good as the new situation with the lower price p_1^1, in other words the amount we would have to pay her to forgo the price change.

One reason for the interest in the duality theory of section 7.1 is that these money measures are quite straightforward to define with the help of the expenditure function. Answer (a) requires us to find the difference between the amount of income the consumer needs to achieve utility level u^0 when price is p_1^1, i.e., $y^1 = y(p_1^1, p_2^0, \ldots, p_n^0, u^0)$, and the amount of income she needs to achieve utility level u^0 when the price is p_1^0, which we already know to be $y^0 = y(p_1^0, p_2^0, \ldots, p_n^0, u^0)$. This money measure is therefore

$$C = y^1 - y^0 = y(p_1^1, \ldots, p_n^0, u^0) - y(p_1^0, \ldots, p_n^0, u^0) \tag{7.23}$$

which is usually called the *compensating variation* in income. Essentially it is the change in income that compensates for the price change by keeping the consumer on her *initial* indifference curve. Because price has fallen, the income change in this case is negative.

Answer (b) requires us to find the difference between the amount of income the consumer needs to reach utility level u^1 when price is p_1^1, which is $y^0 = y(p_1^1, p_2^0, \ldots, p_n^0, u^1)$, and the amount she needs to reach the same utility level u^1 when price is p_1^0, which we have seen to be $y^2 = y(p_1^0, p_2^0, \ldots, p_n^0, u^1)$. This money measure is therefore

$$E = y^2 - y^0 = y(p_1^0, \ldots, p_n^0, u^1) - y(p_1^1, \ldots, p_n^0, u^1) \tag{7.24}$$

which is usually called the *equivalent variation* in income. Essentially, it is the change in income that puts the consumer on her final indifference curve in the absence of the price change.

Fig. 7.6(a) illustrates these measures for the two-good case (ignore (b) for the moment). Assume the price of x_2 is 1, so that the budget constraints are given by

$$B_0: p_1^0 x_1 + x_2 = y^0, \qquad B_1: p_1^1 x_1 + x_2 = y^1,$$
$$B_2: p_1^0 x_1 + x_2 = y^2, \qquad B_3: p_1^1 x_1 + x_2 = y^0.$$

The initial equilibrium is at $x_1 = 0$, $x_2 = y^0$ on indifference curve u^0. The price fall to p_1^1 with income unchanged at y^0 gives budget constraint B_3 and achieves utility level u^1. The income level that makes the consumer indifferent between having price p_1^1 or p_1^0 when utility level is u^0 is clearly y^1 in the figure, which requires a reduction of income of C. The income level that makes the consumer indifferent between having price p_1^0 or p_1^1 when utility is u^1 is y^2, implying an increase in income of E.

Note that there is no reason in general for these two consumer surplus measures to be equal, as the figure shows (but see question 1 of Exercise 7.2). Choice of which measure to use in any applied context will depend on the precise question being asked. We might want to know how much consumers are prepared to pay for some investment which will lower the price, in which case C is the appropriate measure; or we might wish to know how much consumers require to be paid to forgo such an investment, in which case E would be appropriate.

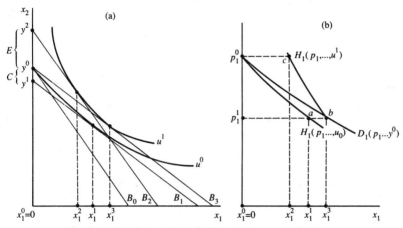

Fig. 7.6: Hicksian and Marshallian consumer surpluses

We now want to relate these consumer surplus measures to areas under demand curves, in order to see how they compare to the earlier *ad hoc* measure proposed by Dupuit and Marshall. We can do this easily by use of Shephard's lemma, derived in section 7.1. Recall that this implies that

$$\partial y(p_1, p_2, \ldots, p_n, u^0)/\partial p_1 = H_1(p_1, p_2, \ldots, p_n, u^0), \tag{7.25}$$

$$\partial y(p_1, p_2, \ldots, p_n, u^1)/\partial p_1 = H_1(p_1, p_2, \ldots, p_n, u^1), \tag{7.26}$$

where H_1 is the Hicksian demand function for good 1. These are illustrated as H_1^0 and H_1^1 in Fig. 7.6(b). Integrating over the expressions in (7.25) and (7.26) gives

$$\int_{p_1^1}^{p_1^0} (\partial y(p_1, p_2, \ldots, p_n, u^0)/\partial p_1)dp_1 = y(p_1^0, p_2, \ldots, p_n, u^0) - y(p_1^1, p_2, \ldots, p_n, u^0)$$

$$= \int_{p_1^1}^{p_1^0} H_1(p_1, p_2, \ldots, p_n, u^0)dp_1 = -C, \tag{7.27}$$

$$\int_{p_1^1}^{p_1^0} (\partial y(p_1, p_2, \ldots, p_n, u^1)/\partial p_1)dp_1 = y(p_1^0, p_2, \ldots, p_n, u^1) - y(p_1^1, p_2, \ldots, p_n, u^1)$$

$$= \int_{p_1^1}^{p_1^0} H_1(p_1, p_2, \ldots, p_n, u^1)dp_1 = E. \quad (7.28)$$

Thus, C is simply the area under the Hicksian demand curve corresponding to indifference curve u^0, between prices p_1^0 and p_1^1; while E is the area under the Hicksian demand curve corresponding to indifference curve u^1, between the same prices. In Fig. 7.6(b) these are shown as areas $p_1^0 a p_1^1$ and $p_1^0 c b p_1^1$ respectively. In the figure, the Marshallian demand curve is labelled D, and so the consumer surplus found by integrating under this demand curve is area $p_1^0 b p_1^1$. We therefore see that the 'Marshallian consumer surplus' does not quite correspond to either of the two measures defined earlier — it overstates C and understates E.

It could be thought (and in the economics literature in the past *has* sometimes been thought) to be a problem that the area under the Marshallian demand curve does not quite give either of the consumer surplus measures (except under special circumstances — see question 1 of Exercise 7.2). However, this is not the case as long as we have the Marshallian demand function for *individual consumers* since in that case, as we argued in section 7.1, we can then derive a Hicksian demand function and thus measure the appropriate consumer surplus. Where a problem does arise is if we only have the *aggregate market* Marshallian demand curve. Only under very special assumptions is it possible to get from this back to the *individual* Hicksian demand functions, and so we could not in general use it to derive the appropriate measures of consumer surplus. To take the area under the market Marshallian demand curve as a measure of consumer surplus involves an unknown error, as well as of course the assumption that it is permissible to add the money measure of welfare of different consumers to arrive at an aggregate measure.

Exercise 7.2

1. Suppose that a consumer's indifference curves are vertical displacements of each other, i.e., in the x_1, x_2-space along a vertical line at any value of x_1 the slope of every indifference curve is the same. Show diagrammatically in this case that all Hicksian demand curves for x_1 coincide with each other and with the Marshallian demand curve for x_1. What are the implications of this for the measurement of consumer surplus? Can you suggest a functional form for the consumer's utility function which corresponds to this case?
2. For the consumer described in question 2 of Exercise 7.1, measure C and E for (a) a fall in p_1 to 60¢, and (b) a rise in p_1 to \$2. Then find the errors involved in taking the Marshallian consumer surplus in each case.
3. A firm selling good x_1 has 1,000 customers, each of whom has the utility function and income of the consumer in the previous question. It currently sells the good at a

price of $1. Its brilliant new marketing manager suggests offering customers the option of paying a lump sum to join a 'discount club', and selling the good at 50¢ per unit to each member of this club. What is the maximum lump sum the firm can charge, consistent with customers joining the club? Will its total revenue increase as a result of the proposal? (Assume the price of x_2 stays fixed at $2 throughout.)

7.3 Summary

Corresponding to the primal problem of maximizing utility subject to a budget constraint, there is a dual problem, of minimizing expenditure subject to a utility constraint. This can be interpreted as the problem of finding the bundle of goods which minimizes the cost of achieving a particular standard of living.

Both problems are solved at a tangency point of a budget line and an indifference curve. If the *achieved* utility level in the primal problem is the *constraint* utility level in the dual, or if the *achieved* expenditure level in the dual problem is the *constraint* income level in the primal, then the solution to the two problems will be identical: the same bundle of goods is chosen in each case.

From the dual problem we derive the consumer's *Hicksian demand functions*. These express quantities demanded as functions of prices and the required utility level. They are distinct from *Marshallian demand functions*, which express quantities demanded as functions of prices and consumer's income.

The horizontal distance between a point on a Hicksian demand curve and a point on the Marshallian demand curve is equal to the *income effect*.

The slope of a Hicksian demand curve shows the *substitution effect*.

The *Slutsky equation* establishes precisely the relation between the slopes of the Hicksian and Marshallian demand curves, and allows us to give a precise statement of the conditions under which the Marshallian demand curve will have a negative slope. For example, we can say that the Marshallian demand curve for a *normal good* must have a negative slope.

By substituting the Hicksian demand functions into the expression for expenditure we obtain *the expenditure function*. *Shephard's lemma* shows that the derivative of the expenditure function with respect to the ith price is the quantity demanded of good i.

An important application of the expenditure function is to the problem of defining a money measure of the change in consumer benefit caused by a price change. We can define two such measures: the *compensating variation*, which asks what change in income would keep the consumer at the *initial* or *pre-change* utility level after the price change; and the *equivalent variation*, which asks what change in income would put the consumer at the *final* or *post-change* utility level at the pre-change prices. Each measure is easily obtained as a difference between two values of the expenditure function.

Using Shephard's lemma, we can show that these money measures of welfare change are given by areas under the corresponding Hicksian demand curves.

This in turn tells us that the area under a consumer's Marshallian demand curve will not be exactly equal to either of these benefit measures, *unless* there is a zero income effect.

However, if we know a consumer's Marshallian demand functions, we should be able to retrieve the utility function parameters and construct the consumer's expenditure function, which in turn will give us equivalent and compensating variations. Unfortunately, this is not in general the case if all we have is the aggregate market Marshallian demand function for a good.

8 The Theory of Supply: Production and Costs

We now want to analyse in some depth the underlying determinants of market supply and the nature of the market supply function. Typically, the seller in a market for goods and services is a *firm*. Thus, the basis of the theory of supply is a *theory of the firm*.

In microeconomic theory we take quite an abstract view of the firm, in order to focus on its essential role in the economic system, that of organizing production and supply. We regard the firm *as if* it were controlled by a single individual, who hires or buys inputs and combines them in certain technological processes, to produce outputs which he then sells. Thus the firm is a buyer in input markets and a seller in output markets. We abstract from the fact that firms may be large, complex, even bureaucratic organizations with a variety of goals — it is still a matter of some debate among economists as to whether anything essential is lost thereby. Our view is that as long as we are chiefly concerned with the analysis of demands for inputs and supplies of outputs (rather than the internal processes of firms) then the simplicity and clarity of this somewhat abstract model, and the fact that its predictions do not seem up to now to have been significantly refuted, more than justify its apparent lack of descriptive realism.

8.1 The production function

The objective of the firm in choosing quantities of inputs to hire or buy, and quantities of outputs to sell, is to maximize profit, defined as the difference between revenue from sales and the cost of inputs used up in production (though see Chapter 13 where we discuss the intertemporal theory of the firm). We proceed as if the firm consisted of two 'departments':

(i) *the production department*, whose role is to acquire inputs and produce any given output *at minimum cost*. Cost minimization is obviously an *implication* of profit maximization (explain why). The production department must also calculate a *cost function*, i.e., a relationship between the rate of output and minimized production cost. It then passes this function on to:

(ii) *the sales department*, whose role is to decide what rate of output will maximize the firm's profit given the conditions on the market it faces. For this it has to estimate the firm's *revenue function*, the relationship between sales revenue and output. Then, putting this together with the cost function provided by the production department allows it to find the profit-maximizing output for the production department to produce.

Given this basic framework, we make a number of assumptions that are essentially simplifying and may be relaxed at some point (possibly as exercises) later. We assume that the firm produces only one output, denoted by q, with two inputs, labour, l, and capital, k. Output is measured in physical units per unit time, for example tonnes per week, gallons per day. Labour input is measured in labour hours per unit time, e.g., a firm with 800 workers each working a 35-hour week would have a labour input of 28,000 labour hours per week. Capital could be thought of as a stock of machines, and the capital input is machine hours per unit time, e.g., a firm with 300 machines working a 35-hour week would have a capital input of 10,500 machine hours per week.

An implication of this way of measuring inputs is that they are perfectly divisible — so that l and k are non-negative real numbers — even if the underlying workers and machines are not. This of course requires further that a worker can be employed for any amount of time up to his or her theoretical maximum (24 hours a day, 168 hours a week, etc.), as can a machine.

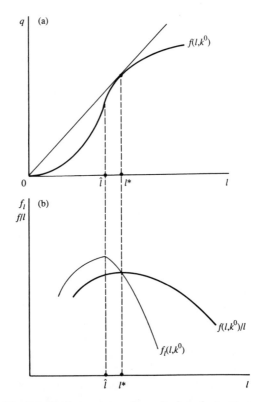

Fig. 8.1: Total, average, and marginal product curves

Finally, the prices of the inputs are taken as given by the firm: there is a given wage rate per hour of labour time, w, and a given *rental rate of capital* per hour of machine time, v. The derivation of this latter price is itself a matter needing some examination: while the firm hires its labour from week to week, it owns its machines, and so the question of what exactly is the cost of one hour of machine time from a stock of machines owned by the firm is worth explicit consideration (for which see Chapter 13 below). For the moment, we simply take it that the answer can be expressed as a rental rate per hour of machine time which can be treated in the same way as a wage rate.

The core of the model is the nature of the technology by which inputs are combined to produce outputs. However, for purposes of economic analysis we are not at all interested in the physical details of these processes, simply in what they imply for the shape of the relationship between inputs and output. The central assumption we make is that there is a well-defined relationship called the *production function*, written

$$q = f(l, k), \qquad (8.1)$$

which tells us how much output results from combining the inputs in the most efficient technological processes known to the firm, i.e., for given quantities of l and k, the largest possible q is produced (again, this is an *implication* of profit maximization – explain why). Then, all we know – or care – about the technology is summarized by the properties of this function, as defined by a number of assumptions.

The first assumption we make is that the function is twice differentiable, and its first-order partial derivatives, denoted by f_l and f_k, are called the *marginal productivities* of *labour* and *capital* respectively. To consider the shape of the function, instead of drawing it in 3 dimensions we take first a 2-dimensional cross-section. Fixing the value of k at k^0, Fig. 8.1(*a*) shows the type of shape we usually assume for the relationship $q = f(l, k^0)$, giving output as a function of labour input only, the 'total product of labour curve'.

Output first increases more than proportionately, then less than proportionately, with labour input. The marginal productivity function $f_l(l, k^0)$ gives the slope of the total product curve at each l-value, and this is graphed in Fig. 8.1(*b*). The point of inflexion of the total product curve occurs at $\hat{l}$ and so this is the point at which $f_l(l, k^0)$ is at a maximum. At l^*, the ray and the tangent to the total curve coincide, and so $f_l(l^*, k^0) = f(l^*, k^0)/l^*$, i.e., marginal product = average product at l^*. As we shall see in section 9.1 below, the essential feature we require of the total product curve is that at *some* point – here, $\hat{l}$ – its slope, though positive, starts diminishing, i.e., $f_l(l, k^0)$ has a *negative slope* over the range $l > \hat{l}$ in Fig. 8.1(*b*). A rationalization of this is the idea that, with k fixed at k^0, increasing l increases the intensity of use of the capital stock, so l has less k to work with, leading increments in l to yield smaller and smaller increments in output. Note that in general changing the fixed amount of capital would shift all three curves in Fig. 8.1.

If we now took a second cross-section through the function, fixing a value of l and observing how q varied as k varied, we would postulate the same *general* sort of relationship as in Fig. 8.1. We can therefore summarize by expressing these properties as:

(i) $f_k, f_l > 0$;

(ii) for any given k, there exists an l-value $\hat{l}$ such that for $l > \hat{l}, f_{ll} < 0$, and likewise for any given l there exists a k-value $\hat{k}$ such that for $k > \hat{k}, f_{kk} < 0$.

Also of interest is the effect of a change in quantity of one input on the marginal product of the other. This is expressed by the second-order cross-partial derivatives $\partial f_l(l, k)/\partial k = f_{lk}$, and $\partial f_k(l, k)/\partial l = f_{kl}$. Since $f(.)$ is assumed twice differentiable, we have that at any point (l^1, k^1), $f_{lk}(l^1, k^1) = f_{kl}(l^1, k^1)$. It is usual to assume that inputs are *cooperant*, meaning that increasing the quantity of one input increases the marginal product of the other, so that we have

(iii) $f_{kl} = f_{lk} > 0$ at every point (l, k). (See question 1 of Exercise 8.1.)

So far we have considered how output varies with one input held fixed and the other allowed to vary. Now we want to consider the possibility of variation in both inputs. Diagrammatically, we do this in terms of the *contours* of the production function. A contour of a function is a set of points which yield the same value of the function. We have already met the concept in Chapter 6, since an indifference curve can be regarded as a contour of the utility function. A contour of the production function is called an *isoquant*, and is defined by the equation

$$f(l, k) = q^0 \tag{8.2}$$

where q^0 is a fixed level of output defining the isoquant. Isoquants are typically assumed to have the general shape shown in Fig. 8.2. This can be expressed as follows. Differentiating through (8.2) gives

$$dq = f_l dl + f_k dk = 0 \tag{8.3}$$

since q^0 is fixed, and so

$$\frac{dk}{dl} = \frac{-f_l}{f_k} \tag{8.4}$$

is the slope of the isoquant. This slope (or rather its negative f_l/f_k) is called the *marginal technical rate of substitution*, and is equal to the ratio of the inputs' marginal products.

The isoquant in Fig. 8.2 shows that as l increases, the slope falls in absolute value but, since it is negative, it increases in algebraic value, and so

$$\frac{d}{dl}\left(\frac{dk}{dl}\right) > 0 \quad \text{or} \quad \frac{d}{dl}\left(\frac{-f_l}{f_k}\right) > 0 \quad \text{or} \quad \frac{d}{dl}\left(\frac{f_l}{f_k}\right) < 0. \tag{8.5}$$

This then implies (carrying out the differentiation)

$$f_{ll}f_k^2 - 2f_l f_k f_{kl} + f_{kk}f_l^2 < 0 \tag{8.6}$$

Now, recalling assumption (i), that $f_k, f_l > 0$, we see from (8.4) that this implies $dk/dl < 0$. Note that $f_{ll}, f_{kk} < 0$, and assumption (iii), that $f_{kl} = f_{lk} > 0$, are *sufficient* for (8.6) to be satisfied. However, we may have $f_{ll} > 0$ or $f_{kk} > 0$ over some range (refer back to Fig. 8.1), in which case the assumption that (8.6) holds for *all* (l, k) does imply further restrictions on the production function.

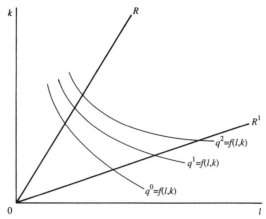

Fig. 8.2: Typical isoquants

Consider now what happens to output as both inputs increase in the same proportion. We refer to this as an *expansion in the scale of production.* Diagrammatically, it means we are moving out along some ray such as *or* in Fig. 8.2. We are then interested in how output increases as we expand scale, and we distinguish:

increasing returns to scale exist when output expands by a larger proportion than the proportionate expansion in inputs, e.g., a 10% increase in each input increases output by 15%;

constant returns to scale exist when output expands by the same proportion as inputs, e.g., a 10% input increase causes a 10% output increase;

decreasing returns to scale exist when output expands by a smaller proportion than the increase in inputs, e.g., a 10% input increase causes an 8% output expansion.

Usually, we assume one of four cases: that a given production function exhibits first increasing, then constant, then diminishing returns to scale; *or* any one of these over its entire domain; *or* first increasing then constant returns; *or* first constant then diminishing returns. We make whatever assumption seems best to fit the problem at hand, or is suggested by whatever facts we may have about the technology of the good in question.

Note also that, in general, the precise nature of the returns to scale may depend on the ray we take along which to expand inputs. For example, in Fig. 8.2, the returns to scale along ray *OR*, where the isoquants are tightly bunched, clearly differ from those along ray OR^1, where the same isoquants are more widely spaced.

This last point leads to consideration of a class of functions for which this is not true: the same returns to scale obtain along any ray. These are the *homogeneous functions.* A production function is called *homogeneous of degree n* when changing inputs by the multiple $\lambda > 0$ changes output by the multiple λ^n, i.e.

$$\lambda^n q^0 = f(\lambda l^0, \lambda k^0) \quad \text{when} \quad q^0 = f(l^0, k^0) \tag{8.7}$$

for any initial value (l^0, k^0). Since inputs are increasing by $(\lambda - 1) 100\%$ while output is increasing by $(\lambda^n - 1) 100\%$, clearly $n > 1$ implies increasing returns to scale, $n = 1$ implies constant returns to scale, and $n < 1$ implies decreasing returns to scale. Moreover, since n is the same whatever the initial point (l^0, k^0), returns to scale are the same along any ray, by definition.

The best-known example of a homogeneous function is the Cobb-Douglas function

$$q = al^{b_1} k^{b_2} \qquad a, b_1, b_2 > 0. \tag{8.8}$$

It is easy to see that this is homogeneous of degree $b_1 + b_2$. Given any l_0, k_0, and $q^0 = al_0^{b_1} k_0^{b_2}$, let $l_1 = \lambda l_0$ and $k_1 = \lambda k_0$. Then

$$q_1 = al_1^{b_1} k_1^{b_2} = a(\lambda l_0)^{b_1} (\lambda k_0)^{b_2} = al_0^{b_1} k_0^{b_2} \lambda^{b_1 + b_2} = \lambda^{b_1 + b_2} q_0. \tag{8.9}$$

Thus $b_1 + b_2 > 1$ implies increasing returns to scale, $b_1 + b_2 = 1$ implies constant returns to scale, and $b_1 + b_2 < 1$ implies diminishing returns to scale. We shall make use of the Cobb-Douglas production function quite extensively in what follows (see also questions 2 and 3 of Exercise 8.1).

To conclude this section, we now have to consider the *time horizon* of the production and supply decisions. To fix ideas, suppose that the firm is considering its possible weekly output level. It hires labour on weekly contracts, so that its labour input can be varied from week to week. However, suppose that it takes one year to bring about a change in its stock of machines, i.e., it takes a year to order and install a new machine, or to sell one of its existing machines. Consider the firm's decision problem at a specific point of time, say the first Monday of the year. In deciding upon its output for that week, the firm has a given labour force and a given stock of machines, say l', k', and so its maximum possible output for the week is $q' = f(l', k')$. It could of course produce $q < q'$ by leaving some of its workers and machines idle, but as long as output has a positive price it would *never pay it to do so*. This is because it must incur the wage bill wl' and the capital cost vk' *regardless* of how much of the inputs it actually uses, and so it is clearly best to use these inputs to the full to obtain as much revenue as possible. In other words, for the current week, all costs are fixed costs, and the additional cost of increasing output from zero all the way up to the 'capacity level' q' is zero.

However, in planning output for the next week, it is possible to vary the labour force by hiring extra workers or not re-employing existing workers, although the capital stock will still have to be taken as fixed. This is in fact true for every following week of this year. Only in planning output for next year is it possible to treat capital stock as variable, because there is time to vary the capital stock by then. This difference is recognized by referring to production in any week *this* year (after the very first week) as being in the *short run*, when capacity must be taken as fixed; and to production next year, or indeed in any future year, as being in the *long run*, when capacity can be treated as variable. Clearly, the set of production possibilities is much greater in the long run than in the short run, and we will have to analyse the firm's production and supply decisions for both these cases. In our notation, we will denote short-run variables by q_0, l_0, k_0, and long-run variables by q_1, l_1, k_1. We will assume that the production function f and input prices w, v are the same in both short and long runs. We consider first the long run, then the short run, and then compare the results for each case.

Exercise 8.1

1. Given that l and k are cooperant inputs, show the effect on the total product curve in Fig. 8.1(a) of an increase in k, paying particular attention to what happens to the slope of the curve at every l-value.

2. (i) If $f(x_1, x_2)$ is homogeneous of degree n, then its partial derivatives, f_1, f_2 are each homogeneous of degree $n - 1$. Show that this holds in the case of the Cobb-Douglas production function $q = al^{b_1}k^{b_2}$.

   (ii) If $f(x_1, x_2)$ is homogeneous of degree n, then the following (*Euler's theorem*) is identically true:

$$f_1 x_1 + f_2 x_2 = nf(x_1, x_2).$$

   In the case where f is a production function, use this to explain the following statement: if the inputs are each paid a price equal to their marginal product, the resulting total payments will
   (*a*) exceed total output if $n > 1$,
   (*b*) just equal total output if $n = 1$,
   (*c*) fall short of total output if $n < 1$.
   Then show that Euler's theorem holds for the Cobb-Douglas production function.

   (iii) The slope of a contour of a homogeneous function $f(x_1, x_2)$ depends *only* on the ratio x_1/x_2, and *not* on the value of the function. Show that this holds for the Cobb-Douglas function. Show also that this implies that along a ray the slopes of all contours, at their points of intersection with the ray, are the same.

3. Given the Cobb-Douglas production function $q = al^{b_1}k^{b_2}$, with $0 < b_1, b_2 < 1$, show that
   (i) marginal products are everywhere positive but diminishing;
   (ii) the inputs are cooperant;
   (iii) the contours have a curvature satisfying (8.6), whether $b_1 + b_2 \gtrless 1$.

8.2 Long-run production and costs

When both labour and capital can be treated as variable, the production department has to choose labour l_1 and capital k_1 to minimize cost $C_1 = wl_1 + vk_1$, taking the input prices as given, and subject to its production function $q_1 = f(l_1, k_1)$.
   This implies the problem

$$\min \; C_1 = wl_1 + vk_1 \qquad \text{s.t.} \, f(l_1, k_1) = q_1$$

where q_1 is some given output level. Necessary conditions are

$$w - \lambda f_l = 0,$$
$$v - \lambda f_k = 0, \qquad (8.10)$$
$$f(l_1, k_1) - q_1 = 0$$

where λ is the Lagrange multiplier. The first two conditions in (8.1) yield the tangency condition

$$\frac{w}{v} = \frac{f_l}{f_k} \qquad (8.11)$$

specifying that when cost is minimized the input price ratio is equated with the ratio of marginal products, or equivalently the marginal rate of technical substitution. The first two conditions of (8.10) could alternatively be rearranged to give

$$\frac{w}{f_l} = \frac{v}{f_k} = \lambda \qquad (8.12)$$

which has an important interpretation. We refer to λ as *long-run marginal cost*, dC_1/dq_1, the rate at which minimized cost changes when output varies, and (8.12) shows that this is given by the common ratio of input price to marginal product when the given output q_1 is produced at minimum cost. To see the intuition behind this, note that since f_l denotes $\partial q/\partial l_1$, and f_k denotes $\partial q/\partial k_1$, we have

$$w \frac{\partial l_1}{\partial q_1} = v \frac{\partial k_1}{\partial q_1} = \lambda. \qquad (8.13)$$

Thus, at the cost-minimizing levels of l_1 and k_1, to a first approximation the cost of a small increase in output is the increase in the wage bill if it is achieved by increasing labour with capital fixed, or the increase in capital costs if it is achieved by increasing capital with labour fixed.

We can solve conditions (8.10) to obtain the cost-minimizing value of l_1 and k_1 as functions of the parameters of the problem, w, v, and q_1. We write these *constant-output* demand functions as $l_1 = l(w, v, q_1)$ and $k_1 = k(w, v, q_1)$. Inserting these into the cost equation $C_1 = wl_1 + vk_1$ gives

$$C_1 = wl(w, v, q_1) + vk(w, v, q_1) = C_1(w, v, q_1).$$

$C_1(w, v, q_1)$ is called the *long-run total cost function*.

Note the close similarity between the cost minimization problem here and the consumer's expenditure minimization problem in Section 7.1 (especially given the assumption that the contours of the production function – the isoquants – have the same general shape as the indifference curves in consumer theory). In fact the two problems have exactly the same structure, and we can exploit that to note that we can apply Shephard's lemma in this case to obtain

$$\frac{\partial C_1}{\partial w} = l(w, v, q_1); \qquad \frac{\partial C_1}{\partial v} = k(w, v, q_1).$$

However, at this point our interest lies far more in the relationship between cost and output than with the input demand functions, and so we assume in what follows that input prices are constant, and suppress them in the notation, to write the long-run total cost function simply as $C_1(q_1)$.

We are particularly interested, for reasons that will become clearer in Chapter 9 below, in the shape of this function, i.e., the precise way in which costs increase as output increases. As can be seen from the way the function is derived, this depends fundamentally (for given input prices) on the form of the production function. We shall clarify this by means of an example in a moment, but first note that one way of describing the shape of the cost function is in terms of *economies and diseconomies* of scale. *Economies of scale* are said to exist when increasing output leads to a less-than-proportionate increase in cost; *diseconomies of scale* exist when increasing output leads to a more-than-proportionate increase in cost. We can denote the proportionate rate of change of output by dq_1/q_1, and the proportionate rate of change of cost by dC_1/C_1. Moreover, we note that

$$dC_1 = C_1'(q_1)dq_1 \qquad (8.14)$$

where $C_1'(q_1)$, the derivative of total cost with respect to output, is long-run marginal cost (LRMC). We also define $C_1(q_1)/q_1$ as long-run average cost (LRAC), and so we have that the derivative of LRAC is given by

$$\frac{d}{dq_1}\left(\frac{C_1(q_1)}{q_1}\right) = \frac{1}{q_1}\left(C_1'(q_1) - \frac{C_1(q_1)}{q_1}\right). \qquad (8.15)$$

Then: (i) *economies of scale* imply

$$\frac{dC_1}{C_1} < \frac{dq_1}{q_1}$$

$$\Rightarrow \qquad C_1'(q_1)\frac{dq_1}{C_1} < \frac{dq_1}{q_1} \qquad \text{using (8.14)}$$

$$\Rightarrow \qquad C_1'(q_1) < \frac{C_1}{q_1}$$

$$\Rightarrow \qquad \frac{d}{dq_1}\left(\frac{C_1(q_1)}{q_1}\right) < 0 \qquad \text{using (8.15);}$$

(ii) *diseconomies of scale* imply

$$\frac{dC_1}{C_1} > \frac{dq_1}{q_1}$$

$$\Rightarrow \qquad C_1'(q_1)\frac{dq_1}{C_1} > \frac{dq_1}{q_1} \qquad \text{using (8.14)}$$

$$\Rightarrow \qquad C_1'(q_1) > \frac{C_1}{q_1}$$

$$\Rightarrow \qquad \frac{d}{dq_1}\left(\frac{C_1(q_1)}{q_1}\right) > 0 \qquad \text{using (8.15).}$$

Thus economies of scale imply that LRMC < LRAC and that the latter is falling, while diseconomies of scale imply LRMC > LRAC and the latter is rising. Note also that if LRAC is at a minimum at an output $\hat{q}_1$, so that $d[C_1(\hat{q}_1)/\hat{q}_1]/dq_1 = 0$, then (8.15) implies

$$C'_1(\hat{q}_1) = \frac{C_1(\hat{q}_1)}{\hat{q}_1},$$

i.e., LRMC = LRAC at its minimum value.

These relationships are illustrated in Fig. 8.3. In (*a*) we have a total cost function $C_1 = C_1(q_1)$, and in (*b*) we have the LRMC function $C'_1(q_1)$ and LRAC function C_1/q_1. The point of inflexion of the function occurs at $\tilde{q}_1$, and this is the value at which C'_1 is at a minimum. LRMC cuts LRAC at $\hat{q}_1$, at which point the slope of a tangent to the total cost function (= LRMC) is just equal to the slope of a ray (= LRAC).

To illustrate the way in which the cost function depends on the production function, we take the Cobb-Douglas function

$$q_1 = l_1^{b_1} k_1^{b_2} \qquad b_1, b_2 > 0.$$

The cost-minimization problem for this function becomes

$$\min \ wl_1 + vk_1 - \lambda(l_1^{b_1} k_1^{b_2} - q_1),$$

yielding the conditions

$$w - \lambda b_1 l_1^{b_1-1} k_1^{b_2} = 0,$$
$$v - \lambda b_2 l_1^{b_1} k_1^{b_2-1} = 0,$$
$$l_1^{b_1} k_1^{b_2} - q_1 = 0.$$

Eliminating λ gives

$$\frac{w}{v} = \frac{b_1 l_1^{b_1-1} k_1^{b_2}}{b_2 l_1^{b_1} k_1^{b_2-1}} = \frac{b_1}{b_2} \frac{k_1}{l_1} \ \Rightarrow \ l_1 = \frac{v}{w} \frac{b_1}{b_2} k_1. \tag{8.16}$$

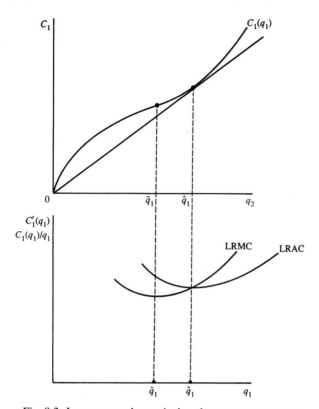

Fig. 8.3: Long-run total, marginal, and average cost curves

(8.16) is the equation of the *output expansion path*, i.e., the locus of all points of tangency of isoquants with cost lines, and it shows that for this production function, the output expansion path is a line through the origin. A corollary of this is that the capital/labour ratio k_1/l_1 is independent of the *scale* of output and depends only on relative input prices, i.e.

$$\frac{k_1}{l_1} = \frac{w}{v}\frac{b_2}{b_1}. \tag{8.17}$$

(8.16) alone does not of course allow us to solve the problem. Substituting into the production function constraint gives

$$\left(\frac{v}{w}\frac{b_1}{b_2}k_1\right)^{b_1} k_1^{b_2} - q_1 = 0 \;\Rightarrow\; k_1 = \left(\frac{w}{v}\frac{b_2}{b_1}\right)^{\beta_1} q_1^{\alpha} \tag{8.18}$$

where $\beta_1 = b_1/(b_1 + b_2)$, $\alpha = 1/(b_1 + b_2)$. This is the constant-output demand curve for capital. We obtain that for labour by substituting from (8.18) back into (8.16) to obtain

$$l_1 = \left(\frac{v\,b_1}{w\,b_2}\right)^{\beta_2} q_1^{\alpha} \tag{8.19}$$

where $\beta_2 = b_2/(b_1 + b_2)$.

We then derive the cost function by substituting for l_1 and k_1 into the cost equation to obtain

$$C_1 = w\left(\frac{v\,b_1}{w\,b_2}\right)^{\beta_2} q_1^{\alpha} + v\left(\frac{w\,b_2}{v\,b_1}\right)^{\beta_1} q_1^{\alpha} = Aw^{\beta_1}v^{\beta_2}q_1^{\alpha} \tag{8.20}$$

where $A = (b_1/b_2)^{\beta_2} + (b_2/b_1)^{\beta_1}$.
Or, taking w and v as fixed, we have

$$C_1 = \beta q_1^{\alpha} \qquad \text{where} \qquad \beta = Aw^{\beta_1}v^{\beta_2}, \tag{8.21}$$

which then implies the LRMC and LRAC functions

$$C_1'(q_1) = \alpha\beta q_1^{\alpha-1}, \qquad C_1(q_1)/q_1 = \beta q_1^{\alpha-1}. \tag{8.22}$$

Clearly, the shapes of the total, marginal, and average cost curves depend on the value of $\alpha = 1/(b_1 + b_2)$. Recall the definitions of increasing, constant, and diminishing returns to scale from the previous section. Then, since

$$C_1''(q_1) = \alpha\beta(\alpha - 1)q_1^{\alpha-2} \qquad \text{and} \qquad \frac{d}{dq_1}\left(\frac{C_1(q_1)}{q_1}\right) = \beta(\alpha - 1)q_1^{\alpha-2}, \tag{8.23}$$

we have that

(i)   with increasing returns to scale, $b_1 + b_2 > 1$, so $\alpha < 1$, therefore from (8.23) marginal and average costs are falling with output, implying economies of scale;

(ii) with constant returns to scale, $b_1 + b_2 = 1$, so $\alpha = 1$, therefore from (8.23) marginal and average costs are constant and (from (8.22)) equal — so there are neither economies nor diseconomies of scale;

(iii) with diminishing returns to scale, $b_1 + b_2 < 1$, so $\alpha > 1$, therefore from (8.23) marginal and average costs are increasing with output, implying diseconomies of scale.

These three cases are illustrated in Fig. 8.4.

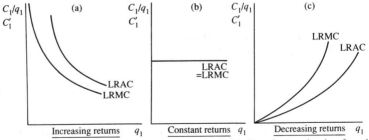

Fig. 8.4: Returns to scale and cost functions for Cobb-Douglas production functions

Note that in the case of the Cobb-Douglas production function, we do not obtain the 'usual' U-shaped average and marginal cost curves. However, the Cobb-Douglas function does provide a simple way of showing the relationship between returns to scale (the shape of the production function) and economies of scale (the shape of the cost function).

Exercise 8.2

1. Illustrate diagramatically the solution to the cost minimization problem analysed in this section. Explain how the cost–output pairs which define the total cost function can be derived from the output expansion path. Compare the analysis here to that given for the consumer in Section 7.1.
2. Show that the elasticity of total cost with respect to an input price is equal to the proportion of total cost paid out to that input [Hint: the elasticity of total cost with respect to the wage rate is $(\partial C_1/\partial w)(w/C_1)$, and likewise for the capital rental; use Shephard's lemma.] Confirm that this holds for the Cobb-Douglas production function.
3. A firm has the production function $q = l^{1/2}k^{1/4}$, with $w = \$10$, $v = \$20$. Derive its long run total, average, and marginal cost functions, and sketch them on a graph.

8.3 Short-run production and costs

The fact that capital cannot be expanded or sold off in the short run has two implications for the cost minimization problem the production department has to solve. First, production is subject to the constraint $k_0 \leq \bar{k}$, where $\bar{k}$ is the fixed amount of capital available to the firm. Note that we allow in principle less than the maximum capital to be *used*. Second, the firm must pay its capital cost $v\bar{k}$ regardless of how much output it produces, and so this represents a fixed cost, which we denote by F. Thus the short-run cost-minimization problem is

$$\min C_0 = wl_0 + F \qquad \text{s.t.} \quad q_0 = f(l_0, k_0) \quad \text{and} \quad k_0 \leq \bar{k}.$$

It is hardly necessary to solve this problem explicitly, however. Since short-run total cost varies only with l_0, the problem is simply to minimize l_0 while producing the required level of output. But, since l and k are substitutes, this in turn implies using as much k as possible, so we should set $k_0 = \bar{k}$. Then the required labour input is found by solving the constraint $q_0 = f(l_0, \bar{k})$ for l_0, so that we have

$$l_0 = f^{-1}(q_0, \bar{k}) \tag{8.24}$$

and the short-run cost total cost function becomes

$$C_0 = wf^{-1}(q_0, \bar{k}) + F = C_0(q_0, \bar{k}) + F. \tag{8.25}$$

For the moment we are not interested in $\bar{k}$, so we suppress it in the notation and write simply $C_0 = C_0(q_0) + F$. The first component of cost is known as total variable cost, the second as total fixed cost. This procedure is illustrated in Fig. 8.5.

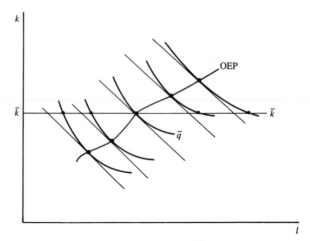

Fig. 8.5: Short-run output expansion path $\bar{k}\bar{k}$ compared to the long run
output expansion path OEP

With capital set at $\bar{k}$ the optimal labour inputs are found from the successive isoquants, and so the horizontal line $\bar{k}$ is the short-run expansion path. Note that it intersects the long-run expansion path on the isoquant corresponding to output $\bar{q}$. At this output, the fixed capital $\bar{k}$ is in fact the cost-minimizing value, and so short- and long-run costs are equal. At outputs $q < \bar{q}$ and $q > \bar{q}$, however, short-run costs, though minimized for the fixed $\bar{k}$, are higher than long-run costs, because $\bar{k}$ is not the appropriate level of capital for those outputs. As $q \rightarrow \bar{q}$, from both above and below, the 'cost penalty' associated with having the 'wrong' k-value, or equivalently the 'wrong' k/l ratio, goes to zero.

Short-run marginal cost (SRMC) is given (from (8.25)) by

$$C_0'(q_0) = w\frac{\partial f^{-1}}{\partial q_0} = w \left/ \frac{\partial f}{\partial l_0} \right. \tag{8.26}$$

Thus short run marginal cost is simply the ratio of the wage rate to the marginal product of labour. With w fixed, the graph of this function as output increases will be the inverse of the graph of the marginal product function (refer back to Fig. 8.1(b)). As output expands, with k fixed, $\partial f/\partial l_0$ increases, reaches a maximum, and then falls, and so $C_0'(q_0)$ will first decrease, reach a minimum, and then rise. The assumption that f_l *must* start diminishing at some level of l and q for any fixed k implies that SRMC *must* start increasing at some q.

We now have three kinds of average cost in the short run. These are:

$$\text{average variable cost,} \quad \text{AVC} \equiv \frac{C_0(q_0)}{q_0} = \frac{wl_0}{q_0} = w \left/ \frac{q_0}{l_0} \right. ;$$

$$\text{average fixed cost,} \quad \text{AFC} \equiv \frac{F}{q_0} ;$$

$$\text{average total cost,} \quad \text{AC} \equiv \text{AVC} + \text{AFC}.$$

The first of these definitions shows that AVC, which is simply the wage bill divided by output, can be expressed as the ratio of the wage rate to the average product of labour. It follows that the graph of this function is the inverse of the graph of the average product function. If average product first increases, reaches a maximum, then falls, AVC will first fall, reach a minimum, and then rise. Moreover,

$$\frac{\partial f}{\partial l_0} = \frac{q_0}{l_0} \implies \text{AVC} = \text{SRMC},$$

and since the first of these equalities holds when average product is at its maximum, the second will hold when AVC is at its minimum.

The second definition shows that the graph of AFC will be a rectangular hyperbola. The third shows that AC is found by adding AVC to AFC at each level of output.

The curves drawn in Fig. 8.6 summarize this discussion. The shape of the short run total cost curve in (a) of the figure reflects the assumed shape of the total product curve in Fig. 8.1 (as (8.25) implies), with an intercept given by fixed cost F. The marginal and average cost curves in (b) of the figure are derived from this, and, as we have just discussed, reflect the properties of the marginal and average product curves in Fig. 8.1. The minimum point of the SRMC curve in (b) occurs at output $\bar{q}_0$, corresponding to the point of inflexion of the total cost curve in (a). At output $\hat{q}_0$, the ray drawn from F to the total cost curve coincides with the tangent to the curve, and so at that output SRMC = AVC. At output q'_0, the ray drawn *from the origin* to the total cost curve coincides with the tangent to the curve, and so at that output SRMC = AC.

The Cobb-Douglas production function again provides a simple example of the derivation of the short-run cost curves. With $k_0 = \bar{k}$, we have

$$q_0 = a l_0^{b_1} \bar{k}^{b_2} \implies l_0 = (a\bar{k}^{b_2})^{-1/b_1} q_0^{1/b_1} = \gamma q_0^{1/b_1} \tag{8.27}$$

where $\gamma \equiv (a\bar{k}^{b_2})^{-1/b_1}$. Then

$$C_0(q_0) = w\gamma q_0^{1/b_1} + F,$$
$$C'_0(q_0) = (w\gamma/b_1) q_0^{(1/b_1)-1},$$
$$\text{AVC} = w\gamma q_0^{(1/b_1)-1}.$$

Thus, given the condition $0 < b_1 < 1$ we have both SRMC and AVC increasing with output. Again, the Cobb-Douglas production function does not give us the 'U-shaped' average and marginal cost curves often assumed.

Exercise 8.3

1. Take the production function and input prices in question 3 of Exercise 8.2, and assume the firm's capital is fixed at 10 units. Derive its short-run total, marginal, average variable, and average total cost functions and sketch them on a graph.
2. Explain carefully why:
   (a) the SRMC curve is just the 'mirror image' of the marginal product of labour curve;
   (b) the AVC curve is just the 'mirror image' of the average product of labour curve;
   (c) the SRMC curve cuts the AVC curve at its lowest point;
   (d) the SRMC curve cuts the ATC (= AVC + AFC) curve at its lowest point;
   (e) SRMC is independent of fixed cost.

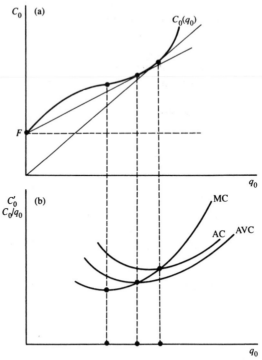

Fig. 8.6: Short-run cost curves

8.4 Relationship between long-run and short-run cost functions

We now have to examine the relationship between long- and short-run costs. Recall we made the assumptions that input prices and the production function are the same in both long and short runs. Then, we can first obtain an intuitive picture of the relationship between the cost curves by a purely diagrammatic argument. Refer back to Fig. 8.5. We see that as output increases – we move through successively higher isoquants – the short-run expansion path and the long-run expansion path converge, meet at output $\bar{q}$, and then diverge. This implies that at outputs above and below $\bar{q}$ short-run total costs are higher than long-run total costs, but that this difference narrows to zero as output converges to $\bar{q}$. In Fig 8.7(a) we show the implications of this for the total cost curves. The short-run curve C_0 begins at $F (= v\bar{k})$, converges to the long-run curve C_1, meets it at output $\bar{q}$ then diverges. Thus the two curves are *tangent* at output $\bar{q}$. However, there is nothing special about the choice of $\bar{k}$ as the fixed capital value. If we had chosen a larger capital stock $\bar{k}^1$, with $\bar{q}^1$ the corresponding output for which it is the cost-minimizing level (isoquant on which short- and long-run expansion paths would interesect in Fig. 8.5), then by the same argument we would have the same general relationship between the new short-run cost curve and the long-run curve (the intercept is now higher because of course $v\bar{k}^1 > v\bar{k}$), as shown in Fig. 8.7(*a*). Indeed, we could take each possible value of $k > 0$ and derive a corresponding short-run cost curve, that will be tangent to the long-run curve at just one point, the output for which it is the cost-minimizing k-value. Thus we would have a *family* of short-run cost curves, with the long-run curve tracing out a lower boundary, or *envelope*, of them all, as Fig. 8.7(*a*) illustrates.

Given the relationship between the short- and long-run total cost curves in Fig. 8.7(*a*), we can derive the relationship between the corresponding sets of marginal and average cost curves, shown in Fig. 8.7(*b*). Since there is only one long-run total cost curve, there is only one LRMC curve and one LRAC curve, derived as shown in Fig. 8.3. Then, corresponding to each short-run total cost curve is a SRMC curve and a SRAC curve. Their relationships to the long-run curves are:

(*a*) The SRMC curve *intersects* the LRMC curve at that output level for which the fixed capital stock underlying the short-run curve is the cost-minimizing value. To see this, note that in Fig. 8.7(*a*), at a point of tangency between short- and long-run total cost curves, their slopes are the same and so the two marginal costs must be equal;

(*b*) The SRAC curve is *tangent* to the LRAC curve at that output level for which the fixed capital stock underlying the short-run curve is the cost-minimizing value. To see this, note that in Fig. 8.7(*a*) a ray drawn from the origin to a point of tangency between a short-run total cost curve and the long-run total cost curve has a slope equal to both SRAC and LRAC at that point. At other outputs, the given short-run total cost curve lies above the long-run total cost curve and so the given SRAC curve must be above the LRAC curve at those points. Thus the point of tangency between short- and long-run *total* curves implies a tangency between short- and long-run *average* curves. This in turn implies that the LRAC curve is the envelope of the entire family of SRAC curves.

Note that, since SRAC and LRAC curves meet at a point of tangency, the sign of the slope of the SRAC curve must be the same as that of the LRAC curve. This in turn implies that, with the exception of only one output level, q_* in Fig. 8.7(*b*), there is a difference between

(i) that output for which a given value of k is the cost-minimizing value;
(ii) that output at which the SRAC curve for a given value of k is at a minimum.

When there are economies of scale, the output level in (i) is less than that in (ii); when there are diseconomies of scale the reverse applies. Only at that output at which LRAC is at a minimum are the two output levels equal (refer to the diagram).

This result can be put in practical terms as follows. Suppose it is desired to produce output q_0 in Fig. 8.7(*b*) at minimum cost. The solution, since there are economies of scale, is *not* to install that level of capacity for which q_0 achieves the minimum *short-run* average cost, but to install a somewhat larger capacity, taking advantage of the economies of scale. In other words it is better to operate a larger plant at less than optimum capacity because of the economies of scale. Likewise, if it is desired to produce output q_0' in Fig. 8.7(b) at minimum cost, we should install a somewhat smaller level of capacity than that for which q_0' minimizes SRAC, because of the diseconomies of scale. Only when there are neither diseconomies nor economies of scale do we install capacity for which the desired output is the level that minimizes SRAC.

We now have to analyse the relationship between short-run and long-run costs more rigorously. Essentially, to provide confirmation of the diagrammatic argument in Fig. 8.7, we have to show the following. Suppose we arbitrarily fix $\bar{k}$, solve the short-run cost-minimization problem, and obtain the cost function $C_0 = C_0(q_0, \bar{k}) + v\bar{k}$ as in (8.25). Expressing C_0 as a function of $\bar{k}$ emphasizes the fact that for any required q, the cost-minimizing l_0 depends on $\bar{k}$. Varying $\bar{k}$ then traces out a whole family of these cost functions. Now suppose we *fix* q_0 at some value and ask what value of $\bar{k}$ *minimizes* C_0 for this q_0 This $\bar{k}$ must obviously satisfy, using (8.25),

$$\frac{dC_0}{dk} = \frac{w \partial f^{-1}}{\partial k} + v = 0. \tag{8.28}$$

But

$$\frac{\partial f^{-1}}{\partial k} = \frac{dl}{dk} = \frac{-f_k}{f_l}. \tag{8.29}$$

Thus (8.28) becomes

$$\frac{w}{f_l} = \frac{v}{f_k} \quad \text{or} \quad \frac{w}{v} = \frac{f_l}{f_k} \tag{8.30}$$

which is the condition for the (l, k) -pair which minimizes long-run costs of producing q_0. Thus the k that minimizes short-run costs of producing any output q_0 is the k that minimizes long-run costs of producing that output, as is the value of l, and so short- and long-run costs coincide at that k.

We can illustrate this with the Cobb-Douglas example used earlier. The short-run cost function (refer to (8.27)) was

$$C_0 = wa^{-1/b_1} k^{-b_2/b_1} q_0^{1/b_1} + vk. \tag{8.31}$$

Taking q_0 as fixed and differentiating gives

$$\frac{dC_0}{dk} = -wa^{-1/b_1} \frac{b_2}{b_1} k^{-(1+b_2/b_1)} q_0^{1/b_1} + v = 0. \tag{8.32}$$

But $q_0 = al^{b_1} k^{b_2}$, and so substituting into (8.32) gives

$$-w\frac{b_2}{b_1} \frac{l}{k} + v = 0 \Rightarrow \frac{w}{v} = \frac{b_1 k}{b_2 l} \tag{8.33}$$

where the right-hand side is the ratio of marginal products. Thus the cost-minimizing k, and implied value of l, lie on the long-run output expansion path. Thus the minimum short-run cost of producing an output q_0 is the point on the long-run cost function corresponding to that output.

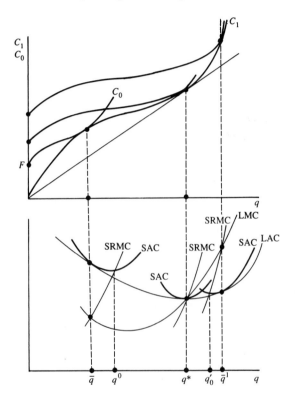

Fig. 8.7: Relationship between long- and short-run cost curves

Exercise 8.4

1. Discuss the nature of the short-run output expansion path, and the relationship between short- and long-run cost curves, in the case where the firm is able to dispose of capital, at the rental rate v, within one week (e.g., by hiring it out), but it still takes a year to expand capacity. [Hint: what exactly *are* fixed costs in the short run, and how does this affect the firm's short run cost-minimization problem?]
2. Suppose that in one year's time:
   (a) both the wage rate and the rental rate on capital are expected to increase in the same proportion; or
   (b) a given combination of inputs will produce 50% more output.

   How would you interpret the envelope of the short run curves in this case? What

becomes of the relation between short- and long run total costs? What will short run costs be like in one year's time?
3. Using the relationship: TC = AC(q) q, differentiate and use the result to show that when short- and long run total cost curves are tangent, SRAC is tangent to LRAC and SRMC = LRMC.

8.5 Summary

Typically the sellers in markets for goods and services are firms. Therefore, to develop a theory of the determinants of supply we need to develop a theory of the firm.

We view the firm *as if* it were controlled by a single decision-taker. His problem is to choose inputs and outputs in such a way as to maximize profit, defined by the difference between revenue and cost.

We separate the problem into two parts. The first, the subject of this chapter, is the problem of choosing the cost-minimizing input quantities to produce each possible output level, and then to use the solution of this to construct the firm's *cost function*, the relation between (minimized) cost and output (as well as input prices, which however are often treated as parameters rather than as variables).

The central relationship is the production function, which summarizes the technological possibilities for combining inputs to produce output.

The assumptions we make about this function are that it is twice differentiable, and that there is always a point at which the marginal product of each input starts to fall as the use of the input increases. Also important is the assumption we make about the nature of returns to scale, which specify how output changes as we increase inputs in the same proportion (i.e., increase the *scale* of production). Essentially, the shape of the production function determines the shape of the cost function.

We also need to distinguish between *the long run*, a period far enough ahead that all inputs can be varied and so the scale of production can be chosen and the short run, in which one input (typically capital) *and the cost associated with it* are fixed, and only one input (typically labour) is variable.

Solving the problem of minimizing cost of producing a given output allows us to state the conditions which must be satisfied by the input quantities the firm chooses, and allows us to derive constant output–input demand functions and total cost functions.

From the latter marginal and average cost functions follow directly. In discussing these, the main interest lies in examining how the assumed properties of the production function determine the shapes of the cost functions.

In general, short- and long-run cost functions differ. That is, in general it is cheaper to produce a *given output level* when we can choose *both* inputs so as to minimize costs, than when one is fixed and we can only choose *one* input. However, it may happen that the fixed level of capital is at the level that *would* be chosen to produce the given output level. Hence at that output level short- and long-run cost functions just touch. This accounts for the 'envelope' relation between the long run (total, average) cost functions and the short run (total, average) cost functions.

9 Competitive Market Supply

Given the cost functions constructed by the 'production department', the 'sales department' must then use these to solve the problem of choosing profit-maximizing output levels. In the short-run, the problem is to choose a current level of output. At the same time, the sales department must *plan* the output it will produce in the long run. When the choice of short-run output is made, the production department will then know how much labour – the variable input – to hire. When the choice of the long-run output plan is made the production department will know how much capital it will need in the future, and so can set in motion *now* the appropriate investment (or divestment) plan.

The most important task of the sales department is to formulate the firm's *revenue function*, $R(q)$, which expresses the relationship between output sold and revenue gained. The nature of this function is determined by conditions on the output market and in particular by the nature of competition. We distinguish three cases:

1. *Perfect competition.* There is a 'very large' number of firms in the market, each so small that its sales department can take market price as a constant, unaffected by its own output. This implies that its revenue function is $R = pq$, where p is the given market price. A further important aspect of perfect competition is that in the long run, new firms can enter the market and compete on the same terms as existing firms – there are no *barriers to entry*. There are therefore two sources of variation in long-run supply: changes in output by existing firms, and changes in the number of firms.
2. *Monopoly.* There is only one firm controlling entire market supply. This case is dealt with in Chapter 10.
3. *Oligopoly.* There is a small number of firms who recognize the interdependence of their decision-taking. This case is dealt with in Chapter 11.

In this chapter, we analyse the determinants of supply in competitive markets. The procedure is to solve for the firm's profit-maximizing output and then to show how this output will vary with market price. The implied relationship between price and the firm's output gives the firm's perceived supply function, i.e., the relation between price and the quantity the firm perceives that it can sell. We then have to consider how *aggregate* market supply will vary with price. In the short run, this is a relatively straightforward matter, but in the long run expectations again become important, as does the possibility of new entry.

9.1 Short-run Supply

Given the short-run cost function (corresponding to the firm's given capital stock $\bar{k}$) derived by the production department, the sales department has to find the level of output that maximizes the firm's profit. In doing this, clearly the conditions on the market it faces will be of central importance. Since in this chapter we assume that the firm sells in a perfectly competitive market, it takes the market price, p, as a constant, unaffected by its own output. Then its total revenue function is given simply by $R(q_0) = pq_0$, and its total profit function, $\pi(q_0)$, by

$$\pi(q_0) = R(q_0) - C_0(q_0) - F = pq_0 - C_0(q_0) - F \qquad (9.1)$$

where F is again fixed cost. Then, maximizing $\pi(q_0)$ with respect to q_0 implies the conditions

$$\pi'(q_0^*) = p - C_0'(q_0^*) = 0, \qquad (9.2)$$

$$\pi''(q_0^*) = -C_0''(q_0^*) < 0 \qquad (9.3)$$

if the profit-maximizing output $q_0^* > 0$ (this qualification is necessary because it may turn out that profit-maximizing output is zero, in which case conditions (9.2) and (9.3) need not hold, as we shall see below). Assuming $q_0^*, > 0$ condition (9.2) says it must equate SRMC with price, and (9.3), the second-order condition, says that SRMC must be increasing at q_0^*.

The solution is illustrated in Fig. 9.1. In (*a*) of the figure are shown the total revenue, cost, and profit functions, and in (*b*) the marginal and average cost functions. The horizontal line at the given value of p could be called the demand curve for the firm: it reflects the firm's perception that variations in its output (over the relevant range around q_0^*) do not affect market price. It could also be called the marginal revenue (MR) curve for the firm, since

$$\text{MR} \equiv R'(q) = p.$$

Marginal revenue – the rate at which revenue varies with output, the slope of the total revenue curve – is constant because extra output can be sold at the same price as existing output.

The significance of the second-order condition (9.3) is brought out in the figure. The first-order condition (9.2) is in fact also satisfied at $\hat{q}_0$, where profit is *minimized*. Hence we need (9.3) to ensure we really do have the maximizing output. From now on we restrict attention to solutions to (9.2) which are true maxima.

We can solve (9.2) for q_0^* in terms of p to obtain the function

$$q_0^* = C_0'^{-1}(p) \equiv S_0(p). \qquad (9.4)$$

This is the firm's *short-run supply function*, for positive levels of output, since it expresses the output the firm wants to produce as a function of the market price. So we see that the firm's short-run supply curve is its SRMC curve (as long as output $q_0^* > 0$), and so is determined by the nature of its production function, and in particular by the marginal product of labour function, $f_l(l, k)$ (refer back to equation (8.26)). The fact that this supply curve has a positive slope is then essentially due to the assumption of diminishing marginal productivity of labour.

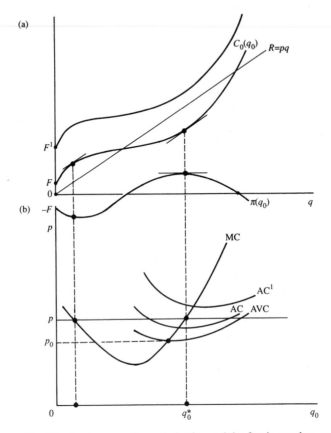

Fig. 9.1: Short-run profit maximization and the firm's supply curve

We now consider the question of the conditions under which the firm will indeed want to produce a positive output rather than zero. Suppose that in Fig. 9.1 fixed costs are actually F^1, so that the firm's cost curves are C^1, SRMC, AVC, and ATC^1 (note the higher fixed cost leaves marginal and average variable cost unaffected). The result is that at every possible output the firm now makes a loss rather than a profit, and q_0^* is now the *loss-minimizing* output. But would the firm do better by producing $q_0 = 0$ rather than q_0^* in this case? To answer this question we note that the fixed cost F^1 must be paid regardless of the level of output. So $\pi(0) = -F$. Then, the firm should produce q_0^* if and only if $\pi(q_0^*) \geq \pi(0)$, i.e., if

$$\pi(q_0^*) = pq_0^* - C_0(q_0^*) - F \geq -F$$

or

$$pq_0 - C_0(q) \geq 0,$$

i.e.

$$p \geq C_0(q_0^*)/q_0^* \equiv \text{AVC}. \tag{9.5}$$

Thus positive output should be produced if and only if price is no less than AVC at the profit-maximizing – or loss-minimizing – output. In Fig. 9.1 this condition is clearly satisfied, because $p > p^0$, but if $p < p^0$, then $q_0 = 0$ would be optimal. This then implies that the full specification of the firm's short-run supply curve is:

$$q_0 = 0 \qquad \text{for} \qquad 0 \leq p < p^0,$$
$$q_0 = S_0(p) \qquad \text{for} \qquad p \geq p^0,$$

where $p^0 = \min \text{AVC}$. Diagrammatically, the firm's short-run supply curve in Fig. 9.1(b) is the interval Op^0 of the vertical axis, and the portion of its SRMC curve above its AVC curve. This implies a discontinuity in the curve, corresponding to the gap between p^0 and b.

The implication of this analysis is that a profit-maximizing firm may continue to produce a positive output in the short run even though it is making losses. This is because it will have to pay its fixed cost whether it produces or not, and so as long as revenue at least covers variable costs it is making *some* contribution to fixed costs by carrying on production. In other words, fixed costs are 'bygones' and can be ignored in the calculation – all that matters are those costs that may vary with output.

We conclude this section by considering some comparative statics. First, what would be the effect on the firm's equilibrium output of a change in the wage rate? Reintroducing w into the cost function, we can rewrite the condition determining the firm's output choice (for $q_0^* > 0$) as

$$p - C_0'(q_0, w) = 0. \tag{9.6}$$

Differentiating (recalling that p is constant) gives

$$\frac{-\partial C_0'}{\partial q_0} dq_0 = \frac{\partial C_0'}{\partial w} dw \tag{9.7}$$

$$\Rightarrow \frac{dq_0}{dw} = -\frac{\partial C_0'/\partial w}{\partial C_0'/\partial q_0}. \tag{9.8}$$

We know from the second-order condition (9.3) that $\partial C_0'/\partial q_0 > 0$, and so if $\partial C_0'/\partial w > 0$ then $dq_0/dw < 0$, or output falls when the wage rate rises. To determine the sign of $\partial C_0'/\partial w$, recall from (8.26) that

$$\frac{\partial C_0}{\partial q_0} = \frac{w}{f_l(l, k)}. \tag{9.9}$$

Hence

$$\frac{\partial}{\partial w}\left(\frac{\partial C_0}{\partial q_0}\right) = \frac{1}{f_l^2}\{f_l - wf_{ll}\} > 0 \tag{9.10}$$

since $f_l > 0$ and $f_{ll} < 0$. Thus an increase in the wage rate raises SRMC and hence reduces the firm's output. Note that the increase in wage rate also increases AVC and so *could* change the firm's output to zero; this would have to be checked in any specific application.

Next consider the effect of a specific tax t per unit of output. In effect, this either changes the revenue function to $R(q_0) = pq_0 - tq_0$, or the cost function to $C_0(q_0) + F + tq_0$. In either case the firm's profit function becomes

$$\pi(q) = pq_0 - C_0(q_0) - F - tq_0$$

and its first-order condition is

$$p - C_0'(q_0^*) - t = 0. \tag{9.11}$$

Then, differentiating through yields

$$-C_0''(q_0^*)dq_0 - dt = 0 \implies \frac{dq_0}{dt} = \frac{1}{-C_0''(q_0^*)} \tag{9.12}$$

implying, since $C_0''(q_0^*) > 0$, that an increase in the tax reduces output. Again, this change could also cause the firm's output to become zero rather than positive (supply the details of the argument).

We can show in Fig. 9.1 that a change in fixed cost leaves the firm's profit-maximizing output unaffected (explain why). Consider now the effect of a profit tax. If levied as a fixed sum, this is clearly equivalent to an increase in fixed cost. Suppose therefore that it is a proportionate profit tax at the rate τ, so that the firm's after-tax profit $\hat{\pi}$, which is of course what it seeks to maximize, is given by the relationship

$$\hat{\pi}(q_0) = (1 - \tau)\pi(q_0) = (1 - \tau)[pq - C_0(q_0) - F].$$

Then the first-order condition is

$$\hat{\pi}'(q_0^*) = (1 - \tau)\pi'(q_0^*) = (1 - \tau)[p - C_0'(q_0^*)] = 0. \tag{9.13}$$

But then $(1 - \tau)$ simply divides out, leaving precisely the same condition as before. Thus, the profit-maximizing firm's short-run output is unaffected by a profits tax, whether lump-sum or proportionate (see question 2 of Exercise 9.1).

Before concluding, we should put the analysis of this section in the context of the decision problem of the firm that we set out in the previous chapter. Given the market price that it expects to prevail in each week of the coming year – the entire short run – the firm determines its profit-maximizing output and labour input for each week. If this price fluctuates from week to week then the firm will be moving along its short-run supply function, determining outputs, and along its short-run labour demand function (still to be analysed – see Chapter 12 below), determining labour inputs. In some weeks, if price falls below minimum AVC, it will shut down altogether.

Clearly, we are assuming that it is costless for the firm to vary output and employment in this way. In reality, however, there are costs involved in hiring and firing workers. In addition, workers may acquire skills specific to the firm which make it worthwhile to hold on to them through temporary drops in demand. On the other side of the coin, it may be possible to store output at reasonably low cost, so that production and employment can take place smoothly over time while sales can vary in response to price changes by running down or building up inventories. Thus the analysis of this section is only the first step in a realistic study of firms' short-run output and supply decisions, though one which is consistent with the approach taken to market supply in Chapters 2–5.

Exercise 9.1

1. Given each firm's short-run supply function as defined in this section, how is the *market* supply function derived? [Hint: how did we derive the market demand function from individual demand functions?] Under what conditions will the discontinuity we noted in the individual firm's supply function occur also in the market supply function? Under what conditions (intuitively) would it disappear?
2. Suppose each firm in the market has the short-run cost function derived in answer to question 1 of Exercise 8.3. What market supply function does that imply? Go through the comparative statics exercises set out in this section for that supply function.

9.2 Long-run supply

As long as we confine attention to the problem of production, the main distinction between short and long runs is the existence of the constraint on capacity and a corresponding fixed cost in the short run. However, when we move on to consider the output supply decision – the sales department's problem – there is another crucial difference. The *short-run problem* is to choose *current* output in the light of the *known* current market price. The long-run decision, on the other hand, is concerned with *planning* future output and input levels on the basis of an *expected* future price, and then deriving from the output plan a current *investment decision*. If the output plan requires a future level of capital greater than that currently available the firm must set about expanding its capital stock now; in the converse case the firm must set about selling off the capital equipment it will not want.

Thus again assume that it takes one year to vary capital stock, and that the sales department is now planning output for one year ahead. Then, formally, its output decision is similar to that made for the short run, except now of course it works with its long-run total cost function $C_1(q_1)$, in which there are no fixed costs. If p_1 is the price the sales department expects to prevail in the future period, the profit-maximizing output in the long run must solve the problem

$$\text{max} \quad \pi_1(q_1) = R_1(q_1) - C_1(q_1) = p_1 q_1 - C_1(q_1),$$

and the solution q_1^*, if positive, must satisfy

$$p_1 - C_1'(q_1^*) = 0, \qquad\qquad (9.14)$$

$$-C''(q_1^*) < 0. \qquad\qquad (9.15)$$

Thus optimal output, if positive, is given by the equality of price with LRMC, which, as (9.15) tells us, must be increasing at the optimal point. Inverting (9.14) gives us the firm's *perceived long-run supply function*

$$q_1^* = C_1'^{-1}(p_1) \equiv S_1(p_1) \qquad\qquad (9.16)$$

for p_1 such that $q_1^* > 0$. How can we identify prices for which output will be positive? We simply note that since in the long run all costs are variable, a zero output implies zero profit. If the firm expected price p_1 to be less than average cost $C(q_1^*)/q_1^*$ at all output levels it would choose zero output, because any positive output yields negative profit. Thus the firm's *perceived long-run supply function* can be written

$$q_1^* = S_1(p_1) \qquad \text{for} \qquad p_1 \geq p_1^0, \qquad\qquad (9.17)$$

$$q_1^* = 0 \qquad \text{for} \qquad p_1 < p_1^0,$$

where p_1^0 corresponds to the minimum LRAC. Fig. 9.2 illustrates this *perceived supply function*. For $p_1 \geq p_1^0$ the firm's supply curve is its LRMC curve; for $p_1 < p_1^0$ it is the vertical axis. Again therefore there is a discontinuity in the firm's supply curve, this time at the minimum point of the LRAC curve.

However, although the firm may *perceive* the function in (9.17) as its long-run supply function, it does not follow that the long-run *equilibrium* of the firm could be at *just any* point on this function (unlike the case of the short run), essentially because in the long-run analysis we have to take account of the *possible entry of new firms into the market*. For this reason, the market long-run supply curve is *not* the horizontal sum of the individual perceived supply curves. To fix ideas, first assume

(a) all firms currently in the market at a given time have identical cost functions;
(b) new firms starting up production have exactly the same production function and face the same input prices as existing firms. So, they must also have the same long-run cost function;
(c) all firms, whether initially in the market or not, have exactly the same price expectations, p_1.

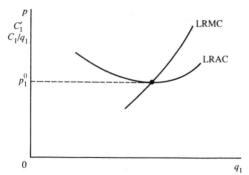

Fig. 9.2: The firm's perceived long-run supply curve

Then, if $p_1 > p_1^0$ in Fig. 9.2, all firms, whether currently in the market or not, will expect that there are positive profits to be made in this market, and so new entry will take place, thus expanding supply next year and driving down price. It may be the case that the equilibrium price next year will still be above LRAC, and, if this creates the expectation that price will be above LRAC in the *following* year, again new entry will take place. Alternatively, price next year may be driven below LRAC, in which case firms will plan to leave the market in the following year. The general point is this: as long as, at each firm's long-run profit-maximizing equilibrium, the expected price differs from LRAC, entry into or exit from the market will take place, and so the market cannot be in long-run equilibrium.

For the market to be in long-run equilibrium, we require that

(a) each firm in the market is planning to produce its profit-maximizing output given the expected price;

(b) there are no profits or losses associated with the expected price, so that no entry or exit is planned;

(c) these plans and expectations held in year t for year $t + 1$ are exactly realized when year $t + 1$ arrives, and so no firm changes its plans and expectations, which are all mutually consistent.

Thus if p_1^* is the long-run equilibrium price, and q_1^* the output of each firm (assumed positive), requirement (a) implies the condition

$$p_1^* = C_1'(q_1^*) \qquad \text{(price = LRMC)}. \qquad (9.18)$$

Requirement (b) implies the condition

$$p_1^* = C_1(q_1^*)/q_1^* \qquad \text{(price = LRAC)}. \qquad (9.19)$$

Requirement (c) implies the condition

$$D(p_1^*) = n^* q_1^* \qquad \text{(demand = supply)}, \qquad (9.20)$$

where $D(p_1^*)$ is the market demand function and n^* is the equilibrium number of firms.

The first two of these conditions imply that each firm must be in equilibrium at the minimum point of its LRAC curve, where LRMC = LRAC. Since all firms have the same cost function, each firm therefore produces the same output. So, market supply is simply this output times the number of firms. Hence we have the third equilibrium condition. Note that this equilibrium condition essentially determines the number of firms, since the equilibrium price is given by the minimum point of each firm's LRAC curve, and so is determined entirely by technology and input prices, independently of demand. In other words, given the assumption of constant input prices, the long-run market supply curve is a horizontal line. Changes in demand change only the size of the market, not long-run equilibrium price.

This is illustrated in Fig. 9.3. In (a) of the figure we show the equilibrium of the firm, in (b) the market equilibrium for an initial demand function $D^0(p)$. Suppose now that there is an unanticipated increase in demand to $D^1(p)$. In the short run, the firms already in the market increase supply along their short-run supply (= SRMC) curves, so that short-run equilibrium price rises to p^1; the supply curve S_0 is the horizontal sum of SRMC curves of firms in the market (recall question 1 of Exercise 9.1). At p^1 positive profits are being earned. Everything now depends on the price expectations formed, both by firms in the market and potential entrants to the market.

Let us assume that they all naively extrapolate the price p^1 as the price which will continue to prevail in the long run. Then existing firms and new entrants will each plan to produce output q^1 in (a) of the figure, and will expand capacity accordingly. The following year, of course, if demand remains at $D^1(p)$ there will be considerable overcapacity, oversupply, price will be driven down, and firms will plan to cut output and reduce capacity. The analysis of the disequilibrium adjustment process is similar to that of the cobweb model analysed in Chapter 5. However, given the conditions of equilibrium in (9.18), (9.19), (9.20), we know that the new equilibrium, if it is reached, must be at price p^* and output n^*q^* in (b) of the figure, since no other point is consistent with equilibrium. But we notice from (a) of the figure that this must mean that, whatever changes may have been made out of equilibrium, any firm that was in the market at the initial equilibrium ends up producing the same output in the new equilibrium. This means that the expansion in supply must have been the result of the entry of new firms, each producing the output q^* (we usually assume that q^* is very small relative to total output, so that n, the number of firms, can be treated as a real number rather than an integer).

What if firms form rational price expectations, rather than naively extrapolating the current period (short-run) equilibrium price? You should apply our earlier model of rational expectations to show that following a demand change the market will move directly to the new long-run equilibrium.

This analysis of course rested on the assumptions that new entrants would have the same production functions as existing firms, and that input prices remain constant as the market output expands. Relaxing either of these assumptions would change the shape of the long-run market supply curve. For example, continue to assume that new firms will have the same production function as existing ones, but that, as the market as a whole expands, it bids up the wage rate. This implies that each firm's marginal and average costs increase. However, the long-run equilibrium conditions (9.18), (9.19), and (9.20) must again hold (note that cost functions of all firms will still be identical since they have the same production functions and face the same input prices). Hence, the long-run market supply curve will be traced out by a sequence of equilibrium positions like those illustrated in Fig. 9.4. With the initial demand function $D^0(p)$, the equilibrium conditions are satisfied at price p^0 and firm's output q^0, with aggregate output n^0q^0, n^0 being the initial number of firms in the market. If demand now increases to $D^1(p)$, in the short-run price will rise along the short-run supply curve (= horizontal sum of SRMC curves), thus giving positive profits. The expansion of capacity and long-run output will, however, now bid up wages and shift each firm's cost curves, as shown in (a) of the figure. The new, long-run equilibrium at price p^1 and total output n^1q^1 again satisfies conditions (9.18)–(9.20), but must involve each firm producing a lower output because its costs have risen. Thus again the expansion in long-run supply as a result of the demand increase is achieved entirely by the entry of new firms.

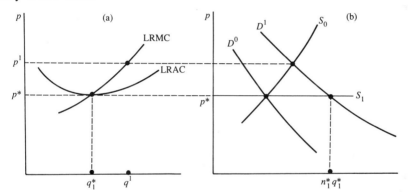

Fig. 9.3: Long-run supply with new entry

This analysis shows that the long-run market supply function is a more complex concept than the short-run market supply function. The latter could be taken to be the sum of firms' individual supply functions (in the absence of changes in input prices), and changes in the output price cause a firm to move along its short-run supply function. In the long run neither of these results carries over (even in the absence of changes in input prices), essentially because of the possibility of new entry. Where new entrants have the same cost function as existing firms, the market supply function is certainly not the sum of firms' perceived supply functions. Moreover, points on the long-run market supply curve represent the outcome of disequilibrium adjustment processes which may be quite complex and drawn out, involving as they do the process by which firms form price expectations and the precise way in which the market operates out of long-run equilibrium. There are plenty of problems there for further study.

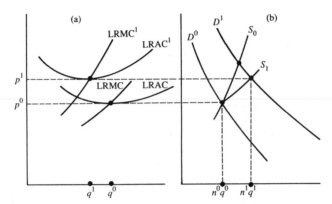

Fig. 9.4: Long-run supply with higher input prices

Exercise 9.2

1. Condition (9.15) requires that the firm must be encountering diseconomies of scale in the neighbourhood of its long-run equilibrium. What are the consequences of assuming that for each firm there are

    (*a*) economies of scale, or
    (*b*) constant returns to scale

for all outputs up to total market demand?
2. Explain carefully why summing firms' perceived long-run supply functions will not in general give the long-run market supply function.
3. Explain carefully what it means for a market to be in long-run equilibrium.
4. Given that all firms (including new entrants) have identical production functions and face the same input prices, explain why a necessary condition for the market long-run supply function to have a positive slope is that input prices increase as the industry expands.
5. A market has the demand function

$$q_D = 1000 - p$$

and each firm (including any new entrant) has the total cost function

$$C_1(q_1) = 3q_1^3 - 4q_1^2 + 2q_1.$$

Solve for long-run equilibrium price, output, and number of firms. [Hint: solve for each firm's output by *minimizing* LRAC, then for the number of firms and market price.]

9.3 Summary

The firm's short-run market supply function has a discontinuity at the price equal to minimum average variable cost. At prices below this its supply is zero. At prices above, its supply is given by the corresponding point on its marginal cost curve. If input prices are constant, the short-run market supply function is the sum of the individual firms' market supply functions.

Lump sum and profit taxes leave market supply unaffected. A tax on a variable input reduces market supply and raises price.

A firm's perceived long-run supply function has a discontinuity at price equal to minimum average cost. At an expected price below this a firm would plan to supply zero output. At an expected price above, a naive firm would plan to produce an output at which price equalled long-run marginal cost. However, we have to allow for the fact that whenever expected price is above long-run average cost, new entry will take place. It follows that in long-run equilibrium, each firm's output is at the minimum point of its long-run average cost curve. If all firms, including new entrants, have the same production functions and face the same input prices, the long-run market supply curve is horizontal at a price equal to minimum long-run average cost. Market demand then simply determines the number of sellers in the market.

We obtain a positively sloped long-run market supply curve if we assume that expansion of market output causes input prices to bid up, and therefore the cost curves of individual firms to rise.

Adjustment between different points on the long-run market supply curve could be quite complicated, depending as it does first on how firms form their price expectations, and secondly upon interactions between output and input markets.

10   Monopoly

A firm which is the sole seller of a good is said to have a monopoly of that good, or to be a monopolist. In analytical terms, this implies that the firm's revenue function is now derived from the market demand function. It would be irrational for the firm to regard itself as able to sell as much as it likes at a given price. Instead, it has to recognize that changes in the quantity it offers for sale will change the price it faces in a way defined by the market demand function. In all other respects the model of the firm is essentially similar to that developed in the previous two chapters: the general nature of the firm's cost function is as we described there, and the firm still seeks to maximize profit. However, the apparently simple change in the firm's revenue function has quite far-reaching consequences.

We shall see that in general a monopoly will earn higher profits than would firms in the same market if it were competitive. The realization of this may well be what causes monopolies to be created, through mergers among firms or acquisitions of firms by others. An important additional reason for the growth of a firm beyond the size consistent with a competitive market, and possibly for the existence of a monopoly itself, is the presence of significant economies of scale. As we shall see (in question 6 of Exercise 10.1), it is quite consistent with monopoly equilibrium that the equilibrium output is at a point on the falling portion of the marginal cost curve, something not consistent with competitive equilibrium.

Given that a monopoly may well be enjoying high profits, an important issue is the question of entry of new firms into the market in pursuit of a share in those profits. Indeed, it has been argued that if a firm would be able to produce with the same production function and input prices as the existing monopoly, and if there are no *sunk costs*, that is, costs which once incurred could not be avoided by shutting down and leaving the market, then the existing monopoly could not earn *any* excess profit. If it did so, a new firm would immediately enter the market and compete them away since, once profits then fell to zero, it could exit costlessly. Though this type of market, known as a *perfectly contestable market*, is something of a special case, it does bring out the importance of the threat of entry to both the short- and long-run pricing behaviour of the monopoly. However, in this book we shall assume entry of new firms is *impossible*, and consider simply monopoly pricing behaviour unconstrained by the threat of entry.

10.1 Monopoly price and output

The firm faces the market demand function $q_D = D(p)$. If, as we assume, this demand function is continuous (indeed differentiable) and monotonic, then we can invert it to obtain the inverse demand function $p = D^{-1}(q) = p(q)$. The reason for doing this is that the firm's cost function, $C(q)$, is defined in terms of output, and so it is convenient to work with output as the independent variable in all the functions we deal with. For convenience also we drop the 'D' subscript on the quantity-demanded variable, which involves the not entirely innocuous assumption that the firm's output, q, is identical to the amounts demanded and actually sold.

We then define the firm's profit function

$$\pi(q) = R(q) - C(q) = p(q)q - C(q), \tag{10.1}$$

where $R(q)$ is the firm's revenue function. We assume the firm's demand function is such that $R''(q) < 0$, for all $q \geq 0$, that is, that the firm's revenue function is strictly concave. The firm's profit-maximizing output q^* is then given by the conditions

$$\pi'(q^*) = R'(q^*) - C'(q^*) = p(q^*) + q^* p'(q^*) - C'(q^*) = 0, \tag{10.2}$$

$$\pi''(q^*) = R''(q^*) - C''(q^*) = 2p'(q^*) + q^* p''(q^*) - C''(q^*) < 0. \tag{10.3}$$

The first condition is usually translated as 'the firm's profit maximizing output equates its marginal revenue R' to its marginal cost C''. The second condition says that at this maximum the derivative of marginal revenue must be less than that of marginal cost.

Fig. 10.1 illustrates this equilibrium. In (a) of the figure we show the total revenue and total cost curves, the latter having the shape consistent with first decreasing, then increasing marginal and average costs. The total profit curve measures the vertical distance at each output between the revenue and cost curves, and the profit-maximizing output is indicated. In (b) of the figure we show the equilibrium equivalently in terms of the demand (or *average* revenue), marginal revenue, marginal and average cost curves. Since marginal revenue $R'(q)$ is the slope of the revenue curve, and marginal cost $C'(q)$ that of the total cost curve, the revenue and cost curves in (a) of the figure are momentarily parallel at q^*. Note also that condition (10.2) is also satisfied at output $\hat{q}$, but this is a profit minimum, not a maximum, and this is captured by the fact that condition (10.3) is not satisfied at $\hat{q}$ (explain why not).

The key to interpretation of this monopoly equilibrium lies in the relation between price and marginal revenue. Thus we have

$$R'(q) = p + qp' = p(1 + qp'/p) = p(1 - 1/e) \tag{10.4}$$

where $e \equiv -(p/q)(dq/dp)$ is own price elasticity of demand at a point on the demand curve.

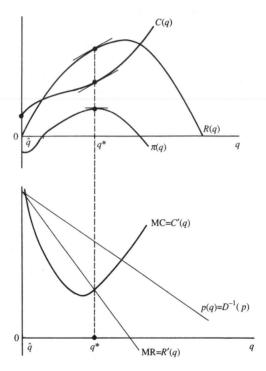

Fig. 10.1: Monopoly equilibrium

Equation (10.2) gives

$$p^*(1 - 1/e) = C'(q^*) \implies (p^* - C'(q^*))/p^* = 1/e. \tag{10.5}$$

The ratio of the excess of price over marginal cost to price, which is defined as the *degree of monopoly*, is determined entirely by the elasticity of demand at the profit-maximizing point. As we shall see in greater detail in Chapter 15, the difference between price and marginal cost is of interest because this reflects the extent to which monopoly causes an inefficient allocation of resources. In a competitive market, equilibrium price is equal to marginal cost for every firm. Recalling that in this case each firm faces a horizontal demand curve, with therefore 'infinite elasticity', we obtain the competitive result from (10.5) by letting $e \to \infty$.

As Fig. 10.1 and equation (10.4) both show, a monopolist's marginal revenue is strictly less than its price wherever output is positive. To see why this is so, take any output level $q_0 > 0$ and consider the change in revenue dR resulting from a small change in output dq:

$$dR = R'(q_0)dq = (p(q_0) + q_0 p'(q_0))dq. \tag{10.6}$$

We can interpret this total differential as saying: when the firm makes a small change in output the resulting change in revenue has two components: the price

p at which the increment in output dq can be sold; and the *loss* of revenue $q_0 p'$ resulting from the fact that in order to sell more output the price must be reduced ($p' < 0$).

Fig. 10.2 illustrates this for a discrete change in output Δq.

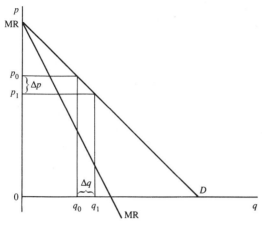

Fig. 10.2: Why marginal revenue is less than price

Against the increase in revenue $p_1 \Delta q$ obtained from selling the output increase must be set the loss of revenue $q_0 \Delta p$ resulting from the fact that previous output q_0 is now sold at a lower price. Thus we have

$$\Delta R = p_1 \Delta q + q_0 \Delta p = \left(p_1 + q_0 \frac{\Delta p}{\Delta q} \right) \Delta q \tag{10.7}$$

and (10.6) is then obtained from (10.7) by letting $\Delta q \to dq$.

Clearly, then, the ultimate reason for the difference between price and marginal revenue is the fact that all output must be sold at the same price. If, in order to sell output dq, the firm did *not* have to reduce price on output q_0, then the marginal revenue of output dq would be the price p_1. This raises the question of the possibility of selling different units of output at different prices, a practice known as *price discrimination*. We shall examine this in some detail in the next two sections.

Exercise 10.1

1. A soft drinks firm is asked to bid for the (monopoly) right to supply drinks at a football match. It estimates the demand function as $p = 500 - 0.05q$, where p is the price per drink, in cents, and q is the number of drinks it will sell. Its total cost function is $C(q) = 100 + q^2$. What is the maximum amount it should bid for the franchise? Illustrate your answer in a diagram.
2. Show that the value of the intercept of a linear demand curve on the q-axis is exactly twice the value of the intercept of its corresponding (linear) marginal revenue curve.

Use (10.6) to explain why the gap between p and MR *must* increase as q increases, and MR must become negative.

3. Suppose a monopolist faces the market demand function $p = aq^{-1}$. Explain why no profit-maximizing output exists in this case. [Hint: what is the elasticity of demand on this demand curve? Sketch the total profit function. Where is the discontinuity?] What would you advise the firm to do in this case? Generalize the discussion to the cases where $p = aq^{-b}$, $b \gtrless 1$.

4. Show that total revenue is maximized at a point on the demand curve where $e = 1$. What is the value of marginal revenue at this point? Use this to explain why a profit-maximizing monopolist with positive marginal cost will always be in equilibrium at a point on the demand curve where $e > 1$.

5. Show that a profit-maximizing monopolist would not change its output in response to the imposition of a tax on profits, whether proportional or lump sum. Show also that imposing a specific tax per unit of output on a monopolist will always raise price and reduce output. [Hint: add tq to the cost function, where t is the tax, derive the profit-maximizing condition, then differentiate totally to solve for dq/dt and dp/dt. Recall the second-order condition (10.3) in signing these derivatives.] Derive and discuss explicit expressions for these effects when the demand function is linear and the total cost function $C(q) = aq^2$.

6. Use condition (10.3) to explain why a monopolist could be in equilibrium at a point on the falling portion of its marginal cost curve. Use the condition to explain diagrammatically what must be true at this point.

7. Suppose that a monopolist wants to maximize *sales revenue*, not profit. It must do this subject to the constraint that its revenue at least covers its total cost. Derive and illustrate the solution to this problem. [Hint: note that there are essentially *two* possible cases, depending on the height of the average cost curve.]

8. Authors of books are paid a 'royalty', consisting of a certain percentage of the price of each book sold. Assume that

    (*a*) a given book has a monopoly of its market;

    (*b*) the author would like to maximize his royalty income from sale of the book;

    (*c*) the publisher would like to maximize profit, defined as revenue − (costs + royalty payment) from sale of the books. Set up a model and solve it to show that as long as marginal cost is positive, the author would always like to sell a larger number of books at a lower price per book than does the publisher.

9. *Monopsony*: A monopsonist is the sole buyer of a good or service. Let a monopsonist have the revenue function $R(l)$, where l is an input, and the (inverse) input supply function $w = w(l)$, $w'(l) > 0$, where w is the input price. Show that the condition on profit-maximizing input purchase l^* is

$$w^*(1 + 1/\eta) = R'(l^*)$$

where w^* is the equilibrium input price and $\eta \equiv w/lw'$ is the elasticity of input supply with respect to w. Illustrate this solution and discuss the parallels with the analysis of monopoly in this section.

10.2 Monopoly price discrimination: two markets

In the previous section we defined price discrimination as the practice of selling different units of the same good at different prices. We now analyse this further. First, note that whenever a given good is being sold at different prices, there is an incentive for *arbitrage*: buyers of the good at the lower price can make a profit by reselling it to buyers of the good at a higher price. For price discrimination to be possible, therefore, the monopolist must be able to prevent this. In all that follows we assume that is the case.

We first take the case in which the market can be divided into two submarkets, with demand functions $p_1(q_1)$ and $p_2(q_2)$. Since there is no physical difference in the units of output sold to each submarket, the firm's cost function is $C(q_1 + q_2)$. The firm's profit function is therefore

$$\pi(q_1, q_2) = p_1(q_1)q_1 + p_2(q_2)q_2 - C(q_1 + q_2) \tag{10.8}$$

and profit-maximizing outputs q_i^* must satisfy:

$$\frac{\partial \pi}{\partial q_1} = p_1^* + q_1^* p_1' - C'(q_1^* + q_2^*) = 0, \tag{10.9}$$

$$\frac{\partial \pi}{\partial q_2} = p_2^* + q_2^* p_2' - C'(q_1^* + q_2^*) = 0, \tag{10.10}$$

implying

$$p_1^* + q_1^* p_1' = p_2^* + q_2^* p_2' \quad \text{or} \quad p_1^*(1 - 1/e_1) = p_2^*(1 - 1/e_2) \tag{10.11}$$

where $e_i \equiv -(p_i/q_i)/p_i'$ is price elasticity of demand in market $i = 1, 2$. From (10.11) we then have

$$\frac{p_1^*}{p_2^*} = \frac{(1 - 1/e_2)}{(1 - 1/e_1)}. \tag{10.12}$$

This tells us first that the equilibrium prices in the two submarkets will be equal only if $e_1 = e_2$ at the equilibrium. Secondly, it tells us that price will be higher in the submarket with the lower elasticity of demand, since for example

$$e_1 < e_2 \implies 1/e_1 > 1/e_2 \implies (1 - 1/e_1) < (1 - 1/e_2) \implies p_1^* > p_2^* \tag{10.13}$$

(recall that at a monopoly equilibrium $e_1, e_2 > 1$).

The results would make complete sense to a businessman. Equation (10.11) says that a given total output should be allocated between the two submarkets in such a way that marginal revenue in one market equals that in the other. To see why this is so, suppose it were not true: say marginal revenue in market 1 were \$10, that in market 2, \$5. Then, without changing total output (and therefore cost), by reducing output to market 2 by one unit and diverting it to market 1 the firm makes a net gain in revenue of approximately \$(10 − 5) = \$5. Thus a necessary condition for a profit-maximizing allocation of output between submarkets is equality of marginal revenues. Then, the optimal total output must satisfy the condition derived earlier: marginal revenue must equal marginal cost.

Given that marginal revenues in the two submarkets are equated, it then follows necessarily that price will be higher in the market with lower demand elasticity. Translated into a businessman's language, this simply says that a higher profit margin will be made in the market in which fewer alternatives exist that consumers can switch to (since it is the possibilities of substitution that determine elasticity of demand).

Exercise 10.2

1. A profit-maximizing monopoly sells its output in the UK and in the US. The demand functions are

$$p_1 = 100 - 2q_1; \quad p_2 = 200 - q_2,$$

   where country 1 is the UK and 2 the US. The cost function is $C(q_1 + q_2) = (q_1 + q_2)^2$, implying that it costs exactly as much to supply a unit of output to the UK as to the US. The price p_1 is expressed in £, the price p_2 in $, and the exchange rate is £r per $. Find the profit-maximizing prices and outputs in the two countries as functions of the exchange rate. What is the effect on these of a devaluation of the £ (rise in r)? Suppose a specific tax t is introduced on output sold to the UK *only*. What is the effect at a given exchange rate on profit maximizing prices and outputs? Explain and illustrate your answer diagrammatically.

2. If you look at the terms on which the leading academic economic journals are sold, you will see that they usually have three prices: the lowest, for sales to students; a higher price for teachers; and the highest for libraries. Use the analysis of this section to explain this. How do the suppliers of the journals prevent arbitrage?

3. Find an example of real-world price discrimination and discuss how it fits in with the analysis of this section.

4. *Two-plant Monopoly*: Suppose a monopolist supplies a single market with output from two separate plants, with cost functions $C_1(q_1)$ and $C_2(q_2)$ respectively, and $C_i'' > 0$, $i = 1, 2$. Analyse the problem of the profit-maximizing allocation of output between the plants, and discuss its similarities with the analysis of this section.

5. *Discriminating Monopsony*: Suppose that a monopsonist (see question 9 of Exercise 10.1) is faced with two sources of supply of an input, with (inverse) supply functions $w_1 = w_1(l_1)$ and $w_2 = w_2(l_2)$, where w_i is the price of input i and $w_i'(l_i) > 0$. The inputs from each source are identical in productivity. Show that the monopsonist will discriminate in price, paying the input with the lower supply elasticity a lower price. Point out the parallels with the analysis of this section. Discuss the application of this analysis to discrimination on the basis of gender or race in labour markets.

10.3 Perfectly discriminating monopoly

Imagine a market in which each buyer consumes just one unit of output per period. Each buyer will have a maximum price, called his or her *reservation price*, such that at any higher price than this, preferred consumption of the good would be zero. Fig. 10.3 illustrates this for the case in which there are 5 buyers, with reservation prices $p_1 > p_2 > p_3 > p_4 > p_5$.

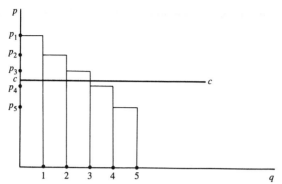

Fig. 10.3: Perfect price discrimination with five buyers

Now suppose this market is supplied by a monopolist. Clearly, if the monopolist sets a price p_n, $n = 1, 2, ..., 5$, it will sell exactly n units of output. For example suppose it sets p_3, so that it sells 3 units. We notice an important and interesting fact about this. Buyer 1 is obtaining a *consumer surplus* of $p_1 - p_3$, because she was prepared to pay a maximum of p_1 for a unit of the good, but she is getting it instead for $p_3 < p_1$. Likewise buyer 2 is obtaining a surplus of $p_2 - p_3$. Only buyer 3, who at p_3 is the *marginal buyer*, is obtaining no surplus.

It could be a source of some frustration to the monopolist, knowing that some buyers are willing to pay more for this consumption than they are in fact being made to pay. Now assume:

(*a*) the monopolist knows exactly which buyer has which reservation price;

(*b*) it is able to prevent arbitrage, i.e., resale of units of the good among buyers.

(*c*) It costs \$*c* to supply each unit of output.

Then clearly, it can increase its profits by taking each buyer separately and offering her the unit of the good at the corresponding reservation price. As compared, for example, to the case in which a single price p_3 is set, earning a total revenue of $3p_3$, the monopolist would receive a profit of $\sum_{n=1}^{3}(p_n - c)$. It would not pay to supply to 4 and 5, because their reservation prices are below the cost of supplying them with the good. But notice the result of this: output, at 3 units, is just where the price to the *marginal* buyer is almost equal to the unit cost of the good. In other words, if this good were supplied under competitive conditions at a price $p = c$, then we would have exactly the same output. The only difference would be that the three consumers would be enjoying surpluses of $p_1 - c$, $p_2 - c$, and $p_3 - c$ respectively. Under perfectly price-discriminating monopoly output would be the same but the monopolist appropriates all the surplus as profit. Thus the two markets differ only in their *distributions of the benefits of trade*, not in the *amount traded*.

This example suggests therefore that under perfect price discrimination monopoly output would be at the competitive level, the only difference being in the distribution of the 'gains from trade' as between producers and consumers. The drawback to the above example is that since it was constructed in discrete terms, with a finite number of buyers, we could not arrive at an exact equality of marginal price (i.e., price to the marginal buyer, 3) with marginal cost (except by accident, if, say, $p_3 = c$). So the purpose of the following generalization is to make everything nicely continuous.

Let $x \geq 0$ denote a consumer's reservation price at or below which she will buy *one and only one* unit of the good, and $n(x)$ the number of consumers with reservation price x, where $n(\cdot)$ is a continuous function. We denote by p^0 the price with the property that

$$x \geq p^0 \Rightarrow n(x) = 0, \qquad x < p^0 \Rightarrow n(x) > 0. \qquad (10.14)$$

In words, at prices of p^0 and above, demand is zero, while below it, demand is positive (p^0 is the intercept of the demand curve on the price axis).

We assume that the monopolist can charge each buyer her reservation price. The monopolist's problem is to choose the *marginal price*, p, which implies choice of total output, given by the sum of purchases by all those buyers whose reservation price is p or higher. The revenue the monopolist will receive from buyers with reservation price x is clearly $xn(x)$, since each buyer buys one unit of the good. Hence, for any marginal price p, total revenue is

$$R(p) = \int_p^{p^0} xn(x)dx, \qquad (10.15)$$

that is, the revenues from all those buyers with reservation prices $p^0 \geq x \geq p$. Total demand and output will be

$$Q(p) = \int_p^{p^0} n(x)dx, \qquad (10.16)$$

that is, purchases by all those buyers with reservation prices $p^0 \geq x \geq p$. It follows that, given the cost function $C(Q(p))$, the monopolist's profit function is

$$\pi(p) = \int_p^{p^0} xn(x)dx - C(Q(p)). \qquad (10.17)$$

It is appropriate to regard this as a function of the marginal price since the lower this price, the greater is output. Essentially, the monopolist is choosing a 'cut off point' in the distribution of buyers: buyers at or above this point are offered the good at their reservation price; buyers below this point are not offered the good at all.

If we maximize $\pi(p)$ with respect to p we obtain the condition

$$\pi'(p^*) = -p^* n(p^*) + C'n(p^*) = 0 \Rightarrow (p^* - C')n(p^*) = 0 \Rightarrow p^* = C' \qquad (10.18)$$

(where we have used the fact that $d \int_p^{p^0} xn(x)dx/dp = -pn(p)$). Thus optimal marginal price p^* is equal to marginal cost. This then simply confirms the idea we obtained from the example with 5 consumers. Intuitively, it pays the monopoly to expand output as long as the price it can obtain from a consumer exceeds the marginal cost, C', of supplying the consumer. The key point here is that offering a lower price to increase output *does not* require the monopoly to reduce price on units of output it is selling at a higher price. This shows that the fundamental reason for the result that equilibrium monopoly price exceeds marginal cost is not simply that there is a single seller, but rather that the seller cannot practise perfect price discrimination. We shall consider this point again in Chapter 15, where we discuss the welfare loss due to monopoly.

Exercise 10.3

1. A monopolist has the total cost function $C = 5Q$ and has a distribution of buyers according to reservation price, given by $n(x) = 100 - 5x$. Find the profit-maximizing marginal price and the corresponding total output and profit.
2. Suppose now that the monopolist in the preceding problem cannot practise price discrimination, but can only set a single price to all consumers. Find the profit-maximizing output in this case, and the corresponding price and total profit. Compare them to your solution to question 1. [Hint: note that the standard demand function is given by $D(p) = \int_p^{p^0} n(x)dx$.] Sketch the $n(x)$, $D(p)$, and marginal revenue functions and explain the relationships between them.

10.4 Regulated monopoly

As we show in Chapter 15, the presumption is that monopoly creates a misallocation of resources (in the absence of perfect price discrimination), and this has led many economists to advocate some form of state control or regulation. Non-economists perceive that monopolies are capable of generating high prices and profits, and this too has led to the call for control of monopoly power. In fact, historically, non-economists' concern with the level of profits and prices was probably the dominant consideration, and this determined the form of regulation of monopoly firms adopted in practice.

Until recently, the main form of control or regulation of monopoly was the so-called 'rate base' system. The regulatory agency decided on an appropriate rate of return on capital, and then sought to control the monopoly's prices so that profits in excess of this were not earned. If we define the firm's cost function to include this rate of return – in effect, in terms of the analysis of Chapter 8, the rental rate of capital would be determined on the basis of this 'allowed rate of return' – and impose the constraint that revenue should be no higher than total costs calculated in this way, then in a single output monopoly this is effectively constraining price to equal *average cost* (explain why). We show in Chapter

15 that allocative efficiency requires price to equal *marginal cost.* Thus, only if marginal and average costs are equal does rate base regulation achieve an efficient resource allocation. Moreover, it has been shown that the monopoly has an incentive to operate inefficiently and increase its costs in order to increase its profits under this form of regulation. The regulator is in fact guaranteeing the monopoly a profit however inefficient it is. Finally, if we add to all this the observation that quite often regulation appears to have acted in the interests of the regulated firm(s), especially in respect of inhibiting the entry of new competition, it is hardly surprising that the conventional form of regulation has been heavily criticized by economists.

Rather than concern ourselves with rate base regulation, let us suppose that the object of regulation is to achieve an efficient allocation of resources, and consider what problems arise and how they might be solved. First, we note that if the regulator knows a monopolist's demand and cost functions, then it is in principle a trivial matter to ensure allocative efficiency: he simply sets a maximum price at the value p^* which is implied by the equation

$$p^* = p(q^*) = C'(q^*). \tag{10.19}$$

That is, he sets the 'marginal cost price' as a maximum (see Chapter 15 for an explanation of why and under what conditions this achieves an efficient allocation of resources). It then follows that the profit-maximizing output for the monopolist is q^* and regulation achieves allocative efficiency. Moreover, the monopolist has every incentive to keep costs down, since inflation of costs will result in lower profit, given the fixed maximum price p^*.

This argument leads to the view that the fundamental problem facing a regulator who seeks to achieve allocative efficiency is that of *asymmetric information.* The regulator does not in fact know the firm's true cost and demand conditions, whereas the firm does. If the firm is required to provide the necessary information to the regulator it has, as we shall see, an incentive to bias this information in a way that is favourable to itself. However, it is possible to set up a system of regulation which not only induces the monopoly to report information truthfully, but can under certain circumstances achieve an efficient allocation of resources. We now analyse a model which explores these points.

Suppose that the regulator knows the monopoly's (linear) demand function $p(q) = a - bq$, but does not have enough information to know for sure its cost function. However, to keep things simple, we assume that the regulator does have enough information to know that the monopoly's cost function is of the form

$$C(q) = \theta q + \frac{c}{2}q^2. \tag{10.20}$$

Moreover, the regulator knows the value of c for sure, but does not know θ.

Let us suppose first that the regulator is quite naive. He asks the monopoly to report a value of θ, and, given that value, sets price equal to the implied marginal cost,

$$\hat{p} = a - b\hat{q} = \hat{\theta} + c\hat{q}. \tag{10.21}$$

where $\hat{\theta}$ is the value of θ reported by the monopoly. This implies

$$\hat{q} = \frac{a - \hat{\theta}}{b + c}; \quad \hat{p} = a - b\left(\frac{a - \hat{\theta}}{b + c}\right) = \frac{ac + \hat{\theta}b}{b + c} \tag{10.22}$$

Then the monopoly has an incentive to give a false report of its true value of θ. We can regard (10.22) as defining its regulated price and output as a function of its cost report $\hat{\theta}$, and so it should clearly choose $\hat{\theta}$ so as to maximize its profit

$$\pi(\hat{\theta}) = \hat{p}\hat{q} - \theta\hat{q} - \frac{c}{2}\hat{q}^2 = R(\hat{q}(\hat{\theta})) - C(\hat{q}(\hat{\theta})). \tag{10.23}$$

Effectively, the monopoly can *choose* its price through its report $\hat{\theta}$ (but note that its actual profit is determined also by the *true* value of θ). We would guess that, as Fig. 10.4 illustrates, the optimal $\hat{\theta}$ is precisely that which will yield a regulated price $\hat{p}$ equal to the monopolist's profit-maximizing price in the absence of regulation, p^0. That is, by claiming its costs are actually higher than they are by the appropriate amount, $\hat{\theta} - \theta$, the monopolist can achieve its true maximum profit. The simplest way to see this is to use (10.23) and the 'function of a function' rule to obtain

$$\frac{d\pi}{d\hat{\theta}} = (R' - C')\frac{d\hat{q}}{d\hat{\theta}} = 0, \tag{10.24}$$

which, since $d\hat{q}/d\theta \neq 0$, implies that output satisfies the profit-maximizing condition $R' = C'$ with respect to *true* marginal cost. To obtain an explicit solution for $\hat{\theta}$, substitute for $\hat{q}$ from (10.22) into (10.23) (do it) to obtain the problem

$$\max_{\hat{\theta}} (a - \theta)\left(\frac{a - \hat{\theta}}{b + c}\right) - \left(b + \frac{c}{2}\right)\left(\frac{a - \hat{\theta}}{b + c}\right)^2, \tag{10.25}$$

the first-order condition for which can be rearranged to give

$$\hat{\theta} = \frac{ab + (b + c)\theta}{2b + c}. \tag{10.26}$$

Alternatively note from Fig. 10.4 that we can express $\hat{\theta} - \theta$ as

$$\hat{\theta} - \theta = p^0 - R'(q^0) = p^0 - p^0\left(1 - \frac{1}{e^0}\right) = \frac{p^0}{e^0} = bq^0 \tag{10.27}$$

where $e^0 = p^0/bq^0$ is elasticity of demand at q^0. Moreover, it is straightforward to confirm that $q^0 = (a - \theta)/(2b + c)$. Thus we have from (10.27)

$$\hat{\theta} - \theta = \frac{(a - \theta)b}{(2b + c)}. \tag{10.28}$$

Then adding $\theta(2b + c)/(2b + c)$ to both sides of (10.28) and rearranging gives (10.26).

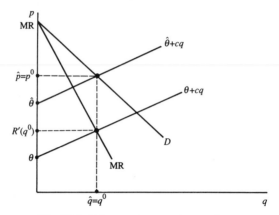

Fig. 10.4: False report to a naive regulator

The regulator may well of course not be so naive. If he can observe the monopolist's true profits then he will be able to detect *ex post* that the monopolist reported false information, since actual profit will be $\hat{p}\hat{q} - (\theta\hat{q} + c\hat{q}^2/2)$ and not $\hat{p}\hat{q} - (\hat{\theta}\hat{q} + c\hat{q}^2/2)$ as he would have expected. Therefore, *ex ante*, he might be able to threaten some punitive action if higher than expected profits are subsequently observed. There are two reasons this might not in practice be enough to induce truthful reporting. The first is that the excess profits might be absorbed within the monopoly, in the form of lavish office suites, generous expense allowances, and various kinds of managerial perquisites. The second is that in reality there is likely to be some uncertainty in the environment the firm faces — for example its demand might turn out to be high or low — and so high profits *ex post* could be attributed to good fortune rather than to a biased cost report. To keep things simple we shall continue to assume complete certainty of demand conditions, but, to retain interest in the problem, we assume that profits *ex post* are not observable.

An alternative way in which the regulator might try to ensure truthful reporting is by carrying out an independent audit of the firm's costs. This could however be expensive, and in fact a costless – to the regulator – alternative may exist, in the form of a *tax incentive mechanism*. The idea is quite simple. The regulator imposes a lump sum tax on the monopolist, which depends on the cost report $\hat{\theta}$ the monopolist makes. By suitable choice of the value of the tax, he can ensure that the true θ is reported, and he can then set the appropriate marginal cost price. We now see how this incentive mechanism can be designed.

Suppose that the regulator knows that the monopolist's cost parameter can take one of only two possible values, θ_1 or θ_2, with $\theta_2 > \theta_1$ as Fig. 10.5 shows. Now if θ_1 were true, the regulator would expect the monopolist to report $\hat{\theta} = \theta_2$, i.e., to lie, since this would result in higher profit than the truth, $\hat{\theta} = \theta_1$. That is, q_2^* is closer to the profit-maximizing output q_1^0 than is q_1^*. However, if θ_2 were true, then the monopolist would clearly tell the truth, because the false report $\hat{\theta} = \theta_1$ would result in price p_1 and lower profit than telling the truth. Thus, in the absence of an incentive mechanism, the monopolist will always report $\hat{\theta} = \theta_2$, whether this is true or not.

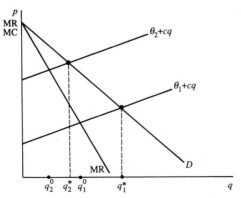

Fig. 10.5: Monopolist always reports θ_2

Fig. 10.6 shows how a tax can be set to induce truthful revelation of costs. The curves show relevant portions of the profit functions $\pi_i = R(q) - \theta_i q - cq^2/2$, labelled according to the value of θ. (Note, these curves are negatively sloped because we are to the right of the profit-maximizing output.) Since $\theta_1 < \theta_2$, π_1 is higher than π_2 at every output, and also, at each q

$$\frac{d\pi_1}{dq_1} = R' - \theta_1 - cq > R' - \theta_2 - cq = \frac{d\pi_2}{dq_2}, \qquad (10.29)$$

which justifies the relative slopes of the curves shown in the figure, i.e., π_1 is flatter than π_2 at each output. The indicated outputs q_1^*, q_2^* are those at which price = marginal cost in each case. We then offer the firm the following scheme: if it reports $\hat{\theta} = \theta_1$, it pays no profit tax, whereas if it reports $\hat{\theta} = \theta_2$, it pays a tax given by T^* shown in the figure. This will induce the firm to report its true θ. If θ_2 is the true cost parameter, the firm will report it, since its profit, even after the tax, is higher than if it reported θ_1 and were made to produce output q_1^* (set price p_1^*). On the other hand, if θ_1 is its true cost parameter it will report θ_1, since if it lied and reported θ_2 its profit, after the tax, would, at π_1^*, be no higher than if it told the truth (if the regulator believed that the firm might lie even when it did not gain from doing so, he may make T^* slightly larger than is shown in the figure).

We see therefore that the condition which defines the optimal tax, paid only when the firm reports $\hat{\theta} = \theta_2$, is

$$R(q_2^*) - \theta_1 q_2^* - cq_2^{*2}/2 - T^* = R(q_1^*) - \theta_1 q_1^* - cq_1^{*2}/2, \qquad (10.30)$$

implying

$$T^* = R(q_2^*) - R(q_1^*) - \theta_1(q_2^* - q_1^*) - c(q_2^{*2} - q_1^{*2})/2. \qquad (10.31)$$

Condition (10.30) is effectively saying that the firm must be no better off by reporting θ_2 when θ_1 is true than by telling the truth.

It might be objected that Fig. 10.6 could represent a special case and an appropriate tax T^* might not in general be available. Two problems could arise. First, could it happen that charging T^* *only* when the firm reports θ_2 might induce it to lie when its cost actually is θ_2, so as to be allowed to produce q_1^*? Fig. 10.6 shows that as long as the condition on the relative slopes of the curve in (10.29) holds, this *cannot* be a problem: the slope of the arc a_1b_1 on π_1 must be smaller in absolute value than the slope of the arc a_2b_2 on π_2, and so there is no way in which point b_2 on π_2 (the profit from reporting θ_1 when θ_2 is true) can be to the left of point c.

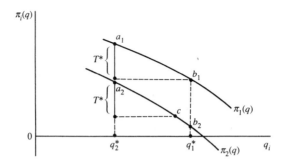

Fig. 10.6: A tax to induce truthful revelation

Secondly, it could happen that θ_2 is sufficiently large, and the profits π_2 sufficiently low at output q_2^*, that imposing the tax T^* on the firm when it reports θ_2 would force it into bankruptcy (note that T^* is determined solely by the π_1 curve). That is, we require T^* to satisfy the no-bankruptcy condition

$$R(q_2^*) - \theta_2 q_2^* - cq_2^{*2}/2 - T^* \geq 0, \tag{10.32}$$

as is clearly the case in the figure. It may happen that it would not. However, it is still possible to work out an incentive scheme to induce truthful revelation, and question 2 of Exercise 10.4 asks you to solve this problem.

An important implicit assumption in this analysis is that the regulator is indifferent to the value of the lump sum tax *as such*: he is concerned only with efficiency of resource allocation and not with tax revenue. If this assumption were relaxed the basic result would not hold: although truthful revelation of costs could still be induced, the regulator would not in general set price equal to marginal cost.

A second important assumption was that marginal cost was increasing, so that marginal cost pricing resulted in profits which could be taxed. If marginal cost was constant or falling with output, marginal cost pricing would imply zero or negative profit. In that case the tax T^* would have to be replaced with a subsidy (see question 3 of Exercise 10.4) and it becomes much harder to accept that the regulator would be indifferent to the value of the incentive payment.

Exercise 10.4

1. A monopolist has the demand function $p = 200 - q/2$, and has the cost function $C = \theta q + q^2/2$.

   (a) Assume $\theta = 20$. The monopolist is faced with a naive regulator. Find the monopolist's optimal report $\hat{\theta}$, and compare the resulting profit with that it would make if it reported truthfully.

   (b) Assume a sophisticated regulator who knows that the monopolist's θ is either 20 or 40. Find the optimal tax to induce truthful revelation. Does it satisfy the no-bankruptcy condition?

2. Suppose that the optimal tax T^* as computed in Fig. 10.6 would not satisfy the no-bankruptcy condition. Show how you would derive an alternative incentive system. [Hint: what would happen if a subsidy were paid when θ_1 is reported?]

3. Consider a monopolist with constant marginal cost, θ. Show how a subsidy incentive system could be designed which would induce truthful revelation of θ.

4. Explain why the following is the appropriate generalization of the case analysed in Fig. 10.6. An incentive scheme to induce truthful revelation of the cost parameter θ is a pair of values $V_1 \gtrless 0$, $V_2 \gtrless 0$, that satisfy the following two pairs of conditions:

   (i) *no bankruptcy*:

   $$R(q_1^*) - \theta_1 q_1^* - cq_1^{*2}/2 + V_1 \geq 0,$$
   $$R(q_2^*) - \theta_2 q_2^* - cq_2^{*2}/2 + V_2 \geq 0.$$

   (ii) *truth-telling*:

   $$R(q_1^*) - \theta_1 q_1^* - cq_1^{*2}/2 + V_1 \geq R(q_2^*) - \theta_1 q_2^* - cq_2^{*2}/2 + V_2,$$
   $$R(q_2^*) - \theta_2 q_2^* - cq_2^{*2}/2 + V_2 \geq R(q_1^*) - \theta_2 q_1^* - cq_1^{*2}/2 + V_1,$$

   where V_1 is a tax or subsidy paid if θ_1 is reported, and V_2 is a tax or subsidy paid if θ_2 is reported (here, q_i^* is the output at which price = marginal cost for θ_i, $i = 1, 2$).

10.5 Summary

A monopoly is the sole seller of a good, and consequently it faces the aggregate market demand function, from which its revenue function is derived. The central feature of monopoly equilibrium is that price exceeds marginal cost, which, since marginal cost is equated with marginal revenue, is in turn due to the fact that price exceeds marginal revenue at positive outputs.

The size of the excess of price over marginal cost is determined by the elasticity of demand at the profit-maximizing point. This must always be greater than one when marginal cost is positive.

Marginal revenue diverges from price because of the assumption that to sell more at any level of output, the firm must reduce its price on *all* units of output.

Price discrimination exists when units of ouptut are sold to different buyers at different prices. For this to be feasible the firm must be able to prevent arbitrage.

Where the market can be divided into different subgroups, profit maximization requires that marginal revenue be equalized across subgroups, which in turn implies that a subgroup will be charged a higher price, the lower its elasticity of demand.

Perfect price discrimination exists when each buyer can be charged her reservation price for each unit of the good. The interesting implication is that if a monopoly *can* practise perfect price discrimination, it will produce where price equals marginal cost.

When monopoly equilibrium results in a price above marginal cost, this implies a misallocation of resources (see Chapter 15). As a result it is argued that monopolies should be regulated. A key problem in regulation is the asymmetry of information between regulator and monopolist, especially in respect of the monopoly's true costs.

If the monopoly is asked to report its costs to the regulator, it has an incentive to bias this information. We saw, however, that it is possible to devise a tax-based incentive scheme which induces the monopoly to report its true costs, allowing the regulator to set price equal to marginal cost. This analysis was conducted in terms of a special case. Whereas it is usually possible, under more general conditions, to devise an incentive mechanism that induces truthful cost reporting, this may require subsidies, in which case the assumption that the regulator is indifferent to the size of the subsidy is harder to sustain.

11 Oligopoly

The defining characteristic of an oligopolistic market is the *perceived strategic interdependence* among sellers. There are few enough sellers in the market for each firm to realize that a change in its price or output will affect the profits of the other firms, who are in turn likely to react. Therefore no firm can predict the profitability of its own price or output decisions unless it knows or can guess the reactions of its rivals.

It is for this reason that quite a wide range of oligopoly models exists. The purpose of these models is to predict the equilibrium outputs and prices in an oligopolistic market. The equilibrium will generally depend on the pattern of reactions the firms are assumed to adopt to each other's decisions. Since different reaction patterns are *a priori* possible, different models, each developing the implications of a particular reaction pattern, follow.

Here we shall make no attempt to survey this range of models. Instead, we shall focus on what appear to us to be the most interesting lines of recent research. This has been heavily influenced by developments in the *theory of games*, a body of theory which seeks to abstract and analyse elements common to all situations of strategic interdependence.

11.1 Cartels and Cournot-Nash equilibrium

For simplicity we focus on a market with the following characteristics. There are two firms, labelled $i = 1, 2$. They produce identical outputs which they sell into a market which has a linear demand function $p = a - b(q_1 + q_2)$, where q_i is firm i's output, $i = 1, 2$. Their cost functions take the form $C_i = c_i q_i^2$, $c_i > 0$, implying that their marginal cost functions are straight lines through the origin, with the equations $C_i' = 2c_i q_i$, $i = 1, 2$. Though this model, which is a *duopoly model*, is quite special in respect of the number of firms and the particular structure of the demand and cost functions, it will serve very well to bring out some of the main general features of modern oligopoly theory.

To make things still more concrete, we shall use a numerical illustration to reinforce the main points. For this, we assume:

$$a = 100, \qquad c_1 = 1,$$
$$b = 1, \qquad c_2 = 2.$$

A very important assumption, which we relax in the next section, is that the firms are in the market for *only one period*. That is, the q_i represent outputs which will be produced and sold in one period, and then the market ceases to exist. This is obviously highly unrealistic. However, until recently the important consequences of moving away from this assumption in a more realistic direction were not fully appreciated.

We must also specify what each firm knows, and whether they can enter into binding agreements with each other. We assume throughout that everything about the situation is *common knowledge* to each firm. This is, each firm knows the market demand function, its own and the other firm's costs, it can observe the other firm's output when this has been chosen and the resulting market price, and moreover, each firm knows that the other firm knows it knows all this.

The question of the possibility of binding agreements is central. Suppose it is possible for the firms to enter into a legally enforceable agreement to produce particular outputs. This means that if one of them violates the agreement, the other is able to take it to court and obtain damages which can be specified in the contract. Then, since the firms seek to maximize their profits, they would be well advised to do the following. They should choose outputs to solve the problem

$$\max_{q_1, q_2} \pi(q_1, q_2) = [p(q_1 + q_2)](q_1 + q_2) - C_1(q_1) - C_2(q_2). \qquad (11.1)$$

That is, they should simply act as a monopolist with two plants, and seek to maximize total profit in the market. Using the earlier assumptions on demand and cost functions, the necessary conditions (which are also sufficient) for this problem are

$$\frac{\partial \pi}{\partial q_i} = a - 2b(q_1^* + q_2^*) - 2c_i q_i^* = 0 \qquad i = 1, 2$$

and this implies the optimum output pair

$$q_1^* = \frac{a - 2bq_2^*}{2(b + c_1)} = \frac{ac_2}{2(b(c_1 + c_2) + c_1 c_2)},$$

$$q_2^* = \frac{a - 2bq_1^*}{2(b + c_2)} = \frac{ac_1}{2(b(c_1 + c_2) + c_1 c_2)}. \qquad (11.2)$$

Using our numerical example, we find:

$$q_1^* = 20 ; \quad q_2^* = 10.$$

As (11.2) shows, the ratio of marginal costs c_1/c_2 determines the ratio of outputs q_1^*/q_2^*, and since in the numerical example c_1 is one-half c_2, we find firm 1's output is twice that of firm 2.

The firms will then earn the joint profit $\pi^* = \pi(q_1^*, q_2^*)$, which in our numerical example is

$$100(30) - (30)^2 - (20)^2 - 2(10)^2 = \$1500.$$

The key point is: by the definition of the problem in (11.1), the largest amount of profit the firms can possibly earn in the market (in the absence of price discrimination – see question 2 of Exercise 11.1 below) follows from producing outputs (q_1^*, q_2^*), and this is π^*.

The firms therefore, if they can, should write a legally binding contract to product outputs (q_1^*, q_2^*) *with appropriate penalties for deviating from these.* When firms are able to enter into such an agreement, the firms are said to form a *cartel,* and we have just derived what could be called the cartel outcome, which is the same as the monopoly outcome.

Why should penalties be necessary, however, and what do we mean by an 'appropriate penalty'? The answer is that quite generally, in a much wider class of cases than our example, it pays any one firm to renege or cheat on the cartel agreement if it can do so unpenalized. To see this, suppose the arrangement is that each firm will under the cartel agreement earn the profit

$$\pi_i^* = p(q_1^*, q_2^*)q_i^* - c_i q_i^{*2} \qquad i = 1, 2. \tag{11.3}$$

That is, it earns the profit generated by selling its own cartel output q_i^* at the price $p(q_1^*, q_2^*) = a - b(q_1^* + q_2^*)$ (= \$70 in our example. This would give firm 1 $\pi_1^* = \$1000$ and firm 2 $\pi_2^* = \$500$). Now suppose firm 1 reasons as follows: if firm 2 in fact produces q_2^*, are we really doing the best we can for our shareholders if we set our output at q_1^*? To answer this, we solve:

$$\max_{q_1} \pi^1(q_1, q_2^*) = p(q_1, q_2^*)q_1 - C_1 = aq_1 - b(q_1 + q_2^*)q_1 - c_1 q_1^2 \tag{11.4}$$

That is, we find the output q_1^+ that maximizes firm 1's profit *given* firm 2 is producing q_2^*. This results in the output (check the derivation):

$$q_1^+ = \frac{a - bq_2^*}{2(b + c_1)}. \tag{11.5}$$

By comparing q_1^+ in (11.5) with q_1^* in (11.2), we see that

$$q_1^+ - q_1^* = \frac{a - bq_2^*}{2(b + c_1)} - \frac{(a - 2bq_2^*)}{2(b + c_1)} = \frac{bq_2^*}{2(b + c_1)} > 0. \tag{11.6}$$

That is, if firm 1 believes that firm 2 will keep to the agreement and produce q_2^*, it will pay firm 1 to cheat and produce $q_1^+ > q_1^*$.

In our numerical example, if firm 2 produces $q_2^* = 10$, and firm 1 produces $q_1^+ = 22.5$ (check the arithmetic using (11.5)), then the market price will be

$$p^+ = 100 - (22.5 + 10) = \$67.50$$

and firm 1's profit will now be

$$\pi^1(q_1^+, q_2^*) = \pi_1^+ = (67.5)(22.5) - (22.5)^2 = \$1012.50.$$

It therefore pays to cheat! (and consumers receive a lower price). This then suggests that if the firms enter into a legally binding agreement, they should recognize that firm 1 has an incentive to cheat and so specify that if firm 1 does not honour the agreement it will have to pay firm 2 damages of some amount no smaller than $\pi_1^+ - \pi_1^*$, since this is the amount it would gain by cheating (in our numerical example this is $12.50).

Of course, the firms should realize that firm 2 also has an incentive to cheat. By going through exactly the same process we find that if it expects firm 1 to produce q_1^*, it would pay firm 2 to produce

$$q_2^+ = \frac{a - bq_1^*}{2(b + c_2)} = q_2^* + \frac{bq_1^*}{2(b + c_2)}. \tag{11.7}$$

In our numerical example, firm 2 produces $q_2^+ = 13.33$, this reduces market price to $66.67 and gives firm 2 a profit of $\pi_2^+ = \$533.33 > \pi_2^*$. Thus in the legally binding agreement damages for firm 2's reneging should be set at no less than $\pi_2^+ - \pi_2^*$ (= $33.33 in the example).

Before we leave the cartel, note finally that the firms could decide to divide the cartel profit π^* differently than was just assumed, but that this can never remove the incentive for at least one of them to cheat. The division of profits (π_1, π_2) must always satisfy

$$\pi_1 + \pi_2 = \pi^*, \tag{11.8}$$

whereas we necessarily have that

$$\pi_1^+ + \pi_2^+ > \pi^*, \tag{11.9}$$

i.e., it is not feasible for *each* firm i to have π_i^+. So even if, say, firm 1 were given π_1^+ instead of π_1^* to stop it cheating, that would leave firm 2 with an even larger incentive to cheat (confirm from the numerical example).

At certain times and in some countries, cartels have been legal, but in most Western industrial countries they are now illegal. *Collusive agreements* of the type which underlay the cartel certainly are not enforceable in the courts, and may be punishable by the anti-trust authorities. It follows that in the present model the agreement to produce outputs (q_1^*, q_2^*) would not be enforceable. *Because this is a one-period market*, once firms produce their outputs that is that, and there is nothing either can do if the other cheats. In that case, would we expect the firms nevertheless to produce (q_1^*, q_2^*)? The argument that they would *not* can be made even stronger by noting that it not only pays to cheat, but a firm loses badly by sticking to the agreement when the other firm cheats. That is, we have for firm 1

$$\pi^1(q_1^+, q_2^*) > \pi^1(q_1^*, q_2^*) > \pi^1(q_1^*, q_2^+) \tag{11.10}$$

and similarly for firm 2. In our numerical example, we can compute

$$\begin{aligned}
\pi^1(q_1^+, q_2^*) &= \$1012.50; & \pi^2(q_1^*, q_2^+) &= \$533.33, \\
\pi^1(q_1^*, q_2^*) &= \$1000.00; & \pi^2(q_1^*, q_2^*) &= \$500.00, \\
\pi^1(q_1^*, q_2^+) &= \$933.33; & \pi^2(q_1^+, q_2^*) &= \$475.00, \\
\pi^1(q_1^+, q_2^+) &= \$937.50; & \pi^2(q_1^+, q_2^+) &= \$500.00.
\end{aligned} \tag{11.11}$$

Note that cheating reduces the *sum* of profits, as must be the case (explain why). Also if *both* cheat so that (q_1^+, q_2^+) is produced, price falls to $64.17.

We can now use these numbers to construct a *payoff matrix*, which is a table showing the profit each firm receives for each possible pair of output choices. Firm 1's output choices are given at the side of the matrix, firm 2's at the top. In each cell, the first entry is firm 1's profit, the second firm 2's. Clearly the matrix is just a useful way of organizing the information given in (11.10) and (11.11).

Firm 2 chooses:

	q_2^*		q_2^+	
q_1^* 1000.00	500.00		933.33	533.33
q_1^+ 1012.50	475.00		937.50	500.00

Firm 1 chooses:

The table makes one thing quite clear: the choice q_i^+ *dominates* the choice q_i^* in the sense that each firm does better choosing q_i^+ for each possible choice by the other firm. Cheating is better if the other firm honours the agreement, but it is also better if the other firm cheats, since that way you reduce the damage that is then inflicted upon you.

If, therefore, the *only* two alternatives open to the firms were the outputs q_i^* and q_i^+, we would predict as the market equilibrium the pair (q_1^+, q_2^+), *in the absence of enforceable, binding agreement*. However, this is worse for the firms than the cartel outcome, since total profits fall from $1500 to $1437.50. Firm 1 could agree to make a side-payment to firm 2 of anything up to $62.50 = \$ (1000 - 937.50)$ to induce it to adhere to the agreement and still be better off than with the output pair (q_1^+, q_2^+). But in the absence of binding agreements there is no way in which the firms can commit themselves to producing (q_1^*, q_2^*), and so the *individually rational* choice for firm $i = 1, 2$ is q_i^+. This is an example of a very important type of game, which occurs in a number of economic and social contexts, known as *the prisoners' dilemma*.

This analysis suggests that if the firms are individually rational profit maximizers they will not choose the cartel outputs. However, does (q_1^+, q_2^+) itself represent an *equilibrium* output pair? The following argument shows that it does not. Suppose that firm 1 has gone through the reasoning so far just as we have. It has come to the conclusion that firm 2 will choose q_2^+ in preference to q_2^*. If firm 2 *does* produce q_2^+ however, is q_1^+ the best output for firm 1? To answer this we have to solve

$$\max_{q_1} \pi^1(q_1, q_2^+) = p(q_1, q_2^+) q_1 - C_1(q_1) = aq_1 - b(q_1 + q_2^+) q_1 - c_1 q_1^2. \quad (11.12)$$

This results in the output level

$$\hat{q}_1 = \frac{a - bq_2^+}{2(b + c_1)} \neq \frac{a - bq_2^*}{2(b + c_1)} = q_1^+. \quad (11.13)$$

Thus, since $q_2^+ > q_2^*$, if firm 1 really expected firm 2 to produce q_2^+, it should produce $\hat{q}_1 < q_1^+$. But then, firm 1 should assume that firm 2 has gone through exactly the same kind of calculation, and will have concluded that *it* should not produce q_2^+ if it expects firm 1 to produce q_1^+. At this point heads begin to ache. Where will this thought-process end?

Game theory suggests the following answer. Note that at each step in the discussion so far we have asked the question: given the output firm i expects firm j to produce, what is i's profit-maximizing output? That is, for *any given* q_j, say $\bar{q}_j$, we solve the problem:

$$\max_{q_i} \pi^i(q_i, \bar{q}_j) \;\rightarrow\; q_i = \frac{a - b\bar{q}_j}{2(b + c_i)} \qquad i, j = 1, 2, \ i \neq j. \qquad (11.14)$$

We call the function obtained in (11.14) (the specific form of which of course results from the particular demand and cost functions we have assumed) firm i's *best response* function, because it tells us i's profit-maximizing response to *any* output j might choose.

We ruled out q_i^* and q_i^+ as possible equilibria essentially because they were not i's best response to the output of firm j that i began with. This suggests what an equilibrium would be, namely a pair of outputs such that each is the best response to the other. For in that case, given the output of firm j that firm i expects, i will produce exactly *that* output that j expects in choosing *its* output. Thus the two outputs are mutually sustaining and consistent. This concept of equilibrium is known as *Nash equilibrium*, after the game theorist John Nash, who first developed it as a solution concept for games of the type of which our oligopoly model is an example.

More formally, we would say that a Nash equilibrium of our duopoly model is a pair of outputs (q_1^N, q_2^N) such that q_i^N solves

$$\max_{q_i} \pi^i(q_i, q_j^N) \qquad i, j = 1, 2, \ i \neq j. \qquad (11.15)$$

In our example this means that the output pair must satisfy the equations

$$q_i^N = \frac{a - bq_j^N}{2(b + c_i)} = \frac{a(b + 2c_j)}{4(b + c_i)(b + c_j) - b^2} \qquad i, j = 1, 2, \ i \neq j. \qquad (11.16)$$

Using our numerical example, we find that:

$$q_1^N = 21.74, \qquad q_2^N = 13.04, \qquad p^N = 100 - 34.78 = \$65.22,$$

$$\pi_1^N = \pi^1(q_1^N, q_2^N) = \$945.25, \qquad \pi_2^N = \pi^2(q_1^N, q_2^N) = \$510.39.$$

We see that, as compared to the cartel outcome, both outputs are larger, market price is lower, firm 1's profit is lower while firm 2's is higher, but of course total profit is less, at \$1455.63. Thus, the firms could increase their joint profits by

$45.37 if they switched to the cartel outputs. Recall that at these outputs firm 2's profit was $500, which is less than that it receives in the Nash equilibrium. Therefore, firm 1 would have to make a side payment of at least $10.39 to firm 2 to induce it to enter into a cartel arrangement, but since its own profit increases by $54.75 there is considerable scope for this. But as we have already argued, in the absence of a binding agreement this would not be an equilibrium.

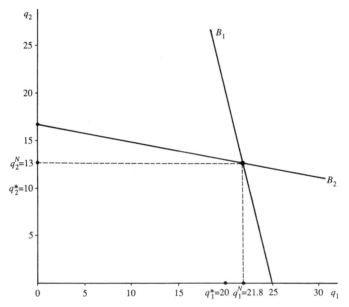

Fig. 11.1: Nash equilibrium in the duopoly example

This solution for the Nash equilibrium is illustrated in Fig. 11.1. The firms' best response functions are

$$q_1 = \frac{100 - q_2}{4}, \quad q_2 = \frac{100 - q_1}{6} \tag{11.17}$$

and these are graphed as B_1 and B_2 respectively. The Nash equilibrium is then the point at which they intersect. The figure shows why (q_1^*, q_2^*) and (q_1^+, q_2^+) (or indeed any other output pair) are *not* equilibria: q_i^* is *not* i's best response to q_j^*, and q_i^+ is not i's best response to q_j^+, as we have already seen. On the other hand, if i expects j to produce q_j^N, then i will want to produce q_i^N, thus confirming j in its choice of q_j^N. This is the essential property of the Nash equilibrium.

The Nash equilibrium as a solution for the duopoly model was proposed by the French mathematical economist Augustin Cournot long before that concept as such was explicitly defined. His explanation of the equilibrium was, however, somewhat different, and was framed in terms of reaction patterns. First, assume that in our model $c_1 = c_2 = 0$, so that each firm's profit is identical to its revenue. Then, suppose that at some initial point in time, firm 1 produces some arbitrary output level q_1^0. Firm 2 then assumes that whatever level of q_2 it chooses, firm 1 will not change its output, and so it solves

$$\max_{q_2} \pi^2(q_1^0, q_2) = [a - b(q_1^0 + q_2)]q_2 \;\rightarrow\; q_2^1 = \frac{a}{2b} - \frac{1}{2}q_1^0, \qquad (11.18)$$

where q_2^1 is what we have called earlier its best response to q_1^0. Firm 1 then observes this value of q_2 and makes the same assumption – that firm 2's output will remain unchanged whatever it chooses – and so it now chooses q_1 to solve

$$\max_{q_1} \pi^1(q_1, q_2^1) = [a - b(q_1 + q_2^1)]q_1 \;\rightarrow\; q_1^2 = \frac{a}{2b} - \frac{1}{2}q_2^1. \qquad (11.19)$$

But then firm 2 make *its* best response to q_1^2, and so on *ad infinitum*. Essentially we are here defining a first-order linear difference equation in outputs

$$q_t = \frac{a}{2b} - \frac{1}{2}q_{t-1} \qquad t = 1, 2, \dots \qquad (11.20)$$

where we ignore the firms' subscripts on output because the best response functions are identical. The solution of this difference equation is

$$q_t = \frac{a}{3b} + \left(q_1^0 - \frac{a}{3b}\right)\left(\frac{-1}{2}\right)^t \qquad (11.21)$$

and so clearly, $\lim_{t \to \infty} q_t = a/3b$, which is precisely the Nash equilibrium in this special case (confirm). In honour of this earlier discovery, what we have called the Nash equilibrium is in oligopoly theory usually called the Cournot-Nash equilibrium.

There are a number of criticisms of the Cournot-Nash equilibrium which we shall consider in the rest of this chapter. However, one criticism of the Cournot model does not carry over to its modern counterpart. In Cournot's story, the market adjustment is actually taking place through 'real time', and at each point the firm choosing its output is making a patently unrealistic assumption – that the other firm will not change its output in response to its own output choice – even when through time the assumption is consistently falsified. The Cournot assumption about what firms assume about their competitors' reactions therefore seems highly implausible.

The rationalization of Nash equilibrium, however, does not involve an actual adjustment process through time. Everything takes place inside the heads of the decision-takers, and only one pair of output choices is finally made. The Nash equilibrium concept is proposed as the appropriate concept for rational, well-informed players to use, because any other choices would involve one player assuming that the other had not thought the problem through as fully as he had. If firm 1 works out that q_1^* is not the best response to q_2^*, it should also acknowledge that firm 2 has also worked that out, and should not expect firm 2 actually to choose

q_2^*. In the light of that, logic compels us to acceptance of the Nash equilibrium concept.

Exercise 11.1

1. For the particular numerical example used in this section, draw a graph showing the market demand and marginal revenue curves and the cartel solution. Show how this determines the individual firms' outputs on the principle that their marginal costs are equated. Then illustrate the Cournot-Nash equilibrium solution. Explain why firm 2 earns higher profit in the Cournot-Nash than in the cartel equilibrium.
2. Suppose that the market can be subdivided into two sub-markets with the linear demand functions $p_k = a - b_k q_k$, $k = 1, 2$, and $b_1 \neq b_2$. Find for the cartel the profit-maximizing prices p_k and the firms' outputs q_i. Illustrate your answer on a diagram. [Hint: total output must be allocated between markets to equalize marginal revenue, and between firms to equalize marginal costs.]

11.2 Collusion as a Nash equilibrium

The single most unsatisfactory assumption underlying the analysis of the previous section was that the market only existed for one period. Typically, firms would see themselves as being in a market for many periods. This implies that we should model oligopoly as a *repeated game*. The game we have just analysed is called the 'one-shot' or *constituent game*, and we define the repeated game as a sequence of repeated plays of this game. We shall assume that the firms regard themselves as playing this game for ever – the sequence of repeated plays is an infinite sequence. Then a very interesting thing happens. The unappealing result of the previous section, that firms would never achieve a collusive equilibrium, disappears. We are able to define circumstances under which collusive outputs can be sustained as a Nash equilibrium of the repeated game.

The reason the 'no collusion' result of the previous analysis is unappealing is simply that it is contrary to observation. As many anti-trust cases and detailed industry studies show, collusion among oligopolists is actually quite widespread, even in countries where it is illegal. Since Nash equilibrium in the one-period game corresponds to no collusion, it is therefore refuted as a prediction of the market outcome. If this meant that the concept of Nash equilibrium *itself* was refuted, that would have quite devastating implications for game theory. However, what we shall see now is that it is the 'one-period' approach which is really at fault.

Though the general analysis of an infinitely repeated game could be quite complex, the point we are interested in can be illustrated in the context of a simple model, that of *balanced temptation equilibrium*. Suppose the firms want to produce the joint-profit-maximizing output pair (q_1^*, q_2^*), and we are at the start of period 1. Such an agreement is not legally enforceable and there is an incentive to cheat. However, the firms can now make the following threats to each other: if firm i cheats in *this* period, then firm j will choose its single period Nash equilibrium output q_j^N in *every future period*. In other words firm j threatens firm i with this punishment, and we now show that under certain conditions it then pays each firm not to cheat. Since the repeated game is exactly the same from whatever point in time we start, this implies that the collusive outputs can be sustained indefinitely.

As we saw in the previous section, if firm 1 cheats when 2 adheres to the agreement it makes an immediate gain in profit of $\pi_1^+ - \pi_1^*$. Now consider what it loses if this induces firm 2 to switch to output q_2^N from next period on. Instead of having profit π_1^* in every future period, firm 1 will now only receive π_1^N (since its best response to q_2^N is of course q_1^N). Thus its cheating will cost it $\pi_1^* - \pi_1^N$ in every future period. Since this represents an infinite annuity, its present value is $(\pi_1^* - \pi_1^N)/r_1$, where r_1 is firm 1's interest rate or cost of capital. It follows that firm 2's threat is sufficient to induce firm 1 not to cheat only if

$$\pi_1^+ - \pi_1^* \le (\pi_1^* - \pi_1^N)/r_1 \tag{11.22}$$

since then the loss from cheating offsets the immediate gain. This can be expressed as a condition on the interest rate:

$$r_1 \le (\pi_1^* - \pi_1^N)/(\pi_1^+ - \pi_1^*). \tag{11.23}$$

We can interpret this as saying that if firm 1 does not discount the future too heavily (r_1 satisfies the inequality in (11.23)), it will not cheat. By the same argument we can define an equivalent condition for firm 2:

$$r_2 \le (\pi_2^* - \pi_2^N)/(\pi_2^+ - \pi_2^*). \tag{11.24}$$

Then, if conditions (11.23) and (11.24) are satisfied, the firms are said to be in a 'balanced temptation equilibrium' (the term was proposed by James W. Friedman, who first carried out this analysis). Whatever the temptations, it pays neither to cheat, given the threat made by the other. Note that this balanced temptation equilibrium is a Nash equilibrium of the repeated game. Firm i's strategy can be described as: produce q_i^* as long as j produces q_j^*, but if j once produces some other output, switch to q_i^N the next period and produce that for ever. This type of strategy is known as a 'trigger strategy' – j's deviation triggers i's reaction – and the key point is that, if (11.23) and (11.24) are satisfied, these strategies are mutually best responses to each other and so form a Nash equilibrium. To convince yourself of this, put yourself in the position of firm 1, and try to find a better strategy *given* that firm 2 *will* play the above trigger strategy.

Using the values of the various profits we derived in the numerical example of the previous section, we have that

$$r_1 \le \frac{1000 - 945.25}{1012.50 - 1000} = 4.38, \qquad r_2 \le \frac{500 - 510.39}{533.33 - 500} = -0.31.$$

This tells us that if firm 2 receives only π_2^* under the collusive agreement, a balanced temptation equilibrium does not exist in this market, since in general we expect $r_2 > 0$. In other words, since firm 2's profit is higher in the Nash equilibrium than under collusion, it *always* pays it to cheat in the first period and then move to the Nash equilibrium. On the other hand firm 1 would have to have a very high interest rate indeed – up to 438% per period – before it would be worthwhile for it to cheat.

Clearly, firm 1 is going to have to make a side-payment to firm 2 if it wants to achieve the cartel outcome. Let π_i^c denote firm i's profit after any such side-payment, where of course we must have

$$\pi_1^c + \pi_2^c = \pi^*. \tag{11.25}$$

Thus the condition for balanced temptation equilibrium *with side-payments* to support the cartel outcome is that there exist a profit pair (π_1^c, π_2^c) satisfying (11.25) and such that

$$r_i \leq (\pi_i^c - \pi_i^N)/(\pi_i^+ - \pi_i^c) \qquad i = 1, 2. \tag{11.26}$$

Given the interest rates r_i, (11.25) and (11.26) imply the necessary condition

$$\sum_{i=1}^{2} \{(r_i\pi_i^+ + \pi_i^N)/(1 + r_i)\} \leq \pi^*. \tag{11.27}$$

If π^* is not large enough to satisfy this inequality then no side-payments exist which can ensure a profit pair (π_1^c, π_2^c) to meet condition (11.26) for the given interest rates.

Let us assume in our numerical example that $r_1 = r_2 = 0.25$, which would be quite reasonable if the period of the constituent game is one year. Then condition (11.27) becomes

$$\frac{1}{1.25}\${(0.25)1012.50 + 945.25 + (0.25)533.33 + 510.39\} = \$1473.68 < \$1500$$

and so side-payments exist which allow a balanced temptation equilibrium. For example, if firm 1 makes a side-payment to firm 2 of $25, we would have:

$$0.25 < \frac{(975 - 945.25)}{(1012.50 - 975)} = 0.79 \; ; \; 0.25 < \frac{(525 - 510.39)}{(533.33 - 525)} = 1.75$$

and so condition (11.26) is readily satisfied.

This discussion raises three issues: first, what if side-payments are not possible? For example, in a country such as the US where collusion is *per se* illegal, large and regular cash transfers between firms in the same market would be strong evidence of collusion. But as we saw, without side-payments, the cartel outcome could not be sustained as a balanced temptation equilibrium.

Secondly, if the purpose of the threat strategy is to enforce collusion, could some punishment other than the Cournot-Nash equilibrium be more effective? As our example shows, relative to the cartel outcome, for firm 2 the Cournot-Nash equilibrium is no punishment at all! Can we therefore define more effective punishment strategies? We shall see that we can.

Finally, we might question whether the threat to choose the Cournot-Nash output *for ever* is really a credible one. Suppose firm 1 has cheated and firm 2 is now punishing it by choosing q_2^N. We might expect that firm 1 would try to persuade firm 2 to begin colluding again, coupled of course with promises not to cheat. If this is thought possible *ex ante*, then the threat underlying the above trigger strategies is not a credible one. Better to make punishment as short and sharp as possible, and then restore collusion after cheating has been shown not to pay.

In the next section, we show, in the context of the special duopoly model of this chapter, that a general approach is available which resolves these issues and generalizes the type of threat strategy we have so far been considering.

Exercise 11.2

1. Explain why the trigger strategies underlying the balanced temptation equilibrium represent a Nash equilibrium of the infinitely repeated game.
2. In (π_1, π_2)-space, illustrate the point (π_1^*, π_2^*) for the numerical example of this section. Then show the set of (π_1, π_2) points achievable by lump-sum redistributions from this point. [Hint: what kind of function from π_1^c to π_2^c is defined by (11.25)?] Show the point (π_1^N, π_2^N), and then define the set of points which (*a*) maximize π_2 for each $\pi_1 \geq \pi_1^N$, and (*b*) satisfy $\pi_2 \geq \pi_2^N$.

11.3 A folk theorem

A folk song is a song which is widely known and sung but whose composer may not be identifiable, perhaps because the origins of the song are lost in the mists of time. A 'folk theorem' has the same connotation, though since game theory is relatively recent there must be some other reason for the inability to attribute the theorem to anyone. The folk theorem we are concerned with in this section generalizes the ideas introduced in our discussion of balanced temptation equilibrium and says that it is possible to devise punishments such that a very wide set of collusive outcomes can be supported by appropriate threat strategies. We shall use our market example to show how these are constructed.

The punishments are based on the idea of one firm 'minimaxing' the other. This term arises because the very worst outcome that one firm can inflict upon the other is found by solving the problem:

$$\min_{q_j} \max_{q_i} \pi^i(q_i, q_j) \qquad i,j = 1, 2, \ i \neq j. \tag{11.28}$$

This means that we first maximize π^i with respect to q_i, taking q_j as fixed, resulting in the best response function derived earlier, which we now write as $q_i = \beta_i(q_j)$, then we minimize $\pi^i(\beta_i(q_j), q_j)$ with respect to q_j.

However, we should recognize that this may result in a value of q_j which is not feasible for firm j. For example, if we take the market model of the previous two sections we find the q_j that satisfies (11.28) is $q_j = a/b$ (see question 1 of Exercise 11.3). That is, firm j produces the output that drives market price down to zero, thus ensuring that the best firm i can do is produce no output and earn zero profit. However, this is costly for j, since its loss is $c_j(a/b)^2$, its total production cost – it earns no revenue. In our numerical example, this would imply that firm 1 makes a loss of $\$(100)^2 = \$10,000$ when it minimaxes firm 2, while 2 would make a loss of $\$2(100)^2 = \$20,000$ when it minimaxes 1.

It seems reasonable therefore to impose a feasibility condition as a constraint on (11.28), and we assume that the firm doing the minimaxing must do no worse than break even, i.e.

$$\pi^j(\beta_i(q_j), q_j) = 0 \qquad i,j = 1, 2, \ i \neq j. \tag{11.29}$$

But that then means that the optimal *constrained* minimaxing outputs q_j^m are determined by the constraints in (11.29) directly. That is, the worst that each firm can do to the other is to drive market price down to the point where it itself just breaks even. Recalling that in our market example the best response functions are $q_i = (a - bq_j)/2(b + c_i)$, (11.29) implies

$$a - b\left(\frac{a - bq_j}{2(b + c_i)} + q_j\right) = c_j q_j \qquad i,j = 1, 2, \ i \neq j \tag{11.30}$$

or, in words, market price equals j's average cost. This then implies that

$$q_j^m = \frac{a(b + 2c_i)}{2(b + c_j)(b + c_i) - b^2} \qquad i,j = 1, 2, \ i \neq j. \tag{11.31}$$

In our numerical example, we find that

$$q_1^m = 45.45; \ q_2^m = 27.27; \ \beta_2(q_1^m) = 9.09; \ \beta_1(q_2^m) = 18.18.$$

That is, when *1 minimaxes 2* we have the output pair (45.45, 9.09), and when *2 minimaxes 1* we have the output pair (18.18, 27.27). When firm *1 minimaxes 2* then the market price and the firms' profits are

$$p = \$45.45; \ \pi^1(45.45, 9.09) = 0; \ \pi^2(45.45, 9.09) = \$247.97,$$

and when *2 minimaxes 1* they are:

$$p = \$54.55; \ \pi^1(18.18, 27.27) = \$661.21; \ \pi^2(18.18, 27.27) = 0.$$

The fact that 2's minimaxed profits are much less than 1's minimaxed profits are of course due to 2's higher average and marginal costs – 1 can afford to force the market price much lower than 2 while still breaking even. We see from (11.31) that

$$\frac{q_1^m}{q_2^m} = \frac{b + 2c_2}{b + 2c_1}, > 1 \quad \text{since } c_1 < c_2. \tag{11.32}$$

Fig. 11.2 illustrates the situation when 1 minimaxes 2. Given 1's output of 45.45, firm 2 has the *residual demand curve* $D'D'$. That is, its output will be added to 45.45 to determine market price, and so firm 2 is in effect a monopolist with respect to $D'D'$. Its profit-maximizing output with respect to this demand curve is then 9.09, found as the intersection of its marginal cost curve (drawn from the 'origin' of 45.45 units of output) with the marginal revenue curve corresponding to $D'D'$. The resulting market price of \$45.45 is just equal to firm 1's average cost at output 45.45.

The folk theorem for repeated games can now be stated in the following particular form. Suppose the collusive agreement between the firms results in some profit $\hat{\pi}_j$ for firm j, such that $\hat{\pi}_j > \pi_j^m$, $j = 1, 2$. Then, if the firms do not discount the future too heavily, this collusive agreement can always be supported as a Nash equilibrium of the repeated game by the following trigger strategy: i will adhere to the collusive agreement as long as j does, but if j once cheats, then i will switch to minimaxing j for ever.

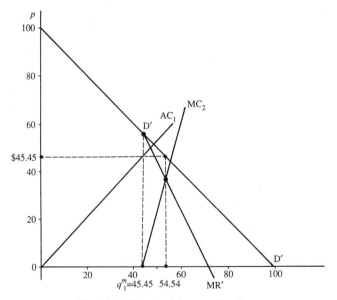

Fig. 11.2: Firm 1 minimaxes firm 2

The condition for this strategy to work is

$$\tilde{\pi}_j - \hat{\pi}_j \leq \frac{\hat{\pi}_j - \pi_j^m}{r_j} \quad \text{or} \quad r_j \leq \frac{\hat{\pi}_j - \pi_j^m}{\tilde{\pi}_j - \hat{\pi}_j} \tag{11.33}$$

where $\tilde{\pi}_j$ is the maximum profit j can make by reneging on the agreement, so that obviously $\tilde{\pi}_j > \hat{\pi}$. π_j^m is again j's minimax profit, and r_j is again j's interest rate. The point is of course that the right-hand side of the second inequality in (11.33) is always positive, and so there must always be *some* set of interest rates that would satisfy the condition. Thus it is always possible, given any interest rate in this set, for i to make it unprofitable for j to cheat. If j believes that i will carry out this threat, his best response is to continue to collude.

If, in our numerical example, we take the cartel profits (π_1^*, π_2^*) as the collusive outcome that the firms seek to support, then we have as the required conditions on the interest rates:

$$r_1 \le \frac{1000 - 661.21}{1012.50 - 1000} = 27.10 \; ; \; r_2 \le \frac{500 - 247.97}{533.33 - 500} = 7.56$$

Thus we see that the cartel outcome can be supported for wide sets of interest rates – anything up to 2710% for firm 1 and 756% for firm 2. In contrast to the case of balanced temptation equilibrium no side-payments are required, because of course the punishment of being minimaxed for ever is much more severe (especially for firm 2) than the punishment of being forced to the Cournot-Nash equilibrium forever.

The idea of punishment by minimaxing, rather than by reverting to the single-period Nash equilibrium, removes some of the criticisms we levelled at balanced temptation equilibrium, but one problem remains. Is it really credible for a firm to threaten to minimax another *for ever*? After all, in our model, this would involve the firm doing the punishing in zero profits for ever, while the firm being punished has positive profits! The temptation to 'kiss and make up' would be irresistible. Recent work in game theory on optimal punishments suggests a resolution of this problem, but to pursue these would take us further than we are able to go in this book. Instead, we note that in any given case, it may only be necessary to punish for a finite, possibly short period of time to wipe out the gains from cheating – the point of a definition such as that in (11.33) is to define the widest possible set of interest rates for which the trigger strategy could support the profit pair $(\hat{\pi}_1, \hat{\pi}_2)$. Thus, let r_j be j's actual interest rate. Then, a policy of minimaxing j for t_j periods and then reverting to a policy which gave j profit of $\hat{\pi}_j$ would just wipe out j's gain from cheating if

$$\tilde{\pi}_j - \hat{\pi}_j \le \hat{\pi}_j/r_j - [\pi_j^m(1 - (1 + r_j)^{-t_j})/r_j - \hat{\pi}_j(1 + r_j)^{-t_j}/r_j]$$

$$= (\hat{\pi}_j - \pi_j^m)(1 - (1 + r_j)^{-t_j})/r_j. \tag{11.34}$$

To see this, note that $\hat{\pi}_j/r_j$ is the present value of what j would receive if it kept to the agreement for ever. The term in square brackets in (11.34) is the present value of a profit stream of π_j^m for t_j periods, starting next period, then $\hat{\pi}_j$ for ever. Thus if t_j is chosen to satisfy (11.34), the one-off profits from cheating will be offset by the present value of the loss of profit from a punishment policy of minimaxing for t_j periods, then reverting to the agreed position.

Suppose in our example that each firm's interest rate is $r = 0.25$ per period (e.g., 25% per year), and the firms wish to sustain the joint profit-maximizing profits (π_1^*, π_2^*). Then using the payoff matrix data and (11.34) we have to solve

$$(1012.50 - 1000) \leq (1000 - 661.21)(1 - (1.25)^{-t_1})/0.25,$$
$$(533.33 - 500) \leq (500 - 247.97)(1 - (1.25)^{-t_2})/0.25.$$

The solutions for t_1 and t_2 which satisfy the *equalities* are actually .04 and .15 respectively, but since each t_i is restricted to be an integer, we can say that the punishment of minimaxing for just one period is much more than enough to wipe out the one-period gain from cheating. The reason for this is of course that, in this example, the loss of profit from being minimaxed is much larger than the gain from cheating. Thus the threat of a short, sharp price war could sustain collusion in this case.

A remaining problem with this approach is that it shows that quite a large set of profit pairs can be sustained by threats of retaliation. Therefore, we still have the problem of providing a model that characterizes a *unique* outcome. This is in the literature still an open question, but the approach through collusion appears to us to be the most promising one.

Exercise 11.3

1. Take the market model we have used in the previous two sections and show that if we place no constraints on a firm's ability to sustain losses, then firm j minimaxes firm i by setting $q_j = a/b$, $i, j = 1, 2$, $i \neq j$.
2. Suppose firms have identical constant marginal costs and face a negatively sloped demand function of the usual kind. What is each firm's minimax output and profit?

11.4 Summary

An oligopolistic market is one in which there are few enough sellers that they perceive their strategic interdependence. If they could make enforceable agreements then, in a one-period market, it would pay them to maximize joint profits, i.e., act as a multi-plant monopolist.

However, in a one-period market – or 'one-shot game' – it pays each firm to cheat on the agreement in the absence of enforceability. Therefore we would not predict collusion as the market outcome. A Nash equilibrium or, in the context of oligopoly, a Cournot-Nash equilibrium, is a pair of outputs which have the property that each is the best response to the other. In a one-period market without enforceable agreements this is the most compelling equilibrium concept if we regard the firms as rational. However, if we view the market as operating over many periods – say, for ever – then the threat of future retaliation may, if the firms do not discount the future too heavily, be enough to deter cheating in any one period. Thus we have an enforcement mechanism and collusion is sustainable as a Nash equilibrium of the repeated game.

One type of future retaliation could take the form of choosing the one-period Cournot-Nash output forever. A wider set of collusive outcomes can be supported however by a more drastic form of retaliation, that of minimaxing the cheat forever. The Folk Theorem shows that any pair of profits greater than the minimax profit pair can be supported in this way.

The threat of retaliation *for ever*, however, may not be credible, the temptation to 'kiss and make up' may be expected to be too strong. In many instances, however, a policy of minimaxing for some period of time, then reverting to the collusive outcome, may be enough to wipe out the gains from cheating.

All this shows that a wide set of collusive outcomes can be supported by threats. We are left still with the problem of predicting a unique oligopoly outcome. However, an approach grounded in the view that firms seek to collude may be more fruitful than one which sees firms acting at arms' length, trying to second-guess each other's reactions.

12 The Labour Market

The most important component of the income of most people is that derived from selling their labour services. The price at which they do this we call the wage, and the basic approach that we adopt in economics is that wages and employment are determined by the forces of supply and demand on the labour market. This chapter therefore first extends the model of consumer choice introduced in Chapters 6 and 7 to analyse labour supply; then extends the model of the profit maximizing firm in Chapters 8 and 9 to analyse labour demand.

There are two basic approaches to the way in which we might model the interaction of supply and demand. One is to treat the labour market as a competitive market, in which buyers and sellers — firms and workers — act as price takers. In that case wages and employment levels are determined very much in line with the analysis of Chapters 2–5. Alternatively, we might regard various kinds of market imperfection, in particular the presence of labour unions, as being of importance, and try to model the effect this will have on the labour market. Elements of this latter approach are developed in the final section of this chapter.

12.1 Labour supply

We consider here the individual's decision on how many hours to work per day, given his or her skills and location. The basic idea is that the worker has available 24 hours in the day, which can be spent either at work or 'at leisure', so, if l denotes the number of hours spent at work and x the number spent at leisure we have the simple identity $x = 24 - l$. The individual is assumed to regard leisure as a good — the more of it the better — and so this identity implies that work is a bad, the less of it the better. The other variable in the analysis is income, y, which represents the ability to purchase all other goods and services and so is a good. Income and work are related by the simple budget constraint

$$y = wl + m = w(24 - x) + m \qquad (12.1)$$

where $m \geq 0$ is exogenous non-wage income that the worker may receive from sources other than labour supply, and w, assumed constant, is the hourly wage rate.

The worker's preferences are expressed by a utility function, which can be defined in either of two equivalent ways: in terms of the *goods*, income and leisure, $u(x, y)$, u_x, $u_y > 0$; in terms of the good and the bad, income and work, $u(24 - l, y) \equiv v(l, y)$, $v_l = -u_x < 0$, $v_y = u_y > 0$.

Since we want to discuss the individual's labour supply, i.e., the chosen value of l as a function of w and m, we choose the second formulation. Then labour supply is given by the solution to the problem

$$\max v(l, y) \quad \text{s.t. } y - wl = m.$$

The simplest way to solve this is just to substitute for y from the constraint into the utility function to give an unconstrained maximization problem in l, $\max v(l, wl + m)$, yielding the condition

$$v_l + v_y w = 0 \implies w = \frac{-v_l}{v_y}. \tag{12.2}$$

This has the usual interpretation: marginal rate of substitution equals price ratio, where the latter is just w because the 'price' of income is 1.

This problem and its solution are illustrated in Fig. 12.1. The budget constraint is drawn for some given $m > 0$, and the object is to reach the highest possible indifference curve. The positive slope of the indifference curves reflects the fact that y is a good while l is a bad; the curvature reflects the assumption that as l increases, the rate at which income must be increased to compensate for an increase in work, dy/dl, also increases (see question 2 of Exercise 12.1).

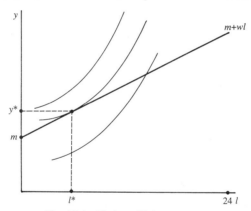

Fig. 12.1: Choice of labour supply

Condition (12.2) gives one equation in one unknown, l, and so we can solve to obtain the optimal l as a function of the parameters of the problem, w and m. We write this function as $S(w, m)$ and call it the individual's Marshallian labour supply function. Of particular interest is the *ceteris paribus* relationship between l and w, the labour supply curve. We would like to know the sign of the slope of this curve: it is usual to suppose that this is positive – the labour supply curve slopes upward – but is this necessarily so?

As we know from Chapter 7, the way to answer this question is to derive the Slutsky equation, since this will identify the income and substitution effects and allow us to see under what conditions they will reinforce or offset each other. This in turn requires us to derive the expenditure function. We do this by forming the dual problem

$$\min \ m = y - wl \quad \text{s.t.} \ v(l, y) = u. \tag{12.3}$$

This says that we want to find the smallest non-wage income that can allow a given utility level to be achieved at a given wage rate when y and l are optimally chosen. Diagrammatically, as Fig. 12.2 shows, we are again finding the lowest possible budget constraint consistent with being on the given indifference curve.

Solving the problem in (12.3), with μ as the Lagrange multiplier, gives the conditions

$$1 - \mu v_y = 0,$$

$$-w - \mu v_l = 0, \tag{12.4}$$

$$v(l, y) - u = 0.$$

As we expect, the first two conditions yield, on eliminating μ,

$$w = \frac{-v_l}{v_y}$$

which is the tangency condition obtained earlier. Solving conditions (12.4) for l and y gives the Hicksian labour supply and income demand functions $l = H(w, u)$, $y = y(w, u)$. In particular, $\partial H / \partial w$ gives the substitution effect on labour supply of a change in the wage rate with utility held constant and, given the assumed slope of the indifference curves, it is always positive (see question 3 of Exercise 12.1).

Then the minimized value of m is given by substituting for y and l in (12.3),

$$m = y(w, u) - wH(w, u) \equiv m(w, u),$$

which is the expenditure function. Moreover, it can be shown (see question 4 of Exercise 12.1) that Shephard's lemma here takes the form

$$\frac{\partial m}{\partial w} = -H(w, u) = -l.$$

In words, to a first approximation, if you work 8 hours a day for a wage of w per hour, and the wage rate rises by 1¢, then your standard of living can be just maintained by reducing your non-wage income by 8¢.

We are now in a position to derive the labour supply Slutsky equation. If the utility level chosen as the constraint in the expenditure minimization (dual) problem is the value of utility achieved at the solution to the utility maximization (primal) problem, then the solution value of l will be identical in each case, so that we have

$$S(w, m) = H(w, u). \tag{12.5}$$

We are going to consider a change in w, allowing l to vary, and taking whatever change in m is required to maintain the equality in (12.5). Accordingly, we substitute $m(w, u)$ into the lefthand side of (12.5), and differentiate totally to obtain

$$dl = \left(\frac{\partial S}{\partial w} + \frac{\partial S}{\partial m} \frac{\partial m}{\partial w} \right) dw = \frac{\partial H}{\partial w} dw. \tag{12.6}$$

Using Shephard's lemma, eliminating dw and rearranging (12.6) gives

$$\frac{\partial S}{\partial w} = \frac{\partial H}{\partial w} + l\frac{\partial S}{\partial m}. \tag{12.7}$$

The slope of the Marshallian labour supply function is then the sum of two effects: the substitution effect, or slope of the Hicksian labour supply function, $\partial H/\partial w$, and an income effect, $l\,\partial S/\partial m$, which represents the effect on labour supply of a change in non-wage income, weighted by the value of labour supply.

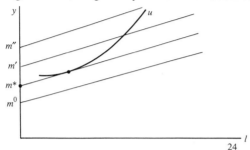

Fig. 12.2: Minimizing m for a given utility level

Now, graphical analysis (see question 3 of Exercise 12.1) shows that in general $\partial S/\partial m \gtrless 0$. Hence, again, although the slope of the Hicksian relationship is unambiguous, the Marshallian supply curve could in general have positive or negative slope. Can the idea of 'normal goods' rescue us from this ambiguity, as it did in Chapter 7 in the case of consumption goods? Unfortunately, not in this case. Thus, define work as a 'normal bad' if $\partial S/\partial m < 0$, i.e., if an increase in non-wage income causes the supply of labour to fall. Then in this case the sign of $\partial S/\partial w$ remains ambiguous. Only in the case $\partial S/\partial m \geq 0$, which we could call the case of an 'inferior bad', is the sign of $\partial S/\partial w$ certainly positive.

It may be hard to see the reason for these definitions of normal and inferior bads, but this is soon resolved by bringing leisure, a good, back into the analysis. We would call leisure a normal good if $\partial x/\partial m \geq 0$, and an inferior good if $\partial x/\partial m < 0$. But recalling that $l = 24 - x$,

$$\frac{\partial x}{\partial m} > 0 \Rightarrow \frac{\partial l}{\partial m} < 0; \qquad \frac{\partial x}{\partial m} \leq 0 \Rightarrow \frac{\partial l}{\partial m} \geq 0 \tag{12.8}$$

So, we can restate our conclusion from the Slutsky equation: if leisure is a normal good, the slope of the (Marshallian) labour supply curve at any point could be positive or negative, while if leisure is an inferior good the slope will certainly be positive.

This result has a simple explanation. If, say, the wage rate rises, then the worker, with the same amount of work, would have a higher income. He or she will then want to enjoy more leisure if, as the evidence suggests, leisure is something the demand for which increases as people become better off. This income effect would then tend to increase leisure and reduce work supplied. On the other hand,

the increased wage rate now means a higher return for extra time spent working
or, equivalently, a higher price of leisure in terms of income forgone, and this will
tend to reduce the demand for leisure and increase the supply of labour. So, the
overall change in labour supply following an increased wage rate could, as the
resultant of these conflicting effects, go in either direction.

   Given that we have a labour supply function for each individual, we can
then define the market labour supply function simply by aggregating those of
the individuals: at each wage rate, aggregate labour supply is the sum of the
individual labour supplies. It might then happen that, even if some individuals
have negatively-sloped labour supply curves, enough others have positively sloped
curves that the market supply curve is also positively sloped. It is also possible,
however, that the negative slopes could dominate, so that both demand and supply
curves in the labour market would be negatively sloped. As we saw in Chapter
5, this does not in itself present problems for the *existence* of an equilibrium, but
may be very important for its *stability*.

Exercise 12.1

1. The worker has the utility function $u = (x - \bar{x})^a (y - \bar{y})^b$, $x \geq \bar{x}$, $y \geq \bar{y}$, where $\bar{x}$
   is interpreted as a minimum required amount of leisure (sleep), and $\bar{y}$ as a minimum
   required income (subsistence). Derive the worker's labour supply function and comment
   on its properties.
2. Explain why l a bad and y a good must imply that indifference curves in the (l, y)-space
   have positive slopes. Explain why the assumed curvature of the indifference curves in
   Fig. 12.1 is consistent with the observation that workers must be paid a higher wage rate
   per hour to induce them to work overtime.
3. Use an indifference map to derive Hicksian and Marshallian labour supply curves [Hint:
   use the type of two-panel construction shown in Fig. 7.2.] Show that the slope of the
   Marshallian supply curve differs from that of the Hicksian supply curve by the income
   effect of a wage rate change. Show also that the income effect may be positive or
   negative.
4. Prove that Shephard's lemma in the labour supply problem takes the form $\partial m / \partial w =
   -l = -H(w, u)$. [Hint: simply adapt the method of proof of Shephard's lemma used in
   Chapter 7.]
5. Show that if t is a proportionate tax on wage income and leisure is a normal good, it
   cannot be asserted that reducing the tax rate will always increase labour supply. [Hint:
   define the after-tax wage rate as $w = (1 - t)w_0$, where w_0 is the fixed pre-tax wage rate,
   and use the Slutsky equation derived in this section.]
6. Suppose the worker has a fixed 'income target' — she wants to earn a given income
   regardless of the wage rate. What kind of labour supply function does this imply?
7. A worker is located at town A, where the wage rate is w_A. Some distance away, at
   town B, the wage rate for the same type of work is $w_B > w_A$. Suppose the worker has
   the preferences described in question 1 of this exercise. Derive an expression for the
   maximum amount of income the worker would pay for the opportunity to relocate in
   town B. [Hint: recall the analysis of consumer surplus in Chapter 7 and apply it to this
   case.]

12.2 Labour demand functions

In Chapter 8 we saw that the solution to the firm's long-run cost minimization problem yielded *constant-output* demand functions for the two inputs l and k (short-run cost minimization yielded only one value of l regardless of w, provided the firm went on producing – the constant output short-run demand curve for labour is vertical). However, it is clear that changes in an input price will change the firm's costs and profit-maximizing output, and so we need to take this into account in the input demand functions. We now derive input demand functions which do this.

The most convenient analytical approach is to replace the two-stage problem we considered earlier – first cost minimization for given output, then profit maximization with a given cost function – by a single maximization problem defined only on the firm's inputs. Thus, given the revenue function $R(q) = pq$ (we again assume a competitive market), we substitute for q from the production function $q = f(l, k)$ to define the problem

$$\max \pi(l, k) = pf(l, k) - wl - vk, \tag{12.9}$$

which is an unconstrained problem in l and k. With k free to vary, we have the *long-run* solution

$$pf_l - w = 0, \tag{12.10}$$

$$pf_k - v = 0. \tag{12.11}$$

These conditions can be solved for the two unknowns l and k in terms of the output and input prices $p, w,$ and v to give the long-run input demand functions of the firm

$$l = l(p, w, v) ; \quad k = k(p, w, v). \tag{12.12}$$

Note that the conditions imply the equality

$$\frac{f_l}{f_k} = \frac{w}{v} \tag{12.13}$$

that we also obtained in Chapter 8 — we have the same expansion path. The conditions now *imply* an output level, which we could locate from the production function, given the solution values of the inputs, i.e.

$$q = f(l, k) = f(l(p, w, v), k(p, w, v)) = S(p, w, v), \tag{12.14}$$

which is in fact the firm's long-run supply function.

Returning to conditions (12.10) and (12.11), we note that pf_l and pf_k are called, respectively, the marginal value product of labour (MVP_l) and the marginal value product of capital (MVP_k). Their interpretations are that in a competitive market, where the firm can vary its output with no effect on price, MVP_l gives revenue resulting from selling the output yielded by a small increment of labour input, and similarly for MVP_k. Conditions (12.10) and (12.11) can then be interpreted as saying: the profit-maximizing firm employs an input up to the point at which its marginal value product is just equal to its price.

To investigate the properties of the input demand functions in (12.12), we recall that f_l and f_k are functions, $f_l(l, k)$ and $f_k(l, k)$, and so totally differentiating through (12.10) and (12.11) gives the system

$$pf_{ll}dl + pf_{lk}dk = -f_l dp + dw, \tag{12.15}$$

$$pf_{kl}dl + pf_{kk}dk = -f_k dp + dv. \tag{12.16}$$

Defining the determinant

$$D = \begin{vmatrix} pf_{ll} & pf_{lk} \\ pf_{kl} & pf_{kk} \end{vmatrix} = p^2(f_{ll}f_{kk} - f_{lk}f_{kl})$$

and using Cramer's Rule then gives the derivatives of the demand functions

$$\frac{\partial l}{\partial p} = \frac{p(f_{lk}f_k - f_{kk}f_l)}{D}; \quad \frac{\partial k}{\partial p} = \frac{p(f_{kl}f_l - f_{ll}f_k)}{D}, \tag{12.17}$$

$$\frac{\partial l}{\partial w} = \frac{pf_{kk}}{D}; \quad \frac{\partial k}{\partial w} = \frac{-pf_{kl}}{D}, \tag{12.18}$$

$$\frac{\partial l}{\partial v} = \frac{-pf_{lk}}{D}; \quad \frac{\partial k}{\partial v} = \frac{pf_{ll}}{D}. \tag{12.19}$$

To be able to sign these derivatives we need to know the sign of the determinant D, which clearly is the same as the sign of the determinant

$$\Delta = \begin{vmatrix} f_{ll} & f_{lk} \\ f_{kl} & f_{kk} \end{vmatrix} = f_{ll}f_{kk} - f_{kl}f_{lk}.$$

Now it can be shown that for conditions (12.10) and (12.11) to define a true maximum we require $\Delta > 0$, which of course is consistent with the assumptions made earlier, that $f_{ll}f_{kk} > f_{kl}f_{lk}$. This assumption is in fact precisely the condition that the production function be *strictly concave* (see question 6 in Exercise 12.2). Given $\Delta > 0$, then $D > 0$, and so we have (check these out carefully for yourself) from (12.17)–(12.19):

(a) $\partial l/\partial p > 0$; $\partial k/\partial p > 0$. An increase in output price increases the firm's long-run demand for both inputs;

(b) $\partial l/\partial w < 0$; $\partial k/\partial v < 0$. The long-run input demand curves are negatively sloped.

(c) $\partial k/\partial w < 0$; $\partial l/\partial v < 0$. The two inputs are complements, which follows from the assumption that they are cooperant, i.e., $f_{kl}, f_{lk} > 0$ (check from (12.18) and (12.19)).

Note that these changes in input demands are derived from the effects of a change in p, w, or v on the profit-maximizing equilibrium of the firm, and so incorporate the effect of a change in the firm's equilibrium output. The effects on output itself are easy to derive from (12.14) (see question 1 of Exercise 12.2).

The last point we shall note about long-run input demands is that the long-run input demand function for the market as a whole *cannot* be derived simply by summing the long-run demand functions of the individual firms that are in the market at any one point in time. The reason for this is exactly that which makes it not possible to derive the long-run market supply function by summing the individual firms' long-run supply functions. A change in an input price will cause entry into or exit from the market, in response to the changed profit expectations, and the resulting change in total input demands must be derived from the new long-run market equilibrium given the change in the number of firms in the market (see question 2 of Exercise 12.2).

Turning now to the *short-run* input demand function, we note that the short-run profit maximizing condition can be rearranged to give

$$pf_l(l, \bar{k}) = w \tag{12.20}$$

and this can be solved for l to give

$$l = f_l^{-1}\left(\frac{w}{p}\right) = l_0(w, p, \bar{k}). \tag{12.21}$$

Thus the firm's short-run demand for the variable input, labour, depends on output price, the wage rate, and the given amount of the fixed factor capital. The properties of this demand function can be obtained by differentiating through (12.20) to obtain

$$pf_{ll}dl = dw - f_l dp - pf_{lk}d\bar{k} \tag{12.22}$$

and so $\partial l/\partial w = (pf_{ll})^{-1} < 0$ since $f_{ll} < 0$ at the profit maximum. $\partial l/\partial p = -f_l/pf_{ll} > 0$; $\partial l/\partial \bar{k} = -pf_{lk}/pf_{ll} > 0$ assuming $f_{lk} > 0$, the inputs are cooperant.

The first two of these signs are just as we would expect. The third tells us that if, somehow, more of the fixed input became available, this would, for fixed p and w, cause an increase in demand for the variable input because it would increase the latter's marginal product. (Note that the price of the fixed input does not appear in the labour demand function. Explain why.)

However (12.21) is not a complete specification of the labour demand function. Recall that if price falls below average variable cost the firm will produce no output in the short run, implying that it employs none of the variable input. Moreover, recall from Chapter 9 that AVC can be expressed as $w/AP_l = w/f(l, k)$. We can therefore define the firm's short-run labour demand function as:

$$l = l_0(w, p, \bar{k}) \quad \text{for } w, p, \bar{k} \text{ such that} \quad w \leq p\frac{f(l_0(w, p, \bar{k}), \bar{k}))}{l_0}, \tag{12.23}$$

where $l_0(.)$ is the function derived from (12.20). This implies a discontinuity in the labour demand function which mirrors that in the short-run supply function. Fig. 12.3 illustrates this for fixed p and $\bar{k}$.

Finally, the aggregation of the individual input demand functions across firms is relatively straightforward for the short run, since the number of firms is fixed at, say, n_0. If all firms have the same production functions *and* quantities of the fixed input then each will have the same demand for labour, $l_0(w, p, \bar{k})$. Hence aggregate market demand for labour is $n_0 l_0(w, p, \bar{k})$. If $S_0(w)$ is the market supply of labour then equilibrium in the labour market is given by the condition

$$S_0(w) = n_0 l_0(w, p, \bar{k}). \tag{12.24}$$

This yields an equilibrium wage rate and level of employment if p is taken as given, or an equilibrium market price and level of employment if w is taken as given. However, if both w and p are free to vary, to determine a full equilibrium we need a condition for equilibrium on the output market. Now, total market supply must be $n_0 q_0 = n_0 f(l_0(w, p, \bar{k}), \bar{k})$. Hence, if $D_0(p)$ is the market demand function for output, the conditions

$$D_0(p) = n_0 f(l_0(w, p, \bar{k}), \bar{k}), \tag{12.25}$$

$$S_0(w) = n_0 l_0(w, p, \bar{k}) \tag{12.26}$$

simultaneously determine an equilibrium wage rate and market price. Thus by aggregating outputs and labour demands over firms, and using the conditions for equilibrium in both output and labour markets, we obtain mutually consistent equilibrium values for market price and the wage rate.

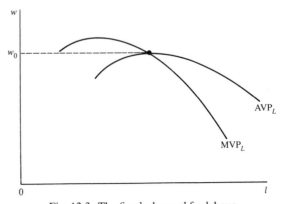

Fig. 12.3: The firm's demand for labour

Exercise 12.2

1. Use equations (12.14) and (12.17)–(12.19) to derive the effects on the firm's output of a change in p, w, and v (all other firms' outputs remaining unchanged).
2. Suppose a market is initially in long-run equilibrium, and then there is a permanent increase in the wage rate. Discuss the subsequent process of market adjustment and the

nature of the new long-run equilibrium. From this explain the nature of the long-run market demand curve for labour. [Hint: review the discussion of the long-run market supply curve in Chapter 9.]

3. Assume the firm has a Cobb-Douglas production function, $q = l^{b_1} k^{b_2}$, $0 < b_1, b_2 < 1$. Derive its short-run labour demand function and discuss its properties. [Hint: use condition (12.20).] Then, assuming that all firms have the same production functions and that the market demand function for output and the supply function of labour are, respectively

$$D_0(p) = p^{-\alpha}; \quad S_0(w) = w^{\beta} \quad \alpha, \beta > 1,$$

solve for the equilibrium price and wage rate, as functions of the number of firms in the market and their given capital stocks. [Hint: use the conditions in (12.25) and (12.26).] Illustrate your answer diagrammatically.

4. Suppose the firm is a profit-maximizing monopolist, facing a market demand function $p = D(q)$, $D' < 0$. Show that the counterparts of conditions (12.10) and (12.11) now involve the elasticity of the market demand function. Define the *marginal revenue product* as marginal revenue $\times$ marginal product and sketch the relationship between this and the marginal value product when the market demand function is linear. Use this to show that an equivalent way of expressing the proposition that a monopoly produces 'too little output', is that it uses 'too little input'.

5. Carry out the comparative statics analysis of conditions (12.25) and (12.26) to show the effects on equilibrium output price and wage rate of

   (a) an exogenous change in fixed capital stock $\bar{k}$;

   (b) the imposition of a specific tax t on output [Hint: define the price consumers pay as $p + t$, with p as the price producers receive, and note that the output demand function is now $D_0(p + t)$, while firms' supply functions remain unchanged];

   (c) the imposition of a tax y on the wage [Hint: define the wage workers receive as $w - y$ and note that the labour supply function is now $S_0(w - y)$, while the firms' supply and demand functions remain unchanged].

6. Given a twice differentiable function $f(x_1, x_2)$, it is strictly concave if $f_{11} < 0$ and its *Hessian determinant*

$$\begin{vmatrix} f_{11} & f_{12} \\ f_{21} & f_{22} \end{vmatrix} = f_{11}f_{22} - f_{21}f_{12}$$

is *positive*, implying $f_{11}f_{22} > f_{21}f_{12}$. Show that the Cobb-Douglas production function $y = l^{b_1} k^{b_2}$ is strictly concave if and only if $b_1 + b_2 < 1$. What problems arise for the competitive firm's long-run demand functions if $b_1 + b_2 \geq 1$? [Hint: what must be true of the firm's long-run equilibrium output in this case?]

12.3 Labour market equilibrium

At the end of the previous section we set out the conditions for the simultaneous equilibrium in an output market and a labour market, to clarify the interdependence between the two. To keep things simple it was useful to take the short run, with a fixed number of firms, and with the firms having identical production functions and fixed capital stocks. In this section, we consider the labour market in isolation, and ignore interaction with the output market. We can think of the market demand function more generally therefore as being the sum of the individual demand functions across all firms that employ the relevant type of labour; likewise the market supply function is the sum of individual supply functions across all individuals who supply that type of labour.

By a 'type of labour' we mean labour services of the same skill level, occupation and location. As with any type of good or service, for different units of labour to be traded on one market and command the same price they must be perfect substitutes in demand and supply. Although in macroeconomic models we might find it useful to talk of 'the' labour market and 'the' wage rate, in reality there are many markets for different types of labour, with a wide range of wage rates. A full explanation of these wage differentials would be the subject of a more specialized book than this. Here we consider simply the nature of the equilibrium in the market for a single type of labour.

If the market is competitive, so that all buyers and sellers, firms and workers, act as price-takers, the competitive market analysis of Chapters 2–5 applies directly. Supply and demand (which is, as we saw in the previous section, *derived* from consumers' demands for outputs) interact to determine the equilibrium wage. In applying the analysis of market stability and adjustment between static equilibria, we should also bear in mind that labour supply curves may well have negative slopes (as we saw in section 12.1).

Labour markets, however, are often not competitive. In particular, workers may be represented by a labour union, which negotiates a wage rate with employers on behalf of its members. If that union controls the labour supply by means of a 'closed shop' agreement, under which all workers must belong to the union, then it can behave as a monopolist. We now analyse the implications of that for the wage rate and employment level.

We treat the union *as if* it were an individual with a utility function

$$u = u(w, L) \quad u_w > 0, \; u_L > 0. \tag{12.27}$$

That is, the union prefers higher wages, w, to lower *and* higher employment levels, L, to lower. Note that L is best thought of as the number of workers, rather than 'effort supply', for simplicity we can assume that each worker works for a fixed number of hours, so L can also be used to denote aggregate labour supply. The market labour demand function is

$$L = D(w). \tag{12.28}$$

A key assumption we make is that the union cannot determine directly the level of employment L, but must accept however many workers the firms want to hire. On the other hand we assume that the union can *set* the wage. Since the union cares about the employment level, it must take account of the fact that choice of w determines L through the labour demand function (12.28). In effect (12.28) represents a constraint on the union. The simplest way to deal with this is to use it to substitute for L in the utility function to obtain the problem

$$\max u(w, D(w)), \tag{12.29}$$

which requires

$$u_w + u_L D'(w^*) = 0 \quad \Rightarrow \quad \frac{u_w}{u_L} = -D'(w^*). \tag{12.30}$$

As usual, we interpret u_w/u_L as the marginal rate of substitution between the wage rate and employment. Thus the optimal wage rate w^* and employment level $L^* = D(w^*)$ are at a tangency of the union's indifference curve u^* with the labour demand curve, $D(w)$, as Fig. 12.4 illustrates.

The underlying determinants of this equilibrium are: the relative strength of the union's preferences for the wage rate as compared to the employment level; and the position and slope of the labour demand function (but see question 3 of Exercise 12.3).

We can note two features of this equilibrium. First, the wage rate is higher and employment lower than it would be under competitive market conditions, when the equilibrium would be at (L_c, w_c) (but again see question 3 of Exercise 12.3). Secondly, there will be an excess supply of labour equal to $L_S - L^*$ at wage rate w^*, which we might call 'involuntary unemployment' in this market. This arises because the firms in the labour market do not *want* to hire more than L^* at the wage w^*, and because (L^*, w^*) is not a competitive market equilibrium.

A central feature of the model we have just analysed was that the union unilaterally set the wage. This is somewhat special. Typically unions and employers bargain over the wage rate, and we now have to examine whether this might cause the equilibrium to differ from that just described.

To model this, let us assume the market is supplied by a monopoly firm. The demand function $D(w)$ is now derived from its marginal revenue product function. That is, if we write the firm's total revenue as a function $R(L)$ of labour only, then profit maximization implies that labour demand satisfies $R'(L) = w$, and we find $D(w)$ by choosing L to satisfy this condition for each w.

We have characterized the union's indifference curves describing its preferences over (L, w) - pairs, but what of the firm's? We can define these by noting that the firm will be indifferent between (L, w)-pairs that yield the same value of profit. Then, a typical profit-indifference curve for the firm will be a set of (L, w)-pairs which, for given π^0, satisfy

$$\pi^0 = R(L) - wL. \tag{12.31}$$

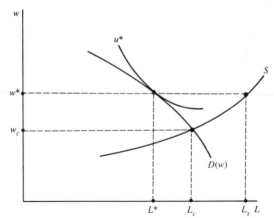

Fig. 12.4: Union sets wage, firm sets employment

To find the slope of the indifference curve we differentiate totally and rearrange to obtain:

$$d\pi^0 = 0 = (R'(L) - w)dL - Ldw \;\Rightarrow\; \frac{dw}{dL} = \frac{R'(L) - w}{L}. \tag{12.32}$$

We then note that at any point on the firm's labour demand curve, since $R'(L) = w$, a profit-indifference curve has a zero slope. Thus, the labour demand curve can be thought of as the locus of points of tangency of the firm's profit-indifference curves with horizontal lines, as Fig. 12.5 shows. In the figure it is *assumed* that these profit-indifference curves are strictly concave over the relevant range. We discuss this assumption further below.

This figure then shows immediately that the equilibrium we identified previously in Fig. 12.4, reproduced in Fig. 12.5, is actually *inefficient*, in the sense that we could *either*

(a) make the union better off and the firm no worse off by increasing L to $\hat{L}$ and reducing w to $\hat{w}$, thus keeping profit constant but putting the union on a higher indifference curve; *or*

(b) make the firm better off and the union no worse off by increasing L to L' and reducing w to w', thus keeping u constant and increasing π (the *lower* the firm's profit-indifference curve, the *higher* the firm's profit: to see this, consider what happens to the firm's profit if L is fixed and w reduced).

The precise point the firm and union will achieve will depend on their relative bargaining strengths. However, we shall not try to model the bargaining situation, but simply assume that wherever they end up, it will be at an *efficient point*; that is, a point which does *not* have the property that one of the parties to the bargain can be made better off with the other no worse off. To characterize such a point, we need to maximize the union's utility for any given level of the firm's profit, i.e., we have to solve the problem

$$\max u(w, L) \quad \text{s.t.} \; \pi^0 = R(L) - wL.$$

Attaching the Lagrange multiplier λ to the profit constraint we have the conditions

$$u_w + \lambda L = 0,$$
$$u_L - \lambda(R' - w) = 0, \tag{12.33}$$
$$\pi^0 - (R(L) - wL) = 0.$$

Eliminating λ from the first two conditions gives

$$\frac{u_L}{u_w} = \frac{-(R' - w)}{L}, \tag{12.34}$$

which implies the kind of tangency point shown in Fig. 12.5. Since $u_L/u_w > 0$, this condition tells us that the bargaining solution must be at a (L, w)-pair at which $R' < w$, that is, it must be a point *off* the labour demand curve, with lower wage and higher employment than would be implied by the point *on* the labour demand curve that yields the same value of profit (refer to Fig. 12.5). Precisely where depends on the profit constraint π^0, which we take to reflect the firm's bargaining power.

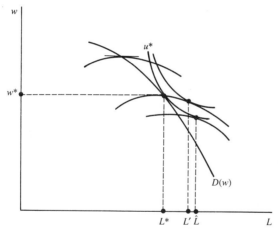

Fig. 12.5: Inefficiency of contracts on the labour demand curve

Note that we had to assume in the above analysis that the firm's profit-indifference curves were nicely concave. To see what is involved in this assumption, take the expression for the slope of one of these curves in (12.32) and note that for the curve to be strictly concave we require

$$\frac{d^2w}{dL^2} = \frac{d}{dL}\left(\frac{R'(L) - w}{L}\right) < 0. \tag{12.35}$$

Carrying out the indicated differentiation (confirm it for yourself) gives

$$\frac{d^2w}{dL^2} = \frac{LR'' - 2(R' - w)}{L^2} \tag{12.36}$$

and so condition (12.35) becomes

$$R'' < 2 \left(\frac{R' - w}{L} \right) = \frac{2dw}{dL}. \tag{12.37}$$

If the revenue function is strictly concave, $R'' < 0$. Then, this condition is certainly satisfied at a point on the labour demand curve, since then $R' - w = 0$. However, at (L, w)-pairs to the right of the labour demand curve we will have $R' - w < 0$, and it is therefore possible that we might have (12.37) violated. The consequences of this are illustrated in Fig. 12.6. The two tangency points shown there each satisfy condition (12.34), but neither is efficient since it is possible to increase the union's utility, without reducing the firm's profit, by increasing L in each case. Clearly, the problem is caused by the convexity of the profit-indifference curve over the indicated range, which in turn is due to the fact that (12.37) is violated. Thus, to ensure a 'well-behaved' solution to the problem, we need to assume that (12.37) is satisfied, which in turn restricts the set of revenue functions we can work with.

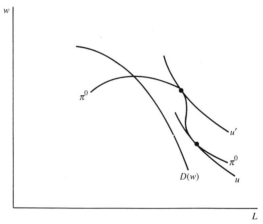

Fig. 12.6: The problem of non-concavity of the profit contours

Exercise 12.3

1. The union's utility function is given by $u = w^{3/4}L^{1/4}$, and the firm's revenue function is given by $R(L) = 500L - 0.25L^2$. Find the efficient wage and employment pair that corresponds to a profit of $100 for the firm. Confirm that this implies a lower wage and higher employment than at the point on the labour demand curve at which profit is also $100.

2. Take a union with the utility function of the previous question, let the firm's revenue function be given by $R = L^\alpha$, $0 < \alpha < 1$ and let the profit constraint be for zero profit. Show that the problem of finding an efficient wage and employment pair has no solution in this case. Explain why this happens.

3. Note that in the analysis of both models in this section, we ignored the labour supply curve. That is, we did not incorporate the constraint that the union, to deliver any given supply of labour, must obtain a wage at least as high as that corresponding to the given

labour supply on the market labour supply curve. Show the consequences of introducing market labour supply into the model. [Hint: assume the market labour supply curve is perfectly elastic at some given wage rate, and see what happens if you introduce this curve at various possible levels in Figs 12.4 and 12.5.]

12.4 Summary

The short-run labour supply decision is modelled as a problem of allocating time between leisure, which is a good, and work, a bad, given that the income derived from work is used to pay for consumption.

We derive the individual's Marshallian labour supply function by maximizing utility subject to a budget constraint. This function may have a positive or negative slope, if leisure is, as we would expect, a normal good.

The reason for this is that an increase in the wage rate increases income, which in turn increases the demand for leisure and so reduces the supply of work; but at the same time the price of leisure and the return to work – the wage rate – has increased and this tends to induce substitution of work for leisure.

Using a dual approach we can derive a Slutsky equation which allows us to give a precise statement of these conflicting effects.

Aggregation over individuals to obtain the market labour supply function may result in a positively sloped supply curve, but then again it may not.

The firm's labour demand function is derived by solving its profit maximization problem. There will be different demand functions depending on whether capital is fixed or variable.

The firm's labour demand also depends on price in the output market. Since this in turn depends on firms' marginal costs and input prices, we require simultaneous determination of prices and wages in a mutually consistent equilibrium.

In the short run this is relatively straightforward because the number of firms is fixed. In the long run, however, the number of firms is free to vary and the market labour demand function is *not* simply the sum of individual firms' long-run labour demand curves. The derivation of the long-run market labour demand function has to be integrated with the derivation of the long-run market output supply function, taking full account of the process of entry or exit of firms.

For simplicity we can construct a model of the labour market in isolation from the output market. If we assume this market is competitive then the earlier analysis of Chapters 2–5 applies in a straightforward way. However, labour markets are often characterized by non-competitive elements, in particular monopoly unions. In a model of a monopoly union which sets the wage but not employment, we saw that the wage would be higher and employment less than under competitive conditions.

If, however, the wage is set by bargaining between firm and union, and they seek an efficient bargain, that is, one which maximizes the union's utility at any given level of the firm's profit, then the chosen wage–employment pair will be off the firm's labour demand curve, with higher employment and a lower wage than in the previous model.

13   The Capital Market

By the capital market we mean the market for loans, or borrowing and lending. We abstract from the details of the institutions such as banks, stock and bond markets, building societies or savings and loan associations and, so on, which make up the market in reality, in order to focus on the basic determinants of the demand and supply of loans.

The price on this market is the interest rate, the premium paid on the amount borrowed or lent over a specified period of time, expressed as a proportion or percentage of that amount. To talk about 'the' interest rate is also a major over-simplification. In reality there is a wide range of instruments for borrowing and lending, which bear different interest rates, or rates of return, because of differences in their *liquidity* – the cost at which they can be converted into money – and in their riskiness – the probability distribution of their possible returns. Here we abstract from these differences and suppose there is just one type of riskless instrument, a loan. There are two sources of supply and demand for these loans. The first of these is the consumer, who, given the interest rate he faces, his preferences for consumption now and consumption in the future, and his income, chooses how much to borrow or lend. The second is the firm. Given the interest rate, the future output price it expects, and its technology of production, it chooses how much it wants to invest, and this, in conjunction with any internal funds it may have, determines how much it wants to borrow or lend. The market price, the equilibrium interest rate, is then found by aggregating supplies and demands for loans across all consumers and firms.

Because of the abstractions we make from the complexities of real capital markets the model we set out here is by no means a complete analysis of interest rate determination. However, it is an important first step in such an analysis.

13.1 Consumer borrowing and lending

Suppose that the consumer knows for certain how much income he will receive this year and in each future year. He also knows how this would be spent on goods and services, and therefore whether it would be preferable to have more money to spend sooner at the expense of less later, or conversely. For example, in their twenties people usually spend more than they earn, because they want to buy houses, cars, and consumer durables, and so tend to borrow quite heavily, while in their fifties people are anticipating retirement and so tend to save. So, borrowing and lending (saving) are the means by which people adjust their *endowed* income stream – the sequence of annual incomes they anticipate – to match their desired or preferred time stream of consumption. We now want to model these decisions.

To simplify notation and to make the main concepts as clear as possible, we assume there are just two time periods, this year, year 1, and next year, year 2. We denote by y_1 and y_2 the amount of money spent on consumption in years 1 and 2 respectively, and we call (y_1, y_2) a *consumption time stream*. The consumer is essentially interested in choosing a consumption time stream – borrowing and lending are simply the means by which this choice is put into effect. The consumer has an *endowed income time stream*, $(\bar{y}_1, \bar{y}_2)$, consisting of the incomes he is sure to receive in each year. Borrowing in year 1 is a way of allowing $y_1 > \bar{y}_1$, at the expense of having $y_2 < \bar{y}_2$, while lending is a way of allowing $y_2 > \bar{y}_2$, at the expense of having $y_1 < \bar{y}_1$. We will make this more precise in a moment.

We take it that consumption in each year is a good, and so the consumer's preferences over income time streams can be represented by the usual kind of indifference curves and utility function $u(y_1, y_2)$. We should note one item of terminology. Given the marginal rate of substitution $-dy_2/dy_1 = u_1/u_2$ at any point (y_1, y_2), we can always find some number ρ such that

$$-\frac{dy_2}{dy_1} = 1 + \rho. \tag{13.1}$$

We call ρ the consumer's *rate of time preference*. Its significance is that it has the interpretation of a kind of subjective *interest rate*. Thus if dy_1 and dy_2 are differentials along an indifference curve, (13.1) implies

$$dy_2 = -(1 + \rho)dy_1, \tag{13.2}$$

so ρ measures the premium the consumer requires to be paid in income *next* period to compensate for a reduction in income this period.

To derive the budget constraint or, as we shall call it, the *wealth constraint* in this model, we first suppose that the consumer can borrow or lend at a given market interest rate, r. Let b denote the amount borrowed or lent, with $b > 0$ if the consumer borrows and $b < 0$ if he lends. If b is borrowed or lent in year 1, then $(1 + r)b$ will be repaid in year 2. Then the consumption expenditures he is able to make in each period are given by

$$y_1 = \bar{y}_1 + b, \tag{13.3}$$
$$y_2 = \bar{y}_2 - (1 + r)b. \tag{13.4}$$

(13.3) says that consumption expenditure in year 1 can consist of initial endowed income plus the proceeds of any borrowing ($b > 0$), *or* less the amount of any lending ($b < 0$). (13.4) says that consumption in year 2 consists of the endowed income less the repayment owed on any borrowing, *or* plus that due on previous lending.

These two constraints can obviously be collapsed into one by solving for b in (13.4), to obtain $b = -(y_2 - \bar{y}_2)/(1 + r)$, and substituting into (13.3) to obtain the wealth constraint. This can be written in either of two equivalent ways:

$$(y_1 - \bar{y}_1) + \frac{(y_2 - \bar{y}_2)}{(1 + r)} = 0 \tag{13.5}$$

or

$$y_1 + \frac{y_2}{(1+r)} = \bar{y}_1 + \frac{\bar{y}_2}{(1+r)}, \tag{13.6}$$

To interpret these constraints, first note that $1/(1+r)$ is usually called a *discount factor*: it tells us how much \$1 to be received or paid in year 2 is worth in year 1, i.e., it gives the *present value* of \$1 receivable next year. It gives us a relative price: one would be prepared to exchange \$1 receivable next year for $\$1/(1+r)$ now, or \$1 now for $\$1(1+r)$ receivable next year.

Then, constraint (13.6) can be interpreted as saying: the consumer can choose an income stream (y_1, y_2) other than the endowment stream $(\bar{y}_1, \bar{y}_2)$, by borrowing or lending, provided that the present value of the chosen time stream is equal to the present value of the endowed income time stream, which we call wealth. Constraint (13.5) expresses the same idea in terms of variations of the chosen time stream around the endowed time stream.

The consumer's optimization problem is then to choose (y_1, y_2) so as to maximize $u(y_1, y_2)$ subject to either constraint (13.5) or (13.6). The conditions are

$$u_1 - \lambda = 0,$$

$$u_2 - \frac{\lambda}{1+r} = 0, \tag{13.7}$$

$$y_1 - \bar{y}_1 + \frac{(y_2 - \bar{y}_2)}{1+r} = 0,$$

where λ is the Langrange multiplier. Eliminating λ gives the condition

$$\frac{u_1}{u_2} = (1+r) \tag{13.8}$$

the usual kind of tangency condition. Recalling the interpretation of the marginal rate of substitution as $1 + \rho$, this condition can be expressed by the statement: *the consumer borrows or lends up to the point at which his time preference rate ρ is just equal to the market interest rate r.*

Fig. 13.1 illustrates this solution. The line W^0 represents the wealth constraint. Note that this line *must* pass through the point $(\bar{y}_1, \bar{y}_2)$ for *all* values of the interest rate, since $y_1 = \bar{y}_1$ and $y_2 = \bar{y}_2$ necessarily satisfy (13.5) and (13.6) for all r. The slope of the line is $-(1+r)$. Hence, changing the interest rate would cause the line to rotate through the point $(\bar{y}_1, \bar{y}_2)$. Changing $\bar{y}_1$ or $\bar{y}_2$ with r unchanged shifts the line parallel to itself.

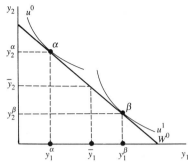

Fig. 13.1: Borrowing and lending solutions

Then the figure illustrates two types of solution to the choice problem. If the consumer's preferences are such that u^0 is one of his indifference curves, then the optimal choice is point α, with consumptions y_1^α, y_2^α. This implies that he chooses to lend the amount $y_1^\alpha - \bar{y}_1$ in year 1, being repaid the amount $y_2^\alpha - \bar{y}_2 = -(1+r)[y_1^\alpha - \bar{y}_1]$ in year 2. If on the other hand u^1 and not u^0 is one of his indifference curves, then the solution is at β with consumptions y_1^β, y_2^β. In this case he chooses to borrow $y_1^\beta - \bar{y}_1$ in year 1 and so must repay $\bar{y}_2 - y_2^\beta = (1+r)[y_1^\beta - \bar{y}_1]$ in year 2.

The general form of this equilibrium is found by solving conditions (13.7) for the chosen comsumptions y_1 and y_2, as functions of the parameters of the problem. Thus we obtain the Marshallian consumption demand functions $y_1 = D_1(r, \bar{y}_1, \bar{y}_2)$, $y_2 = D_2(r, \bar{y}_1, \bar{y}_2)$. Then, as usual we are interested in the *ceteris paribus* relationships and, in particular, the effect of changes in r on current consumption, since this determines how borrowing or lending varies with the interest rate. Thus since, from (13.3)

$$b = y_1 - \bar{y}_1,$$

we have

$$db = dy_1.$$

Hence, if an increase in the interest rate reduces current consumption, this must reduce borrowing, if the consumer is at a borrowing equilibrium, or increase lending, if the consumer is at a lending equilibrium. To determine the effect of interest rate changes on borrowing or lending it suffices to determine their effect on current consumption.

It is easy to show diagrammatically (see question 2 of Exercise 13.1) that in general, as we would expect by now, $\partial D_1 / \partial r \gtrless 0$. Thus again we need a Slutsky equation, and therefore an expenditure function, to investigate this derivative futher. Recall that, diagrammatically, we find an expenditure function by first finding the lowest possible budget line that is just tangent to a given indifference curve. In the present case, the intercept of the budget (i.e., wealth) constraint is determined by both $\bar{y}_1$ and $\bar{y}_2$, and so to express the problem mathematically we need only consider minimizing one of them. It is more convenient to choose $\bar{y}_2$

for this: i.e., from the wealth constraint in (13.6) we have

$$\bar{y}_2 = (y_1 - \bar{y}_1)(1 + r) + y_2$$

and so we form the expenditure minimization problem as

$$\min \bar{y}_2 = (y_1 - \bar{y}_1)(1 + r) + y_2 \qquad \text{s.t. } u(y_1, y_2) = u. \qquad (13.9)$$

The reason for choosing $\bar{y}_2$ is simply that it is easier to differentiate with respect to r than it would be in the corresponding expression for $\bar{y}_1$ and, in any case, we wish to focus on the demand for y_1 (see question 3 of Exercise 13.1).

Necessary conditions for a solution to the minimization problem are

$$(1 + r) - \mu u_1 = 0,$$
$$1 - \mu u_2 = 0, \qquad (13.10)$$
$$u(y_1, y_2) = u,$$

where μ is the Lagrange multiplier associated with the utility constraint. Again, we have the tangency condition

$$\frac{u_1}{u_2} = (1 + r),$$

and solving conditions (13.10) for y_1 and y_2 as functions of the parameters gives the Hicksian consumption demand functions $y_1 = H_1(r, u)$, $y_2 = H_2(r, u)$. Thus, the *minimized* value of $\bar{y}_2$ is given by

$$\bar{y}_2 = (H_1(r, u) - \bar{y}_1)(1 + r) + H_2(r, u) = \bar{y}_2(r, u), \qquad (13.11)$$

which is the expenditure function we want. It can be shown (see question 4 of Exercise 13.1) that Shephard's lemma in this case takes the form

$$\frac{\partial \bar{y}_2}{\partial r} = H_1(r, u) - \bar{y}_1 = y_1 - \bar{y}_1 = b. \qquad (13.12)$$

This has an interesting interpretation: if the consumer is a borrower ($b > 0$), an increase in the rate of interest increases the income required to maintain a given utility level, while if he is a lender ($b < 0$), an increase in the rate of interest reduces the income required to maintain a given utility level. This accords perfectly well with common sense: an increase in the interest rate makes borrowers worse off and lenders better off.

We can now derive the Slutsky equation for y_1. If the value of utility achieved at the solution to the utility maximization problem is taken as the constraint in the (dual) expenditure minimization problem in (13.9), we have

$$y_1 = D_1(r, \bar{y}_1, \bar{y}_2) = H_1(r, u). \qquad (13.13)$$

We want to consider a change in r, with $\bar{y}_1$ and u fixed, and with whatever change in $\bar{y}_2$ is required to maintain the equality in (13.13). Hence we substitute $\bar{y}_2(r, u)$ into D_1 in (13.13) and differentiate to obtain

$$\frac{\partial D_1}{\partial r} + \frac{\partial D_1}{\partial \bar{y}_2}\frac{\partial \bar{y}_2}{\partial r} = \frac{\partial H_1}{\partial r}. \tag{13.14}$$

Then using (13.12) in (13.14) and rearranging gives

$$\frac{\partial D_1}{\partial r} = \frac{\partial H_1}{\partial r} - (y_1 - \bar{y}_1)\frac{\partial D_1}{\partial \bar{y}_2} \tag{13.15}$$

which is the Slutsky equation we want. It can be shown (see question 5 of Exercise 13.1) that $\partial H_1/\partial r < 0$, consistent with the fact that the interest rate is the price of current consumption. Moreover, it seems reasonable to suppose that current consumption is a normal rather than inferior good, i.e., $\partial D_1/\partial \bar{y}_2 > 0$, so that an increase in wealth increases current consumption. In that case, the sign of $\partial D_1/\partial r$ depends on whether the consumer is a lender or a borrower. Thus we can say from (13.15), *given* that current consumption is a normal good,

(i)   if the consumer is a borrower, so that $y_1 - \bar{y}_1 > 0$, then $\partial D_1/\partial r < 0$ unambiguously;

(ii)   if the consumer is a lender, so that $y_1 - \bar{y}_1 < 0$, then $\partial D_1/\partial r \gtrless 0$.

To interpret these conclusions in terms of borrowing and lending, we can say that as long as current consumption is normal, borrowing will fall when the rate of interest rises, while saving may increase, decrease, or stay the same depending on the relative strengths of the income and substitution effects.

The intuitive reason for these results is easy to see. Suppose that the interest rate rises. To a borrower, this results in a fall in real income, as well as an increase in price of current consumption, and so, if current consumption is a normal good, both effects work to reduce current consumption and hence borrowing. For a lender, the price of current consumption has increased, tending to reduce current consumption and thus increase lending; however, real income has also increased, tending to increase current consumption (if it is normal). The two effects therefore work against each other.

If we now consider the individual's demand for current consumption, and the implied borrowing or lending, as a function of the interest rate, we can represent it as a curve such as that drawn in Figure 13.2. At relatively low interest rates the individual will be a borrower ($y_1 > \bar{y}_1$), and borrowing falls unambiguously as r increases. At interest rates above $\bar{r}$, he is a net lender ($y_1 < \bar{y}_1$), and, although in the figure lending is shown as increasing with the interest rate, we have just seen that in fact a portion of this curve could bend back toward the r-axis.

By summing the individual Marshallian demand functions we obtain the aggregate consumer demand for current consumption, and by subtracting aggregate consumers' income in period 1, we obtain aggregate net borrowing or lending. If the economy consisted *only* of consumers, then in equilibrium we would have to have that aggregate net borrowing or lending must be exactly zero: the sum of borrowing must equal the sum of lending, or equivalently current consumption demand must equal aggregate consumers' income. Thus we would have the equilibrium condition

$$\sum_{h=1}^{H} D_1^h(r^*, \bar{y}_1^h, \bar{y}_2^h) = \bar{Y}_1 \equiv \sum_{h=1}^{H} \bar{y}_1^h \tag{13.16}$$

where there are H consumers, the hth consumer's demand function for current consumption is $D_1^h(r, \bar{y}_1^h, \bar{y}_2^h)$, $\bar{Y}_1$ is aggregate endowed income in period 1, and r^* is the equilibrium interest rate. Note that even though consumers may have vastly different general preferences as between current and future consumption, at a market equilibrium each consumer's time preference rate ρ^h is equal to the market interest rate r^* and so they are all equal to each other.

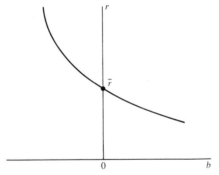

Fig. 13.2: Capital market equilibrium

However, condition (13.16) is not the end of the story, since there are also firms in the economy, and their lending and borrowing will also have an important effect on the market. In the next section we analyse how this is determined.

Exercise 13.1

1. A consumer has the utility function $u = y_1^a y_2^{1-a}$. Derive her Marshallian consumption demand functions, her demand for borrowing/lending function, and comment on their form.
2. Use Fig. 13.1 to show that when the interest rate rises, consumer's Marshallian demand for consumption y_1 may go up or down or stay the same. [Hint: note that increasing r implies rotating the wealth constraint in Fig. 13.1 clockwise through the point $(\bar{y}_1, \bar{y}_2)$.]
3. Use Fig. 13.1 to show the effect on consumption demands of reducing $\bar{y}_2$ with r and $\bar{y}_1$ held fixed. Put this together with your analysis in question 2 to show that if y_1 is a normal good, an increase in the interest rate unambiguously reduces borrowing but the effect on lending may be in any direction. Relate this to the Slutsky equation in (13.15). Now show what can be said if there is a reduction in the interest rate.

4. Prove the form of Shephard's lemma given in (13.12). [Hint: simply adapt the proof given in section 7.1 to the present case.]
5. Use a diagram to show that the derivative of the Hicksian consumption demand function $\partial H_1/\partial r$ must be negative. [Hint: recall that the Hicksian demand function is given by taking a sequence of tangent lines to a given indifference curve.]

13.2 Firms' borrowing and lending

In Chapter 12 we developed conditions which determined long-run input demands of a competitive firm, which for convenience we repeat here. Given the firm's production function $f(k, l)$ and the price p it expects to prevail 'next year' – the 'long run' – its profit-maximizing problem is

$$\max pf(k, l) - wl - vk,$$

giving the conditions

$$pf_l = w,$$

$$pf_k = v,$$

from which we obtain the long-run input demand functions $l(p, w, v)$, $k(p, w, v)$. We could now analyse the demand for capital, k, in exactly the same way as we analysed that for labour, but this would be so repetitive that we leave it as an exercise. Instead, we pursue two closely related questions. First, what is the relation between the firm's demand for physical capital for use in production, and its borrowing or lending? And secondly, what exactly is the price of capital input v, which we referred to in Chapters 7 and 8 as the 'rental rate of capital'? To answer both these questions we need to look more explicitly at the *intertemporal* nature of the firm's capital input decision.

We can recast the firm's decision problem in the following, probably descriptively more accurate way. It expects price p next period, it has the production function $f(k, l)$, and it expects to be able to hire whatever labour it needs at the wage rate w. *Given* the capital it will have available next period, it will choose l next period to solve

$$\max \pi(k, l) = pf(k, l) - wl \tag{13.17}$$

with the necessary condition

$$pf_l - w = 0$$

yielding the demand function for labour $l(p, w, k)$. Substituting for l in (13.17) with p and w assumed constant gives next period profit as a function of k only:

$$\pi(k) = pf(l(k), k) - wl(k). \tag{13.18}$$

The firm must choose *now* how much k to have next period, given the current market price of a unit of the capital good, which we denote by s. In other words, the firm is going to have to go out and buy the capital equipment on the relevant market, which we call the *capital goods* market and assume perfectly competitive. We assume that one unit of the capital good yields one unit of the capital input, k. Next period, it will use the capital, and this use will cause the capital to lose some value, i.e., to depreciate. We let $(1 - \delta)s$ denote the price at the end of next period of one unit of *used* capital, where $0 < \delta < 1$ is the *rate of depreciation*. Then, next period, the firm will have a *net worth* of

$$V(k) = \pi(k) + (1 - \delta)sk \qquad (13.19)$$

consisting of the profit it makes from production *plus* the value of the capital stock it will own. However, this is all in terms of money next year. The firm is going to have to spend sk *this* year to provide the future capital. To make next year's net worth comparable to this year's outlay we need to express $V(k)$ as a *present value*, $\hat{V}(k) = V(k)/(1 + r)$. Then, the firm's objective in choosing the capital stock is to maximize its net present value, i.e., to choose k to solve

$$\max \hat{V}(k) - sk \equiv \frac{pf(l(k), k) - wl(k) + (1 - \delta)sk}{1 + r} - sk, \qquad (13.20)$$

which implies the condition:

$$\frac{(pf_l - w)l'(k)}{1 + r} + \frac{pf_k + (1 - \delta)s}{1 + r} - s = 0. \qquad (13.21)$$

Note first that since l will be optimally chosen, the first term in (13.21) vanishes (this is an example of the important *envelope theorem*). Then rearranging the condition we have

$$pf_k = (r + \delta)s. \qquad (13.22)$$

We then see that we should define the price of a unit of capital, v, as $(r + \delta)s$. The intuitive interpretation of this is straightforward: the opportunity cost of buying a unit of capital at price s and using it for one period is the interest forgone on the money 'tied up' in the capital, rs, plus the fall in value of capital, δs. The term 'rental rate' then follows from noting that if one rented (leased) the unit of capital in a competitive rental market in which the rental charge just covered the leasing company's opportunity cost, the one-period charge would be $(r + \delta)s = v$.

Condition (13.22) now defines a demand function for the capital good $k(r, p, \delta, s)$. From our earlier analysis, we know that increasing v reduces demand for k, and so, since v increases with r, δ, and s, it is straightforward to see that k is inversely related to these parameters.

We now have to see how a firm's demand for *money* capital – its borrowing – is related to its demand for physical capital, k. First, note that the above analysis implicitly assumed that the firm possessed no existing capital stock – the capital k is being 'newly acquired'. In this case k is its investment in new capital. If the firm possesses some initial capital stock, k_0, its net investment demand will be

$$I = k(r, p, \delta, s) - k_0 = I(r, p, \delta, s, k_0), \qquad (13.23)$$

that is, its total desired capital stock *minus* its existing capital. The dollar cost of this investment will be sI. The firm may also have some initial retained profit π_0 that it can use to fund investment, and so the firm's demand for borrowing will be

$$b = sI - \pi_0 = sI(r, p, \delta, s, k_0) - \pi_0. \tag{13.24}$$

Note that b might be negative, implying the firm is a lender, if it has more than enough retained profit to finance its investment programme. Note finally that we do not have the possibility that $\partial b/\partial r > 0$, as was the case with the consumer. As long as the demand for physical capital is inversely related to the rental rate, the demand for money capital will be inversely related to the interest rate.

Exercise 13.2

1. The firm has the production function $f(l, k) = l^{b_1} k^{b_2}$, $b_1 + b_2 < 1$, and has a zero initial capital and no retained profit. Derive its borrowing demand function (13.24).
2. Show that $\partial b/\partial r < 0$, but that $\partial b/\partial s \gtrless 0$. [Hint: note that $dI = dk$, and use (13.22).] Explain these results.

13.3 Capital market equilibrium

If we denote the individual firm by the superscript $f = 1, ..., F$, then from (13.24) we have the firm's net demand for loans – its borrowing if positive, its lending if negative, is given by

$$b^f = sI^f(p, r, \delta^f, s, k_0^f) - \pi_0^f \tag{13.25}$$

where we assume that output price and price of the capital good (as well as of course the market interest rate) are the same for all firms, but that depreciation rates, initial capital, and retained profits may all be firm-specific. The aggregate net demand for loans across all consumers and firms, which we denote by B, is then given by

$$B = B(r, \{\bar{y}_1^h\}, \{\bar{y}_2^h\}, p, \{\delta^f\}, s, \{k_0^f\}, \{\pi_0^f\})$$
$$= \sum_{h=1}^{H} [D_1^h(r, \bar{y}_1^h, \bar{y}_2^h) - \bar{y}_1^h] + \sum_{f=1}^{F} [sI^f(r, p, \delta^f, s, k_0^f) - \pi_0^f] \tag{13.26}$$

where the notation in the $B(.)$ function expresses the idea that in general the aggregate net demand for loans depends on the *distributions* of endowed incomes, depreciation rates, initial capital, and retained profits. At an equilibrium of the capital market we must of course have $B = 0$ since the supply and demand for loans must be equal. This implies that the equilibrium market interest rate must satisfy

$$\sum_{h=1}^{H} D_1^h(r^*, \bar{y}_1^h, \bar{y}_2^h) + s \sum_{f=1}^{F} I^f(r^*, p, \delta^f, s, k_0^f) = \sum_{h=1}^{H} \bar{y}_1^h + \sum_{f=1}^{F} \pi_0^f. \tag{13.27}$$

The right-hand side of (13.27) shows the total supply of income available for loans, from all consumers and firms, while the left-hand side shows the total demand for income to be used for consumption and investment. Although for an individual consumer (or firm) consumption (or investment) can differ from endowed income (retained profit) because of borrowing and lending, across the entire economy borrowing must equal lending, or, equivalently, consumption plus investment must equal the aggregate income of consumers and firms. Essentially, equation (13.27) constitutes the explanation this model gives for what determines the market interest rate.

Exercise 13.3

1. Explain carefully why we would expect the market interest rate to be affected by:
   (i)   firms' expectation of future output price;
   (ii)   the distribution of consumers' future endowed incomes;
   (iii) the distribution of firms' depreciation rates;
   (iv)   the distribution of firms' initial capital stocks.
2. What would be the effect on the equilibrium market interest rate of:
   (i)   an increase in every consumer's future endowed income;
   (ii)   an exogenous reduction in every firm's initial capital stock;
   (iii) an increase in s, the market price of the capital good;
   (iv)   an increase in every firm's retained profit π_0^f;
   (v)   an increase in every consumer's current endowed income.
   (Where necessary, assume that current consumption is a normal good in respect of increases in endowed income.)

13.4 Summary

For a consumer, borrowing or lending is a means of transforming an endowed consumption time stream into a preferred consumption time stream.

We can apply the standard model of consumer choice to show how borrowing/lending is determined by the interaction of preferences, endowed incomes, and the interest rate, which is the market price in this context.

The assumption that current consumption is a normal good is sufficient to imply that borrowing is inversely related to the market interest rate, but in the case of lending (saving), we find that income and substitution effects work against each other and lending could fall as the interest rate rises.

To derive the firm's borrowing or lending, we first have to determine its demand for physical capital as an input. We see that this will depend on its expectation of the future output price. The analysis also shows that the appropriate price of a unit of capital is the interest rate *plus* the depreciation rate multiplied by the market price of a unit of capital. This is the 'rental price of capital'.

The firm's demand for physical capital is inversely related to the interest rate.

We find the firm's demand for investment, in money terms, by subtracting its existing capital from its desired capital and multiplying this difference by the market price of the capital good.

Its borrowing or lending is then given by the difference between its demand for investment and any internal funds it may have, in the form of retained profits it has not paid out to shareholders.

The net market demand for loans is then found by aggregating over all consumers and firms. Since in the aggregate borrowing must equal lending in equilibrium, this implies that the equilibrium market interest rate is determined by the condition that the aggregate net demand for loans be zero. This in turn can be translated into the condition that aggregate consumption and investment demand be equal to aggregate endowed income, where the latter consists of consumers' income *plus* firms' retained profits.

14 General Equilibrium

In microeconomics we visualize an economy as a system of interrelated markets. Each good or service, input or output has its own market, on which its price and quantity traded are determined. In principle, the outcome of any one market will depend on prices on *all* other markets, though in practice the interdependence between particular markets may be weak. For example, conditions on the market for automobiles will have a strong impact on the price of steel and the wages of steelworkers, but are unlikely to affect *strongly* the price of potatoes (though it is also hard to maintain that there would be no effect at all).

General equilibrium analysis is concerned with the interactions among markets at the level of the economy as a whole. Within a given time period the economic system somehow solves simultaneously a complex set of problems: households consume particular amounts of goods and services and supply inputs to firms; firms employ particular quantities of inputs to produce outputs which are then allocated to households. How are all these quantities determined? How are all the separate independent decisions of consumers and firms reconciled and made consistent? At this stage of the book you should be able to anticipate the answer: all this is achieved through the market mechanism. We now want to explore this answer.

It is useful to begin by breaking up the system of mutual interaction into two parts. First, we examine how, with output prices fixed, the production decisions of firms determine outputs, the allocation of an input, and the distribution of income among individuals who supply the input. Secondly, we examine how, with output prices now variable, the preferences and incomes of consumers determine prices. To rationalize this separation we begin by making the assumption that the economy in question has access to perfectly competitive world markets in the goods it produces and consumes. That is, these goods have given 'world prices', which can be taken as exogenous to this economy. This allows us to solve for household incomes and firms' outputs independently of demands, but of course begs the question of how output prices are determined. To close the model, in sections 14.2 and 14.3 we shall introduce consumer preferences and demands and drop the assumption of exogenous output prices.

14.1 Production and income distribution in an open economy

We take an economy with two consumers and two outputs, x_i, $i = 1, 2$. Each consumer has a given, fixed amount of labour L to supply, so that the total supply of labour in the economy is $2L$. Each output is produced by one firm, which seeks to maximize profit. There is a competitive world market for each good and an established world price, p_i, $i = 1, 2$. Therefore, firm i will take p_i as the price of its output, whether it sells in the home market or exports: it could not induce any consumer to pay more for the good than p_i, because she could always buy on the world market at that price; and it would not sell for less than p_i on the home market because it could export the good at that price.

We assume that capital stock is fixed and that labour is the only variable input, with l_i the amount of labour used by firm i. Its output is given by the production function

$$x_i = f^i(l_i) \qquad i = 1, 2 \tag{14.1}$$

with $f_l^i > 0$, $f_{ll}^i < 0$ (we have suppressed capital in the production function since it is assumed constant throughout).

Since output prices are given, there is only one price to be determined in this economy, that is the wage rate, w. We assume that the labour market is perfectly competitive – firms and workers act as price takers – and, since labour supply is fixed independently of the wage rate at $2L$, the equilibrium wage will be determined essentially by labour demand.

Each firm will choose its labour input by maximizing profit:

$$\pi_i = p_i x_i - w l_i = p_i f^i(l_i) - w l_i \qquad i = 1, 2, \tag{14.2}$$

implying the conditions

$$p_i f_l^i(l_i) = w \qquad i = 1, 2, \tag{14.3}$$

and solving for l_i gives the labour demand function

$$l_i = l_i(w, p_i) \qquad i = 1, 2. \tag{14.4}$$

Thus, for given p_i and L the equilibrium wage rate w^* in this economy is determined by the equilibrium demand = supply condition:

$$l_1(w^*, p_1) + l_2(w^*, p_2) = 2L. \tag{14.5}$$

Fig. 14.1 illustrates this equilibrium. (14.3) implies of course that the firm's labour demand curve is its marginal value product of labour (MVP_l^i) curve (recall Chapter 12). In the figure, we measure l_1 *rightward* from the origin, and l_2 *leftward* from the point $2L$, which is the total supply of labour. Thus, any point on the horizontal axis represents an allocation of the given supply of labour between the two firms. The left-hand vertical axis measures MVP_l^1, the righthand vertical axis MVP_l^2, and the shapes of the curves reflect the assumption of diminishing marginal productivity of labour (make sure you can explain why the MVP_l^2 curve is drawn the way it is). The equilibrium w^* is found at the intersection of the two marginal value product curves. Only at that point does the wage induce two labour demands l_i^*, $i = 1, 2$, which exactly sum to the given labour supply. Thus we see that the allocation of the given supply of labour between firms is determined first by output prices and secondly by technology, which determines marginal products of labour.

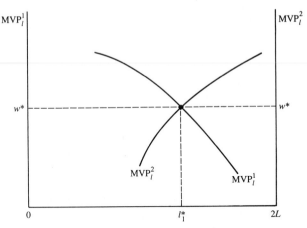

Fig. 14.1: Labour market equilibrium in the two-sector model

The figure also clearly brings out a very important feature of this market equilibrium: total labour supply is allocated between the firms in such a way that marginal value products of labour in the two firms are equalized (this can also be readily seen from condition (14.3)). If labour were allocated so that this were not the case, as at l_1' in Fig. 14.1, then it would pay the firm with the higher MVP$_l$ to bid labour away from the firm with the lower, since the former could compensate the latter for its loss of labour and still increase its profit.

This important implication of condition (14.3) can be put in yet another way. The equilibrium wage in a competitive labour market measures the value of the marginal unit of labour to each firm and so, in effect, any one firm is being charged for its use of labour at the *marginal opportunity cost of labour* – the value of the output the marginal unit of labour would generate if it were diverted to an alternative use in another firm.

The derivation of the comparative statics results for this model is straightforward and is left as an exercise. An increase in an output price increases the wage rate, increases output and employment in its own sector, and reduces those in the other. An increase in labour supply reduces the wage rate and increases output and employment in both sectors (see question 5 of Exercise 14.1).

Now consider the distribution of income in this economy. Given the assumption that each individual is endowed with the same amount of an identical type of labour, the wage income of each of them is w^*L. (Note that we assume that a worker can supply labour to more than one firm.) However, this is not the end of the story, because in this equilibrium each firm is making non-negative profits, i.e.

$$\pi_i^* = p_i f^i(l_i^*) - w^* l_i^* \geq 0 \qquad i = 1, 2. \tag{14.6}$$

To see this, note that (14.3) implies that $p_i f_l^i(l_i^*) = w^*$ and so

$$\pi_i^* = p_i[f^i(l_i^*) - f_l^i(l_i^*)l_i^*] = p_i l_i^* [f^i(l_i^*)/l_i^* - f_l^i(l_i^*)] \qquad i = 1, 2. \tag{14.7}$$

But the term in brackets in (14.7) is the difference between the average and marginal products of labour, and, if the firm is producing positive output, given the assumption that marginal product of labour is diminishing ($f^i_{ll} < 0$), this term must be positive or, if l^*_i happens to be at the point of maximum average product of labour, zero. Thus, $\pi^*_i \geq 0$, $i = 1, 2$.

In a general equilibrium model, it is important to recognize that profit represents income to the owners of the firm, who can, in this economy, only be our two individuals. Moreover, what we are here calling 'profit' could also be regarded, at least in part, as a return to the capital which each firm uses with labour to produce output. To keep things simple, we assume that individual j owns a share θ_{ij} in firm i, with of course $\theta_{i1} + \theta_{i2} = 1$ since ownership of a firm is exactly shared between the two individuals. We could suppose that these shares reflect past investment the individuals made in each firm to provide the capital stock it now has. This implies that the total income of individual j is

$$y^*_j = w^*L + \theta_{1j}\pi^*_1 + \theta_{2j}\pi^*_2 \qquad j = 1, 2 \qquad (14.8)$$

where the asterisks emphasize that this is income *at the equilibrium of the economy*.

As we already pointed out, the wage incomes of the two individuals will be the same, on our present assumptions, but clearly their profit incomes could be very different. For example, if 1 owned larger shares in *both* companies than 2, her income would be higher. For a given pattern of shareholdings, the income distribution will be determined by firms' profits, determined in turn by technology and the output price. Thus this model gives only a partial explanation of the differences in individual incomes: it says that they are due in part to differences in individuals' ownership shares of firms, which can be thought of as 'initial wealth endowments', but we have no explanation in this model of these differences in endowments.

To see what determines the level of total income in this economy, we sum the two incomes in (14.8) to obtain

$$y^*_1 + y^*_2 = w^*2L + \pi^*_1 + \pi^*_2 \qquad (14.9)$$

(using the fact that $\theta_{i1} + \theta_{i2} = 1$). Then, substituting for π^*_i from (14.6) gives

$$y^*_1 + y^*_2 = w^*2L + p_1 f^1(l^*_1) + p_2 f^2(l^*_2) - w^*(l^*_1 + l^*_2)$$
$$= p_1 f^1(l^*_1) + p_2 f^2(l^*_2) = p_1 x^*_1 + p_2 x^*_2, \qquad (14.10)$$

where we use the fact that $l^*_1 + l^*_2 = 2L$ and x^*_i is the output of good i in equilibrium. (14.10) is a statement which will again be encountered in Chapter 16. It says that the value of total income in the economy must be equal to the value of total output – what firms receive as revenue they pay out as incomes. Then clearly the level of income in this economy is determined on the one hand by the given output prices, and on the other by the technology underlying the production functions, which determine the productivity of labour.

We have so far determined what could be called the *production allocation* in this economy – the four quantities x_i^*, l_i^*, $i = 1, 2$. Before going on to look at the *consumption allocation*, we consider a particular representation of this production allocation in terms of the *production possibility curve*. First, we derive an expression for the set of feasible output pairs in this economy, which we call the *production possibility set*. Given the production functions $x_i = f^i(l_i)$, $i = 1, 2$, we invert them to obtain the amount of labour a firm requires as a function of the output it produces, $l_i = \phi^i(x_i)$. Then the constraint that the labour used by firms cannot exceed the amount available,

$$l_1 + l_2 \leq 2L, \tag{14.11}$$

becomes by substitution a constraint on feasible outputs,

$$\phi^1(x_1) + \phi^2(x_2) \leq 2L. \tag{14.12}$$

The set of output pairs which satisfy this constraint is the production possibility set. The *upper boundary* of this set is given by the equation

$$\phi^1(x_1) + \phi^2(x_2) = 2L, \tag{14.13}$$

and when graphed as in Fig. 14.2, it gives the production possibility curve. The intercept x_1^0 is found from (14.13) by setting $x_2 = 0$ and solving the equation $\phi^1(x_1^0) = 2L$, and similarly for x_2^0. Since ϕ^i is the inverse function to f^i, we have $d\phi^i/dx_i \equiv \phi_x^i = 1/f_l^i$, $i = 1, 2$. Then the slope of the production possibility curve at a point is found by differentiating (14.13) totally (recalling L is a constant) to obtain

$$\phi_x^1 dx_1 + \phi_x^2 dx_2 = 0 \implies \frac{dx_2}{dx_1} = -\frac{\phi_x^1}{\phi_x^2} = -\frac{f_l^2}{f_l^1}. \tag{14.14}$$

Thus the slope of the production possibility curve is the ratio of marginal products. We call this slope the *marginal rate of transformation* between outputs. Intuitively, we move around the production possibility curve by withdrawing a little labour from firm 1, say $-dl_1$, thus causing a fall in output of $dx_1 = -f_l^1 dl_1$; and then by putting this labour into firm 2, causing an increase in output of $dx_2 = f_l^2 dl_1$. Thus, the ratio $dx_2/dx_1 = -f_l^2 dl_1/f_l^1 dl_1$ is exactly that in (14.14). The production possibility curve is drawn concave to the origin because $f_{ll}^i < 0$: the proof of this is left as an exercise (see question 6 of Exercise 14.1). Finally, note that a point below the production possibility curve is feasible, in that it implies an output pair which uses less than the total amount of labour available to the economy – (14.11) and (14.12) are satisfied as strict inequalities at such points. These points correspond therefore to underemployment of the economy's resources, whereas points on the production possibility curve correspond to full employment.

Now recall from (14.3) that each firm is in equilibrium with $p_i f_l^i = w^*$, implying that at the equilibrium resource allocation (l_i^*, x_i^*), we must have

$$\frac{p_1}{p_2} = \frac{f_l^2}{f_l^1}. \tag{14.15}$$

That is, as Fig. 14.2 illustrates, at the equilibrium of the economy the marginal rate of transformation, or slope of the production possibility curve, $dx_2/dx_1 = -f_l^2/f_l^1$ is equal to $-p_1/p_2$.

This suggests a futher interesting interpretation of the equilibrium. Define the function $V = p_1x_1 + p_2x_2$, where V is obviously the value of national output at the given world prices. Then solve the problem

$$\max_{x_1,x_2} V = p_1x_1 + p_2x_2 \qquad \text{s.t. } x_1 = f^1(l_1),\ x_2 = f^2(l_2), \qquad l_1 + l_2 = 2L.$$

Then the solution to this problem will be given by the conditions in (14.3), with the wage rate w set equal to the Lagrange multiplier λ associated with the constraint that labour demand = labour supply. The illustration of this solution in (x_1, x_2)-space is precisely that given in Fig. 14.2. Question 3 of exercise 14.1 asks you to consider this interpretation further.

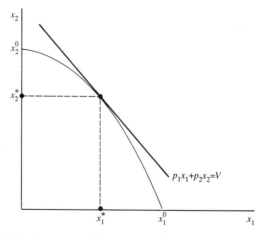

Fig. 14.2: Production possibility curve and value-of-output maximization

In the market economy, we saw that the consumers would have incomes $y_j^* = w^*L + \theta_{1j}\pi_1^* + \theta_{2j}\pi_2^*$. They are also faced with given prices of the goods, and so we can specify for each consumer a budget constraint:

$$p_1x_{1j} + p_2x_{2j} = y_j^* \qquad j = 1,2 \tag{14.16}$$

where x_{ij} is j's consumption of good i, $i,j = 1,2$. Moreover, each consumer will have a utility function $u^j(x_{1j}, x_{2j})$, and so each will choose consumptions x_{ij}^* which maximize utility subject to (14.16). So, in this market economy with *fixed output prices*, that is how consumptions are determined. Consumers simply buy their preferred consumption bundles with the income they receive from their labour supply and shares in firms' profits.

The question then naturally arises: are the consumers' chosen consumptions *feasible*? After all, we have imposed no 'supply = demand' type of equilibrium condition on them. We now show that they are. Denote the chosen consumptions by x^*_{ij}, $i, j = 1, 2$, and note that they must satisfy the consumers' budget constraints, so that we have

$$p_1 x^*_{1j} + p_2 x^*_{2j} = y^*_j = w^* L + \theta_{1j} \pi^*_1 + \theta_{2j} \pi^*_2 \qquad j = 1, 2. \qquad (14.17)$$

Then adding these budget constraints, recalling that $\theta_{i1} + \theta_{i2} = 1$, $i = 1, 2$, gives

$$p_1(x^*_{11} + x^*_{12}) + p_2(x^*_{21} + x^*_{22}) = w^* 2L + \pi^*_1 + \pi^*_2 \qquad (14.18)$$

where of course $x^*_{i1} + x^*_{i2}$ is total demand for good $i = 1, 2$. Then substituting from (14.2) the definitions for the π^*_i and rearranging gives

$$p_1(x^*_{11} + x^*_{12}) + p_2(x^*_{21} + x^*_{22}) = p_1 x^*_1 + p_2 x^*_2 \qquad (14.19)$$

which says that the value of total consumption in the economy equals the value of total production or national output. We can express this as

$$p_1(x^*_{11} + x^*_{12} - x^*_1) + p_2(x^*_{21} + x^*_{22} - x^*_2) = 0, \qquad (14.20)$$

which tells us that if the total consumption of one of the goods exceeds the economy's production of it, then the converse must be true for the other good. (14.20) is in fact a balance of trade condition for the economy. Thus assume

$$e^*_1 \equiv (x^*_{11} + x^*_{12} - x^*_1) < 0 ; \quad e^*_2 \equiv (x^*_{21} + x^*_{22} - x^*_2) > 0.$$

Then we can regard e^*_2 as the economy's *imports* of good 2, since the excess of consumption over production must be obtained on the world market; and we can regard e^*_1 as the economy's *exports* of good 1, the excess of production over consumption is sold on the world market. Condition (14.20) then says that $p_1 e^*_1 + p_2 e^*_2 = 0$, that is, that the value of exports equals the value of imports and the economy's trade with the rest of the world is in balance.

The essential point of the analysis is that the exact consumption bundles chosen by the households are of no concern as long as they satisfy the individual budget constraints. The economy has available an income of $p_1 x^*_1 + p_2 x^*_2$ to spend on consumption. This is allocated among consumers as the incomes y^*_j. As long as each consumer spends exactly y^*_j on consumption, the economy's aggregate consumption expenditure will equal its income, which is all that is required.

Exercise 14.1

1. Take an economy with a large number of consumers each supplying a fixed amount of labour; a large number of goods, each of which has a given world price; and a large number of firms producing each good. Labour is the only input. Derive the conditions which hold at a market equilibrium of this economy, and show that they are essentially

similar to those given in (14.3) and (14.5). Obtain the counterparts of (14.8) and (14.10) for this economy. [Hint: this problem is essentially an exercise in the use of subscripts and the $\sum$-notation.]

2. In the two-good, two-consumer model of this section, the production functions are

$$x_1 = 10l_1^{1/2} ; \quad x_2 = 8l_2^{1/2}.$$

The consumer's utility functions are

$$u^1 = x_{11}^{1/2} x_{21}^{1/2} ; \quad u^2 = x_{12}^{1/4} x_{22}^{3/4}.$$

Each consumer supplies 50 units of labour. The given world prices of the goods are

$$p_1 = \$20, \qquad p_2 = \$10.$$

Consumer 1 owns all of firm 1, consumer 2 owns all of firm 2. Solve for the equilibrium wage rate, outputs, consumptions, exports, and imports in this economy. Draw the counterparts of Figs 14.1 and 14.2 for this case. Explain why the production allocation, though not the consumption allocation, is independent of the consumers' shareholdings. Explain what 'ultimately determines' the production and consumption allocations.

3. For the economy given in the previous question, solve the problem;

$$\max_{x_1, x_2} V = p_1 x_1 + p_2 x_2 \qquad \text{s.t. } x_1 = f^1(l_1), \ x_2 = f^2(l_2); \quad l_1 + l_2 = 2L$$

and show that exactly the same production allocation is obtained as in question 2, with the Lagrange multiplier $\lambda^* = w^*$. [Hint: substitute for x_1 and x_2 into V and solve with respect to l_1 and l_2; *or* substitute for l_1 and l_2 in the labour supply constraint.] Explain why it could be argued that consumers are made as well off as possible by choosing a resource allocation which maximizes the value of national output at the given world prices.

4. What would be the effects, in the economy of question 2, of the imposition of a minimum wage rate of $\$12$? Illustrate in the counterparts of Figs 14.1 and 14.2. Calculate the resulting falls in the consumers' utilities. In this kind of economy, what are the only ways in which *all* consumers can be made better off?

5. Derive the comparative statics effects of a change in one of the world output prices, and of an increase in each consumer's labour supply. [Hint: find the effects on w^* by differentiating totally through (14.5) and rearranging. Use (14.3) and (14.4) to give the required restrictions on signs of partial derivatives.]

6. Prove that the production possibility curve derived in this section is strictly concave to the origin.

14.2 Preferences and prices: a pure exchange economy

We now move to an economy which is closed, in the sense that it does not
have access to outside markets for goods. Output prices are to be determined
endogenously. We approach the problem in two steps. In this section, we study
a *pure exchange economy*, that is, an economy without production. In the next
section we reintroduce production. Output supplies are now assumed to consist
of fixed amounts of the goods, distributed between the two consumers as *initial
endowments*. Economic activity consists only of the exchange of these two goods
between the consumers. However, because we want to say something about
market economies, we assume that exchange, even in this very simple economy,
takes place through markets. Thus, the consumers face prices p_1, p_2, which they
take as outside their influence or control. Consumer j has initial endowments
$(\bar{x}_{1j}, \bar{x}_{2j})$, and the utility function $u^j(x_{1j}, x_{2j})$, $j = 1, 2$. The consumption x_{ij} may of
course differ from the endowment $\bar{x}_{ij}$, but the key step in the model is to specify
exactly how, for each consumer, these quantities must be related.

We define as the consumer's wealth v_j the value of her initial endowments at
given market prices, i.e., $v_j \equiv p_1\bar{x}_{1j} + p_2\bar{x}_{2j}$. The budget constraint is then defined
by the restriction that the value of the consumer's consumption must equal her
wealth, that is

$$p_1x_{1j} + p_2x_{2j} = v_j \implies p_1(x_{1j} - \bar{x}_{1j}) + p_2(x_{2j} - \bar{x}_{2j}) = 0 \qquad j = 1, 2. \qquad (14.21)$$

The second form of the constraint in (14.21) gives a good idea of its meaning. If
consumer j initially owns $\bar{x}_{ij}$ of good i, and ends up consuming x_{ij} of it, then she
must have made a net trade of $(x_{ij} - \bar{x}_{ij})$ – a purchase if this is positive, a sale if this
is negative. Hence, (14.21) is simply saying that the value of what she sells must
equal the value of what she buys, at *whatever* market price happens to prevail.
The consumer then chooses consumptions x_{ij} to solve

$$\max_{x_{ij}} u^j(x_{1j}, x_{2j}) \qquad \text{s.t. } p_1x_{1j} + p_2x_{2j} = v_j \qquad j = 1, 2,$$

which yields Marshallian demand functions $x_{ij} = D_{ij}(p_1, p_2, v_j)$, $i, j = 1, 2$.

Equilibrium prices are those which equate total consumers' demands to the
given available supply of each good. Thus they must satisfy, for each good i,

$$D_{i1}(p_1, p_2, v_1) + D_{i2}(p_1, p_2, v_2) = \bar{x}_{i1} + \bar{x}_{i2} \qquad i = 1, 2. \qquad (14.22)$$

Two equations in two unknowns – should be easy! However, there is a little twist
to the story. Let us show this by an example.

Consumer j has the utility function $u^j = x_{1j}^{a_j} x_{2j}^{1-a_j}$, with $0 < a_j < 1$, $j = 1, 2$.
From Chapter 6 we know that this will yield the Marshallian demand functions
$x_{1j} = a_jv_j/p_1, x_{2j} = (1 - a_j)v_j/p_2, j = 1, 2$. Thus prices must satisfy the counterparts
of (14.22):

$$a_1v_1/p_1 + a_2v_2/p_1 = \bar{x}_{11} + \bar{x}_{12}, \qquad (14.23)$$

$$(1 - a_1)v_1/p_2 + (1 - a_2)v_2/p_2 = \bar{x}_{21} + \bar{x}_{22}. \tag{14.24}$$

Now multiplying through (14.23) by p_1 and (14.24) by p_2, and substituting the definitions of v_1 and v_2

$$a_1(p_1\bar{x}_{11} + p_2\bar{x}_{21}) + a_2(p_1\bar{x}_{12} + p_2\bar{x}_{22}) = p_1(\bar{x}_{11} + \bar{x}_{12}), \tag{14.25}$$

$$(1 - a_1)(p_1\bar{x}_{11} + p_2\bar{x}_{21}) + (1 - a_2)(p_1\bar{x}_{12} + p_2\bar{x}_{22}) = p_2(\bar{x}_{21} + \bar{x}_{22}). \tag{14.26}$$

It is now straightforward to show that either one of these equations can be rearranged so that it is identical to the other (do it!). That is, we have in effect only one equation to solve for the two unknowns, and so the system is underdetermined – it appears that we cannot find equilibrium prices!

This conundrum (which is *not* just a result of the example we have chosen) was solved by the mathematical economist Leon Walras, who noticed the following. If we add the consumers' budget constraints in (14.21) we obtain

$$p_1(x_{11} + x_{12}) + p_2(x_{21} + x_{22}) = v_1 + v_2 = p_1(\bar{x}_{11} + \bar{x}_{12}) + p_2(\bar{x}_{21} + \bar{x}_{22}) \tag{14.27}$$

or, rearranging,

$$p_1[x_{11} + x_{12} - (\bar{x}_{11} + \bar{x}_{12})] + p_2[x_{21} + x_{22} - (\bar{x}_{21} + \bar{x}_{22})] = 0. \tag{14.28}$$

The terms in square brackets are the *net market demands* of the goods, the difference between total demand and total supply. Then (14.28), which is called *Walras's Law*, is saying that the sum of net market demands, valued at whatever prices generated those demands, must be exactly zero.

The reason Walras's Law must hold is easy to see. If, for each consumer, the value of a positive excess demand must be exactly offset by the value of a negative excess demand, in accordance with (14.21), then this must be true when we aggregate the net demands across consumers. The not so obvious implication is that we have not *two* demand = supply conditions to determine equilibrium prices, but only *one*. As (14.28) shows, if one market is in equilibrium with zero net demand, then so *must* the other be. This is why (14.23) and (14.24) reduce to the same condition. There is in fact only one independent condition, which we could take *either* as (14.23) *or* as (14.24).

But what about our two unknown prices? The answer is that we can determine only *one price*, namely the number of units of one good which is exchanged for one unit of the other, or the *relative price* of one good in terms of the other.

Thus, in (14.21) divide through each consumer's budget constraint by p_2. With $r \equiv p_1/p_2$, the budget constraints then become

$$rx_{1j} + x_{2j} = r\bar{x}_{1j} + \bar{x}_{2j} \equiv v_j' \qquad j = 1, 2. \tag{14.29}$$

Consumer j's Marshallian demand functions will now be $D_{ij}(r, v_j')$, and the equilibrium value of r is determined by the single supply = demand condition

$$D_{11}(r, v_1') + D_{12}(r, v_2') = \bar{x}_{11} + \bar{x}_{12}. \tag{14.30}$$

For example, in the special case in which $u^j = x_{1j}^{a_j} x_{2j}^{1-a_j}$, $0 < a_j < 1$, $j = 1, 2$, we have the condition

$$a_1 v_1'/r + a_2 v_2'/r = \bar{x}_{11} + \bar{x}_{12}. \tag{14.31}$$

Substituting for v_1' and v_2' in (14.31) gives

$$a_1(r\bar{x}_{11} + \bar{x}_{21})/r + a_2(r\bar{x}_{12} + \bar{x}_{22})/r = \bar{x}_{11} + \bar{x}_{12} \; \rightarrow \; r^* = \frac{a_1\bar{x}_{21} + a_2\bar{x}_{22}}{(1-a_1)\bar{x}_{11} + (1-a_2)\bar{x}_{12}} \tag{14.32}$$

where r^* is the equilibrium relative price. (14.32) tells us that this equilibrium relative price will be higher

(a) the higher the x_{1j}-exponents a_1 and a_2 in the consumers' utility functions, i.e., the stronger their preference for good 1 relative to good 2;
(b) the greater their initial endowments – the supply – of good 2, $\bar{x}_{21}$ and $\bar{x}_{22}$;
(c) the smaller their initial endowments – the supply – of good 1, $\bar{x}_{11}$ and $\bar{x}_{12}$;

all of which seem intuitively plausible.

Underlying the mathematics we have been doing is the question of what we mean by a 'price'. It is natural to think of the price p_1 and p_2 as expressed in terms of 'units of account' – so many £, or $, or DM per unit of goods 1 and 2 respectively. What we have just seen is that the type of general equilibrium model we have so far been dealing with in this chapter *cannot solve for prices in this sense*.

When we take the price ratio $p_1/p_2 = r$, we are expressing prices as rates of exchange – the number of units of good 2 one unit of good 1 will exchange for. We are in effect choosing good 2 as the unit of account or *numeraire* (Walras's term) in terms of which to express prices – the price of good 2 in this numeraire is of course 1. For example, if a can of beer costs $2, a pizza costs $10, and we adopt beer as the numeraire, then the relative price of a pizza is 5 cans of beer. We have just seen that the model gives us an equilibrium solution for prices expressed in this way. The model solves for equilibrium prices as rates of exchange. It follows that *any* prices in terms of dollars (or any other unit of account) are consistent with this equilibrium as long as their ratio is r^*, i.e., all the (p_1, p_2)-pairs that satisfy $p_1 = r^* p_2$ are consistent with equilibrium (except, trivially, $p_1 = p_2 = 0$). We express this by saying that in this type of model *the absolute level of prices is indeterminate*.

This equilibrium solution is illustrated in Fig. 14.3, which makes use of a very important construction known as the Edgeworth box diagram. The horizontal side of the box represents $\bar{x}_1 = \bar{x}_{11} + \bar{x}_{12}$, the total supply of good 1 to this economy. The vertical side of the box represents $\bar{x}_2 = \bar{x}_{21} + \bar{x}_{22}$, the total supply of good 2. Consumer 1's consumptions are measured from the south-west corner, origin 0_1, consumer 2's from the north-east corner, origin 0_2. A point in the box has therefore *four* coordinates, $(x_{11}, x_{21}, x_{12}, x_{22})$ with the property that $x_{11} + x_{12} = \bar{x}_1$, $x_{21} + x_{22} = \bar{x}_2$, i.e., the point is a feasible consumption allocation for the economy. The point $\bar{x} = (\bar{x}_{11}, \bar{x}_{21}, \bar{x}_{12}, \bar{x}_{22})$ in the box is the initial endowment point, where consumers are located before they enter the markets. The point $x^* = (x_{11}^*, x_{21}^*, x_{12}^*, x_{22}^*)$ is the equilibrium in the economy. The slope of the line marked E is equal to r^*, the equilibrium relative price ratio. This implies that

consumer 1 sells $\bar{x}_{21} - x_{21}^*$ of good 2 and uses the proceeds to buy $x_{11}^* - \bar{x}_{11}$ of good 1;

consumer 2 sells $\bar{x}_{12} - x_{12}^*$ of good 1 and uses the proceeds to buy $x_{22}^* - \bar{x}_{22}$ of good 2.

We then see that the quantities of each good bought and sold are equal, which we require for x^* to be an equilibrium.

Fig. 14.3 also shows an important property of this equilibrium, namely that the consumers' marginal rates of substitution are equal. This is because the line E is in fact each consumer's budget constraint, and since each consumer chooses a consumption bundle at the point of tangency of the line and an indifference curve, their indifference curves must also be mutually tangent, that is

$$\frac{dx_{21}}{dx_{11}} = \frac{-u_1^1}{u_2^1} = \frac{-p_1}{p_2} = -r^* = \frac{-u_1^2}{u_2^2} = \frac{dx_{22}}{dx_{12}} \tag{14.33}$$

at the equilibrium consumption allocation $(x_{11}^*, x_{21}^*, x_{12}^*, x_{22}^*)$.

To see that E is indeed the budget constraint for each consumer, note that it is a straight line with slope r^* passing through each consumer's initial endowment point, and so the points on it satisfy the equation

$$r^* x_{1j} + x_{2j} = r^* \bar{x}_{1j} + \bar{x}_{2j} \qquad j = 1, 2. \tag{14.34}$$

This fact about the line E also explains why in this model we can determine *only* an equilibrium relative price r, and not absolute prices p_1, p_2. The consumer's budget constraint, using (14.21) can be written out as

$$x_{2j} = \left(\bar{x}_{2j} + \frac{p_1}{p_2} \bar{x}_{1j}\right) - \frac{p_1}{p_2} x_{1j} = \left(\bar{x}_{2j} + r\bar{x}_{1j}\right) - r x_{1j}. \tag{14.35}$$

Then clearly, neither parameter of this line – the intercept, in brackets, and the slope $r = p_1/p_2$ – changes if p_1 and p_2 change proportionately. Thus the consumer's demand stays unchanged when prices change equiproportionately – only changes in *relative* prices change demands. Hence the consumers' demand functions are actually functions of *relative* price only, and so the supply = demand conditions can serve only to determine the equilibrium value of this relative price (see also question 3 of Exercise 14.2).

Finally, we can suggest how the model might be extended to determine also the absolute equilibrium price levels. Let p_1^* and p_2^* be equilibrium market prices in units of account, say \$. We know that one condition they must satisfy is

$$\frac{p_1^*}{p_2^*} = r^* \implies p_1^* = r^* p_2^* \tag{14.36}$$

where r^* is determined by condition (14.30).

Economic activity in this economy consists of each individual selling on the market an amount of one good and buying an amount of the other. Let us think of this sequentially. First, consumer 2 goes into the market for good 2 and buys $x_{22}^* - \bar{x}_{22}$ units of it at a price p_2^* (consistent with Fig. 14.3), and so she will require an amount of money equal to $\$M = \$p_2^*(x_{22}^* - \bar{x}_{22})$ to do this. Since it is consumer 1 who has sold to consumer 2, he receives these $\$M$, goes into the market for good 1, and spends $p_1^*(x_{11}^* - \bar{x}_{11})$ on buying his desired quantity of good 1. How do we know that this costs exactly $\$M$? Because

$$x_{11}^* - \bar{x}_{11} = -(x_{12}^* - \bar{x}_{12}) \qquad \text{(i.e., demand = supply for good 1)}$$

and

$$p_1^*(x_{11}^* - \bar{x}_{11}) = -p_1^*(x_{12}^* - \bar{x}_{12}) = p_2^*(x_{22}^* - \bar{x}_{22})$$

(from consumer 2's budget constraint).

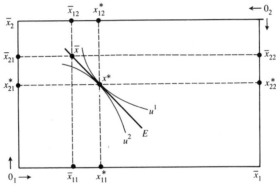

Fig. 14.3: Market equilibrium in a 2×2 pure exchange economy

Thus we have that the supply of money in existence in this economy must be large enough to enable these transactions to be carried out at these *absolute* prices p_1^*, p_2^*, that is, the supply of money must be M^*, where this satisfies

$$p_1^*(x_{11}^* - \bar{x}_{11}) + p_2^*(x_{22}^* - \bar{x}_{22}) \equiv 2M^*. \qquad (14.37)$$

The left-hand side of this identity is the value of goods bought ($\equiv$ the value of goods sold). The '2' on the right-hand side reflects the fact that $\$1$ in this economy is used twice: once by consumer 2 to buy good 2, then by consumer 1 to buy good 1. Then, it would be said that the 'velocity of circulation' of money in this economy is 2. Using (14.36) we can rewrite (14.37) as

$$p_2^*[r^*(x_{11}^* - \bar{x}_{11}) + (x_{22}^* - \bar{x}_{22})] \equiv 2M^*. \qquad (14.38)$$

The term in square brackets can be thought of as the 'real value of trade' in this economy: it is determined by *relative* prices, and is independent of the absolute price level. Denote it by T^*. Then, we have finally the identity

$$p_2^* \equiv \frac{2M^*}{T^*}, \qquad (14.39)$$

which shows that the absolute price level (fixed once p_2^* is known) is proportional to the 'money supply', M^*.

The identity in (14.39) is a version of the identity which underlies the so-called 'quantity theory of money'. The supply of money has no effect on *real* economic activity (in this case trade in goods). It determines in a simple proportional way only the absolute level of prices. These ideas will be encountered again, in extended form, in Chapter 17.

Exercise 14.2

1. Assume we have a pure exchange economy with a large number of consumers and a large number of goods. Derive Walras's Law in this case. Set out the conditions to determine equilibrium relative prices. How many should there be? Derive the counterpart of the identity in (14.38) for this economy [Hint: again this problem is largely an exercise in use of subscripts and $\sum$-notation.]
2. In the type of 2-consumer, 2-good economy considered in this section, the consumers' utility functions are, respectively,

$$u^1 = x_{11}^{1/2} x_{21}^{1/2}; \qquad u^2 = x_{12}^{1/4} x_{22}^{3/4}.$$

Consumer 1 is endowed with 20 units of good 1 and 10 units of good 2, consumer 2 with 10 units of each good.
   (*a*) Find the equilibrium relative price.
   (*b*) Confirm that individuals' marginal rates of substitution are equal at the equilibrium allocation.
   (*c*) If the money supply is \$1000, what are absolute equilibrium prices in this economy?
   (*d*) Sketch the counterpart of Figure 14.3 for this case.
3. A function $y = f(x_1, x_2)$ is homogeneous of degree n if $\lambda^n y = f(\lambda x_1, \lambda x_2)$. On the basis of the discussion of equation (14.25), explain why Marshallian demand functions are homogeneous of degree zero in prices.
4. Explain why it might be said in connection with the model of this section, 'money is a veil'.
5. Suppose that consumers *simultaneously* buy and sell goods in this economy. What will be the velocity of circulation of money? Write down the counterpart of (14.38) to determine absolute prices.

14.3 Preferences, production, and prices

The assumption that supplies of goods take the form of given initial endowments is clearly unrealistic, but allowed us to develop in a simple context some important ideas, in particular Walras's Law, and the distinction between the determination of relative and absolute prices. We now construct a model in which supply is the result of production; in effect we combine the models of the previous two sections to determine endogenously the wage rate *and* prices. As in the first section, we have two firms, each producing one output, with labour hired on a competitive market. There are two consumers, each of whom supplies L units of labour independently of the wage rate, owns a share in each firm, and chooses a consumption bundle which maximizes utility subject to a budget constraint. The income defining this constraint consists of wage income plus shares in companies' profits. We could also assume consumers possess initial endowments of goods, but to lighten notation we assume these are zero. In the labour and output markets, buyers and sellers act as price takers.

For any given wage rate and prices, firms choose outputs and labour inputs to solve

$$\max_{x_i, l_i} \ p_i x_i - w l_i \qquad \text{s.t. } x_i = f^i(l_i),$$

which gives supply functions for outputs $x_i = S_i(p_i, w)$, and labour demand functions $l_i = l_i(p_i, w)$, $i = 1, 2$. This implies that the profits π_i can be written

$$\pi_i = p_i S_i(p_i, w) - w l_i(p_i, w) = \pi^i(p_i, w) \qquad i = 1, 2 \qquad (14.40)$$

to emphasize that, via the maximization process, profits are essentially a function of prices and the wage rate. Two important things to note about these functions are:

(a) output x_i and labour demand l_i are unaffected by equiproportionate changes in prices and the wage rate, since the condition that determines profit-maximizing l_i and x_i is

$$p_i f^i_l(l_i) = w \ \Rightarrow \ f^i_l(l_i) = w/p_i$$

and so only changes in prices *relative to* the wage rate cause changes in output and employment. That is, the labour demand and output supply functions are homogeneous of degree zero (recall question 3 of Exercise 14.2).

(b) A given, equal proportionate change in prices and the wage rate increases profits in the same proportion. To see this, replace w and p_i in (14.40) by λw and λp_i, $\lambda > 0$, and note that profit changes to $\lambda \pi_i$. Thus $\lambda \pi_i = \pi^i(\lambda p_i, \lambda w)$ and the profit function is homogeneous of degree one in prices and the wage.

Turning to the consumers' choices of consumption bundles, the budget constraints and their properties are the crux of the analysis. As in section 14.1, these are written as

$$p_1 x_{1j} + p_2 x_{2j} = y_j \equiv wL + \theta_{1j} \pi_1 + \theta_{2j} \pi_2 \quad j = 1, 2. \qquad (14.41)$$

The first thing to note is that the budget constraint is again unaffected by equiproportionate changes in prices and the wage rate. If w and the p_i are multiplied by $\lambda > 0$, then, as we just saw, π_1 and π_2 increase to $\lambda \pi_1$ and $\lambda \pi_2$, and so y_j increases to λy_j. Then the budget constraint $\lambda p_1 x_{1j} + \lambda p_2 x_{2j} = \lambda y_j$ clearly defines exactly the same set of consumption bundles (x_{1j}, x_{2j}) as that in (14.41). It follows that when we solve

$$\max \ u^j(x_{1j}, x_{2j}) \qquad \text{s.t. } p_1 x_{1j} + p_2 x_{2j} = y_j$$

to obtain the Marshallian demand functions $x_{ij} = D_{ij}(p_1, p_2, y_j)$, these demand functions are homogeneous of degree zero in prices and the wage rate – equiproportionate changes in these variables leave demands unchanged.

Secondly, if we sum the consumers' budget constraints we have

$$p_1(x_{11} + x_{12}) + p_2(x_{21} + x_{22}) = w2L + \pi_1 + \pi_2 \qquad (\text{using } \theta_{i1} + \theta_{i2} = 1)$$

$$= p_1 x_1 + p_2 x_2 + w(2L - l_1 - l_2)$$

$$(\text{using the definition of profit})$$

or

$$p_1(x_{11} + x_{21} - x_1) + p_2(x_{21} + x_{22} - x_2) + w(l_1 + l_2 - 2L) = 0. \qquad (14.42)$$

The terms in brackets in (14.42) are the net market demands for the two goods and labour, respectively. Thus (14.42) is the appropriate statement of Walras's Law in this model. Note again its implications: if any two of three markets in the economy are in equilibrium, with zero net market demands, then *so must the third be*.

So, just as in section 14.2, we have that only *relative* prices matter, and we have one fewer supply = demand condition – because of Walras's Law – than the number of markets in the economy. To proceed we simply choose a numeraire in terms of which to express relative prices and use two market clearing conditions to determine these. Let us choose labour as the numeraire. In effect we are setting $w = 1$, and solving for the relative prices $p_1/w \equiv r_1$, $p_2/w \equiv r_2$. With this change in notation we can write the supply = demand conditions as

$$D_{11}(r_1, r_2, y_1') + D_{12}(r_1, r_2, y_2') - S_1(r_1, 1) = 0, \qquad (14.43)$$

$$D_{21}(r_1, r_2, y_1') + D_{22}(r_1, r_2, y_2') - S_2(r_2, 1) = 0. \qquad (14.44)$$

Note that the incomes $y_j' \equiv y_j/w$ are not separate independent variables but themselves simply functions of r_1 and r_2:

$$y_j' = L + \theta_{1j}(r_1 S_1(r_1, 1) - l_1(r_1, , 1)) + \theta_{2j}(r_2 S_2(r_2, 1) - l_2(r_2, 1)), \qquad (14.45)$$

where the terms in brackets are the profit functions expressed in *relative prices* ($w = 1, r_i = p_i/w$).

Thus (14.43) and (14.44) give us two equations in the two unknowns, r_1 and r_2. The underlying parameters which determine this solution are: the fixed labour supplies L; the initial endowments of shares θ_{ij}; the parameters of the utility and production functions, which determine the demand and supply functions.

Thus, to the question: what determines relative prices and the corresponding resource allocation in this model? we would give the above list. However, it is possible to make the supply of labour endogenous (see question 2 of Exercise 14.3) and so we can reduce the fundamental determinants of a competitive market resource allocation to just two sets of parameters: initial endowments of wealth, which we take to be exogenously determined, perhaps by past saving and inheritance; and the nature of preferences and technology in the economy, which again we take as exogenous data.

An Example

We now set up a particular model of this economy to reinforce these points. Firm i's production function is

$$x_i = l_i^{\beta_i} \qquad 0 < \beta_i < 1, \qquad i = 1, 2.$$

Consumer j's utility function is

$$u^j = x_{1j}^{a_j} x_{2j}^{1-a_j} \qquad 0 < a_j < 1, \qquad j = 1, 2.$$

Note that these are the only functions we need to specify. Everything else follows from these functions and the values of L and θ_{ij}, $i, j = 1, 2$.

Since we set $w = 1$, we write the firms' profit functions as

$$\pi'_i = r_i l_i^{\beta_i} - l_i \qquad i = 1, 2,$$

where the prime denotes the fact that profits are measured with labour as the numeraire. Maximizing π'_i with respect to l_i yields the labour demand function (check the derivation)

$$l_i = (\beta_i r_i)^{1/1-\beta_i} \qquad i = 1, 2$$

and, by substituting into the production function, the output supply function

$$x_i = (\beta_i r_i)^{\beta_i/1-\beta_i} \qquad i = 1, 2.$$

Hence the firm's profit function is

$$\pi'_i = r_i x_i - l_i = \gamma_i r_i^{\sigma_i} \qquad i = 1, 2$$

where $\gamma_i \equiv (\beta_i^{-1} - 1)\beta_i^{\sigma_i}$; $\sigma_i \equiv 1/(1 - \beta_i)$; $i = 1, 2$ (be sure you agree with the manipulations of the exponents). Note that each firm's profit is an increasing function of the relative output price, as we would expect.

We know that for the type of utility function specified here consumer j's demand functions are

$$x_{1j} = a_j y'_j / r_1; \qquad x_{2j} = (1 - a_j) y'_j / r_2 \qquad j = 1, 2$$

where

$$y'_j = L + \theta_{1j} \gamma_1 r_1^{\sigma_1} + \theta_{2j} \gamma_2 r_2^{\sigma_2} \qquad j = 1, 2.$$

Thus we have, as the market clearing conditions, the counterparts of (14.43) and (14.44)

$$a_1 y_1'/r_1 + a_2 y_2'/r_1 - x_1 = 0 \Rightarrow a_1 y_1' + a_2 y_2' - r_1 x_1 = 0,$$

$$(1 - a_1) y_1'/r_2 + (1 - a_2) y_2'/r_2 - x_2 = 0 \Rightarrow (1 - a_1) y_1' + (1 - a_2) y_2' - r_2 x_2 = 0.$$

Substituting for the y_j' and x_i and rearranging gives

$$(a_1 + a_2)L + \{(a_1 \theta_{11} + a_2 \theta_{12}) \gamma_1 - \delta_1\} r_1^{\sigma_1} + (a_1 \theta_{21} + a_2 \theta_{22}) \gamma_2 r_2^{\sigma_2} = 0,$$

$$(2 - (a_1 + a_2))L + \{1 - (a_1 \theta_{11} + a_2 \theta_{12})\} \gamma_1 r_1^{\sigma_1} + \{(1 - (a_1 \theta_{21} + a_2 \theta_{22})) \gamma_2 - \delta_2\} r_2^{\sigma_2} = 0,$$

where $\delta_i \equiv \beta_i^{\beta_i/1 - \beta_i}$.

As a first step to interpreting these equations, let us assume that the consumers have identical preferences, so that $a_1 = a_2 = a$. Then, recalling that $\theta_{i1} + \theta_{i2} = 1$, $i = 1, 2$, the equations become quite simply

$$2aL + (a\gamma_1 - \delta_1) r_1^{\sigma_1} + a\gamma_2 r_2^{\sigma_2} = 0,$$

$$2(1 - a)L + (1 - a)\gamma_1 r_1^{\sigma_1} + ((1 - a)\gamma_2 - \delta_2) r_2^{\sigma_2} = 0.$$

Thus the share ownership parameters drop out and do not determine the resource allocation. The reason for this is that share ownership determines the distribution of income, and this affects the resource allocation only if consumers have different preferences: if their preferences are the same it does not matter how income is divided between them (though of course this does determine the final utility levels of the consumers).

If we define the variables $Z_i \equiv r_i^{\sigma_i}$, we note that the above equations are linear in the Z_i. We can write the system as

$$(a\gamma_1 - \delta_1)Z_1 + a\gamma_2 Z_2 = -2aL,$$

$$(1 - a)\gamma_1 Z_1 + ((1 - a)\gamma_2 - \delta_2)Z_2 = -2(1 - a)L.$$

We can then solve for the Z_i using Cramer's Rule:

$$r_1^{\sigma_1} \equiv Z_1 = 2a\delta_2 L/|A| \Rightarrow r_1^* = (2a\delta_2 L)^{1/\sigma_1}/|A|^{1/\sigma_1},$$

$$r_2^{\sigma_2} \equiv Z_2 = 2(1 - a)\delta_1 L/|A| \Rightarrow r_2^* = (2(1 - a)\delta_1 L)^{1/\sigma_2}/|A|^{1/\sigma_2}.$$

Thus we have explicit solutions for the equilibrium relative prices, with the determinant $|A|$ given by

$$|A| = (1 - a)\delta_1 \beta_2^{\sigma_2} + a\delta_2 \beta_1^{\sigma_1} > 0$$

(it takes a little algebra to establish that but you should work through it none the less).

To explore the influence of the parameters on these equilibrium relative prices we take only the expression for r_1^*, the discussion for r_2^* being entirely symmetrical. Note that in the expression for Z_1, the term $a\delta_2$ occurs in both numerator and denominator. Hence we obtain

$$Z_1 = 2L/(\beta_1^{\sigma_1} + (1 - a)\delta_1\beta_2^{\sigma_2}/a\delta_2).$$

Recall that $\delta_i = \beta_i^{\beta_i/1 - \beta_i}$, $\sigma_i = 1/(1 - \beta_i)$, and so the denominator simplifies in the following way:

$$Z_1 = 2L/\beta_1^{\sigma_1}(1 + (1 - a)\beta_2/a\beta_1).$$

Thus

$$r_1^* = Z_1^{1/\sigma_1} = \{2L/\beta_1^{\sigma_1}(1 + (1 - a)\beta_2/a\beta_1)\}^{1-\beta_1}.$$

The following parameter effects then follow immediately:

(a) An increase in the labour supply L increases r_1^* (and r_2^*). This may seem counterintuitive: an increase in labour supply will reduce the wage rate and firms' costs, and so we would expect prices to fall. Remember, however, that this is a general equilibrium model with labour as the numeraire in the *relative* prices r_i. An increase in labour supply reduces the wage rate relative to output prices, hence the r_i^* increase.

(b) An increase in a, representing an increase in the strength of consumers' preferences for good 1, increases r_1^* (and reduces r_2^*). To see this result, which is perfectly intuitive, note that an increase in a reduces $(1 - a)\beta_2/a\beta_1$, and therefore reduces the denominator in the expression for r_1^*.

(c) An increase in β_2 reduces r_1^*, and this effect will be greater, the stronger the preference for good 2. Thus, in the expression for r_1^*, an increase in β_2 increases $(1 - a)\beta_2/a\beta_1$, therefore increases the denominator and reduces r_1^*. Intuitively, an increase in productivity in production of good 2 reduces its marginal cost, reduces its price, therefore reduces demand for good 1, therefore reduces *its* price. An increase in productivity reduces the prices of outputs relative to the wage. A given amount of labour buys more goods.

(d) The effect of an increase in β_1 is to reduce r_1^*, again as we would expect. The way in which β_1 enters into the expression for r_1^* is obviously quite complicated and so this effect has to be derived more formally. Question 5 of Exercise 14.3 guides you through this derivation.

The assumption that the consumers' preferences are identical clearly simplified these results a great deal. If we relax that assumption then the distribution of income influences equilibrium prices. Thus the solutions for these prices will depend on the specific shareholdings the consumers are endowed with as well as the equilibrium profits of firms. However, the procedure we have just used can still be adopted to solve for the Z_i and hence the r_i^*. Rather than pursue that here, we leave it as an exercise (see question 4 of Exercise 14.3).

Returning now to the general model for which the equilibrium conditions are (14.43) and (14.44), we note that the following conditions are satisfied at the equilibrium resource allocation x_{ij}^*, x_i^*, l_i^*.

(i) Since each firm is maximizing profit and faces the same equilibrium wage rate, we have

$$r_1^* f_I^1 = r_2^* f_I^2 \Rightarrow f_I^2 / f_I^1 = r_1^* / r_2^*. \tag{14.46}$$

That is, the property, discussed at some length in section 14.1, of equality of marginal value products across firms, carries over to the present model.

(ii) Since each consumer is maximizing utility and faces the same equilibrium output prices, we have

$$-\frac{dx_{21}}{dx_{11}} = \frac{u_1^1}{u_2^1} = \frac{r_1^*}{r_2^*} = \frac{u_1^2}{u_2^2} = -\frac{dx_{22}}{dx_{12}}. \tag{14.47}$$

That is, marginal rates of substitution are equalized across consumers, a result we discussed in section 14.2.

Putting (14.46) and (14.47) together shows that a third condition is also satisfied at the equilibrium:

$$\frac{u_1^j}{u_2^j} = \frac{r_1^*}{r_2^*} = \frac{f_I^2}{f_I^1} \qquad j = 1, 2. \tag{14.48}$$

That is, each consumer's marginal rate of substitution is equal to the economy's marginal rate of transformation. The meaning and importance of this result is at the moment not at all obvious. In Chapter 15 however, we shall explore this in some depth.

Exercise 14.3

1. Assume that the economy consists of a large number of consumers, each of whom supplies L units of (identical) labour; and a large number of outputs, each of which is supplied by a large number of firms. Extend the model of this section to this case. What are the 'ultimate determinants' of relative prices in this economy?

2. Suppose that instead of supplying a fixed amount of labour regardless of the wage, households choose their labour supplies. Extend the model of this section to take account of that. [Hint: include the consumer's labour supply as an argument in her utility function, derive a labour supply function, show that it is homogeneous of degree zero in prices and the wage, and include the labour supply functions in the labour market equilibrium condition.]

3. Suppose that in the economy of this question there is a given stock of money, and that consumers first receive payments of wages and profits from firms, and then spend these incomes on goods. Show how the absolute level of equilibrium prices, in units of account, is determined in this economy. [Hint: refer back to the discussion of this point in section 14.2.]

4. In an economy of the type considered in this section, the consumers' utility functions are

$$u^1 = x_{11}^{1/2} x_{21}^{1/2} ; \quad u^2 = x_{12}^{1/4} x_{22}^{3/4}.$$

Each consumer supplies 50 units of labour. The firms' production functions are

$$x_1 = l_1^{1/2} ; \quad x_2 = l_2^{1/4}.$$

Solve for equilibrium relative prices on the assumptions that:

(*a*) consumer 1 owns all of firm 1, and consumer 2 owns all of firm 2;

(*b*) each consumer owns half of each firm;

(*c*) consumer 1 owns all of both firms.

Discuss and explain the differences in equilibrium prices in these three cases. Compare the values of the consumers' utilities in the three cases and explain the differences.

5. Given the expression

$$Z_1 = 2L/\beta_1^{\sigma_1}(1 + (1 - a)\beta_2/a\beta_1)$$

and that $r_1^* = Z_1^{1/\sigma_1}$, where $\sigma_1 = 1/(1 - \beta_1)$, show that an increase in β_1 reduces r_1^*. [Hint: note that the sign of dr_1^* is the same as the sign of $d\log r_1^*$. Since $\log r_1^* = \log Z_1/\sigma_1 = (1 - \beta_1)\log Z_1$, we have $d\log r_1^* = (1 - \beta_1)d\log Z_1 - \log Z_1 d\beta_1$. So we need to prove that $d\log Z_1/d\beta_1 < 0$. Note: $\log Z_1 = \log(2L) - \log[\beta_1^{\sigma_1}(1 + (1 - a)\beta_2/a\beta_1]$. So you just have to show that $d\log[\beta_1^{\sigma_1}(1 + (1 - a)\beta_2/a\beta_1]/d\beta_1 > 0$. The key is to show that $d\beta_1^{\sigma_1}/d\beta_1 > 0$.]

6. Explain why we can interpret the relative prices r_i in this section as the number of hours (or other time unit) of work it takes to earn enough to buy one unit of good i.

14.4 Summary

General equilibrium analysis is concerned with the way in which markets interact to determine an allocation of resources and a set of prices in the economy as a whole.

Taking output prices as fixed (e.g., on 'world markets'), and assuming consumers and firms act as price-takers – markets are perfectly competitive – an equilibrium allocation of labour between firms implies a wage rate equal to the marginal value product of labour in each firm.

The distribution of income is determined by consumers' shares in the profits of firms, assumed exogenously given, and firms' profits. The level of income is determined by the productivity of labour in producing output and the values of the exogenous output prices. The production possibility curve shows the set of output pairs available to the economy when it uses available resources to their fullest extent. Its slope is the marginal rate of transformation between outputs.

Solving the problem of maximizing the value of the economy's output at world prices, subject to a constraint represented by the production possibility curve, gives the same resource allocation as that achieved by the competitive equilibrium.

Consumers' choices of consumption bundles in this economy determine its exports and imports. As long as consumers' choices satisfy their budget constraints, the economy's trade will be in balance.

In a pure exchange economy, where consumers have fixed initial endowments of goods, the relative prices of goods are determined by these endowments and by consumers' preferences. This model brings out clearly the importance of Walras's Law, and of the fact that the condition that supply and demand for each good be equal determines only relative and not absolute prices.

Walras's Law says that at any prices, as long as consumers' choices satisfy their budget constraints, the values of net demands for goods sum to zero. The implication of this is that if all markets but one are in equilibrium with zero net demand, this one market must be in equilibrium. The further implication of this is that there is one fewer supply = demand condition to determine prices than there are markets in the economy.

The apparent under-determination of the unknown equilibrium prices is resolved by noting that one good can be chosen as the numeraire, its price set to 1, and any other price expressed in terms of the number of units of the numeraire good which can be exchanged for one unit of the good in question. The system of supply = demand conditions then solves for these relative prices or rates of exchange.

The absolute prices in units of account or 'money' can be found by noting that the supply of money must be proportional to the absolute price of the numeraire, if the equilibrium trades are to be financed. Equivalently, absolute prices are at whatever general level is necessary to permit the equilibrium trades to be financed by the given supply of money. These trades are, however, determined independently of the money supply.

When we move to a general model in which outputs and their prices are determined endogenously, we see that the equilibrium resource allocation and prices depend on preferences, productivities and the initial endowments of shares (assuming labour supplies are endogenous). The central features of the two simpler models continue to hold: the supply = demand conditions determine *relative* prices, Walras's Law holds and only absolute price levels are determined by the money supply.

At the competitive market equilibrium resource allocation, because consumers and firms all face the same relative prices, marginal rates of substitution are equal across consumers, marginal value products of labour are equal across firms, and the marginal rate of transformation between outputs is equal to the common consumers' marginal rate of substitution. These results will be shown in the next chapter to have some importance for the evaluation of the optimality of a competitive equilibrium resource allocation.

15   Welfare Economics

Welfare economics is that area of economics which attempts to assess the efficiency of resource allocation systems and to provide a basis for rational economic policy. Both these concerns are reflected in this chapter. First, we examine two propositions, known as the *fundamental propositions of welfare economics*, which attempt to characterize in as precise a way as possible the efficiency properties of the competitive market mechanism. Then, we go on to consider the economic analysis of a range of issues – monopoly, taxation, pollution, tariffs, depletion of natural resources – which provides the basis for the policy prescriptions economists have made in these areas. A notable feature of this chapter is the change from the general equilibrium level at which the two fundamental propositions are discussed, to the partial equilibrium level at which the specific policy issues are analysed. Section 15.2 is very important in providing the link between these two levels.

15.1 The two fundamental propositions of welfare economics

Adam Smith's famous proposition concerning the 'invisible hand' can be put for our purposes in the following way: an economic system in which individual decision-takers (consumers and firms) pursue their own selfish interests (utility maximization, profit maximization) without regard for the 'wider public interest' can, surprisingly perhaps, achieve an optimal allocation of resources in the economy as a whole, when their decisions are co-ordinated through a competitive market mechanism. Put more succinctly: a competitive market equilibrium achieves a Pareto efficient resource allocation. This requires us to explain what is meant by a 'Pareto efficient resource allocation'.

A resource allocation is a set of quantities: of consumption of each good and supply of each input by each consumer; and use of each input and output of each good by each firm. Given some initial resource allocation, if it were possible to change it in such a way that at least one consumer became better off and no other consumer became worse off, it seems a reasonable value judgement to say that such a change would be worth making. A *Pareto efficient allocation* is then a resource allocation which cannot be improved upon in this way: it is only possible to make one consumer better off by making another worse off. The concept bears the name of the Italian economist and sociologist Vilfredo Pareto, who is credited with having formulated it.

To interpret this mathematically, we take the two-consumer, two-good, one-input model of section 14.3. We wish to find a Pareto efficient allocation of resources in this economy, or, more generally, to characterize such a resource allocation by the conditions it must satisfy. Given the utility functions of the individuals, $u^j(x_{1j}, x_{2j})$, $j = 1, 2$, we find a Pareto efficient allocation by maximizing the utility of one individual, say 1, for any given level of utility of the other, 2. This means that we find a resource allocation such that 1 cannot be made better off given 2's utility level. The choice of the given level of 2's utility is essentially arbitrary, as long as it is feasible for this economy. This suggests that choosing different utility levels for 2 will generate different Pareto efficient allocations, so that the criterion that a resource allocation be Pareto efficient is not strong enough to define a *unique* resource allocation for the economy. We shall now show this in an example.

Take a special case of our economy in which the aggregate outputs of the two goods are each fixed at 10 units. Each individual has the utility function $u^j = x_{1j}^{1/2} x_{2j}^{1/2}$, $j = 1, 2$. Set 2's utility level arbitrarily at $\bar{u}_2$. Then we can formulate the problem of finding a Pareto efficient resource allocation in this economy as

$$\max_{x_{ij}} u^1$$

$$\text{subject to } u^2 = \bar{u}_2 \text{ and} \tag{15.1}$$

$$x_{11} + x_{12} = 10,$$

$$x_{21} + x_{22} = 10.$$

That is, we maximize 1's utility for a given level of 2's utility *and* subject to the constraint that the consumptions we choose are feasible for this economy. Note that we must require that $\bar{u}_2 \leq 10 (= 10^{1/2} 10^{1/2})$ as well as being non-negative (explain why).

The Lagrange function for this problem is

$$L = x_{11}^{1/2} x_{21}^{1/2} + \lambda \left[x_{12}^{1/2} x_{22}^{1/2} - \bar{u}_2 \right] + \rho_1 [10 - x_{11} - x_{12}] + \rho_2 [10 - x_{21} - x_{22}] \tag{15.2}$$

and the necessary conditions are

$$\frac{\partial L}{\partial x_{11}} = (1/2) x_{11}^{-1/2} x_{21}^{1/2} - \rho_1 = 0,$$

$$\frac{\partial L}{\partial x_{21}} = (1/2) x_{11}^{1/2} x_{21}^{-1/2} - \rho_2 = 0; \tag{15.3}$$

$$\frac{\partial L}{\partial x_{12}} = \lambda (1/2) x_{12}^{-1/2} x_{22}^{1/2} - \rho_1 = 0,$$

$$\frac{\partial L}{\partial x_{22}} = \lambda (1/2) x_{12}^{1/2} x_{22}^{-1/2} - \rho_2 = 0; \tag{15.4}$$

together with all three constraints in (15.1). Taking the ρs to the right-hand sides of these conditions, taking the ratios of conditions (15.3) and then of conditions (15.4), and cancelling terms wherever possible gives

$$\frac{x_{21}}{x_{11}} = \frac{\rho_1}{\rho_2} = \frac{x_{22}}{x_{12}}. \tag{15.5}$$

We then have (15.5) and the three constraints in (15.1) to determine our four unknowns ((15.5) is in fact the condition that the consumers' marginal rates of substitution are equal, as we shall see more generally below).

Using the feasibility constraints we can re-write (15.5) as

$$\frac{(10 - x_{22})}{(10 - x_{12})} = \frac{x_{22}}{x_{12}} \tag{15.6}$$

and cross-multiplication gives

$$x_{12} = x_{22}. \tag{15.7}$$

Substituting from (15.7) into the utility constraint gives

$$x_{12}^{1/2} x_{22}^{1/2} = \bar{u}_2 \implies x_{22} = \bar{u}_2 = x_{12} \tag{15.8}$$

and so from the feasibility constraints we have

$$x_{11} = x_{21} = 10 - \bar{u}_2. \tag{15.9}$$

Thus, given any numerical value for the utility constraint $\bar{u}_2$ in the interval [0, 10], we immediately have from (15.8) and (15.9) the Pareto efficient values of the consumption allocations, x_{ij}.

Substituting from (15.9) into the expression for 1's utility function gives

$$u^1 = (10 - \bar{u}_2)^{1/2}(10 - \bar{u}_2)^{1/2} = 10 - \bar{u}_2. \tag{15.10}$$

This equation gives us the maximum achievable level of 1's utility for each possible level of $\bar{u}_2$ in [0, 10]. It is graphed in Fig. 15.1, and the line is usually referred to as the *utility possibility frontier*, since it shows the upper bound of the set of utility pairs which are possible or feasible in this economy. Clearly, the point (0, 10) corresponds to 2 receiving all of the available quantities of the goods; (10, 0) to 1 receiving everything, and the remaining points on the line to divisions of the goods between them.

Though this example is obviously very special (which makes it easy to derive the explicit expression for the utility possibility frontier in (15.10)), we can illustrate some important general points about Pareto efficient allocations and their resulting utility distributions with it.

First, the negative slope of the line indicates that, in the set of Pareto efficient allocations, 1 can only be made better off by making 2 worse off, as we would expect.

Second, as we suggested earlier, there is no unique Pareto efficient allocation: there is an infinity of them, with potentially large variation in the distribution of utility between the two individuals. To choose some specific allocation from this 'menu' of Pareto efficient allocations would require some explicit ranking of utility distributions, which in turn requires us to make 'interpersonal comparisons of utility' – we must be prepared to evaluate gains in utility to one individual against losses in utility to the other.

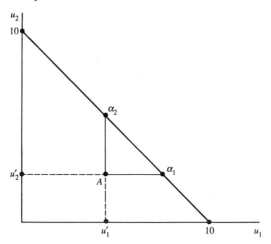

Fig. 15.1: Utility frontier for a special case

What is the significance of a utility distribution such as that at point A in Fig. 15.1? It is clearly Pareto inefficient, since at least one individual, and possibly both, can be made better off by moving to the line segment $\overline{\alpha_1 \alpha_2}$. But what causes this inefficiency? A first guess might be that the individuals are receiving in total less than the available amounts of the two goods at that point. Though a possible explanation, there is a more interesting one: such a point corresponds to non-satisfaction of the necessary condition (15.5). To see this consider Fig. 15.2, which shows an Edgeworth box with sides corresponding to available total consumptions of each good, and with 1's indifference curves drawn with respect to origin 0_1 and 2's with respect to 0_2. The diagonal line in the box, along which $x_{11} = x_{21}$ and $x_{12} = x_{22}$, is the *contract curve*, the locus of tangencies of the indifference curves. To see this, note that individual j's marginal rate of substitution in this example is

$$-\frac{dx_{2j}}{dx_{1j}} = \frac{u_1^j}{u_2^j} = \frac{(1/2)x_{1j}^{-1/2}x_{2j}^{1/2}}{(1/2)x_{1j}^{1/2}x_{2j}^{-1/2}} = \frac{x_{2j}}{x_{1j}} \qquad j = 1, 2$$

and condition (15.5) is then simply requiring the equalization of these marginal rates of substitution. Thus the set of consumption allocations lying along the contract curve in Fig. 15.2 generates the set of utility pairs along the utility possibility frontier in Fig. 15.1 – we simply read off the utility values of the indifference curves tangent to each other on the contract curve. Then point A in Fig. 15.1 could be generated by a point such as A' in Fig. 15.2, which has the sum of the individuals' consumptions equal to the total available, but is not on the contract curve – the individuals' marginal rates of substitution are unequal at A'. Then, by moving along u_1' to α_2', or along u_2' to α_1', or to anywhere on the line segment $\overline{\alpha_1' \alpha_2'}$, we make at least one individual, and possibly both, better off. This is the counterpart of moving from A to $\overline{\alpha_1 \alpha_2}$ in Fig. 15.1.

Of course, point A in Fig. 15.1 *could* have been generated simply by throwing away consumption: if we put individual 1 at point α_2' and 2 at α_1' in Fig. 15.2, this would generate the same utility pair as point A' but would involve the 'waste' of $x_{12}' - x_{11}'$ units of x_1 and $x_{22}' - x_{21}'$ units of x_2. This illustrates very well that failure to satisfy the necessary conditions for Pareto efficiency is equivalent, in terms of the *achieved distribution of utilities*, to simply throwing away resources.

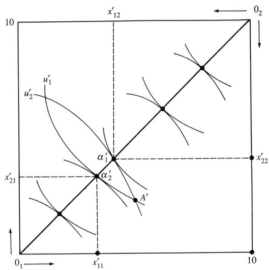

Fig. 15.2: Pareto inefficiency in consumption allocation

This simple example should have brought out the central points concerning the meaning of Pareto efficiency. We now return to our model of the economy with production, and the first fundamental theorem of welfare economics.

We wish to find a Pareto efficient allocation of resources in this economy. Given the labour input requirements of the firms, l_i, $i = 1, 2$, we impose the constraint

$$\sum_{i=1}^{2} l_i = 2L \tag{15.11}$$

where $2L$ is the total available labour supply. We also have to accept that the consumption of each good by each consumer x_{ij}, $i, j = 1, 2$ must in aggregate equal the available supply of the good, the output of the firm concerned:

$$\sum_{j=1}^{2} x_{ij} = x_i \qquad i = 1, 2. \tag{15.12}$$

The technological possibilities of transforming labour into outputs are described by the production functions

$$x_i = f^i(l_i) \qquad i = 1, 2, \tag{15.13}$$

which are also constraints on choice of resource allocation. (15.11)–(15.13) are then the basic constraints on any resource allocation mechanism: employment cannot exceed available labour supply, consumption cannot exceed production, and input–output possibilities are technologically determined by the production functions.

To find a Pareto efficient allocation, again we set consumer 2's utility at $\bar{u}_2$, and solve the problem

$$\text{max } u_1 = u^1(x_{11}, x_{21}) \tag{15.14}$$

$$\text{s.t. } u_2 = u^2(x_{12}, x_{22}) = \bar{u}_2 \tag{15.15}$$

and subject also to the constraints (15.11)–(15.13). The Lagrange function is

$$L = u^1(x_{11}, x_{21}) + \lambda(u^2(x_{12}, x_{22}) - \bar{u}_2) + \mu \left(2L - \sum_{i=1}^{2} l_i\right)$$

$$+ \sum_{i=1}^{2} \rho_i \left(x_i - \sum_{j=1}^{2} x_{ij}\right) + \sum_{i=1}^{2} \beta_i \left(f^i(l_i) - x_i\right) \tag{15.16}$$

and the first-order conditions with respect to the variables x_{ij}, l_i, x_i are

$$\partial u^1/\partial x_{i1} - \rho_i = 0 \qquad i = 1, 2, \tag{15.17}$$

$$\lambda \partial u^2/\partial x_{i2} - \rho_i = 0 \qquad i = 1, 2, \tag{15.18}$$

$$\beta_i f_l^i - \mu = 0 \qquad i = 1, 2, \tag{15.19}$$

$$\rho_i - \beta_i = 0 \qquad i = 1, 2, \tag{15.20}$$

together with the constraints (15.11)–(15.13) and (15.15). Taking the ratio of the two conditions in (15.17), and the similar ratio of the two conditions in (15.18), we obtain

$$\frac{\partial u^1/\partial x_{11}}{\partial u^1/\partial x_{21}} = \frac{\rho_1}{\rho_2} = \frac{\partial u^2/\partial x_{12}}{\partial u^2/\partial x_{22}}. \tag{15.21}$$

That is, a necessary condition for Pareto efficiency is that the consumers' marginal rates of substitution be equal. Taking the two conditions in (15.19) and substituting for the β_i from (15.20) gives

$$\rho_1 f_l^1 = \mu = \rho_2 f_l^2 \implies \rho_1/\rho_2 = f_l^2/f_l^1. \tag{15.22}$$

Recall that in Chapter 14 we called the ratio f_l^2/f_l^1 the marginal rate of transformation between outputs x_1 and x_2. Then (15.21) and (15.22) together imply that a necessary condition for Pareto efficiency is that consumers' marginal rates of substitution equal the marginal rate of transformation.

Thus, to summarize, it is necessary for a resource allocation x_{ij}, l_i, x_i to be

Pareto efficient that it satisfy constraints (15.11)–(15.13) and (15.15), that the allocation of consumptions between individuals be such that their marginal rates of substitution are equal, and the allocation of labour between firms – and the corresponding output pair – be such that the marginal rate of transformation equal this common marginal rate of substitution.

Fig. 15.3 is the generalization of Fig. 15.2 for this economy, and illustrates a Pareto efficient allocation. The indifference curve $\bar{u}_2$ corresponds to the required utility level for consumer 2. The slope of the line R is the ratio ρ_1/ρ_2, the common marginal rate of substitution, P is the production possibility curve, and the slope of the line T is $\rho_1/\rho_2 = f_l^2/f_l^1$, the marginal rate of transformation. The quantities x_{ij}^*, x_i^* represent a Pareto efficient allocation (and l_i^* the corresponding efficient labour allocations). The construction under the production possibility curve is an Edgeworth box.

The assumptions we have made on the consumers' utility functions and firms' production functions ensure that conditions (15.17)–(15.20) are *sufficient* as well as *necessary* for a Pareto efficient resource allocation. Thus any allocation that satisfies them is Pareto efficient. But recall from Chapter 14 that at the equilibrium of a competitive market economy these conditions are indeed satisfied: because consumers face the same price ratio and maximize utility, an equilibrium resource allocation must be one in which their marginal rates of substitution are equal; because firms face the same wage rate for labour, face the same output prices as consumers, and maximize profit taking the wage and prices as given, their employment and output decisions imply equality of marginal value products of labour with the wage, and this implies satisfaction of the condition of equality of the marginal rates of substitution with the marginal rate of transformation. Thus the competitive market allocation is Pareto efficient.

Note that the choice of consumer 2's level of utility $\bar{u}_2$ in constraint (15.15) was again essentially arbitrary, but that the basic form of the conditions (15.17)–(15-20) is unaffected by this: choice of $\bar{u}_2$ would determine the precise numerical solution to the optimization problem but not the necessary conditions for that solution. Let $\bar{u}_2^0$ denote the lowest utility level it would be possible to generate for individual 2 in this economy and $\bar{u}_2^m$ the highest. Then as $\bar{u}_2$ is varied between $\bar{u}_2^0$ and $\bar{u}_2^m$, with the above resource allocation problem being solved at each $\bar{u}_2$ level, and the corresponding *optimized* values of u_1 calculated, we would trace out a set of Pareto efficient utility pairs or welfare distributions (u_1, u_2) just as we did earlier for the simple example. This would again give a negatively-sloped utility possibility curve in (u_1, u_2)-space. Question 1 of Exercise 15.1 asks you to do this in a numerical example of the type of economy we have just been considering. Note that in general we would obtain a curve and not a line.

The proposition that a competitive market economy achieves a Pareto efficient resource allocation could be expressed by saying that it achieves a utility distribution which is a point on the economy's utility possibility frontier. Since points on this frontier differ substantially in the utility distribution they imply, this tells us that market allocations can produce widely differing utility distributions. In

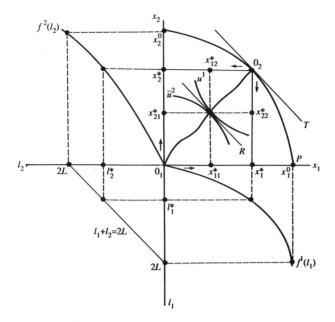

Fig. 15.3: Pareto efficient resource allocation in a 2-good, 2-consumer, 1-input economy

the preceding chapter, we saw that the distribution of income and utility among individuals in the economy was determined by their initial endowments of wealth. Then, very unequal initial wealth distributions can be expected to result in very unequal income distributions in the final market equilibrium. Thus, even though a market equilibrium may be Pareto efficient, it may produce a state of society which, in Amartya Sen's words, is 'perfectly disgusting'.

This observation provides the motivation for the *second* fundamental theorem of welfare economics. Suppose that 'society' or 'policy makers' decide that the distribution of income and welfare resulting from a competitive market equilibrium is unacceptable. The temptation might be to intervene directly in the operation of the market mechanism with measures such as minimum wage laws and price controls to attempt to improve the welfare of particular individuals. The second fundamental theorem provides the basis for an argument that this is not the best policy. It states:

> Any Pareto efficient resource allocation can be achieved as a competitive market equilibrium relative to an appropriate distribution of initial endowments.

We illustrate the sense of the theorem in Fig. 15.4. Suppose the initial endowments in a 2-person, 2-good pure exchange economy (studied in Chapter 14) are as shown by $\bar{x}$ in the figure, and the competitive market allocation is at x^*. As we would expect, since individual 2 starts off with most of the endowments, she ends up in equilibrium consuming most of the goods. A policy maker would like to achieve a more egalitarian position. The first point to note is that it makes sense to choose such a position *on* the contract curve, i.e., within the set of Pareto efficient allocations, since we know that both individuals could be made better off if a position off the contract curve – a Pareto inefficient position – were chosen. So suppose that the policy maker would like to move to point $\hat{x}$ on the contract curve. By virtue of being on the contract curve, $\hat{x}$ must be a point of tangency of indifference curves, and so there must be a unique common tangent line such as $\hat{T}$ shown in the figure. This line passes through the point $\tilde{x} = (\tilde{x}_{11}, \tilde{x}_{21}, \tilde{x}_{12}, \tilde{x}_{22})$. Let $\hat{p}_1$ and $\hat{p}_2$ be prices of the two goods having the property that their ratio $\hat{p}_1/\hat{p}_2$ is equal to the slope of the line $\hat{T}$. Then if the policy maker redistributes initial endowments, taking $\bar{x}_{12} - \tilde{x}_{12}$ away from individual 2 and giving it to 1, and announces market prices $\hat{p}_1$, $\hat{p}_2$ at which the individuals can buy and sell, they will arrive at point $\hat{x}$ as a market equilibrium. In effect, the line $\hat{T}$ represents a budget constraint for each consumer. Since such a line corresponds uniquely to every point on the contract curve, an appropriate redistribution of endowments would allow every such point to be achieved by trade at the prices implied by the slope of the corresponding line. In terms of the utility possibility frontier, the theorem is saying that any point on this frontier can be reached as a market equilibrium by choosing appropriate initial endowments.

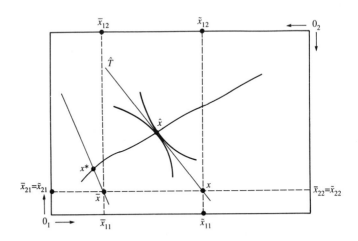

Fig. 15.4: Illustration of the second fundamental theorem

A proof of this theorem requires more advanced methods than those used in this book. Here we simply note that the theorem implies a separation between income distribution goals and the market mechanism. If it is desired to change the final distribution of welfare, this can be done by 'lump sum' redistribution of initial wealth endowments rather than by causing prices to diverge from their equilibrium levels. Lump sum redistributions preserve Pareto efficiency, price distortions create inefficiency.

The two theorems taken together present a strong argument in favour of the market mechanism. However, though obviously the theorems are true within the context of the model of the competitive market economy, there is room for genuine debate about how closely a real market economy approaches this competitive ideal. A number of types of 'market failure', some of which we shall analyse later in this chapter, may cause an actual market system to produce Pareto inefficient outcomes. The applicability of the second fundamental theorem can be limited by the fact that there do not exist sufficient possibilities of *purely* lump-sum redistribution – the fact that, in the model we have considered, the origin of the initial wealth endowments is totally unexplained should lead us to the approach the application of its conclusions with some caution. It may be that in a more general, fully specified model there is very little that is truly an exogenous initial endowment, and what there is may not be observable by a policy maker. The separation between income redistribution and efficient resource allocation then breaks down.

Exercise 15.1

1. Solve the necessary conditions for Pareto efficiency, given some arbitrary $\bar{u}_2$, for the following economy, and derive an explicit function which gives the utility possibility frontier. Discuss the shape of this frontier.

   (a) Consumers have identical utility functions $u^j = x_{1j}^{1/2} x_{2j}^{1/2}$, $j = 1, 2$.

   (b) Firms have identical production functions $x_i = 10 l_i^{1/2}$, $i = 1, 2$.

   (c) $L = 50$.

   [Hint: use (15.21), (15.22) and constraints (15.11)–(15.13) and (15.15) to obtain 6 equations in the 6 unknowns x_{ij}, l_i. Solve first for x_{i2} as functions of $\bar{u}_2$, then for x_{i1} as functions of $\bar{u}_2$. Then substituting into 1's utility function gives u^1 as a function of $\bar{u}_2$.]

2. In the pure exchange economy used as an example in this section, suppose that 1's initial endowments are 2 units of each good, and 2's initial endowments are 8 units of each good. What will be the utility distribution at the corresponding market equilibrium resource allocation? Suppose that a policy maker wishes to achieve a utility distribution in which 1's utility is 6 and 2's is 4. Find a redistribution of initial endowments and an equilibrium price ratio that will support this utility distribution.

15.2 Partial equilibrium welfare economics

We have just discussed the two fundamental theorems of welfare economics at the general equilibrium level. In the remainder of this chapter we go on to analyse a number of policy issues at the partial equilibrium level, i.e., the level of an individual market or firm, ignoring the interactions with the rest of the economy. The reason for doing this is simply that the main ideas and insights can be most clearly brought out in this way. The cost is that the results are not really rigorously established and a number of important problems and qualifications have been swept under the carpet. The purpose of this section is to give at least some idea of the main assumptions that underly the rather simple analyses of the rest of this chapter.

We make extensive use of the idea that the area under the market demand curve gives a money measure of the (gross) aggregate benefits consumers derive from their consumption of a good. This idea goes back to the consumer surplus definitions of Dupuit and Marshall (cf. section 7.2). Thus let $p(q)$ be the (inverse) market demand function, giving price p as a function of total output q. Then we define aggregate consumer benefit from consumption q as

$$B(q) = \int_0^q p(x)dx. \tag{15.23}$$

Intuitively, price p is thought of as a money measure of the benefit derived from the marginal unit of consumption, and (15.23) is then assumed to follow.

The market demand function is an *aggregate Marshallian* demand function – the market demand curve is the horizontal sum of individual Marshallian demand curves. There are therefore two issues to be considered: what is involved in taking the area under an individual's *Marshallian* demand curve as a benefit measure? and what is involved in *aggregating* these measures across consumers?

In section 7.2 we considered three measures of the benefit a consumer derives from a change in price: the CV, the amount by which the consumer's income has to change after the price change to hold her to the *initial* utility level; the EV, the amount by which the consumer's income has to change to give her, with prices unchanged, the utility level which would have been achieved after the price change; and the Marshallian consumer surplus. Each of these was shown to be given by the area under a corresponding demand curve, Hicksian in the first two cases, Marshallian of course in the third. The natural question is: which is the right one to take? The general answer unfortunately is that it all depends on the problem at hand.

For example, suppose that a new road is to be built which will reduce prices of the goods bought by the people served by the road, and will be financed by a lump sum payment by each individual. If consumers' CVs are each greater than their payments, then their utilities must increase as a result of the road construction, and it therefore would be welfare-improving to build the road. In fact, since such decisions are usually based on likely changes to *existing* utility levels, the CV will often be the natural measure to take. It is, on the other hand, very difficult to find cases in which the Marshallian consumer surplus is an appropriate measure of welfare change, though that will have to stand as an unsupported assertion here.

As we saw in section 7.3, the Marshallian consumer surplus differs from the CV (and EV) to an extent depending on the income effect. The Slutsky equation for good i is

$$\frac{\partial D_i^h}{\partial p_i} = \frac{\partial H_i^h}{\partial p_i} - x_{ih} \frac{\partial D_i^h}{\partial y_h} \tag{15.24}$$

where x_{ih} is consumption of good i by consumer h, p_i is its price, D_i^h and H_i^h are Marshallian and Hicksian demands respectively, and y_h is income. We express this in elasticity terms by multiplying through by $-p_i/x_{ih}$ and defining ε_{ih}^H, ε_{ih}^D, and η_i^h as, respectively, own price elasticities of the Hicksian and Marshallian demands and income elasticity, to obtain

$$\varepsilon_{ih}^D = \varepsilon_{ih}^H + \left(\frac{p_i x_{ih}}{\sum_i p_i x_{ih}} \right) \frac{y_h}{x_{ih}} \frac{\partial D_i^h}{\partial y_h} = \varepsilon_{ih}^H + s_i^h \eta_i^h, \qquad (15.25)$$

where we use the fact that $y_h = \sum_i p_i x_{ih}$, and s_i^h is the share of the ith good in consumer h's total expenditure. Thus the difference between the elasticities of the two demand curves depends on how important the good is in the consumer's budget, and how income elastic it is. A good such as housing is likely to have both large s_i^h and large η_i^h, whereas for many other goods one or both of these could be small. If we know the consumer's Marshallian demand curve and the values of s_i^h and η_i^h then it is straightforward to calculate the difference between CV and Marshallian consumer surplus.

For example, suppose that a consumer spends 10% of her income on a good and the income elasticity of her demand for it is 2. Then, if the price elasticity of her Marshallian demand is, say, 1.5, the elasticity of her Hicksian demand is $\varepsilon_{ih}^H = 1.5 - (0.1)(2) = 1.3$. Suppose that its price is currently $10 and that this falls by $1, or 10%. Then the quantity change along the Hicksian demand curve is 13% and that along the Marshallian is 15%. Suppose she currently consumes 10 units of the good. Then consumption along the Hicksian demand curve changes by 1.3 units, and along the Marshallian by 1.5 units. As Fig. 15.5 shows, the difference in the two measures of consumer surplus is, using the standard definition of the area of a triangle (and assuming sufficient linearity of the demand curves), $(1/2)\$1(1.5) - (1/2)\$1(1.3) = \$0.75 - \$0.65 = \$0.10$. The total CV is equal to $10 + $0.65 = $10.65, and so the proportionate error involved in taking the Marshallian measure is $0.1/10.65 \approx 0.009$ or 0.9%. Pretty negligible. On the other hand, note that if we are interested *only* in the *triangular areas* above the quantity changes (which, in many applications, as we shall see later, is the case – these are the so-called *deadweight loss triangles*) then the proportionate error is rather more significant at just over 15%.

Recall, however, the point from section 7.3, that if we have a Marshallian demand function for an *individual consumer*, which is consistent with an underlying utility function and has been estimated subject to the restrictions implied by the model of consumer demand, no approximation is necessary: we can derive the expenditure function and hence CV, or EV, whatever we want. The ultimate difficulty, then, is that the only market demand function we usually have is the *aggregate market function* (though not always – labour supply studies often have individual household data). The conditions under which we can deduce from this the aggregate Hicksian demand function are very restrictive and unlikely to be met.

On the other hand, the Slutsky equations in (15.24) and (15.25) hold for each consumer, and so, if we can form some idea of the likely typical magnitude of the term $s_i^h \eta_i^h$, then we can get an idea of the degree of approximation in using the area under the aggregate Marshallian demand curve – in the case of goods with low expenditure shares for most consumers and moderate income elasticities the approximation is likely to be quite close (see question 1 of Exercise 15.2). It should be emphasized, however, that (15.24) and (15.25) refer to a change in only one price. If several prices change simultaneously, it can be shown that further problems arise with the Marshallian measure which do not arise in the case of CV (or EV).

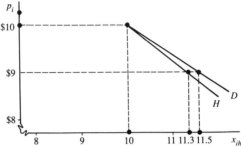

Fig. 15.5: Approximating CV by Marshallian consumer surplus

Now let us assume that, because income effects are negligible, the Marshallian consumer's surplus is a good approximation for CV and EV of an individual consumer, or, equivalently, the area under her Marshallian demand curve is just about equal to the areas under the Hicksian demand curves for the good in question. To add these areas across consumers to obtain 'aggregate consumer benefit' then implies the value judgement that the social significance of the marginal $1's worth of income is the same for every consumer, or, put differently, the distribution of income is judged to be socially optimal.

To see this, we introduce the idea of a social welfare function (swf). This is a function

$$W = W(u_1, u_2, \dots, u_H) \tag{15.26}$$

which represents the preference ordering of the 'policy maker' over utility levels of individuals, where $u_h = u^h(x_{1h}, \dots, x_{nh})$, $h = 1, \dots, H$. That is, we assume that the policy maker is willing and able to make *interpersonal comparisons of utility* to arrive at a ranking of individuals' utility levels well-behaved enough to be represented by the above function (this would, for example, allow a choice to be made from the points on the economy's utility frontier).

Suppose now that we want to determine the social benefit associated with a marginal increment of output dq_i of good i. Totally differentiating through (15.26) (assuming W *is* differentiable) gives

$$dW = \sum_{h=1}^{H} W_h \frac{\partial u_h}{\partial x_{ih}} dx_{ih} \tag{15.27}$$

where $\sum_h dx_{ih} = dq_i$, and $W_h \equiv \partial W/\partial u_h$. Since q_i is sold on a market, we know from the conditions for consumer equilibrium that

$$\frac{\partial u_h}{\partial x_{ih}} = \lambda_h p_i \qquad h = 1, \ldots, H \qquad (15.28)$$

where $\lambda_h = \partial u_h/\partial y_h$ is the marginal utility of income to consumer h. Substituting this into (15.27) gives

$$dW = \sum_{h=1}^{H} W_h \lambda_h p_i dx_{ih} = p_i \sum_{h=1}^{H} \sigma_h dx_{ih} \qquad (15.29)$$

where $\sigma_h \equiv W_h \lambda_h$ is the marginal *social* utility of income to consumer h, and represents the policy maker's evaluation of a marginal \$1 of income to individual h (i.e., $\partial W/\partial y_h = (\partial W/\partial u_h)(\partial u_h/\partial y_h)$ by the function of a function rule). If incomes are so distributed that the σ_h are equal for all h, at say σ, then

$$dW = \sigma p_1 \sum_{h=1}^{H} dx_{ih} = \sigma p_i dq_i \quad \Rightarrow \quad \frac{dW}{dq_i} = \sigma p_i. \qquad (15.30)$$

The expression dW/dq_i is the marginal social benefit of output q_i, and (15.30) shows that on the assumption of equality of the σ_h this is simply proportional to the market price, regardless of how the consumption is distributed amongst consumers. If the σ_h are *not* equal, from (15.29) we obtain

$$\frac{dW}{dq_i} = p_i \sum_{h=1}^{H} \sigma_h \frac{dx_{ih}}{dq_i} \qquad (15.31)$$

and the marginal social benefit of output q_i depends on precisely how the output is distributed among consumers. For example, it will be higher the greater the increases in consumption of consumers with high values of σ_h – the 'socially more deserving' consumers. If $\sigma_h = \sigma$, all h, we can write

$$\frac{1}{\sigma} \frac{dW}{dq_i} = p_i. \qquad (15.32)$$

The expression $(1/\sigma)(dW/dq_i)$ can be interpreted as the money measure of the marginal social benefit of output (since $1/\sigma$ is in units of \$/'social utility units' and can be thought of simply as a conversion factor to change dW/dq_i from 'social utility units'/'units of q_i' to \$/units of q_i, which is how p_i is measured). Thus we have that on the assumption of equal σ_h, price measures marginal social benefit in money terms regardless of how output is distributed among consumers.

How could this equality of marginal social utilities of income σ_h be achieved? One way would be by lump sum redistributions of income which are designed to achieve what the policy maker regards as the optimal income distribution. To see how this works, first note that, given the consumer's expenditure function $y_h^* = y_h(p_1, ..., p_n, u_h)$ as derived in section 7.1, we can invert this to derive a function $u_h = v^h(p_1, ..., p_n, y_h)$, $h = 1, ..., H$ known as the *indirect utility function*, so-called because it gives utility not directly as a function of quantities, but indirectly via the consumer's maximization process, as a function of prices and income. (See question 2 of Exercise 15.2.) A lump sum redistribution is a set of income payments I_h, one for each consumer, such that $\sum_{h=1}^{H} I_h = 0$, so that income is simply shifted around among consumers. Then an *optimal* income redistribution policy for a policy maker with the swf in (15.24) is a set of values I_h^* which solve the problem

$$\max_{I_h} \; W = W(u_1, \ldots, u_H) = W(v^1(p_1, \ldots, p_n, y_1 + I_1), \ldots,$$

$$v^H(p_1, \ldots, p_n, y_H + I_H))$$

$$\text{s.t.} \sum_{h=1}^{H} I_h = 0.$$

Setting up the Lagrange function $W + \sigma \sum_{h=1}^{H} I_h$ and solving gives

$$W_h \frac{\partial u_h}{\partial I_h} - \sigma = 0 \qquad h = 1, ..., H, \tag{15.33}$$

$$\sum_{h=1}^{H} I_h^* = 0. \tag{15.34}$$

But, since σ is the same for all h, (15.33) simply amounts to equalizing $W_h \partial u_h / \partial y_h$ across consumers, or, in our earlier notation, setting σ_h equal to σ for all h.

If we can assume that the income distribution is optimal, then we can simply add money measures of benefits across consumers because the social value of $1 accruing to each consumer is the same. Or, put differently, the approach of taking the area under a market demand curve as a money measure of social benefit implicitly assumes that the income distribution has been optimized (with respect, note, to the policy maker's distributional preferences, not necessarily to anyone else's).

Before leaving the question of measuring consumer benefit, we should distinguish between *gross* and *net* benefit. The net benefit is of course consumer surplus: the money measure of the benefit a consumer derives in excess of the amount she pays to consume the good. The gross measure, already defined by (15.23), is consumer surplus plus expenditure on the good, $p_i x_{ih}'$. This gives the total area under the consumer's demand curve corresponding to consumption x_{ih}'. Fig. 15.6 shows the relation between them (a similar figure can be drawn for the aggregate market demand curve on the assumptions we have made so far).

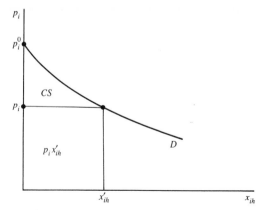

Fig. 15.6: Gross and net consumer benefit

As well as using the area under a market demand curve as a measure of social benefit, we shall also use the area under the market supply curve (for a competitive market) or marginal cost curve (for a monopoly) as a measure of social cost. To fix ideas, take the case of a monopolist producing output q_i in the short run, and buying the labour input in a competitive market. Then as we saw in Chapter 8, its marginal cost $MC_i = w/MP_l^i$, where w is the market wage and MP_l^i the marginal product of labour in producing q_i. To the first order, $dq_i = MP_l^i dl_i$. Also, taking some other competitive industry j, we have $w = p_j/MP_l^j$ as the condition for profit maximization in that industry, and again to the first order $dq_j = MP_l^j dl_j$. Putting all this together

$$MC_i = wdl_i/dq_i = (p_j/MP_l^j)(dl_i/dq_i) = (p_j dq_j/dq_i)(dl_i/dl_j). \tag{15.35}$$

But if the increment in labour dl_i is drawn from industry j we have $dl_i = -dl_j$, and so

$$MC_i = -p_j dq_j/dq_i \ (> 0), \tag{15.36}$$

so that marginal cost in i measures the value of the output lost in j by the diversion of resources from j to i. That is, marginal cost in i *measures the marginal social opportunity cost* of output q_i. It follows that the total social opportunity cost of output q_i is

$$C_i(q_i) = \int_0^{q_i} MC_i(x)dx, \tag{15.37}$$

which of course is the area under the monopolist's marginal cost curve (note that it excludes fixed costs: explain why). A similar argument leads to the same interpretation of the monopoly's long-run marginal cost.

In the case of a competitive market in the short run the above argument can be used for each firm – the area under its marginal cost curve gives the social opportunity cost of its output. Let q_f denote the output from firm $f = 1, \ldots, F$, and $C_f(q_f)$ its total variable cost. Then $q = \Sigma_f q_f$ is total market output and

$$C(q) = \Sigma_f C_f(q_f) \ \Rightarrow \ dC = \Sigma_f C_f' dq_f. \tag{15.38}$$

But since the C'_f are equal across firms (each firm equates its marginal cost to price), denoting this common value by $C'(q)$ we have

$$dC = C'(q) \sum_f dq_f = C'(q)dq \;\Rightarrow\; \int_0^q \frac{dC}{dq}dq = \int_0^q C'dq = C(q) \qquad (15.39)$$

and so we have that total social costs are given by the area under the 'industry marginal cost' or *market supply curve*.

For a competitive market in the long run the argument is a little more complicated, because, as we saw in Chapter 9, the derivation of the long-run market supply curve is more complicated. The result can, however, still be established.

Note also that it must be assumed that the wage or, more generally, the input price really does measure the input's marginal value product in an alternative use. Two cases in which that would not be true are: first, if the alternative use of the input is by a monopolist, then its wage is equal to its marginal *revenue* product, which is below its marginal *value* product, and so this input's true opportunity cost is understated (this is another way of expressing the resource misallocation effect of monopoly); secondly, if, because of frictions and imperfections in the way the input market works, the input would be unemployed if released from the market in question, then the input price *overstates* its social opportunity cost.

Putting now the measure of *gross* consumer benefit together with the measure of social costs gives the *net social benefit measure*

$$S(q) = \int_0^q p(x)dx - \int_0^q C'(x)dx = \int_0^q (p - C')dx. \qquad (15.40)$$

It follows that the output q^* which maximizes net social benefit $S(q)$ must satisfy

$$S'(q^*) = p(q^*) - C'(q^*) = 0, \qquad (15.41)$$

that is, it equates price with marginal cost. Fig. 15.7 illustrates this partial equilibrium analysis of the welfare-maximizing output. $S(q)$ in (15.40) is clearly given by the area between the demand and industry marginal cost curves, and this area is maximized at q^*. If output is set at q', S is less than it can be by the area abc; intuitively, the value of a marginal bit of output, $p(q')$, exceeds the opportunity cost of the resource required to produce it, $C'(q')$, and so net benefit is increased by increasing output. At output q'', S is again less than it might be: the area bde must be subtracted from the area $p^0bC'_0$ to obtain net social benefit, because it corresponds to *negative* net benefit. Over the output range $q'' - q^*$, social opportunity cost, $\int_{q^*}^{q''} C'(q)dq$, exceeds social benefits, $\int_{q^*}^{q''} p(q)dq$, and net benefit is actually increased by reducing output over this range. The condition in (15.41), illustrated in Fig. 15.7, is often called the 'marginal cost pricing rule'.

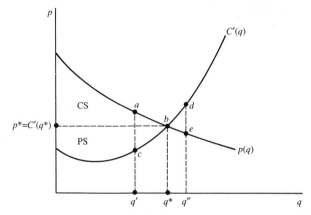

Fig. 15.7: The marginal cost pricing rule

The expression for $S(q)$ in (15.40) can be put slightly differently. Adding and subtracting expenditure or revenue $p(q)q$ gives

$$S(q) = \left\{ \int_0^q p(x)dx - p(q)q \right\} + \left\{ p(q)q - \int_0^q C'(x)dx \right\} = CS + PS \qquad (15.42)$$

The first term is aggregate consumer surplus, the second term is called producer surplus, since it is the excess of revenue over total variable costs to producers (further subtracting fixed costs would give the firm's profits). In effect, we are dividing the net benefit area in Fig. 15.7 into two parts, CS and PS, and we could just as well describe the marginal cost pricing rule as resulting from the maximization of the sum of consumer and producer surplus, the two components of the aggregate net social benefit.

Consider now the net social benefit of the output q_f of a single firm in a competitive market. Its social opportunity costs are $\int_0^{q_f} C'_f(x)dx = C_f(q_f)$, but what about the benefits? We give the following intuitive argument for asserting that we can represent these simply by its revenue, pq_f. The market price p measures the marginal social benefit of output of this good. In a competitive market, the output of any one firm is so small relative to the total that variations in it can be regarded as leaving market price unchanged, and the output q_f itself can be regarded as marginal to the entire market. Thus, to a first approximation pq_f measures the benefit generated by the firm's output. It follows that net social benefit $S(q_f) = pq_f - \int_0^{q_f} C'_f(x)dx$, the firm's producer surplus (which differs from its profit by the amount of fixed cost, if any). Note that maximizing $S(q_f)$ gives $p = C'_f(q_f)$, confirming that outputs are optimal in competitive markets.

As the last topic of this section, we consider the question of the measurement of social benefits over time (we shall need this for the analysis of optimal exploitation of a natural resource in section 15.8 below). Suppose that a market will exist over a sequence of periods $t = 0, 1, \ldots, T$, where $t = 0$ is the present and T may or may not be finite. If q_t is total output in period t, we can define

$$S_t(q_t) = \int_0^{q_t} (p_t(x) - C_t'(x))dx \qquad (15.43)$$

as net social benefit in period t (where the notation allows the possibility that both demand and industry marginal cost may vary over time). The question then arises: how are the $S_t(q_t)$ to be aggregated over time: do we simply take $\sum_{t=0}^{T} S_t(q_t)$, or are things more complicated than that?

Recall from Chapter 13 that at the equilibrium of the (competitive) capital market an interest rate r is established at which each consumer will value future flows of consumption or income. That is, given an income flow $y_0, y_1, \ldots, y_T$, a consumer values it at its present value $v = \sum_{t=0}^{T} y_t(1 + r)^{-t}$. Since the net benefits S_t are equivalent to consumers' income, it therefore seems only consistent to say that a stream of net benefits should be valued at its present value

$$V = \sum_{t=0}^{T} \delta^t S_t \qquad \text{for } \delta \equiv (1 + r)^{-1}, \qquad (15.44)$$

and this of course will differ from simple summation when $r > 0$. This is the approach usually adopted in economics and we use it in section 15.8 below.

This procedure has been the subject of much controversy, however, and we mention here just two of the main points. The effect of discounting on benefits (or costs) accruing far into the future is very dramatic. For example, for $r = 0.1$, $\delta^{60} = .003$, so \$1 worth of benefit accruing in 60 years time is virtually worthless today at an interest rate of 10%. But this means that benefits accruing to future generations, or costs being imposed upon them, say by nuclear energy, resource depletion or environmental pollution, are given hardly any weight in current decisions. Many people, including some economists, have argued that this is excessively myopic and that δ should be very close if not equal to 1 ($r \approx 0$) for such decisions.

Secondly, the model of Chapter 13 is based on the assumption that the capital market is perfectly competitive, with just a single equilibrium price r, paid by all buyers and received by all sellers. In actual fact real capital markets have a wide variety of interest rates and other rates of return, lending rates differ from borrowing rates, and different borrowers and lenders are in equilibrium at different rates. The major reasons for this are the existence of differences in tax rates among consumers and firms, uncertainty, and differences in liquidity of various types of asset. At the very least, this makes it difficult to define a single value for r, but, more fundamentally, the whole approach to the valuation of benefits over time has to be reformulated. In this book, however, we adopt the competitive market model as a first approach.

Exercise 15.2

1. The consumers in the market can be partitioned into 3 types, of each of which there are 1000 consumers:

   Type 1: consume 10 units of the good each, spend 1% of their income on it, have (Marshallian) price elasticity of 1 and income elasticity of 2;

   Type 2: consume 20 units of the good each, spend 5% of their income on it, have price elasticity of 2 and income elasticity of 1;

   Type 3: consume 40 units of the good each, spend 10% of their income on it, have price elasticity of 0.5 and income elasticity of 1.

   Calculate the proportionate error involved in taking the aggregate Marshallian consumer surplus as an approximation to the sum of individual CVs for a 10% price change. The price of the good is $10.

2. A consumer has utility function $u = x_1^a x_2^{1-a}$ and faces the usual budget constraint. Her Marshallian demands are therefore $x_1 = ay/p_1$, $x_2 = (1 - a)y/p_2$. Show that exactly the same indirect utility function is derived by

   (a) substituting the Marshallian demands into the utility function and expressing it as a function of prices and income; or

   (b) finding the expenditure function and then inverting it to obtain utility as a function of prices and income.

   Then show by differentiation that

$$\frac{\partial u}{\partial p_1} = \frac{-\partial u}{\partial y} x_1 \quad \text{(Roy's Identity)},$$

   where $\partial u/\partial y$ is the derivative of the indirect utility function with respect to y, i.e., the marginal utility of income.

15.3 The welfare loss due to monopoly

In Chapter 10 we saw that a profit-maximizing monopolist will set output at the level at which marginal revenue equals marginal cost, implying, since demand elasticity is finite, that price exceeds marginal cost. We now show that this implies a loss of economic welfare, and consider how to measure this loss.

Given our partial equilibrium net benefit measure:

$$B(q) = \int_0^q p(x)dx - C(q), \tag{15.45}$$

this is maximized at the output q^* at which

$$B'(q^*) = p(q^*) - C'(q^*) = 0, \tag{15.46}$$

$$B''(q*) = p'(q^*) - C''(q^*) < 0. \tag{15.47}$$

We assume (15.47) is always satisfied and concentrate on (15.46). Since the monopolist sets $q = \hat{q}$ such that

$$p(\hat{q})(1 - 1/e) = C'(\hat{q}) \tag{15.48}$$

where e is price elasticity of demand, finite and > 1, we must have $\hat{q} < q^*$ and $p(\hat{q}) > p(q^*)$: the monopoly produces too little output at too high a price. The corresponding loss of economic welfare, given our net benefit measure, is

$$\begin{aligned} \text{LM} &= B(q^*) - B(\hat{q}) \\ &= \int_0^{q^*} p(q)dq - C(q^*) - \int_0^{\hat{q}} p(q)dq + C(\hat{q}). \\ &= \int_{\hat{q}}^{q^*} p(q)dq - [C(q^*) - C(\hat{q})] \end{aligned} \tag{15.49}$$

LM, the monopoly welfare loss, is given by the loss of consumer benefit resulting from the lower output, *less* the saving in resource costs. Since

$$C(q) = \int_0^q C'(x)dx \tag{15.50}$$

we can rewrite (15.49) as

$$\text{LM} = \int_{\hat{q}}^{q^*} [p(q) - C'(q)]dq. \tag{15.51}$$

This tells us that the monopoly welfare loss is the area between the demand and marginal cost curves over the output interval $[\hat{q}, q^*]$.

Suppose for example that the demand and marginal cost functions are both linear, given by $p = a - bq$ and $C' = \alpha + \beta q$ respectively (a, b, α, β all positive, $a > \alpha$). Then inserting these into (15.51) gives

$$\begin{aligned} \text{LM} &= \int_{\hat{q}}^{q^*} [(a - \alpha) - (b + \beta)q]dq \\ &= [(a - \alpha)q - (1/2)(b + \beta)q^2]_{\hat{q}}^{q^*} \\ &= (a - \alpha)(q^* - \hat{q}) - (1/2)(b + \beta)(q^{*2} - \hat{q}^2) \end{aligned} \tag{15.52}$$

This is illustrated in Fig. 15.8. The value of LM is given by the shaded triangle, the area of which is $(1/2)[p(\hat{q}) - C'(\hat{q})](q^* - \hat{q})$, that is, $1/2 \times$ base $\times$ height. Question 1 of Exercise 15.3 asks you to show that the expression in (15.52) is equivalent to this expression for the area of the shaded triangle in the figure.

If the demand and marginal cost functions are non-linear, then the counterpart of the shaded triangle in Fig. 15.8 would naturally have two curved sides. If the curvature is not too great, the formula for the area of a triangle could still be used to approximate the monopoly welfare loss. Otherwise, the integral in (15.51) should be explicitly evaluated (see question 4 of Exercise 15.3).

Note that the fact that the monopoly may be making 'excessive profits' is not in itself of direct relevance for the measure of monopoly welfare loss. The essential point is that, in order to maximize profit, the monopoly restricts output below the level at which net benefits are maximized. If output were expanded from $\hat{q}$ to q^*, the value of output forgone elsewhere in the economy as a result is $\int_{\hat{q}}^{q^*} C'(q)dq$, the value of the output increase to consumers is $\int_{\hat{q}}^{q^*} p(q)dq$, and, since the latter exceeds the former, there would be a net welfare gain.

As we saw in Chapter 10, if the monopoly could practise perfect price discrimination it would in fact make this output increase, since it could then appropriate the entire gain in consumer benefit, $\int_{\hat{q}}^{q^*} p(q)dq$, for itself, and profit would increase by the increase in net benefit. There would then be no monopoly welfare loss, and we would say that output was at its socially optimal level – the precise division of net benefits between consumers and the owners of the monopoly firm is irrelevant from the point of view of economic efficiency. If a policy maker disliked the resulting distribution of income, however, she could simply impose a lump sum profit tax on the monopoly, which would leave its output unchanged.

In reality perfect price discrimination is not feasible. The fact that consumers will be charged a uniform price implies that marginal revenue will be less than price at positive output levels, and monopoly in a market will give rise to a welfare loss.

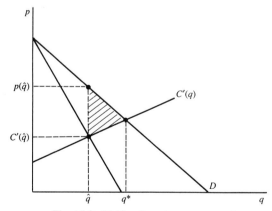

Fig. 15.8: Welfare loss due to monopoly

Exercise 15.3

1. Show that the expression (15.52) is equivalent to the $1/2 \times$ base $\times$ height expression for the monopoly welfare loss.

2. In the case of linear demand and marginal cost functions, show how the monopoly welfare loss varies with the slopes of these functions.

3. Suppose that the market demand function has a constant elasticity of 1, and marginal cost is constant. Show that there is no finite welfare loss measure. [Hint: what is true of the monopoly equilibrium in this case? Refer back to question 2 of Exercise 10.1.]

4. A monopoly has the demand function $p = 10q^{-1/2}$, and the marginal cost function $C' = q^2$. Find the welfare loss due to the monopoly. How closely is this approximated by the linear formula $(1/2)[p(\hat{q}) - C'(\hat{q})](q^* - \hat{q})$?

15.4 The excess burden of a tax

In Chapter 5 we saw that the imposition of a tax on a good drives a wedge between the price consumers pay and the price producers receive. Moreover, relative to the market equilibrium with no tax, consumers have lower consumption at a higher price. Thus, there is a welfare loss associated with the imposition of a tax, and this is generally referred to as the 'excess burden' of the tax.

Let q^* represent the market equilibrium output in the absence of a tax, and $q(t)$ the market equilibrium output when the specific tax t is imposed (refer back to Chapter 3 to see how this output is calculated). Then the excess burden of the tax is given by

$$\text{EB}(t) = \int_{q(t)}^{q*} (p(q) - C'(q))dq. \tag{15.53}$$

The interpretation is similar to that of the monopoly welfare loss. Raising the tax t reduces output and so causes a welfare loss to consumers given by the corresponding area under the demand curve. However, the fall in output releases resources which are used to produce outputs in other parts of the economy, and the value of these is the corresponding area under the supply or industry marginal cost curve. The net welfare loss is the difference between these two quantities, therefore, as expressed by (15.53).

Exercise 15.4

1. The market demand function for a good is $p = a - bq$, and its supply function is $p = \beta q$. Find an expression for the excess burden of a specific tax as a function of the tax rate, t. Comment on the nature of this function. On a supply-demand diagram, show the triangle corresponding to the excess burden, and show that the expression for the area of this triangle is equivalent to that you obtain from the definition in (15.53).

15.5 Optimal indirect taxation

Suppose that a policy maker wishes to raise a given amount of tax revenue, T, by imposing specific taxes on just two goods in the economy (there are others which remain untaxed). He wishes to do so in a way which causes the *smallest possible loss of economic welfare.* We can formulate this problem as that of *minimizing the sum of excess burdens or welfare losses* in the two markets, subject to the constraint that the amount of tax raised is just equal to that required, given also that equilibrium outputs on the markets are functions of the tax rates. That is, we have the problem

$$\min_{t_1, t_2} EB_1(t_1) + EB_2(t_2) = \int_{q_1(t)}^{q_1^*} \{p_1(q_1) - C_1'(q_1)\} dq_1$$

$$+ \int_{q_2(t)}^{q_2^*} \{p_2(q_2) - C_1'(q_2)\} dq_2$$

$$\text{s.t. } t_1 q_1(t_1) + t_2 q_2(t_2) = T$$

where q_i^*, the equilibrium outputs at zero taxes, can of course be taken as fixed. Note that the equilibrium output q_i of good $i = 1, 2$ is assumed to depend only on its *own* tax rate, and not on that of the other good. This is equivalent to assuming that one good's price does not appear in the demand or supply function of the other good (see question 1 of Exercise 15.5). Using the Lagrange approach to solve this problem gives as first-order conditions

$$-(p_i - C_i')q_i' + \lambda[q_i + t_i q_i'] = 0 \qquad i = 1, 2, \tag{15.54}$$

$$t_1 q_1 + t_2 q_2 = T, \tag{15.55}$$

where λ is the Lagrange multiplier associated with the tax constraint and $q_i' \equiv dq_i/dt_i$ (in deriving these conditions we used the fact that the derivative of an integral with respect to the value of its lower limit is the negative of the integrand at that value, together with the 'function of a function' rule). Recall that $p_i - C_i' = t_i$, $i = 1, 2$. That is *the difference between price and marginal cost (the price producers receive) is the tax rate.* Then dividing through (15.54) by q_i' and rearranging gives

$$t_i = \frac{\lambda}{(1 - \lambda)} \frac{q_i}{q_i'} \qquad i = 1, 2. \tag{15.56}$$

Finally, dividing through (15.56) by p_i gives

$$\frac{t_i}{p_i} = \frac{\alpha}{e_i} \qquad i = 1, 2 \tag{15.57}$$

where $e_i \equiv -[q_i/q_i' p_i]^{-1}$ is the own price elasticity of demand (since $dq_i/dp_i \equiv dq_i/dt_i \equiv q_i'$), and $\alpha \equiv -\lambda/(1 - \lambda)$. It can be shown that $\lambda < 0$ (see question 2 of Exercise 15.5) and so (15.57) tells us that the optimal tax rate on a commodity is inversely proportional to its own price elasticity of demand, with a higher tax rate the lower the demand elasticity. Intuitively, relatively higher taxes should be placed on goods for which given proportionate price increases lead to smaller proportionate reductions in output. This 'rule for the optimal indirect taxation' is known as the *Ramsey rule* after the English mathematician F. P. Ramsey who first derived it.

We can gain a little more insight into the rule by rearranging (15.54) to obtain

$$\frac{(p_1 - C_1')q_1'}{t_1 q_1' + q_1} = \lambda = \frac{(p_2 - C_2')q_2'}{t_2 q_2' + q_2}. \tag{15.58}$$

We can rewrite these ratios in a more suggestive notation as

$$\frac{dB_1/dt_1}{dT_1/dt_1} = \frac{dB_2/dt_2}{dT_2/dt_2}. \tag{15.59}$$

The numerator of each ratio is the *marginal welfare change* due to a variation in the tax rate on a good: $p_i - C_i'$ (> 0) is the marginal welfare change resulting from a small change in output and q_i' (< 0) is the rate of change of output with respect to the tax rate, and so the 'function of a function' rule leads to this interpretation. The denominator of each ratio is the marginal tax revenue, i.e., the derivative of total tax revenue $T_i = t_i q_i(t_i)$ with respect to t_i. Then, the condition is saying that the ratios of these derivatives have to be equalized across goods. For suppose not: suppose for example that the ratios have the values -5 for good 1 and -2 for good 2. This tells us that if we reduced the tax rate on good 1 by enough to reduce tax revenue by \$1, we would reduce the excess burden by about \$5; increasing the tax rate on good 2 by enough to make up the \$1 in tax revenue increases the excess burden by about \$2. Thus, with no change in total tax revenue, we have reduced the aggregate excess burden by about \$3. Such possibilities always exist when the ratios in (15.59) are unequal.

Exercise 15.5

1. The market demand functions for two goods are $D_i = a_i - b_i p_i$, and the supply functions are $S_i = \beta_i x_i$, $i = 1, 2$, where p_i is the price consumers pay for good i, x_i is the price producers receive, and $x_i = p_i - t_i$ where t_i is the specific tax on good i. Use the analysis of this section to find expressions for the tax rates t_i^* which minimize aggregate excess burden in the two markets. [Hint: review section 3.2, which shows how to determine equilibrium market output as a function of the tax rate.]

2. Show that $\lambda < 0$ if $T > 0$ and $dT_i/dt_i > 0$. [Hint: first show that $T > 0 \Rightarrow \lambda \neq 0$; then that given $q_i' < 0$, $\lambda < 0$. Use (15.54) in each case.] Explain why we can take it that $dT_i/dt_i > 0$ at the optimum.

3. In problem 1, as in all of this section, we have assumed that the aggregate tax requirement T *can* be met by taxing these two goods. Find a condition on T such that no taxes t_1, t_2 exist which can satisfy the requirement; i.e., the tax requirement is too high for the constraint to be met. [Hint: review the analysis in Chapter 5 of the *tax revenue maximizing* tax rate in a market.]

15.6 External effects: the case of pollution

The two fundamental theorems of welfare economics we considered in section 15.1 were proved for an economy in which every good or service was traded on a market. All economic interactions between individuals are reflected in market prices. For example, suppose that a large proportion of the buyers in a market for a good decide to increase their demands for that good, thus pushing up its price. All buyers now face higher prices for the good, and those buyers whose demands remained unchanged may well feel that those 'responsible' for the price increase have inflicted a cost on them. Nevertheless, the new equilibrium will satisfy the conditions for economic efficiency and there is no reason for a policy maker to intervene *on efficiency grounds* to change the resource allocation.

By *external effects* we mean economic interactions between individuals which take place outside the market mechanism, and which are not reflected in market prices. Their result is that the market mechanism does not allocate resources efficiently: it would be possible to make everyone better off by some change in the allocation produced by the market.

To analyse this type of issue we take a specific model: we have two firms, each of which sells in a perfectly competitive market, but the output of one firm increases the cost of the other. For example, a sawmill located on a river upstream from a brewery discharges its effluent into the river and causes the brewery to incur higher costs of cleaning up the water before it can use it to brew beer. Firm 1 (the sawmill) has cost function $C_1(q_1)$, firm 2 (the brewery) has cost function $C_2(q_1, q_2)$, with the external effect of sawmill pollution captured by the restrictions $\partial C_2/\partial q_1 > 0$, $\partial^2 C_2/\partial q_1 \partial q_2 > 0$, that is, increased output (and therefore effluent) of the sawmill raises both total and marginal costs of the brewery.

Since each firm sells in a perfectly competitive market it faces a given price p_i for its output, $i = 1, 2$. Firm 1's profit-maximizing output q_1^* then satisfies the condition

$$p_1 = C_1'(q_1^*) \tag{15.60}$$

and firm 2's the condition

$$p_2 = \frac{\partial C_2}{\partial q_2}(q_1^*, q_2^*). \tag{15.61}$$

Note that firm 2's equilibrium output depends on that of firm 1, and we can clarify this by differentiating totally through (15.61) to obtain

$$\frac{dq_2^*}{dq_1^*} = \frac{-\partial^2 C_2/\partial q_1 \partial q_2}{\partial^2 C_2/\partial q_2^2} < 0 \tag{15.62}$$

where the negative sign follows from the restriction we placed on the numerator and the fact that the second-order condition on firm 2's maximization — that marginal cost be increasing — implies that the denominator is positive. Thus, the higher firm 1's equilibrium output, the lower firm 2's equilibrium output, because its marginal cost increases with firm 1's output (sketch the diagram).

Consider now the economically efficient output levels. We can define net social benefit as a function of the firms' outputs (recall the discussion in section 15.2) as

$$B(q_1, q_2) = p_1 q_1 + p_2 q_2 - C_1(q_1) - C_2(q_1, q_2), \tag{15.63}$$

and maximizing this gives the conditions

$$\frac{\partial B}{\partial q_1} = p_1 - C_1'(\hat{q}_1) - \frac{\partial C_2(\hat{q}_1, \hat{q}_2)}{\partial q_1} = 0, \tag{15.64}$$

$$\frac{\partial B}{\partial q_2} = p_2 - \frac{\partial C_2(\hat{q}_1, \hat{q}_2)}{\partial q_2} = 0. \tag{15.65}$$

Condition (15.64) tells us immediately that firm 1's socially optimal output $\hat{q}_1 < q_1^*$, its 'privately optimal' output. To see this, note that p_1 is the same in both cases, and, since $\partial C_2/\partial q_1 > 0$, the output which satisfies $p_1 = C_1' + \partial C_2/\partial q_1$ must be smaller than that which satisfies $p_1 = C_1'$ when C_1' is increasing in q_1. It then follows that $\hat{q}_2 < q_2^*$. Thus we can say that the market equilibrium implies a resource allocation in which firm 1's output is too large and firm 2's output too small relative to the economically efficient resource allocation.

The reason for this difference should be clear from the formulations of the two maximization problems. In choosing its profit-maximizing output firm 1 ignored the cost it inflicted on firm 2, through its pollution of the river. It did this because there is no market price it has to pay for its use of the river as an input in its production process – the firm is treating this as a free good but it is not, because firm 2 has to incur costs to clean it up. The social benefit maximization of (15.63) takes this 'social cost' into account and thus results in a different resource allocation.

Having identified the way in which the market fails to achieve an efficient resource allocation, we are led naturally to the question of how we might remedy this failure. One approach was suggested by A. C. Pigou. Suppose we have a policy maker who knows the cost functions and prices which appear in (15.63). He can therefore solve (15.64) and (15.65) for the optimal output pair $(\hat{q}_1, \hat{q}_2)$. If he had the powers to do so, he could then simply instruct firm 1 to produce output $\hat{q}_1$ (ensuring firm 2 produces $\hat{q}_2$ — explain why). However, suppose that the only power the policy maker possesses is that of imposing taxes. Then, if he sets a tax per unit of firm 1's output given by

$$t^* = \frac{\partial C_2(\hat{q}_1, \hat{q}_2)}{\partial q_1}, \tag{15.66}$$

he will achieve the efficient resource allocation. This is because firm 1 will then solve the problem

$$\max_{q_1} \; p_1 q_1 - C_1(q_1) - t^* q_1 \tag{15.67}$$

which must yield as a solution precisely the output $\hat{q}_1$ (show why).

In practice the requirement that the policy maker has all the information necessary to compute the optimal Pigovian tax t^* is very strong and unlikely to be fulfilled. This raises the question of whether a less centralized solution is possible, which in turn leads to a line of argument first proposed by R. H. Coase. He begins with the question: *why* does the interaction between the two firms exist as an external effect and not as a market transaction? Firm 2, the brewery in our example, must realize that the activities of firm 1 are imposing costs on it. Why then do negotiations not take place which would 'internalize' the externality in the absence of a Pigovian tax?

To present Coase's analysis of this issue it is useful to express the profits of both firms as functions of q_1 only. Firm 1's profit is simply $\pi_1(q_1) = p_1 q_1 - C_1(q_1)$. To find a similar expression for firm 2, note first that we can solve (15.61) for firm 2's profit-maximizing output as a function of firm 1's output choice, $q_2 = q_2(q_1)$, and inserting this into the expression for firm 2's profit gives

$$\pi_2 = p_2 q_2(q_1) - C(q_2(q_1), q_1) = \pi_2(q_1). \qquad (15.68)$$

Note that (15.62) gives the form of the derivative $q_2'(q_1)$. Since q_1^* maximizes firm 1's profit, it satisfies $\pi_1'(q_1^*) = 0$. The interesting fact is that the socially optimal output level $\hat{q}_1$ satisfies the condition $\pi_1'(\hat{q}_1) + \pi_2'(\hat{q}_1) = 0$. To see this, write the condition out explicitly as

$$\pi_1' + \pi_2' = p_1 - C_1'(q_1) + \left(p_2 - \frac{\partial C_2}{\partial q_2}\right) q_2'(q_1) - \frac{\partial C_2}{\partial q_1}$$
$$= p_1 - C_1'(q_1) - \frac{\partial C_2}{\partial q_1}(q_2(q_1), q_1) = 0, \qquad (15.69)$$

where the term $(p_2 - \partial C_2/\partial q_2)q_2'(q_1)$ vanishes because, for every q_1, firm 2 sets its output such that price is equal to marginal cost. But (15.69) is identical to (15.64) and results in the same output solutions $\hat{q}_1$, $\hat{q}_2 = q_2(\hat{q}_1)$.

This is illustrated in Fig. 15.9. The functions $\pi_1'(q_1)$ and $-\pi_2'(q_1)$ are shown by the corresponding curves $(-\pi_1'(q_2) > 0$, since $\pi_2'(q_2) < 0)$, and q_1^* and $\hat{q}_1$ must lie at the points shown. Thus the figure again shows that the socially optimal level of firm 1's output is below the profit-maximizing level.

Coase now argues as follows. Suppose it is absolutely clear that firm 1 has the legal right to pollute the river. Since both firms are rational, they should realize that it is in firm 2's interest to pay firm 1 to reduce q_1 as long as $-\pi_2' < \pi_1'$, since the resulting saving in cost to firm 2, $-\pi_2'$, is then greater than the resulting loss of profit to firm 1. For example, if firm 2 paid firm 1 the amount r^* per unit of output reduced from q_1^* to $\hat{q}_1$, then *both* firms gain, firm 2 by the area

$$G_2 = -\int_{\hat{q}_1}^{q_1^*} \pi_2'(q_1) dq_1 - r^*(q_1^* - \hat{q}_1) > 0 \qquad (15.70)$$

and firm 1 by the area

$$G_1 = r^*(q_1^* - \hat{q}_1) - \int_{\hat{q}_1}^{q_1^*} \pi_1'(q_1) dq_1 > 0. \qquad (15.71)$$

Of course some other payment function might be adopted, since the only condition a payment for a reduction in q_1 must satisfy is that it exceed π_1' and be less than $-\pi_2'$. What is clear though is that it never pays firm 2 to bribe firm 1 to reduce its output below $\hat{q}_1$, since $\pi_1' > -\pi_2'$ over this range. Thus the specific payment function adopted will determine the division of the 'gains from trade' between the two firms but not the final output arrived at, $\hat{q}_1$.

Coase then goes on to raise another interesting possibility. Suppose firm 2 takes firm 1 to court, and establishes that firm 1 has no right to pollute the river, but rather may do so only with the consent of firm 2. If, now, firm 2 maximizes its profit *given* its effective control over the output of firm 1, it will set $q_1 = 0$, since this minimizes its own cost and maximizes its profit. But in that case, q_1 is *too low* relative to the socially optimal level $\hat{q}_1$ — some amount of pollution is socially desirable, essentially because firm 1's output is valuable and has a marginal *social* value, over the interval $[0, \hat{q}_1]$, above its marginal *social* cost.

In this case, if the firms are rational, firm 1 will be prepared to pay firm 2 to allow it to increase its output. As long as $\pi_1' > -\pi_2'$, there will be some amount that firm 1 would find it worthwhile to pay, and firm 2 to accept, to permit an increment in q_1. However, firm 1 would want to expand output only to $\hat{q}_1$, since beyond that point the payment firm 2 would require exceeds the marginal profitability of output to firm 1. Thus both firms can gain by firm 1's expansion of output to $\hat{q}_1$. For example, referring again to Fig. 15.9, if firm 1 agreed to pay firm 2 r^* per unit of output, then firm 1 would gain

$$G_3 = \int_0^{\hat{q}_1} \pi_1'(q_1)dq_1 - r^*\hat{q}_1 > 0 \tag{15.72}$$

and firm 2 would gain

$$G_4 = r^*\hat{q}_1 - \int_0^{\hat{q}_1} (-\pi_2'(q_1))dq_1 > 0. \tag{15.73}$$

Again, the precise payment function adopted would determine the distribution of gains but not the output which it is optimal for firm 1 to produce.

The interesting result of this analysis, then, is that regardless of whether firm 1 or firm 2 has the legal right to decide on the level of output of firm 1, if the firms rationally negotiate an output level, this will be the socially optimal $\hat{q}_1$. In that case, we can say that the 'externality problem' between these two firms no longer exists. In effect, a price has been set for firm 1's use of the river, a market in this use has been established, and there would be no external effect. We then say that the externality has been *internalized*.

It would be quite wrong to conclude that the 'Coase Theorem', a special version of which has just been presented, 'proves' that external effects cannot exist.

Rather, it gives us a deeper explanation of why externalities exist. Suppose that in our example, it was unclear whether firm 1 had the right to pollute without firm 2's consent. Then there would be no clear basis for negotiation over who should pay whom for reducing or increasing pollution. Though, from the point of view of the economic analysis, who pays whom is relevant only for the distribution of gains and not for the efficiency of the equilibrium output, it is clearly important for the parties concerned! Thus a necessary condition for the externality to be eliminated is a clear definition of the legal rights – what are often called *the property rights* – in the situation.

Given that these legal rights are well defined, the parties then have to negotiate and reach an agreement. This will involve them in costs, usually termed *transactions costs*, and, in a given situation, it may be that these costs are greater than the perceived gains from trade. So in that case the externality will continue to exist. The Coase Theorem does, however, yield a testable prediction: if property rights are well-defined and transactions costs are negligible, then rational individuals will negotiate an allocatively efficient solution to the externality problem. It also yields some policy recommendations: first, given the difficulty of meeting the informational requirements for a Pigovian tax, it may be more effective to direct policy at clarifying property rights and reducing transactions costs so that decentralized negotiations can take place; secondly, the mere observation of something like pollution need not imply an inefficient allocation – the parties concerned may have negotiated an efficient level of the pollution-generating activity such as $\hat{q}_1$ in our example – and to impose a Pigovian tax in this situation would actually lead away from the efficient allocation. Note, however, that the whole analysis implicitly assumes that only users of the river have an interest in its pollution, something which an 'environmentalist' might strenuously contest.

Exercise 15.6

1. Firm 1 has the cost function $C_1 = 0.5q_1^2$. Firm 2 has the cost function $C_2 = 0.5q_1q_2^2$. The price in firm 1's perfectly competitive market is \$10. The price in firm 2's perfectly competitive market is \$5.

   (a) Find the privately optimal resource allocation and compare it to the socially optimal resource allocation.
   (b) Find the optimal Pigovian tax.
   (c) Identify the gains and losses defined in (15.70)–(15.73) and explain why they would come about.
   (d) Find the resource allocation that would result if the two firms had negotiated the efficient resource allocation and then the Pigovian tax found in (b) were imposed.
   (e) Illustrate your answer by drawing the counterparts of Fig. 15.9 for this case.
2. Suppose now that, rather than selling into a perfectly competitive market, firm 1 is a monopolist with demand function $p_1 = 100 - 6q_1$. Find its profit-maximizing output and compare it to those outputs found in question 1(a). Explain why monopoly might in this situation be 'better' than competition.

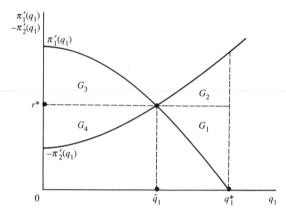

Fig. 15.9: Internalizing the externality

15.7 Free trade and the welfare cost of a tariff

When a country faces given world prices for the outputs it produces, Pareto efficiency requires that consumers pay and producers receive these prices for consumption and production within the economy. If we take the case of just a single good, the resulting demands and supplies will be $D(p_w)$, $S(p_w)$, where p_w is the given world price. Suppose for this good $D(p_w) > S(p_w)$, implying there must be imports of $M(p_w) = D(p_w) - S(p_w)$. A *tariff* is a tax imposed on imports by the government: each unit of the good imported is subject to a tax of t, so that it will cost a consumer $p_w + t$ to buy. This implies that the price received by a domestic producer of the good also rises to $p_w + t$, since the domestic output of the good and imports are assumed perfect substitutes – a producer would not be maximizing profit if he sold at a price less than $p_w + t$. Fig. 15.10 illustrates what happens in this case. Initially, demand was q_D^*, domestic supply was q_S^*, and imports were $m^* = q_D^* - q_S^*$. Imposition of the tariff t effectively raises the domestic market price to $p_w + t$, causing demand to contract to $\hat{q}_D$, domestic supply to expand to $\hat{q}_S$, and imports to shrink to $\hat{m}$. We now show that the welfare loss resulting from imposition of the tariff is given by:

$$\text{WL}(t) = \int_{\hat{q}_D}^{q_D^*} \{p(q) - p_w\}dq + \int_{q_S^*}^{\hat{q}_S} \{C'(q) - p_w\}dq \tag{15.74}$$

that is, the sum of the triangles a and b in the figure. The first term in this sum represents the net loss to consumers arising from the price increase from p_w to $p_w + t$; the loss of benefit from consuming $(q_D^* - \hat{q}_D)$ units of the good, *less* the saving in expenditure $p_w(q_D^* - \hat{q}_D)$. Note that this latter is also a saving to the economy, since it represents the reduction in the value of goods which have to be exported to finance the import quantity $(q_D^* - \hat{q}_D)$.

The second term represents the excess of the cost of producing the increased domestic supply, $\hat{q}_S - q_S^*$, over the saving in cost to the economy of importing that quantity. From the economy's point of view, the value of the resources used in increasing domestic production, $\int_{q_S^*}^{\hat{q}_S} C'(q)dq$, exceeds the cost to the economy of importing that amount of output, $p_w(\hat{q}_S - q_S^*)$. The economy would do better to import this amount of the good and use the corresponding resources to produce something else.

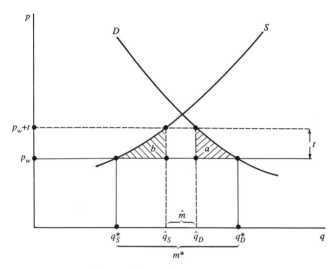

Fig. 15.10: Welfare costs of a tariff

Given these welfare losses, why would the tariff be imposed in the first place? One possible explanation is the need to raise tax revenue: in this case government revenue is $t\hat{m} = t(\hat{q}_D - \hat{q}_S)$ where $\hat{m} \equiv (\hat{q}_D - \hat{q}_S)$ is post-tariff imports. But note that the same revenue could be achieved with a smaller welfare loss if a tax were placed on *all* sales of the good, and not just imports. To see this, suppose that the price of the good is raised to $p_w + t'$, where t' is a specific tax paid on all units of the good sold and not just imports. Then, domestic producers continue to receive p_w net of tax, and so do not change their production. Hence there is no welfare loss corresponding to area b in Fig. 15.10: total cost to the economy of meeting its consumption demand is still minimized. Moreover, if the revenue $T = t\hat{m}$ is to be raised, then since *total* output, and not just imports, is being taxed, the tax rate must be smaller than the tariff, therefore there will be a smaller increase in price to consumers and a correspondingly smaller consumer welfare loss. Thus, a tax rather than a tariff is a more efficient way of raising a given amount of revenue for the government.

A more probable reason for the imposition of a tariff can be seen by noting that producers' revenues rise from $p_w q_S^*$ to $(p_w + t)\hat{q}_S$, while their costs rise by $C(\hat{q}_S) - C(q_S^*) = \int_{q_S^*}^{\hat{q}_S} C'(q)dq$. As Fig. 15.10 shows, this represents a rise in profit in the market shown by area $p_w \alpha \beta (p_w + t)$. Hence it is clearly in producers' interests to campaign, lobby, pay political contributions, and incur other costs in order to obtain the 'protection' of a tariff. What the analysis shows is that this gain to producers is obtained at the cost of an overall welfare loss to the economy as a whole.

Exercise 15.7

1. The demand function in a market is given by $q_D = 100 - 2p$; the supply function by $q_S = 1/2p$. The world market price is $10 per unit. The government imposes a tariff of $10 per unit on imports of the good.
   (a) Calculate the aggregate welfare loss due to the tariff;
   (b) calculate the gain in profit producers make as a result of the imposition of the tariff;
   (c) find the tax which would be levied on *all* sales of the good and which yields the same revenue as the $10 tariff. Calculate the welfare loss caused by this tax and compare it to the welfare loss in (a).

15.8 Optimal depletion of a non-renewable resource

It is usual to distinguish between two types of 'natural resources': first those such as fish, game, and forests which are capable of reproducing themselves and are thus called 'renewable': secondly those such as coal, oil, and minerals which exist in fixed finite amounts, and so are called 'non-renewable'. Here we model the problem of the socially optimal utilization or depletion of the second type of resource.

We assume that we are at year 0, the present, and we envisage a fixed time horizon T sometime in the future beyond which the world ceases to exist. We let $t = 0, 1, \ldots, T$ denote the years from now until then. We have at present some known, given, stock of a resource, R, and the fundamental constraint we face is that the cumulative consumption of the resource over the years from 0 to T cannot exceed R. Let q_t be the amount of the resource extracted and used up (we ignore above-the-ground storage) in year t, and suppose that the cost per unit of the resource extracted is c, and that there is a known demand function $p(q_t)$ which is the same in every year.

If we apply the partial equilibrium welfare analysis set out in section 15.2 to the extraction and consumption of the resource in *any one given year t*, we would conclude (outline the argument) that optimal output is that level $\hat{q}_t$ that satisfies the 'price = marginal cost' condition

$$p(\hat{q}_t) = c, \tag{15.75}$$

implying, on our present assumptions, equal extraction and consumption in each year. However, this does not take account of the fact that the resource is not quite like an ordinary manufactured good: there is a fixed stock of it available and consumption of some of it in one year reduces the amount available for consumption in subsequent years.

In fact two cases are possible. Suppose that, given $\hat{q}_t$ as defined by (15.75), we have

$$(T+1)\hat{q}_t \leq R.$$

That is, we can consume the resource at the rate $\hat{q}_t$ in every year $t = 0, 1, \ldots, T$ and never run out (except possibly at the end of time). Then, essentially, the resource can be treated as if it *were* an ordinary manufactured good, since we would not fully deplete it. However, if $(T+1)\hat{q}_t > R$, then consuming at $\hat{q}_t$ would mean that we would run out of the resource before T, say at date $\hat{t} < T$. This then raises the question: would it not be better to give up some consumption before $\hat{t}$ in order to have some consumption after $\hat{t}$? This is essentially the question of the *optimal depletion path* of the resource, and we now turn to the solution of the problem, *given* the assumption that $(T+1)\hat{q}_t > R$, so that resource depletion *is* an issue.

The problem is to choose a time path of extraction and consumption $q_0, q_1, \ldots, q_T$, subject to the fundamental resource constraint

$$\sum_{t=0}^{T} q_t = R \tag{15.76}$$

(question 2 of exercise 15.8 asks you to explain why this constraint can be expressed as an equality). In any given year t, the net social benefit from producing and consuming the resource is

$$B(q_t) = \int_0^{q_t} p(x)dx - cq_t. \tag{15.77}$$

The key element of the problem, however, is its intertemporal aspect. We assume that the 'social planner' discounts future benefits using a discount factor $\delta^t \equiv (1+r)^{-t}$, where $r > 0$ is the 'social interest rate', so that he wishes to maximize the *present value* of net social benefits given by

$$V = \sum_{t=0}^{T} \delta^t B(q_t). \tag{15.78}$$

Then maximizing V subject to the resource constraint in (15.76) gives the first-order conditions

$$\delta^t(p_t^* - c) - \lambda^* = 0 \qquad t = 0, \ldots, T, \tag{15.79}$$

$$\sum_{t=0}^{T} q_t^* = R,$$

where λ^* is the Lagrange multiplier associated with (15.76), and, note, its value is independent of t. Rearranging (15.79) gives

$$p_t^* = c + \lambda^*/\delta^t \qquad t = 0, \ldots, T. \tag{15.80}$$

This tells us that in every period, the price for sales of the resource should be above the marginal extraction cost and, since δ^t must fall through time, p_t^* is growing through time. The difference $(p_t^* - c) = \lambda^*/\delta^t$ is called the *royalty*, which we denote by ρ_t^*, and (15.80) shows that when the resource *is* relatively scarce $\rho_t^* > 0$, all t; that ρ_t^* increases through time, and will be higher the smaller the quantity of the resource (since this drives up the value of λ^*) and the more heavily future benefits are discounted (the smaller is δ). This royalty can be thought of as a cost which is additional to the cost of actually making the resource available to consumers, c, and which arises out of the fact that an increment of consumption at one point in time permanently lowers the available stock of the scarce resource. λ^* can be interpreted as the marginal value of the resource 'in the ground'.

Further insights can be gained if we note that (15.79) implies that for years t and $t - 1$

$$\delta^t \rho_t^* = \delta^{t-1} \rho_{t-1}^* \qquad t = 1, \ldots, T \qquad (15.81)$$

and so, subtracting ρ_{t-1}^* from both sides and rearranging gives

$$\frac{\rho_t^* - \rho_{t-1}^*}{\rho_{t-1}^*} = \frac{\delta^{t-1}}{\delta^t} - 1 = r \qquad t = 1, \ldots, T \qquad (15.82)$$

where r is the interest rate in the discount factor formula $\delta = (1+r)^{-1}$. Thus, along the optimal extraction path for the resource, the proportionate rate of change in the royalty ρ_t^* is equal to the interest rate the planner uses in discounting the future. The higher this interest rate – the greater the planner's preference for benefits now rather than in the future – the faster the royalty increases. This fundamental relationship in natural resource economics is called 'Hotelling's Rule' after the mathematician H. Hotelling who first postulated it.

(15.80) shows that the market price will be rising through time, essentially because the future is being discounted, $\delta < 1$ or $r > 0$. For this reason therefore the optimal extraction path is one of steadily declining output of the resource, since the demand curve is the same in each period and price is rising. Future consumers do less well than present consumers, an inescapable consequence of discounting.

Exercise 15.8

1. The market demand function for consumption of a natural resource is given by $p = 100 - 2q_t$ in each period t. The economy exists for exactly 3 periods, i.e., $t = 0, 1, 2$. The marginal extraction cost of the resource is $10 per unit in each period. There are 80 units of the resource available. The social planner's interest rate $r = 0.1$.

   (a) Show that depletion of the resource *is* an issue.
   (b) Find the optimal time paths for price and consumption of the resource.
   (c) Find the time path of the royalty and confirm that it satisfies Hotelling's Rule.
   (d) Show the effects on the depletion path of the resource if the planner's interest rate were (i) $r = 0.3$, (ii) $r = 0$. What must be true of the royalty rate in the latter case?
   (e) Show the effects on the depletion path of a new discovery which increases the available resource stock to 120.

2. The natural way to express the constraint on cumulative consumption of the resource is $\sum_{t=0}^{T} q_t \leq R$. State the assumption we made which allows us to write this constraint as an equality.

15.9 Summary

Welfare economics is concerned with assessing the optimality of resource alloca-tion systems and with providing an analytical basis for rational economic policy.

The first fundamental proposition of welfare economics says that the resource allocation corresponding to the equilibrium of a competitive market system is Pareto efficient. All possibilities of making some consumers better off and none worse off have been exhausted. The economy is on its utility possibility frontier.

The reason is that all consumers and firms face the same prices and take them as given in maximizing utility or profits. The first-order conditions their choices satisfy then correspond to those that characterize a Pareto efficient allocation. Since Pareto efficient allocations can vary widely in the distributions of income and utility they imply, a market allocation, though efficient, may not necessarily be thought desirable on distributional grounds.

The second fundamental proposition of welfare economics says that any Pareto efficient resource allocation can be attained as a competitive market equilibrium given an appropriate initial distribution of wealth endowments. This implies that income distributional concerns can be met by lump-sum wealth redistribution, leaving the market mechanism free to allocate resources.

The two propositions taken together give a precise understanding of the sense in which, and conditions under which, Adam Smith's famous invisible hand proposition could hold.

However there is room for debate about the validity of these propositions in any real market economy, since some of the required conditions – price-taking behaviour, no external effects, identical prices to all buyers and sellers of a given good, for example – may not be satisfied. Moreover, the scope for purely lump-sum redistributions may be so restricted that the separation between economic efficiency and income distribution implied by the second fundamental proposition breaks down.

Many issues of economic policy are analysed at the level of partial rather than general equilibrium. It is very convenient to use areas under (Marshallian) market demand curves and supply or industry marginal cost curves as measures of social benefits and costs. The former is an acceptable approximation if individual income effects are small and the income distribution is considered to be socially optimal. The latter is based on the important idea that the social opportunity cost of a given increment in one output is the value of the loss(es) of other output(s) resulting from the resource transfer. For this we require that input prices correctly measure marginal value products of the inputs.

In evaluating consumer benefits over time, a first approach is to take their discounted present values using a discount factor based on 'the' market interest rate. This is, however, quite controversial; two reasons being the consequently low weight given to benefits and costs accruing to future generations and the extent to which real capital markets diverge from the theoretical model.

Using the partial equilibrium approach it can be shown that the optimal output level in a market is that which equates price and (industry) marginal cost. Since a monopoly produces output below this level, it generates a welfare loss which can be measured in a straightforward way.

Similarly the imposition of a specific tax on a good creates a welfare loss known as the 'excess burden' of the tax. This again can be measured in a straightforward way. An optimal set of specific taxes on goods is one which minimizes the sum of excess burdens subject to generating the desired tax revenue. In the simplest possible framework of two goods with zero cross-elasticities of demand, the optimal rate of tax on a good is inversely proportional to its demand elasticity – less elastic goods are more heavily taxed. This is the famous 'Ramsey rule' of optimal taxation.

External effects are interactions among the decisions of economic agents which are not reflected in market prices and which tend to result in non-Pareto- efficient resource allocations. For example, pollution can be regarded as a case in which a producer uses an input for which it does not pay, and so in its decision-taking ignores the costs it imposes on others. One approach to 'correction' of externalities is to impose a tax as a substitute for a market price. Coase's approach suggests that external effects exist because of poor definition of property rights or the transactions costs of reaching agreement to 'internalize' the externalities. This leads to alternative possible policies. It also suggests that mere observation of an apparent external effect might not be enough to establish market failure, since the parties involved may have negotiated the optimal level of the activity.

A tariff on imports creates two types of welfare loss: the first arising from a loss of consumer surplus; the second resulting from higher costs of producing the good domestically rather than importing some amount of it. As a means of raising tax revenue, a tariff is inferior to a specific tax on total sales of the good. An explanation of the popularity of tariffs lies in the increase in producer surplus that they generate, even though there is a net loss in social benefit to the economy as a whole.

Given a non-renewable natural resource that would be exhausted before the end of time if consumed at the rate implied by equating its price to its marginal extraction costs, the problem is to determine the optimal time path of its consumption or depletion. This is done by maximizing the present value of net social benefits subject to the constraint that the sum of consumption over time equals the amount of the resource available. The optimal path is characterized by *Hotelling's Rule*: the royalty (excess of price over marginal extraction cost) should increase at a rate equal to the interest rate. This implies a rising price and falling consumption over time.

PART II: MACROECONOMICS

Macroeconomics is concerned with the determination of the overall level of activity in the economy. The issues addressed include the determination of total output, employment and price changes (inflation) as well as the volume of trade between countries. By their nature, these issues are complex. In principle, we should study *complete* general equilibrium models along the lines of those in Chapter 14 in order to 'build' on microeconomics. However, such models would need to be considerably more complex than those of Chapter 14, and probably more complex than is required to understand some basic forces at work in the economy. In several approaches to macroeconomics there is an attempt at crude aggregation of markets into broad groups. We often identify *the goods market*, *the labour market*, and *the asset market*, and then build simple general equilibrium models to see how these quite different markets interact. This is the approach taken in many first texts on macroeconomics and is broadly the approach taken here.

In Chapters 16 and 17, we set out some institutional and other background material to help set the scene. These chapters are not designed to be comprehensive, but they are useful in providing a view of the phenomena we are attempting to explain. Chapters 18 to 20 develop the so-called IS-LM model. Here, we look at an economy in which prices are fixed and adjustments out of equilibrium involve only *quantity* changes – a contrast with the market analysis of Chapter 2. The IS-LM model links two broad 'markets': those for goods and money. In Chapter 20, we take a first look at the effects of government policy in the IS-LM model, and in particular the factors which determine the effectiveness of policies designed to increase national output. In essence, the IS-LM model is simply a way of modelling the *demand* side of the economy and Chapter 21 makes this explicit. A simple *supply* side analysis is carried out in Chapter 22, and then an *equilibrium* analysis in which both output and the price level are determined. The supply side story of Chapter 22 has two key features. First, it is built on a model of competition in the labour market. Secondly, it is built on perfect information on the part of agents about market opportunities. In Chapter 23 we study the consequences of labour market *imperfections* and in Chapter 24 we model the role of *expectations* in influencing labour market outcomes. Chapters 25 to 27 build on the analysis by considering trade in goods, and flows of assets between countries. Chapters 28 and 29 present *simple* models of dynamics. In Chapter 28 the emphasis is on price dynamics (inflation), building on the work of Chapter 24. In Chapter 29 the emphasis is on 'long-run' issues such as capital accummulation and growth of output.

16   National Accounts

National accounts in general give a broad picture of the total output and income generated in a country, the level of total spending, the general level of prices, and the net position of a country in its dealings with other countries. The national income accounts bring together the aggregate output, income, and expenditure of a country. This turns out to be an important cornerstone of the conceptual framework of macroeconomics. The next section studies this in some detail. We then look at the methods of and difficulties in producing an aggregate price index. Section 16.3 studies the accounts as they relate to an open economy. That is, we examine the implications of trade with other countries for the national accounts.

16.1 National income accounts – output, income, and expenditure

The national income accounts show the flows of income, value of output, and expenditure in a particular period, usually over a quarter or a full year. As we shall see, output, income, and expenditure are closely linked by the production and sales process. In complex economic systems a series of accounting conventions serve to maintain an equality between measured or recorded output, income and expenditure. We can use an example to illustrate the tripartite relationship.

16.1.1 National output

Consider first the value of output produced in the economy during a period, say a year. Producers of output may be classified by the types of goods they produce. *Primary producers* such as mining and timber companies produce an output which is almost exclusively destined for use in a further production process. *Intermediate producers* will use the output of the primary producers to manufacture products which will be used in yet a further stage of production. Producers of *final products* will use as input the output from primary and intermediate producers to manufacture goods for sale to consumers. For example, primary producers, say extractors of coal and iron ore, sell their output to manufacturers of steel, an intermediate product, who in turn sell their output to manufacturers of automobiles, a final product purchased by consumers.

To illustrate how national output is calculated suppose that there are just three firms in the economy, p, i, and f, representing respectively a primary, intermediate and final good producer. What is the total value of output in the economy? Table 16.1 summarizes the transactions between these three firms (sectors). Suppose firm p produces an output of value $1500, $1000 of which is bought by firm i and $500 by firm f. Firm i uses its purchase to produce an output worth $2500 which is sold to firm f which uses this along with the $500 of output bought directly from firm p to produce an output worth $5000. Total gross output of all firms in the economy is the sum of $1500, $2500, and $5000, that is, $9000. However, this sum does not represent the economy's *net output* because, for example, $1000 worth of firm i's output was bought in and is already accounted for as the value of output of firm p. The net value of firm i's output or the *value added* by firm i is actually $2500 − $1000 = $1500, the output of firm i less the value of inputs bought from firm p.

The final product sold to consumers in this example has a gross output value of $5000. The value added by the final stage of the manufacturing process is obviously this value of gross output *less* the value of inputs bought from firm p and from firm i, which we assume were used up entirely in the manufacture of f's product. Thus the $500 of purchases from firm p and the $2500 of purchses from firm i must be deducted from $5000 to find the value added by firm f, which is therefore $2000.

To summarize, suppose we denote the value added in each industry by v_p, v_i, and v_f respectively, and gross output of each industry by q_p, q_i, and q_f. Also, let q_{pi} denote the amount of the primary producers' outputs going to intermediate producers, q_{if} the value of intermediate producers' outputs going to final producers, and q_{pf} the value of primary producers' outputs going to final producers. Total gross output of all industries is therefore $q_p + q_i + q_f$ and value added by each industry is

$$v_p = q_p,$$
$$v_i = q_i - q_{pi},$$
$$v_f = q_f - q_{pf} - q_{if},$$

giving total net output or value added as $v_p + v_i + v_f$.

The total of value added in the economy from the production of all goods and services is known as *Gross Domestic Product* or GDP. In our example we have $q_p = 1500$, $q_i = 2500$, and $q_f = 5000$ but

$$v_p = 1500,$$
$$v_i = 2500 - 1000 = 1500,$$
$$v_f = 5000 - 500 - 2500 = 2000,$$

so GDP = $v_p + v_i + v_f = 5000$.

It is no coincidence that GDP, at $5000, is exactly equal to the sales of firm f. Note that $q_p = q_{pi} + q_{pf}$. Then summing the above expressions gives

$$v_p + v_i + v_f = q_p + q_i + q_f - (q_{pi} + q_{pf}) - q_{if}$$
$$= q_f.$$

Thus, in measuring national output, we need only consider sales of goods to final consumers, since these embody the value of output at all previous stages of production.

16.1.2 National income

What is total *income* in this economy? Income is received by all those involved in the production process. Clearly, firm p received a total of $1500 from the sale of its output, which will find its way into the pockets of its workers, managers, entrepreneurs, and shareholders. In the accounts for firm p, the $1500 will be the sum of wages, salaries, (retained) profits, and dividends (distributed profits). Although firm i sold its output for $2500, its net revenue after the purchase of inputs from firm p is $1500, which again will be the income for all those associated with firm i. Finally, firm f generated $5000 in revenue, $500 of which went to firm p and $2500 of which went to firm i. The net income for firm f therefore is $2000, which is distributed to all those associated with firm f in the form of wages, salaries and dividends. The total income of the economy is $1500 + $1500 + $2000 = $5000, which is identical to total value added. This clearly makes sense. Total receipts not spent on raw material inputs must accrue as income to somebody associated with the firm.

16.1.3 Gross income and net income – the problem of depreciation

It is useful at this point to consider an important complication to the analysis so far. To see the origin of this problem we return to our three-industry example.

Consider the $2500 spent by firm f on the output of firm i. Suppose that this purchase had been of a machine. Then the value added by firm f depends on how much of the machine is 'used up' during the final production process. We take two extremes. In the first, the machine is entirely 'used up' in this period and is scrapped as soon as final output is produced. This case is similar to the use of a raw material as an input, and $2500, the entire value of the machine, must be deducted from the value of gross output of firm f. This leaves value added by this firm as $2000. Value added in this case for the whole economy is given by column (a) in Table 16.1. Of course, this is the case we have already considered.

In the second extreme, the machine bought by firm f is exactly as productive after the production of the final product as it was before, in which case we can argue that none of it has been 'used up' in production. Value added by firm f is therefore revenue *less* cost of raw material, or $5000 − $500 = $4500. The value added by firms i and p are unchanged and so the value added by the economy has increased as shown in column (b). In column (a) the economy's value added is $5000 which, because all inputs have been fully used, is simply the value of the final product of firm f. In column (b) total value added is the sum of the value of the output of firm f *and* that of firm i, since none of the output of firm i is used up in the production process. The output of firm i is now not treated as an input for accounting purposes but rather represents a final purchase, not by consumers, but by firm f.

Table 16.1 An example of gross output and value-added

Producing Company	Received by	Gross Output (Dollars)	Value Added (Dollars)	
			(a)	(b)
p	i	1000	1000	1000
p	f	500	500	500
i	f	2500	1500	1500
f	Consumers	5000	2000	4500
Total			5000	7500

These two cases are of course extremes. In general the value of a machine used during a production process will depreciate to some extent but not completely and, as our example suggests, this *depreciation* reduces net value added. We show how depreciation is treated in the national accounts in section 16.2.4 below.

16.1.4 National expenditure

Finally, we come to expenditure as the third approach to national income accounts. Under case (*b*) of Table 16.1 there are two final expenditures, namely the spending by consumers (amounting to $5000) and that by firm *f* ($2500). This then amounts to the value of GDP, $7500. National accounts typically distinguish between expenditures made by consumers and those made by firms. Firms purchase machinery and other capital equipment which is meant to provide a service over a number of years. Expenditure of this kind is referred to as investment. Thus in our example we see that the sum of consumption and investment expenditure is equal to GDP. Denoting GDP by Y, consumers' expenditure by C, and investment by I we reach the following accounting identity,

$$Y \equiv C + I. \tag{16.1}$$

Moreover, we observe from our example that whilst total income held by individuals is $7500, only $5000 is spent by private consumers. The difference, $2500, is defined as personal savings, S. We therefore have the further identity,

$$Y \equiv C + S. \tag{16.2}$$

Combining (16.1) and (16.2) gives

$$S \equiv I. \tag{16.3}$$

That part of personal income not consumed is saved and these funds are used for investment. Notice that expressions (16.1) to (16.3) are accounting identities, necessarily true given our definitions.

It appears from this simple example that there are three ways of arriving at the total value of the economy's output. First, we may sum up the value added contributed by each firm. Secondly, we may sum up the personal incomes from all sources. Finally, we may add together all expenditures on goods. This output–income–expenditure equivalence holds in more complicated economies than that summarized in our example, as we shall see. However, in practice there are differences in the sums arrived at via each route because of measurement errors. The three approaches are clearly set out in the national accounts of each country. Section 16.4 lists some of the official statistical sources.

16.2 Extensions to the accounting framework

It may not be self-evident that the output–income–expenditure equivalence holds in general, since actual economies are much more complicated than our example. We now turn to some complications and show how these are treated in the accounting framework.

16.2.1 Investment and inventories

The total amount of investment expenditure, including both new or net investment and replacement investment, is referred to as *gross domestic fixed capital formation* (GDFC) or simply gross investment. This amounts to expenditure on machinery, buildings, industrial plant, and so on. There is one further type of 'investment' recorded separately in the expenditure flows which businesses may plan for or which may occur incidentally. Additions to *inventories*, i.e., stocks of unsold goods and raw materials, are recorded because they constitute unsold output, and as such they cause recorded outputs to exceed recorded expenditures (purchases). Treating inventories as 'expenditure' or as a type of investment by firms brings output and expenditure back into balance. For many firms inventory accumulation is a conscious decision based on an anticipation of increases in demand so that in times of buoyant demand, supply of goods will be more readily forthcoming and production delays will not prevent demand from being satisfied. Hence treating additions to inventories as a form of investment and reductions as disinvestment is reasonable. Note that national accounts record *changes* in inventories and not their *levels*. This is because the level of inventories at the begining of a period reflects past expenditure. It is the current period's addition to or reduction in inventory holding which constitutes the firm's 'expenditure' or investment decision.

There are two further things to note about change in inventories. First, as this discussion suggests, the entry for such changes in the accounts may be positive (investment) or negative (disinvestment). Secondly, the change in the value of inventory at market prices is a *real* change. If the actual amount of inventories held during a period was unchanged but their market value increased by an increase in their prices, this would not show up as an increase in inventories in the national accounts. Only if the physical amount changes is this recorded. This is to ensure that the *flow* of production or output in a period is matched by the *flow* of expenditure, which now includes inventory accummulation.

16.2.2 The government sector

The government affects the income–expenditure flows in two ways; by creating an additional source of expenditure and by collecting taxes. These activities of the government result in three main consequences for the system of accounts.

The most obvious consequence is that there is now an additional customer for national output. Thus government expenditure is added to consumers expenditure and investment expenditure. Denoting government expenditure by G gives the following identity to replace (16.1),

$$Y \equiv C + I + G. \tag{16.4}$$

Note that G includes both investment spending and consumption spending by *all* levels of government.

A second consequence of government activity is the effect of taxation on the amount available for purchasing goods by private citizens. Consumption takes place out of *disposable* or post-tax income. Denote the total amount of direct or income tax collected by government by T. Then disposable income, Y_d, is given by

$$Y_d \equiv Y - T. \tag{16.5}$$

Consumers now consume and save out of disposable income, so that $Y_d \equiv C + S$. Using this in (16.5) and rearranging gives, instead of (16.2),

$$Y \equiv C + S + T. \tag{16.6}$$

Combining (16.4) and (16.6) gives, in place of (16.3),

$$S + T \equiv I + G. \tag{16.7}$$

The third consequence of government activity is the levying of taxes on goods and services. Expenditure taxes or indirect taxes create a difference between the factory prices of goods and their selling prices. Firms' output therefore may be valued in two different ways. Valued at *market prices*, the value includes the effects of taxes on goods. Valued at *factor cost* the tax component of the market price is deducted. Thus, it is gross domestic product valued at factor cost which corresponds to the income received by all individuals concerned with producing output. So linking expenditure-based GDP estimates at market prices with income-based estimates will require a factor cost adjustment to the former.

Finally, it is worth noting that there are some goods which receive subsidies from the government to bring down their market price, so the adjustment from market prices to factor cost values is made by deducting taxes *net* of subsidies. Hence,

GDP at factor cost = GDP at market prices − taxes on expenditure

+ subsidies.

16.2.3 The external sector

We give fuller consideration to the consequences of assuming an open economy with trade and payment flows in section 16.3 below. However, there are some immediate consequences for a country's income and expenditure account arising out of its trade with other countries. First, not all of the expenditure by domestic residents is on domestically produced goods. Thus income and expenditure will diverge (the latter exceeding the former) to the extent that goods are imported. Secondly, some domestically produced goods will be purchased by foreign residents, and the extent to which income will exceed domestic expenditure is accounted for by the value of exports. Taking these factors into account gives a new identity for GDP at market prices appropriate for the open economy. The identity (16.4) must be adjusted to account both for the added expenditure on domestically produced goods and services from abroad — exports (X), and for the extent to which C, I, and G contain purchases from abroad — imports (Q). We have

$$Y \equiv C + I + G + X - Q. \tag{16.8}$$

The term $(X - Q)$ is often referred to as the balance of trade surplus or net exports (NX).

Finally, domestic residents may receive income from abroad not through export sales but by virtue of owning property or having other assets overseas. Examples are, rental income from property, interest income from assets, and dividends received from shares. Similarly, overseas residents may receive income from the domestic economy. The net effect of these two flows is known as *net property income from abroad*, and is added to gross domestic product to obtain *gross national product* (GNP).

It is worth reviewing the relationships between the various flows discussed so far. You can do this by refering to Table 16.2 which provides a summary using data from four OECD countries.

It is, of course, not very informative to compare these absolute magnitudes between countries, because in their present form they are denominated in different currencies and we also have not allowed for population differences. Some interesting facts emerge, however, if we take particular ratios of some of the numbers and compare these between countries. Table 16.3 gives ratios of the more important expenditure components to GDP. These ratios are fairly similar across countries over short periods of time with the exception of US foreign trade ratios.

Notice that consumption represents the largest single component of GDP in each country. The lower consumption ratios in Australia and Canada are matched by higher investment ratios. Government expenditure ratios are, interestingly, quite uniform. Foreign trade is clearly less important for the US than for Canada and the UK. A GNP/GDP ratio in excess of 100 per cent reflects the extent to which foreign earnings augment personal incomes. Thus the past roles of the UK and the US as exporters of capital, and of Canada and Australia as importers, shows up in the table.

Table 16.2 GDP and GNP in four OECD countries, 1987

	Australia[1] ($m)	Canada ($m)	UK (£m)	USA ($m)
Consumers expend.	169994	318434	256759	2983670
+ Investment exp.	69871	114378	70767	772635
+ Government exp.	52413	106193	85804	831989
+ Increase in inventory	−721	1954	627	37072
= Total domestic expenditure	291557	540959	413957	4625366
+ Exports	48737	144213	108108	332034
= Total final expenditure	340294	685172	522065	4957400
− Imports	51721	140284	112211	484494
= GDP at market prices[2]	291887	544859	409854	4472906
+ Net income from abroad	−10512	−16607	869	29482
= GNP at market prices	281375	528252	410723	4502388
− Taxes net of subsidies	36673	58011	60578	334540
= GNP at factor cost	244702	470241	350145	4167848

(1) Fiscal year 1 July 1987 to 30 June 1988.

(2) Sums may not add to the figures shown because of statistical error.

Source: OECD National Accounts, ii, 1960–1987, Paris 1989. All figures are in local currency.

In Tables 16.2 and 16.3 the emphasis is on the expenditure element of the national accounts, while in Table 16.4 we present the corresponding income flows which go to produce GDP at factor cost. Table 16.4 is very much simplified. National accounts in each country go into greater detail on the sources of income, and countries tend to differ in the details of their breakdown of incomes.

Table 16.3 Ratios of expenditure components to GDP in 1987

	Australia	Canada	UK	USA
Consumption/GDP	0.58	0.58	0.63	0.67
Investment/GDP	0.24	0.21	0.17	0.17
Government exp./GDP	0.18	0.19	0.21	0.18
Exports/GDP	0.17	0.26	0.26	0.07
Imports/GDP	0.18	0.26	0.27	0.11
GNP/GDP	0.96	0.97	1.00	1.01

Source: Table 16.2.

Table 16.4 Income flows and GDP at factor cost in four OECD countries, 1987

	Australia ($m)	Canada ($m)	UK (£m)	USA ($m)
Wages, salaries, etc.	145,557	295,665	226,764	2,697,055
+ Gross trading profits and surpluses	61,644	127,851	79,938	897,961
+ Other incomes	48,013	63,332	42,574	543,350
= GDP at factor cost	255,214	486,848	349,276	4,138,366
+ Taxes net of subsidies	36,673	58,011	60,578	334,540
= GDP at market prices	291,887	544,859	409,854	4,472,906

Source: As for Table 16.2. All in local currencies. Sums may not add to the figures shown because of statistical error.

16.2.4 Depreciation

We encountered two extremes of depreciation, or capital consumption (the two terms are interchangeable in the national accounts), in our example of Table 16.1. In column (a) the purchase of the machine by firm f from firm i was regarded in a similar way to the purchase of raw materials. That is, the machine was regarded as having been 'used up' in one period during the production process. In column (b), on the other hand, the machine was regarded as being left with its productivity unimpaired at the end of the production process, and so for accounting purposes, can be regarded as final expenditure. In practice we observe cases which lie between these two extremes. The extent to which a piece of equipment is thought to have been used up, that is the amount of depreciation, is the value of the input the machine contributes to output. In terms of our example we can think of it as the amount of input firm f buys from firm i. Value added is therefore net of any depreciation recorded during the period.

Table 16.5 GNP and national income in four OECD countries, 1987

	Australia ($m)	Canada ($m)	UK (£m)	USA ($m)
GDP at factor cost	255,214	486,848	349,276	4,138,366
+ Net income from abroad	−10,512	−16,607	869	29,482
= GNP at factor cost	244,702	470,241	350,145	4,167,848
− Depreciation	48,013	63,302	48,238	551,464
= Net national product *or* National Income	196,689	406,939	301,907	3,616,384

Source: As for Table 16.2. All in local currencies. Sums may not add to the figures shown because of statistical error.

By convention, the amount recorded in firms' accounts for depreciation is taken as being determined by the 'economic life' of the machine. At the end of the planned economic life of the machine the accumulated totals recorded for depreciation will equal the initial purchase cost. This reflects the fact that it is very difficult to measure with precision the exact fall in value of all of a firm's capital actually resulting from its use in production in any given period.

Net national product or *National Income* is the aggregate resulting when depreciation is deducted from GNP at factor cost. Continuing our illustration from Tables 16.2 and 16.4, we present these new relationships in Table 16.5.

Notice that the notes to Tables 16.2, 16.4, and 16.5 refer to a *statistical error*. This arises because some data are collected on an expenditure basis and some on an income basis. The size of the numbers involved inevitably creates a margin for measurement error. These errors are usually recorded in the national accounts when an expenditure-based estimate occurs in an income table and vice versa.

16.2.5 Constant price series — accounting for inflation

There is, of course, no such thing as *the* price of national output in the same sense as there is a price of a single good. However, there is a need to distinguish between those increases in the value of output which stem from increases in the quantity or volume of goods produced and those increases which are the result of output selling for higher prices. Since *changes* in price are of importance, we do not need to worry about the conceptual difficulties of an absolute price *level*. Price indices take the value of output in one year, the *base year*, and calculate the value of output in a subsequent year at the prices which prevailed in the base year. This is done in such a way that any difference in the value of output in any two years evaluated at a common base year set of prices can be interpreted as a difference in the volume or quantity of output. An increase in the volume of output resulting from this exercise is often referred to as a *real* increase as opposed to a money value or *nominal* increase.

In many countries, the popular reporting of inflation — increases in the general level of prices — is expressed in terms of increases in a consumers' price index. In the UK, the *index of retail prices* is the most commonly quoted price index, while in the US it is the *consumer price index*. For national income accounting purposes, however, the construction of a *real* series for GDP or GNP involves the construction of a somewhat different price index known as the *implicit deflator* for GNP or GDP.

Consider the following calculations. Suppose we calculate at the current date, date 1, the total value of goods and services currently produced in the economy. In fact we calculate the current value of gross domestic product GDP_{11} at market prices,

$$\text{GDP}_{11} = \sum_{i=1}^{n} p_i^1 x_i^1 \qquad (16.9)$$

where p_i^1 is the market price of good i in the current period and x_i^1 is the quantity of good i produced in the current period. There are n goods produced as final output. Similarly, we may calculate the total value of goods produced in an earlier period, period 0 as,

$$\text{GDP}_{00} = \sum_{i=1}^{n} p_i^0 x_i^0 \qquad (16.10)$$

Period 0, which need not be immediately prior to period 1, will be taken as the base year. Now define the *value index* as

$$V_{00}^{11} \equiv \frac{\text{GDP}_{11}}{\text{GDP}_{00}} = \frac{\sum_{i=1}^{n} p_i^1 x_i^1}{\sum_{i=1}^{n} p_i^0 x_i^0}. \qquad (16.11)$$

Clearly, V_{00}^{11} gives the ratio of values of output between periods 0, the base period, and period 1, the current period, and is a measure of how the total money value of output has changed between periods. If $V_{00}^{11} > 1$, for example, the total money value of output has increased between periods.

Now suppose we calculate the total volume of current (period 1) output measured at the prices prevailing in the base period,

$$\text{GDP}_{01} = \sum_{i=1}^{n} p_i^0 x_i^1. \tag{16.12}$$

A measure of the change in the volume of output is then the base-weighted *volume index*,

$$V_{00}^{01} \equiv \frac{\text{GDP}_{01}}{\text{GDP}_{00}} = \frac{\sum_{i=1}^{n} p_i^0 x_i^1}{\sum_{i=1}^{n} p_i^0 x_i^0}. \tag{16.13}$$

This is the *Laspeyres volume index*, which is a measure of the change in the volume of output at constant prices between periods. The change in prices, on the other hand, will be measured by

$$\text{GDPD} \equiv \frac{V_{00}^{11}}{V_{00}^{01}} = \frac{\sum_{i=1}^{n} p_i^1 x_i^1}{\sum_{i=1}^{n} p_i^0 x_i^1}, \tag{16.14}$$

which is the *Paasche price index* known as the GDP deflator. A similar technique is used in those countries which calculate mainly GNP figures, with obvious changes in definitions. The calculation may also be made using factor cost rather than market prices. Changes in the GDP or GNP deflator are common measures of inflation. Notice that, since the GDP deflator measures the change in purchasing-power in the sense of the money value of the post-price-change bundle of goods, it has an affinity with the *equivalent variation* measure of welfare change discussed in section 7.2.

In the base year, of course, GDPD is equal to unity, or 100 to use the scale employed in most national accounts, and the base year is changed from time to time to reflect shifts in patterns of expenditure and the introduction of new products.

We make use of the published GDP deflator to remove the effects of inflation from the current price series to obtain a constant price or real series. This is necessary to see how any single economy is doing in terms of real growth.

Table 16.6 shows GDP at current market prices for the UK in 1980, 1984 and 1987. The GDP deflator is also shown. This permits the calculation of the final row of GDP at constant market prices for these years. The base year for the GDP deflator is 1980.

Table 16.6 Nominal and real GDP in the UK, 1980, 1984, 1987 (£ billions)

	1980	1984	1987
GDP at current market prices	230.1	322.1	409.8
GDP deflator (market prices)	100.0	131.8	151.5
GDP at constant (1980) prices	230.1	244.4	270.5

Source: UK National Accounts, 1989 Ed., Central Statistical Office, London.

The equation for calculating the last row of Table 16.6 is simply,

$$\text{GDP in period } t \text{ at 1980 prices} = \frac{\text{GDP in period } t}{\text{GDP deflator in period } t} \times 100.$$

Table 16.7 shows GDP in our four OECD countries for two years, 1984 and 1987, expressed in constant 1980 prices. The last two rows of Table 16.7 show the percentage change in real GDP (the economy's growth rate) and in the implicit price deflator (economy's inflation rate).

Table 16.7 GDP in four OECD countries at 1980 prices

	Australia ($bn)	Canada ($bn)	UK (£bn)	USA ($bn)
GDP in 1984	144.23	337.90	243.56	2886.88
GDP in 1987	170.90	380.10	270.50	3301.30
%Δ in real GDP, 1984–87	18.5	12.5	11.1	14.4
%Δ in the price level, 1984–87 (Inflation rate)	19.5	10.9	16.1	5.8

Source: OECD National Accounts, ii, 1960–1987, Paris, 1989.

We see in Table 16.7 that the fastest-growing economy between 1984 and 1987 was Australia, but Australia also had the highest inflation rate on this OECD estimate. The slowest growing economy was the UK and the US had the lowest rate of inflation of these four countries.

Although we have presented changes in the GDP (GNP) deflator as being our preferred measure of inflation, many countries use an alternative measure designed to reflect the prices of a typical basket of goods bought by consumers. The *consumer price index* (CPI) or *index of retail prices* is a *Laspeyres price index* which keeps the quantities of a particular basket of m consumer goods fixed at some base level,

$$\text{CPI} = \frac{\sum_{j=1}^{m} p_j^1 x_j^0}{\sum_{j=1}^{m} p_j^0 x_j^0}. \tag{16.15}$$

(Compare with (16.13).) Note two key differences between this price index and the GDP (GNP) deflator. First, the CPI is a base-weighted (Laspeyres) index whereas the GDPD is a current-weighted (Paasche) index. (Thus, the CPI has an affinity with the *compensating variation* discussed in section 7.2.) Secondly, the composition of the basket of goods included in the CPI is quite different from that of the GDP deflator. The CPI excludes prices of government purchases and capital goods but includes the prices of imported goods. This is simply because the CPI is designed to capture changes in prices faced by households. Neither index is a perfect measure of inflation, and the choice of index depends on the context and the purpose for which the measure of inflation is required. For example, if import prices rise rapidly the CPI will grow more quickly than the GDP deflator and may be the preferred measure of inflation for cost of living calculations.

Notice that the rates of growth of nominal and real GDP and the rate of inflation are linked by the formula,

Growth rate of nominal GDP = Growth rate of real + Inflation rate.

To see this mathematically, denote by Y_M nominal GDP; Y is real GDP and P is the price level. Then,

$$Y_M = PY \tag{16.16}$$

so,

$$dY_M = P\, dY + Y\, dP \tag{16.17}$$

and dividing through by Y_M

$$\frac{dY_M}{Y_M} = \frac{dY}{Y} + \frac{dP}{P} \tag{16.18}$$

or

$$\hat{Y}_M = \hat{Y} + \hat{P} \tag{16.19}$$

where $\hat{Y}_M$ is the growth rate of nominal output, $\hat{Y}$ the growth rate of real output, and $\hat{P}$ the growth rate of prices, or the inflation rate. This is an example of a general relationship between proportional rates of change known as the 'hat' calculus. (See the Mathematical Appendix.)

16.3 The open economy

All economies engage in some sort of trade or have some financial dealings with other economies, although the degree of 'openness' or the degree of trade dependence of economies may vary considerably. We noted this in our examination of

Table 16.3. We can further illustrate these differences by constructing a measure of openness to trade from the national accounts of Table 16.2. An obvious measure is the percentage of the total value of trade (exports *plus* imports) to GDP. Table 16.8 gives the results.

The results of this calculation suggest that the UK and Canada are economies which depend on trade to a large extent while the US is not nearly such an open economy, so that issues such as the balance of payments, the level of exchange rates, and the effect of interest rates on international capital flows will be far more important for macroeconomic policy in Canada and the UK than in the US, with Australia in an intermediate position.

Table 16.8 The 'openness' of four OECD countries in 1987

Australia	34.42
Canada	52.21
UK	53.75
US	18.25

Note: Openness = [(Value of exports + Value of imports)/GDP] × 100.

We have already shown the impact of trade flows on our income and expenditure definitions. However, there are several other important consequences of trade in an open economy. We briefly set out here some of the key concepts and the types of statistics they generate. Our main concerns are the balance of trade, the balance of payments, and the exchange rate.

16.3.1 The balance of trade

The balance of trade in goods and services has already been discussed briefly in the context of equation (16.8). The *balance of trade surplus* or *net exports* is given by,

$$NX \equiv X - Q, \tag{16.20}$$

which may be positive (a surplus) or negative (a deficit). From the data in Table 16.2 we see that in 1987 Canada ran a trade surplus whilst the other countries ran deficits in that year. If the value of net exports in any year is zero, then the economy is in trade balance, with receipts from the sale of domestically produced goods to other countries just offset by the outflow of expenditure by domestic residents on goods produced abroad.

16.3.2 The balance of payments

Table 16.9 produces summary balance of payments accounts for our four OECD countries in 1987. The accounts differ between countries in their detailed presentation, but Table 16.9 gives the broad view of the main features of such accounts.

In general, balance of payments accounts are produced in 'double-entry' format so that any inflow such as export receipts will be matched by an outflow, say an addition to foreign currency reserves. The sign conventions adopted ensure that the total of all flows on the account cancel each other out so that the account is in balance.

Table 16.9 Summary balance of payments for four OECD countries, 1987[1]

	Australia ($m)	Canada ($m)	UK (£m)	USA ($bn)
1 Current account:				
2 *Visible balance*	−742	11613	−9659	−160
3 *Invisible balance*	−11666	−22172	8055	6
4 Current balance (2 + 3)	−12408	−10559	−1604	-154
5 Net capital transactions	9204	18089	−2684	80
6 Official financing:				
7 *Change in reserves*[1]	−441	−3527	−11350	8
8 *Exceptional financing*[2]	−206	−283	15382	49
9 *Balancing item*[3]	3848	−3712	256	18

(1) (+) indicates a drawing on reserves; (−) indicates additions to reserves.

(2) (+) indicates net borrowing.

(3) Net errors and omissions.

Source: IMF Balance of Payments Yearbook, 1988.

This does not, of course imply that a country's balance of payments position is always neutral, since some of the flows recorded, such as changes in reserves, will arise precisely because of shortfalls or surpluses in transactions relating to trade and capital flows.

In the UK, the two types of flows are referred to as *above the line* (rows 1–5 in Table 16.9) and *below the line* (rows 6–8), referring to actual transactions and accommodating transactions, respectively. In the accounts presented in Table 16.9 'official financing' is below the line and the underlying balance of payments is the sum of the current account and capital account surpluses. The current account records all flows arising from the sale and purchase of goods ('visibles') and services ('invisibles' such as insurance services, tourism services, and so on). The capital account records flows of assets, investment funds, short-term capital movements, and so on. The 'above the line' balance of payments is the sum of rows 4 and 5. Any deficit arising on the current account may be partially or totally offset by a surplus on the capital account.

From the point of view of the overall balance of payments, an excess of purchases over receipts on the current account is not a problem if there is an excess of investment earnings over investment outflows on the capital account. (Of course, there may be other reasons for a country to be concerned about persistent deficits on the current account.) If the net effect of the current balance and the capital account is zero, then there are no 'below the line' consequences and the balance of payments is truly in balance. In practice, a precise balance is a rarity.

To consider the consequences of imbalance for below the line transactions we take the case of a balance of payments deficit. A deficit indicates that insufficient income is received from all sources to finance expenditures and other outflows. Crudely, the demand for domestic currency for the purposes of conversion into foreign currency for the purchases abroad cannot be met from the inflow of domestic currency received from foreigners. The shortfall must be met from elsewhere. In the short run at least, the shortfall may be met by drawing on official reserves, or from borrowing, either from other countries or from the IMF. These transactions are known as 'official financing', and the sign conventions are that a drawing on reserves, say, is entered as a positive flow as far as the balance of payments is concerned. Thus we observe, generally, that countries with deficits on the balance of payments will have a positive entry under official financing, reflecting an inflow of funds onto the account from various funding sources, including foreign currency reserves.

16.3.3 Exchange rates

Trade between countries creates the need to exchange currencies. A British seller of goods in the US will want to convert the dollars received into pounds sterling. A US exporter to the UK will want to do the reverse. Thus currencies are demanded and supplied according to the role they play as a prerequisite for the purchases of goods, services and assets. The demander of Canadian goods in the US is also, at some point, a demander of Canadian dollars. The suppliers of Canadian dollars are those who are giving up their holdings to take up quantities of other currencies.

Markets for foreign exchange naturally come into existence, and exchange rates are the relative prices of currencies established on these markets. The exchange rate between the US dollar and the pound sterling, for example, arises in the market where the two currencies are exchanged. The higher is the price of one currency in terms of another, the greater must be the demand for the former relative to that of the latter. Ultimately, of course, the strength of demand for a currency in foreign exchange markets is a reflection of the strength of demand for a country's goods, services, and assets by foreign customers and investors.

The currencies of all our four OECD countries may be freely exchanged against each other, generating a system of cross exchange rates. A look through a newspaper such as the *Financial Times* or *Wall Street Journal* will show a set of exchange rates, some of which are of more importance than others, reflecting differences in the importance of trade between various countries. Particularly important exchange rates are those involving the US dollar, the Deutschmark (DM) and the Japanese yen, simply because these three economies have the greatest impact on the world economy.

Table 16.10 Exchange rates of national currencies, average of daily rates, 1987

	One unit of:					
Buys:	$Aus.	$Can.	£UK	$US	DM	Yen
$Aus.	1.00	1.07	2.34	1.43	0.80	0.0099
$Can.	0.93	1.00	2.18	1.33	0.74	0.0092
£UK	0.43	0.46	1.00	0.61	0.34	0.0042
$US	0.70	0.75	1.64	1.00	0.56	0.0069
DM	1.25	1.34	2.93	1.79	1.00	0.0123
Yen	101.23	108.46	237.18	144.62	80.98	1.0000

Source: OECD Economic Outlook, No. 43, June 1989.

Table 16.10 shows some exchange rates between Australia, Canada, UK, US, Germany and Japan for 1987. Since all the exchange rates vary from day to day, the figures shown are averages of all daily rates throughout the year.

These exchange rates can be used, among other things, for comparing values of goods in one country in terms of another country's currency. For example, we may express all the values of GDP at market prices in Table 16.2 in a common currency, say the US dollar, to get a better idea of how they compare. The resulting figures are given in Table 16.11.

Table 16.11 GDP and GDP per capita at market prices in four OECD countries, 1987

	Australia	Canada	UK	USA
GDP at market prices ($USm)	204,321	408,644	672,160	4,472,906
GDP per capita ($US)	12,719	15,916	11,842	18,514

16.3.4 Making international comparisons

It is often desirable to attempt to compare the relative prosperity of different countries over a period of time. Two of the adjustments to data already discussed are prerequisites for this type of exercise: constant price series and conversion to common currency units using exchange rates. A third adjustment is necessary, however, for practical comparisons. The data in Table 16.11 illustrate the need for this final adjustment. According to the figures in the first line the US in 1987 had more than ten times the output of Canada, nearly twenty times that of Australia, and nearly seven times that of the UK. The reason this does not reflect the differences in the income of people in these countries is the failure to allow for population differences. We would generally expect differences in output between countries at a similar stage of economic development to be explained partly by population differences. Per capita figures are therefore the appropriate basis for comparison. The second row of Table 16.11 therefore gives GDP per head for our four OECD countries for 1987.

The ranking of countries after appropriate adjustments is clear enough. Even so, the figures are really only crude estimates because of the various transformations the data have been put through. More importantly, great care has to be taken in trying to interpret these figures as a measure of relative overall well-being of individuals. They do not, for example, take into account income disparities across the population within each country.

Exercise 16.3

1. Consider an economy with three firms: A, B, and C. Firm A has an annual output worth 5000, and sells its output to firms B, C and to consumers. Firm B buys 200 of A's output and firm C buys 2000, the rest of firm A's output is bought by consumers. Firm B produces an annual output worth 500 which it sells directly to consumers. Firm C produces an annual output of 6000, of which 3000 worth is bought by A, and the rest by consumers.
   (i) Calculate national income by value added, assuming that all inputs are used up in the production process.
   (ii) Calculate national income if firm C sets aside only 500 for 'depreciation'.
2. Consider the example in Table 16.1. Construct a value added column, (c) in which 50 per cent of the value of the machine bought by firm f from firm i is written off during the period's production.
   (i) What is gross domestic product in this case?
   (ii) What is national income in each of the three cases (a), (b) and (c)?
3. Use the identities (16.4) to (16.8) to justify the identity

$$S + T + Q \equiv I + G + X.$$

4. Calculate GDP in question 1 by expenditures. How is GDP in question 1 affected if firm A buys an imported good of value 1000, and firm C exports part of its output of value 1500, all other things unchanged? What is the trade balance?
5. Calculate the GDP deflator and the consumer price index for 1989 from the following data, using 1988 as the base year.

	1988	1989
Food	5 @ 14	6 @ 30
Housing	3 @ 10	5 @ 20
Fun	4 @ 5	2 @ 6

6. Construct a table showing wages as a proportion of GDP in Australia, Canada, UK, and the US in 1987. Are there any important differences?

16.4 Statistical Sources

A. *OECD Economic Outlook* (monthly), Paris.
B. *OECD Main Economic Indicators* (monthly), Paris.
C. *OECD National Accounts 1960-1980*, Paris 1989.
D. *IMF Balance of Payments Statistics Yearbook*, Washington.
E. Australia, *Australian National Accounts, National Income and Expenditure* (annual), Australian Bureau of Statistics, Canberra.
F. Canada, *National Income and Expenditure Accounts*, Statistics Canada, Ottawa.
G. *UK Economic Trends* (annual), Central Statistical Office, London.
H. *US Survey of Current Business*, Department of Commerce, Washington.

17 Asset Markets and Labour Markets

This chapter introduces some key concepts concerning money and labour, and their roles in the economy. As with the previous chapter, the emphasis is on the definition and measurement of various aggregates rather than on the economic theories which attempt to explain the behaviour of money markets and labour markets. In the money market, we describe briefly a hypothetical banking system and suggest how that system can vary the amount of cash in circulation. We then look at why the banking system holds deposits denominated in foreign currencies, and the effect these foreign currency *reserves* have on domestic money arrangements. In the labour market, we study the ways in which we can measure *unemployment* and its impact.

17.1 Money and banks

The concept of 'the money supply' will figure prominently in our later analysis. It is useful at this point to introduce the methods of accounting for the money supply, and thereby describe the composition of this important economic aggregate. At the detailed level there are some differences in the way the monetary accounts are constructed in different countries and these differences reflect, in part, differences in institutional arrangements. We aim to keep the discussion as general as possible.

We take as given the existence of a central monetary authority appointed by the government to manage the day-to-day functioning of the monetary system and to enact the government's policy on monetary matters. In many countries the monetary authority is the central bank, such as the Bank of England in the UK or the Federal Reserve System in the US. The monetary authority's role in monetary policy is discussed in Chapter 20. In this section, the central bank's role is simply that of providing a particular volume of money or 'liquidity' to the economy. To see how an economy makes use of this money and to attempt to define the money supply we focus on the activities of commercial banks and their customers.

Money in general is any medium of exchange or means of payment acceptable in settlement of a debt or obligation. In a pure *barter* economy in which, say, cotton is exchanged for wool directly, transactions may be cumbersome. I may want to sell cotton but do not want any of the products on offer in exchange by potential buyers of cotton, at least not today. If another commodity is generally acceptable in exchange for both cotton and for wool, and for all other products, it may become a medium of exchange. In most developed economies notes and coin have become money in this sense, though social custom and availablity of suitable commodities have made 'commodity money' take a variety of forms in many societies at various times. A successful and universally acceptable medium of exchange will also become:

1. a unit of account in which 'prices' are denominated and transactions recorded;
2. a store of value so that purchases may be made in the future

A medium of exchange will also have the physical properties of being durable and portable.

A highly valued commodity, such as gold cast into convenient shapes and sizes will serve as money, but *token money* with legal status, such as notes and alloy coins will also suffice. The important thing here, of course, is that the production and distribution of token money must be possible only with the authority of the State. Central banks alone have the authority to produce and distribute token money. For the moment we will think of token money as being the medium of exchange. There are others, such as personal cheques, which we come to presently.

Banks have two prime functions associated with two (not necessarily distinguishable) types of customers. First, banks take deposits from customers who open personal accounts. The advantages of putting money in a bank are mainly safety and convenience, though as we shall see, banks also offer investment opportunities. Secondly, banks use deposited money to make advances to customers who pay interest on the amount advanced. Banks are able to offer this second service because during an interval of time they can be sure that not all their depositors are going to withdraw their entire deposits. Subject to ensuring enough *till money* to satisfy over-the-counter demands for cash, banks can lend money to borrowers on the basis of deposits. On their monthly statements, of course, depositors see the full value of their deposits recorded.

17.1.1 Credit creation

At any one time the total money supply in a simple economy is the sum of cash in people's pockets (cash in circulation) *plus* bank deposits. With no borrowing or lending this total volume of money would be the sum of cash inside and outside banks. If banks lend to borrowers then the total value of *deposits* in banks will exceed the value of *cash* held by banks. This arises in the following way. Consider the balance sheet of a hypothetical bank shown in Table 17.1. A bank accepts a deposit from customer A of value $100. The bank's *assets* are the $100 cash and its *liabilities* are the $100 that the bank 'owes' customer A. This is shown in the first line of Table 17.1. Now customer B asks for a loan of $50 from the bank, which takes the form of a credit to a new account in B's name from which withdrawals can be made. The new balance sheet of the bank is given by the second line of Table 17.1. Its assets are the $100 cash *plus* the loan of $50 and the liabilities amount to the total which might be withdrawn – $100 by customer A *plus* $50 by customer B. Customer B then spends the $50 as a cash withdrawal with store C, which decides to open an account at the same bank and deposits $50. In line 3 of Table 17.1 the total assets of the bank have remained the same at $150 but are now composed of $50 of customer A's original deposit *plus* store C's deposit of $50 *plus* the outstanding loan to individual *B* of $50. The liabilities, also of $150, are the total value of deposits – $100 due to customer *A* and $50 due to store *C*.

Table 17.1 The balance sheet of a hypothetical bank

	Assets	Liabilities
1.	*A*'s cash	$100 deposits
2.	*A*'s $100 cash *plus* $50 loan to *B*	$150 deposits
3.	$50 of *A*'s cash *plus* $50 loan to *B plus* $50 of *C*'s cash	$150 deposits

For simplicity we have confined our attention to a single bank which we might think of as representing the entire commercial banking sector. The key to this example is that by making a loan the bank has increased the recorded deposits by 50 *per cent*. Further loans would increase the total value of deposits still further. This constitutes an increase in the money supply because bank deposits are regarded as funds available for settlement of debts or as means of payments.

In practice, because there are many potential sources of means of payment, there are several corresponding definitions of the money supply. However, one definition fits our story very well, and is known as M1;

$$M1 \equiv \text{Notes and coin in circulation } + \text{ Private 'sight' deposits.}$$

Thus in our example of line 3 in Table 17.1, the M1 money supply is composed of any notes and coin held by customers *A* and *B* and by store *C plus* the sum of their bank deposits. We use the term 'sight deposit' because the M1 definition of the money supply conforms to the narrowest of definitions, which includes cash and 'near cash'. Sight deposits may be converted into cash so quickly that they are regarded as 'near cash'. Note that it would not matter for our purposes if customer *B* had made the purchase at store *C* with a cheque drawn on the bank. The effect is the same. The spending power is immediate and the bank simply debits *B*'s overdraft account and credits store *C*'s account. No cash has changed hands but it is *as if* it had.

Banks also offer deposits in interest-bearing accounts or 'time' deposit accounts. These deposits earn interest but may not be converted into a means of payment (turned into cash or have cheques drawn upon them) so easily. These accounts generally do not enter into the narrow, M1, definition of the money supply, but will enter into 'broader' definitions such as M2 and M3 as we shall see. Modern banking practices and competition among financial institutions are causing breakdown of this simple classification of accounts, with cheque accounts earning interest and time-deposit accounts that have immediate access. However, with penalties, minimum amounts to be held in accounts, and bank charges, it is still generally the case that the less liquid (not so easily converted into cash) deposits attract higher interest.

17.1.2 The money multiplier

There is a limit to a bank's ability to 'create' credit or 'money' as in Table 17.1, and since this example can be used to mimic the entire banking sector there is a limit to how much the initial stock of notes and coin can be multiplied up into a larger value of total deposits. The limit is determined both by individuals' cash requirements and the proportion of the value of deposits banks must retain to satisfy cash demands comfortably. The latter proportion is sometimes known as the bank's *reserve ratio*, given by the ratio of cash reserves to total deposits. We may calculate the limit of credit creation for a simple banking system using some straightforward algebra.

Denote by R_c the total value of cash reserves held by the banking sector and let α be the reserve ratio – a fraction determined by the banks or the monetary authority as being appropriate to satisfy the day-to-day cash requirements of banks. Then, if the total value of deposits is D, we have

$$R_c = \alpha D. \tag{17.1}$$

In our example of Table 17.1, the bank retained $50 out of the $100 cash deposit by customer A and lent the other $50 to customer B. If this represented the bank's maximum permitted loan given a deposit of $100 then $\alpha = 0.5$. In practice, α can be considerably less than this before banks are in danger of being unable to satisfy cash demands by customers. Suppose that customers as a whole require an amount C in their pockets as cash in circulation and that this is a proportion, β, of total deposits. Then we have

$$C = \beta D. \tag{17.2}$$

The *monetary base*, or total of notes and coin, is sometimes referred to as the stock of *high-powered* money, H, and is clearly the sum of cash held by banks and by individuals, so

$$H = C + R_c. \tag{17.3}$$

Finally we use the definition of M1 which in this notation becomes (using (17.2))

$$M1 = C + D = (1 + \beta)D. \tag{17.4}$$

Using (17.1) and (17.2) in (17.3), solving for D in terms of H and then substituting into (17.4) gives

$$M1 = \frac{(1 + \beta)}{(\alpha + \beta)}H = mH. \tag{17.5}$$

Since $0 < \alpha < 1$ and $0 < \beta < 1$, $(1 + \beta) > (\alpha + \beta)$ and so M1 $> H$. The money supply exceeds the stock of notes and coin by a factor $(1 + \beta)/(\alpha + \beta)$, known as the *money multiplier*. A stock of high-powered money can be multiplied up via credit-creation through lending by a factor that depends negatively on the reserve ratio, α, and negatively on the desired proportion of deposits held as cash by the public, β. These effects are apparent by differentiating m with respect to α and β,

$$\frac{\partial m}{\partial \alpha} = -\frac{(1 + \beta)}{(\alpha + \beta)^2} < 0,$$

$$\frac{\partial m}{\partial \beta} = -\frac{(1 - \alpha)}{(\alpha + \beta)^2} < 0.$$

The more cash is required to be held by banks and the more cash is desired by the public, the less banks are able to lend and the lower is the money multiplier. Table 17.2 gives M1 figures for Australia, Canada, the UK and the US in 1987.

Table 17.2 Money supply M1 outstanding in four OECD countries, end 1987

	Aus $m	Can $bn	UK £	US $bn
M1	31,218	37.61	91,866	766.4

Source: OECD Main Economic Indicators, June 1989.

Broader definitions of money supply which take into account deposits held in financial institutions other than banks, holdings of foreign currencies, credit card accounts and so on are presented in the financial statistics for most countries. For example M3 is composed of M1 *plus* time deposits held in all currencies and is a widely quoted statistic. Most often it is not the actual size of the outstanding *stock* of money but its rate of growth which reflects the monetary authority's attitude towards monetary policy. In later chapters we will see how changes in the money supply affect the key variables in the economy.

17.2 Money and the open economy

The banking system retains accounts denominated in foreign currencies, which constitute the *foreign currency reserves*. These reserves facilitate trade and are 'demanded' by those who need foreign currencies to purchase foreign-produced goods – importers of goods. Foreign currencies and the domestic currency are traded on the *foreign exchange market*, and the workings of this market are discussed briefly in Chapter 25. The price of one currency in terms of another is called the *exchange rate* and routinely central banks (under the direction of the government) intervene in the foreign exchange market to buy or sell foreign currencies as required to maintain the exchange rate at a particular level.

Reserves are part of the economy's money supply. To see this, we start by noting that (assuming that the central bank deals in foreign exchange on behalf of the government) the central bank has two primary types of asset: foreign currency reserves R_f and domestic credit D (including loans to the government and the banking sector). The central bank's only liability in the simple case is its outstanding issue of notes and coin (high-powered money), M. Hence, using the accounting identity from the central bank's balance sheet that assets must equal liabilities, we have

$$M \equiv R_f + D. \tag{17.6}$$

Thus in the case of an intervention to buy domestic currency with foreign currency so as to support a particular exchange rate the banking system experiences a *fall* in R_f leading to a fall in M from (17.6).

17.3 The mechanics of monetary policy

In the previous sections we identified the government as having ultimate responsibility for the volume of currency in an economy, even though up to some limit (determined by the money multiplier) the banking system itself can create credit. How does the government, through the central bank, get money into circulation? After all, when the government announces a ten per cent increase in the money supply this does not happen by everybody receiving a ten percent increase in the value of their bank deposits over night! There are several ways in which the government can introduce new currency into the economy, but they can be subsumed into two broad methods. One is by operating on the goods market and the other is by operating on the asset market.

17.3.1 Goods and money

Since the government always has the ability to 'print money' it can, in principle, pay for direct purchases of goods and services with new money. It simply makes a new deposit in its own bank account and writes cheques on that account. In this way an increase in government expenditure is matched by an increase in the money supply. This is an example of an *accommodating monetary policy*, since it is the change in the money supply which 'finances' the increase in government spending.

The consequences of this type of policy are studied later, where we see that there are a number of good reasons why this apparently easy way out of budget problems should be avoided.

17.3.2 Money and assets

Governments not only 'produce' government services and create money, they also borrow from banks and the public by creating and selling government stock or 'bonds'. In a similar way to firms issuing equity, the government issues bonds to raise funds, but in so doing it incurs a commitment to pay interest (rather than dividends) on the bond, and to buy back the bond at its *maturity date*. The amount of interest paid on the bond is generally fixed when the bond is issued, as is the maturity date. Since bonds compete with other financial assets the rate of return must be sufficient to compensate investors for their forgone interest on alternative assets taking into account the differences in risk attached to high risk assets and low risk government bonds.

When the government buys back bonds from the public it can do so by using *new* money, and so an increase in the money supply is acheived by an *open market operation* to redeem bonds.

Exercise 17.3

1. Which of the following could be regarded as 'money': (i) credit cards, (ii) postage stamps, (iii) overdraft accounts, (iv) department store charge-cards?
2. What is meant by: (i) high-powered money, (ii) the money multiplier?
3. Explain in words why the money multiplier is a decreasing function of α (the reserve ratio) and β (the proportion of deposits held as cash).

17.4 Employment and unemployment

Employment and unemployment figures are among the most important and widely-quoted of economic aggregates. Registered unemployment rates and registered vacancies give a picture of the 'tightness' of the labour market. An increase in the number of vacancies and a fall in the number of unemployed would generally indicate a tighter labour market in which jobs are easier to find. A fall in the number of vacancies and an increase in unemployment indicates a 'slack' labour market with jobs being relatively scarce.

The *working population* refers to all those of working age in an economy. Not all of this group will be *participants* in the labour market in the sense that some will be unable or unwilling to be considered for work. The economically active population consists of those who do decide to participate by looking actively for work, and the ratio of the economically active population to the working population is known as the *participation rate*. Full-time students in university are part of the working population but are not, at least in many countries, labour market participants. Of those participating in the labour market, some will be in employment and some will be registered as unemployed. Countries differ in details about which groups are included in the unemployment figures and the practice changes in many countries from time to time. Table 17.3 shows the

OECD estimates of unemployment rate in Australia, Canada, UK and US in the 1980s.

The number of people registered as unemployed is an often-quoted statistic and a politically sensitive one. Unemployment, particularly on a large scale, is seen as a waste of resources and as a social evil. However, the stock of unemployment or the *unemployment rate* (the ratio of the registered unemployed to the working population) tells only part of the story of how the labour market is working. There are three broad reasons for this.

First, the seriousness of a particular unemployment level can only be appreciated in the context of the job opportunities available. If we have data on the number of job openings (*vacancies*) or 'help-wanted ads' then the seriousness of the problem might be more accurately reflected in the ratio of numbers unemployed to the number of vacancies. If 2 million unemployed are faced with 100,000 vacancies one day and with 250,000 the next, one could argue that conditions have improved and that in a short interval of time, the extra vacancies will be filled and unemployment will fall.

Table 17.3 Unemployment rate in four OECD countries, 1980–1988

	1980	1985	1988
Australia	6.0	8.2	7.2
Canada	7.4	10.4	7.7
UK	6.4	11.2	8.3
USA	7.0	7.1	5.4

At least two factors may foil this interpretation. The extra 150,000 vacancies may be in industries requiring skills which are not readily available among the unemployed. Retraining and re-education takes time and the vacancies may remain unfilled. In addition, the new vacancies may tempt new people into the labour force so that the net effect on unemployment is eliminated or reduced. Finally, the working population itself may be growing. Some European countries, for example, have recently experienced employment growth but with little impact on unemployment. Generally, however, a substantial increase in aggregate vacancies will indicate an easing of labour market conditions.

Secondly, unemployment and vacancy totals refer to *stocks*, whereas the rate at which people enter and leave the unemployment pool will also have a bearing on labour market prospects faced by any individual. The unemployment 'problem' is quite different if total unemployment is 2 million and people spend on average two weeks unemployed, than if 2 million are unemployed with an average unemployment spell of one year. Thus it is important to look at labour market flows to see where the burden of unemployment falls. Fig. 17.1 shows various labour market stocks as boxes and the flows as arrows between them. New hires may be taken directly from the new entrants to the labour market or from the pool of unemployed. Retirees are those who leave the labour force permanently, while *discouraged workers* may leave either temporarily or permanently.

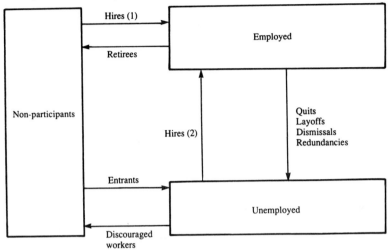

Fig. 17.1: Labour market stocks and flows

The flows into unemployment are accounted for by quits, temporary layoffs, dismissals and redundancies (permanent layoffs).

Taken together the flows into and out of unemployment add an important dimension to the unemployment problem. If a higher rate of unemployment is accompanied by large flows of entrants and new hires, then the unemployment problem will be regarded as less serious than if the flows were sluggish. High rates of flows out of unemployment will be associated with relatively short unemployment durations.

17.4.1 'Types' of unemployment

It is sometimes helpful to think of unemployment in terms of the underlying *cause*. For example, *frictional* unemployment is used to describe the status of people currently between jobs. The source of these frictions may be (temporary) difficulties of skill-matching, job-search, or relocation. Governments may feel that little can or should be done to reduce frictional unemployment since it is largely due to inherent factors such as information or relocation costs, or due to individual choice. If looking for a job is relatively costless then it will be rational to reject low-paying jobs in the expectation of discovering a high-paying job. For as long as jobs are being turned down, an individual is unemployed.

To some extent, frictional unemployment may be *voluntary* and of short duration, but often workers are rationed and are prevented from finding employment by unavailability of jobs. Three further 'types' of unemployment can therefore be identified. First, unemployment may arise because of the *seasonal* nature of work. Occupations in fisheries, agricultrure, hotel and catering, and construction are most affected by the seasonal pattern of demand.

Secondly, reduced demand of a more persistent and general nature may be responsible for unemployment. In the following chapters, we look at how market economies are capable of generating situations such as this. This issue was addressed by the English economist John Maynard Keynes, and his approach associates high unemployment with low aggregate demand for output. In this type of economy, wages do not adjust to clear the labour market in the way that microeconomic analysis suggests they should. This type of model is outlined in Chapters 18 and 19. However, we show in subsequent chapters that, under certain assumptions about the workings of labour markets, this type of unemployment is unlikely to be sustained over a long period. The implication of this is that expansions of demand alone are unlikely to have a permanent effect on unemployment, but are more likely to result in inflationary pressures. Associated with this is the idea of the *natural rate of unemployment*.

The natural rate of unemployment is identified with an economy in equilibrium in which inherent frictions produce an irriducible amount of unemployment. However, the natural rate goes beyond simply frictional unemployment, and can include unemployment resulting from changes in the structure or pattern of output. If a sector or region of an economy experiences a permanent decline for its main output, then certain groups of workers, often with very specific skills, will become unemployed. Other jobs may arise elsewhere, but these workers lack the appropriate skills to fill the new vacancies. With *structural* unemployment there is a *mismatch* between the skills available and the skills required in an economy. Structural changes may displace workers for a long time, though different countries have different experiences. For example, Japan seems able to adapt to new demends on its industry more speedily than the UK.

These distinctions between types of unemployment are only a simple conceptual tool. Both conceptually and empirically the distinctions are far from watertight. For example structural changes may often be most dramatic when part of a general, cyclical decline in the economy.

17.4.2 The natural rate of unemployment

In Chapters 21 to 24 we refer to the level of equilibrium output for the economy as a whole in relation to the interplay of aggregate demand and supply, as in the analysis of a single market. At its simplest, the 'natural' rate of unemployment is that rate of unemployment which corresponds to equilibrium in the goods and labour markets when both are permitted to 'clear'. In other words, the natural rate of unemployment is that which occurs in 'equilibrium' and consists frictional and structural unemployment. However, it may be that labour markets do not have an equilibrating mechanism which tends towards market clearing, and economies may experience high levels of unemployment even in 'equilibrium'. This 'persistent' unemployment arises in the following way.

17.4.3 Persistent unemployment

Persistent or long-term unemployment occurs among groups of workers who may initially not have appropriate labour market skills or who are located in low-growth regions. If jobs are not found, these workers may become 'discouraged' and will reduce their job-search intensity. Employers will discriminate against those who have been out of work for longer periods in preference for workers with more stable employment records.

In some European countries 'long-term' unemployment (the proportion of unemployed who have been without work for one year or more) is high relative to other countries such as the US and Canada. For example in 1988, long-term unemployed accounted for less than one per cent of the labour force in the US and Canada, but 4 per cent in the UK and 7 per cent in Italy (OECD Employment Outlook, 1990, Chart 1.2). Moreover, this unemployment does not seem to respond to variations in demand. In other words, the rise in long-term unemployment appears to correspond to an increase in the 'natural rate'. Changes in skill requirements and an increase in the mismatch between the skills possessed by workers and those required by firms appear to be an explanation.

Theories which attempt to explain persistent unemployment are considered in Chapter 28.

Exercise 17.4

1. Which of the following would increase the natural rate of unemployment:
    (*a*) an increased preference for leisure by workers
    (*b*) introduction of labour-saving technology
    (*c*) a reduction in exports?

17.5 Summary

Money is a particular commodity which acts as a medium of exchange, and as a unit of account.

The government of a country (through a central bank) controls the amount of money in circulation.

Money finds its way into the economy either by replacing assets or, directly, by the government purchase of goods and services.

Unemployment should be thought of in terms of *flows* of workers. The pool of unemployed does not stay the same and there is a continual flow of newly unemployed people and people who are successful in finding jobs.

It is sometimes helpful to think of unemployment in terms of its *cause*. A distinction can be made between *frictional, seasonal, structural* and *demand-deficient* or *cyclical* unemployment.

18 Equilibrium Income: Goods and Money Markets

It is clear from the discussion of Chapter 16 that national income and the components of expenditure on output using that income are precisely linked by a series of accounting identities. However, accounting identities are true by definition, and these relationships as they stand give no clue as to what causes national income to be one particular value rather than another. In other words, the accounting identities themselves do not provide a theory of income determination. In this chapter we will look at income determination in a very simple economy. Since the well-being of a large number of people depends on it, the determinants of the level of income in the economy are of great importance. In the so-called IS-LM model developed in this chapter it is expenditure or demand which determines equilibrium income, but before we examine the reasons for this we will define what we mean by equilibrium.

In very broad terms, an equilibrium in an economic system will be characterized by a state of rest with no tendency for the system to change. The implication, of course, is that out of equilibrium the economy will be changing in some way, and this in turn requires us to specify the behaviour of the economy and how adjustments take place. Mathematically, this means that in addition to an equation or system of equations which hold in equilibrium (the *equilibrium conditions*) we must also specify how the various components of the equilibrium conditions are determined (the *behavioural equations*). In the main we will be interested in the equilibrium position. Our discussions of disequilibrium will be brief and directed towards a fuller understanding of the equilibrium itself. For example, we might be interested in a disequilibrium position to see if and how equilibrium is restored. An important feature of an equilibrium as we have defined it is that decision-makers in the economy will have no reason to change their plans – all demands will, in aggregate, be met and no unexpected additions to or subtractions from stocks of output will be made. In other words there will be neither excess demand nor supply. Hence in equilibrium, output (supply) equals expenditure (demand). But is this not simply what the national accounts told us, and therefore is the economy not in equilibrium all the time, by definition?

The answer is, no. In the national accounts, it is actual or outturn output and expenditure which are equal. Equilibrium, on the other hand requires that *planned* expenditure be equal to output. To see the significance of this distinction, suppose that recorded planned expenditures fell short of output produced. All demands can be met, but producers have unplanned or unexpected additions to their inventories. The national accounts will tell us that total expenditure plus the addition to inventories (a form of investment expenditure by firms, remember) equals output produced. However, this is not an equilibrium. Since the inventory accumulation was unplanned and therefore forced upon firms, they will reduce their planned production for the future. If producers change their plans, then the economy is not in equilibrium.

The equality of demand and supply will therefore characterize an equilibrium, but in fact it is demand which is emphasized in this and the following chapter. This is facilitated by assuming a fixed price level, or alternatively, a horizontal aggregate supply curve. Denoting the economy's price level by P and income (or output) by Y we may draw the economy's demand and supply curves as in Fig. 18.1. The curves of immediate interest are AD_0 and AS. The downward-sloping demand curve is justified by appeal to the usual microeconomic argument – less is demanded at a higher price. We show this more formally at the beginning of Chapter 21. The horizontal supply curve is justified by the assumption that the economy is operating at less than full capacity and that extra output can be generated at no extra unit costs, all the resources (including labour) being readily available. This assumption is relaxed in Chapter 22, by which time we will have a thorough understanding of the demand side. That output is *demand determined* in this model should be clear from Fig. 18.1. If demand for the economy's output suddenly expanded to AD_1, perhaps because of a 'consumer boom', equilibrium output (or income) would increase from Y_0 to Y_1.

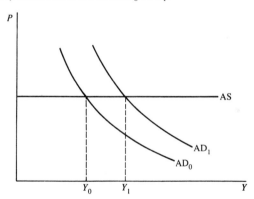

Fig. 18.1: Output determined by demand

Alternatively we may interpret our example of unplanned inventory accumulation using Fig. 18.1. Suppose that producers expect the economy's demand conditions to be represented by curve AD_1 and produce Y_1 in expectation of being able to sell all of the output. However, aggregate expenditure plans are represented by AD_0, and since at the going price, this implies that Y_0 of output is demanded, all demands are realized. Unsold goods amount to $Y_1 - Y_0$ and actual expenditure is Y_0. In the next period, producers will perceive their over-optimistic forecast of demand, because they see that inventories have unexpectedly increased and they will plan to reduce their output so that just Y_0 is available for purchase. If demand conditions remain unchanged at AD_0 then the producers' actions will be justified and all plans will be realized with no unexpected outcomes. Hence with demand at AD_0, Y_0 is an equilibrium level of income or output.

This example suggests that the responses of producers to excess supply is to reduce output, thus taking the economy conveniently towards equilibrium. The question is whether and how the economy achieves equilibrium. The answer to this is first presented in the context of the income–expenditure model in the next section. The remainder of the chapter is devoted to the development of an important model of income determination known as the IS-LM model.

18.1 The simple income–expenditure model

It is apparent from Fig. 18.1 and the discussion surrounding it that equilibrium is given by the equality of output available for supply and planned demand. In the simplest model of the determination of income there are only two types of agent who spend on the economy's output: consumers and investors. The total amounts these groups plan to spend in any period will be denoted by C and I respectively. We see then that plans are satisfied if the total of planned expenditures is equal to output available:

$$Y = C + I. \tag{18.1}$$

As a model, equation (18.1) is incomplete since the 'forces' which determine the key variables Y, C, and I are unspecified. In other words we have no behavioural equations. We rectify this now by asking what factors would explain planned consumption and investment. Clearly, if we can explain C and I, (18.1) gives us an explanation of the equilibrium value of Y. A large number of factors may explain C and I not all of which will be entirely 'economic' in nature. For example, business men's investment plans may be determined by 'confidence' and 'expectations' about future market prospects which although important are very hard to pin down and quantify. Initially, we treat investment as *exogenous*: a parameter fixed at a particular, known value.

In the simplest model presented here, we will also assume that income is disposed of only by consumption spending and by saving, and so, that part of income not consumed is saved.

$$Y = C + S. \tag{18.2}$$

Moreover, it makes sense to assume that in addition to consumption increasing when income increases, savings too will increase, so that saving, in aggregate, is also a 'normal' good. This places some restrictions on the form of the consumption–income relationship.

Mathematically we may write consumption as a function of income:

$$C = C(Y) \qquad 0 < C'(Y) < 1. \tag{18.3}$$

The equation in (18.2) is known as the *consumption function* and its derivative is known as the *marginal propensity to consume*. The marginal propensity to consume itself may be a function of income. For example, the amount by which consumption increases as income increases may fall with income so that more of additional income is saved. At the least we would not expect the marginal propensity to consume to *increase* with income, since intuitively more of society's wants are already satisfied at higher income levels, so that further increases in income would not be accompanied by an increase in the extra income consumed. Hence we may have $C''(Y) \le 0$. Because of its simplicity, one formulation which satisfies this condition with equality and also satisfies (18.3) is in very common use, namely the linear form,

$$C(Y) = a + bY \qquad a > 0, \ 0 < b < 1, \tag{18.4}$$

where b is called the marginal propensity to consume. The positive intercept, a, reflects the fact that consumption must take place even at zero income simply for survival. In this case the economy is drawing on its accumulated wealth in order to satisfy its subsistence requirements. Occasionally we find some formulations of the consumption function for which $a = 0$. One further useful definition is the *average propensity to consume*, APC, which is simply,

$$\text{APC}(Y) = \frac{C(Y)}{Y} \tag{18.5}$$

or, using the formulation in (18.4)

$$\text{APC}(Y) = \frac{a}{Y} + b, \tag{18.6}$$

from which we find that

$$\frac{d[\text{APC}(Y)]}{dY} = -\frac{a}{Y^2}. \tag{18.7}$$

Hence the average propensity to consume is constant or decreasing in income depending on whether a is equal to or greater than zero. For the moment this is as much as we need to know about the consumption function.

Using (18.3) in (18.1) gives us a relation which determines equilibrium income, Y^*

$$Y^* = C(Y^*) + I. \tag{18.8}$$

Clearly, using the general form of $C(Y)$ gives us only an *implicit* solution for Y^* and although we will be able to make progress using general forms presently, for the moment we wish to explore *explicit* solutions for equilibrium income. We can obtain an explicit solution by using (18.4) in (18.8) and solving for Y^* to give

$$Y^* = \frac{a + I}{(1 - b)}. \tag{18.9}$$

Equation (18.9) says that equilibrium income is some multiple $((1 - b)^{-1} > 1)$ of the sum of exogenous expenditures. Once Y^* is determined in (18.9) we may find an expression for consumption. This is found by substituting (18.9) into (18.4) to give

$$C^* \equiv C(Y^*) = \frac{a + bI}{(1 - b)}. \qquad (18.10)$$

We can regard the equilibrium values Y^* and C^* as functions of the parameters a, I, and b. Of interest then is what effect a change in one of these parameters has on the equilibrium. We establish the comparative-static effects by evaluating the derivatives of the functions $Y^* = Y^*(a, b, I)$ and $C^* = C^*(a, b, I)$. In the case of equilibrium income we have, from (18.9)

$$\frac{dY^*}{da} = \frac{dY^*}{dI} = \frac{1}{(1 - b)} > 0 \qquad (18.11)$$

and

$$\frac{dY^*}{db} = \frac{I}{(1 - b)^2} > 0. \qquad (18.12)$$

Hence, an increase in planned exogenous consumption or investment increases equilibrium income, as does an increase in the marginal propensity to consume. There are two things to note about (18.11) in particular. First, an increase in exogenous planned expenditure increases equilibrium income by the same amount regardless of the source of the expenditure. This may not be true in all models, but it is in this particular one. Secondly, the change in income resulting from a change in investment (say) is not only positive but greater than one. This implies that a $1m increase in planned investment leads to an increase in income of *more than* $1m! The reason for this is the two-way nature of income and expenditure determination in a world with a fixed price level. Income (output) is determined by demand (expenditure) and demand determined by income — via the consumption function in this case. Thus the initial increase in investment $\Delta I = $1m$ will have the initial effect of raising output by $1m but this implies an increase in income of $1m out of which consumers will spend an extra $b($1m)$. This *induced* increase in planned demand will call forth greater output and income of $b($1m)$, out of which consumers will spend $b^2($1m)$ and so on. The total added expenditure is then $\Delta I(1 + b + b^2 + ...) = \Delta I/(1 - b))$. The value $(1 - b)^{-1}$, known as the multiplier, captures all of these 'knock-on' effects which cause the initial increase in investment to be multiplied up. Not all multipliers take such a simple form, as we shall see, but all perform a similar role. The new value of income, Y_1^*, is found by adding the change in income to the original value Y_0^*; so

$$Y_1^* = Y_0^* + dY^*$$
$$= Y_0^* + \frac{dI}{(1 - b)}$$
$$= \frac{a + I + dI}{(1 - b)}.$$

The determination of income using (18.1) and (18.4) is illustrated graphically in Fig. 18.2, with total expenditure $E \equiv C + I$ plotted against income, Y.

An alternative characterization of equilibrium in this simple economy may be obtained directly from equation (18.1), since by rearranging, we have

$$Y - C = I. \tag{18.13}$$

Now, in this simple model, that part of income which is not consumed is saved (this is true at *any* level of income) so that (18.2) may be written

$$S = I. \tag{18.14}$$

In view of (18.3), however, S will depend on Y and since (18.1) only holds at Y^*, so will (18.13) and (18.14), hence we should write (18.14) as

$$S^* \equiv S(Y^*) = I. \tag{18.15}$$

These relationships are shown diagrammatically in Fig. 18.2. Equilibrium is where total planned expenditure, $C + I$ cuts the 45° (income = expenditure) line.

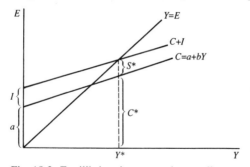

Fig. 18.2: Equilibrium income and expenditure

In Fig. 18.3 we show this equilibrium in terms of savings rather than expenditure. The savings function is, using (18.2),

$$S(Y) = Y - C(Y)$$

or, in our special (linear case)

$$S(Y) = -a + (1 - b)Y, \tag{18.16}$$

where $(1 - b)$ is referred to as the *marginal propensity to save* and its inverse is the multiplier referred to earlier.

It is apparent from Fig. 18.3 that investment exceeds savings at income levels below Y^* while savings exceed investment at income levels above Y^*. Although obviously true in this linear case, this will also be true in general (subject to some qualifications to be discussed presently) as long as $S'(Y) > 0$. Hence a savings function satisfying this condition will be such that

$$S(Y) < I \qquad \text{for } Y < Y^*, \tag{18.17}$$

$$S(Y) > I \qquad \text{for } Y > Y^* \tag{18.18}$$

(and of course $S(Y) = I$ for $Y = Y^*$). The question we want to answer is what adjustments take place if either (18.17) or (18.18) hold since neither constitutes an equilibrium, in which planned demand would just equal planned supply.

Take (18.17). This is a case of *excess demand*, because desired investment demand is in excess of the funds available for this use. In a world where prices respond to excess demand by increasing, there is a tendency towards the elimination of the excess demand. As long as demand exceeds supply, prices rise, reducing demand and stimulating output. (See Chapter2.) In the present model prices are fixed and hence are not able to fulfil this role as adjustment mechanisms. The only variable free to adjust is output. In the case of an excess demand perceived by producers, output will be increased in an attempt to satisfy the demand in future periods. Hence Y increases, thus increasing S. This process continues until (18.15) holds. At Y^* there is neither excess demand nor supply and thus no adjustments take place. In the case of (18.18) there is excess supply, inventories build up and producers respond by lowering output; Y falls until the excess supply is eliminated. In this model output adjusts but does so in a well-behaved way so that the equilibrium level of income Y^* is stable. But notice that output responds to demand in this model, increasing to meet an excess demand and falling in the case of deficient demand. Hence we say that in models of this type, output and income are *demand-determined*. If planned expenditures of one sort are deficient and not compensated for by expenditures of another type, demand and hence output fall. Thus only if planned savings (a withdrawal from expenditure) are 'mopped up' by investment (an injection of expenditure) do we have an equilibrium. This matching of 'withdrawals' and 'injections' is a corollary of the notion that equilibrium income equals the planned expenditure that would take place at that equilibrium income level. These general properties of equilibrium will hold even in more complicated models.

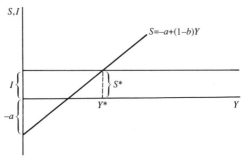

Fig. 18.3: Equilibrium savings and investment

We have shown why Y^* is an equilibrium in this economy and that it is *stable*, for if income is other than Y^* forces operate which take the economy towards Y^*. Two issues concerning Y^* have not been tackled, but they are of some importance. First consider the *existence* of an equilibrium, Y^*. In this model with fixed, positive investment and an increasing linear savings function with a negative intercept, equilibrium is more or less guaranteed (and is guaranteed if income may be indefinitely large). However, there are cases where the equilibrium need not exist. We give two examples. In the first investment is *induced*, that is, it depends on income, and in the second the savings function takes a specific form.

First suppose, that the investment function takes the form

$$I(Y) = \alpha + \beta Y \qquad \alpha, \beta > 0. \tag{18.19}$$

Such a formulation is not wildly unrealistic and need not cause problems of existence of Y^*: everything depends on the values of α and β. The details are left as an exercise, but Fig. 18.4 graphs an example where at no positive value of Y will $S = I$. Clearly if investment *is* induced we want it to be well-behaved.

In the second example, the savings function is increasing but non-linear and investment is fixed. In fact the savings function has the properties

$$S = S(Y), \qquad S'(Y) > 0, \qquad S''(Y) < 0 \tag{18.20}$$

and further $S(Y)$ approaches I as Y tends to infinity. This is graphed in Fig. 18.5. Thus with I as an asymptote to $S(Y)$ there is no finite value of Y^*. Although this case cannot be ruled out as a theoretical possibility, saving being bounded above as income increases could be regarded as representing unreasonable savings behaviour. Recall that the savings function is

$$S(Y) = Y - C(Y) \qquad \text{for all } Y,$$

so that

$$S'(Y) = 1 - C'(Y) \tag{18.21}$$

(the marginal propensity to save is one *minus* the marginal propensity to consume, as above in the linear example where $C'(Y) = b$), but also

$$S''(Y) = -C''(Y). \tag{18.22}$$

Hence, the assumption that $S''(Y) < 0$ implies $C''(Y) > 0$, whereas earlier in this section we argued that $C''(Y) \le 0$ was more reasonable. Clearly $C''(Y) \le 0$ implies $S''(Y) \ge 0$, in which case this particular non-existence problem will not apply.

The second issue concerning equilibrium is its *uniqueness*. Without going into detail, Fig. 18.6 gives an example of a savings function which produces more than one possible equilibrium. This may be considered further as an exercise.

Broadly speaking we will be interested only in well-behaved functional forms which do not produce these difficulties.

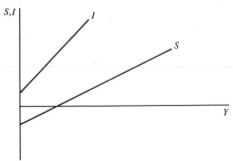

Fig. 18.4: Induced investment and the non-existence of equilibrium

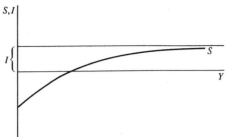

Fig. 18.5: Concave savings and non-existence of equilibrium income

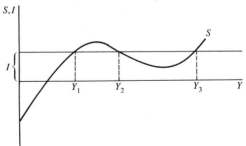

Fig. 18.6: Non-uniqueness of equilibrium

Exercise 18.1

1. Use (18.10) to establish the effects of changes in I, a, and b on C^*. Provide economic explanations of these results and depict the various changes in a diagram similar to Fig. 18.2.
2. Suppose an economy is described by the following system:

$$C = 100 + 0.8Y,$$

$$I = 500.$$

Find

(i)   the value of equilibrium income,
(ii) the value of equilibrium consumption,
(iii) the value of equilibrium savings,
(iv) the value of the multiplier.

What happens to this equilibrium if I increases by 100?

3. Suppose investment is induced and takes the form of equation (18.19) while consumption takes the form of (18.4). What restrictions on values of a, b, α and β produce

(i)   the non-existence of equilibrium as illustrated in Fig. 18.4,
(ii) an equilibrium value of Y?

Do the functions (18.4) and (18.19) allow for the possibility of multiple equilibria?

4. Continuing with question 3, find an expression for a stable equilibrium income level, assuming one to exist, and find an expression for the multiplier associated with an increase in a, autonomous consumption.

5. In Fig. 18.6, equation (18.15) is satisfied at three income levels, Y_1, Y_2, Y_3. All are possible equilibria but not all are stable. Why?

18.2 The IS relation

The simple income–expenditure model is a useful vehicle for introducing the idea of equilibrium. However, we have still left unspecified the determinants of aggregate investment. Of the many possibilities that present themselves our preferred specification of investment is one which makes it a function of the *rate of interest*. This is in line with the microeconomic story of investment of Chapter 13. Much investment is financed either by companies issuing shares or by borrowing from banks. It is this latter aspect which motivates our modelling of investment. If, for simplicity, we assume that the economy has a single rate of interest, r (which essentially assumes away interest rate differentials arising from banks' profit margins or from the existence of a wide variety of financial assets of different maturities and risks), we could argue that, other things remaining constant, an increase in r lowers investment because it increases the cost of borrowing. The 'other things' being kept constant here are the rates of return to various available investment projects. For example, investment need not fall when interest rate rises if the rates of return or profitability of all projects increase also. This caveat aside these arguments allow us to write

$$I = I(r), \qquad I'(r) < 0. \tag{18.23}$$

Now consider what impact the introduction of (18.23) has on the determination of equilibrium income. For a given interest rate r, we know that equilibrium income Y^* is given by (substituting (18.23) into (18.8))

$$Y^* = C(Y^*) + I(r) \tag{18.24}$$

or alternatively by (substituting (18.23) into (18.15))

$$S(Y^*) = I(r). \tag{18.25}$$

Our formulation of the investment function appears to have made no improvement to the model – investment is still exogenous because the interest rate is arbitrarily fixed. Equation (18.25) only tells us what equilibrium income will be *if* the interest rate is at a particular value. We have, so far, one equation with two unknowns. Presently, we will look elsewhere for an explanation of the interest rate. For the moment it is useful to see how equilibrium income changes with changes in the interest rate. Equation (18.25) is an implicit function for Y^* in terms of r, and taking the total differential of (18.25) gives

$$S'(Y^*)dY^* = I'(r)dr \qquad (18.26)$$

so that

$$\frac{dY^*}{dr} = \frac{I'(r)}{S'(Y^*)} < 0. \qquad (18.27)$$

Thus the relationship between the interest rate and the level of income which clears the goods market has a negative slope. The intuition here is that a higher interest rate lowers planned investment, which lowers demand, output, and income. The relationship between r and Y^* implicitly given by (18.25) is known as the IS curve because it gives a locus of (Y, r) points which satisfy the equality of saving (S) with investment (I). An example of an IS curve is drawn in Fig. 18.7.

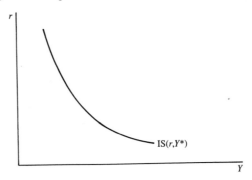

Fig. 18.7: An IS curve

Following our earlier procedure, by assuming sufficiently simple forms for $S(Y)$ and $I(r)$ we may solve (18.25) explicitly. Let $S(Y)$ take the form in (18.16) and let $I(r)$ take the linear form,

$$I(r) = e - hr \qquad e, h > 0 \qquad (18.28)$$

where e and h are constants. Substituting (18.16) and (18.28) in (18.25) and solving for Y^* in terms of r gives

$$Y^* = \frac{a + e - hr}{(1 - b)} \qquad (18.29)$$

and the form of (18.27) may be verified by differentiating Y^* with respect to r to give

$$\frac{dY^*}{dr} = \frac{-h}{(1-b)} = \frac{I'(r)}{S'(Y^*)} < 0. \tag{18.30}$$

Notice that in this case, the IS function (18.29) is a straight line as illustrated in Fig. 18.8, and we are able to ask what happens to the position of this line as $a, e, h,$ and b change.

For example, take an increase in h from h_0 to h_1, $h_1 > h_0$, then by inspection of Fig. 18.8 the intercept on the r-axis falls while that on the Y-axis remains unchanged, and so IS pivots down around the point $(a+e)/(1-b)$. Mathematically we present this movement as follows. Observe that at any r a fall in h lowers income. Thus if we differentiate (18.29) with respect to h, r constant, we find

$$\frac{dY^*}{dh} = \frac{-r}{(1-b)} \leq 0 \qquad r \geq 0, \tag{18.31}$$

which establishes that equilibrium income at a given positive interest rate falls as h increases. However, it tells us more, since the amount by which income falls depends on r. As an extreme, $dY^*/dh = 0$ where $r = 0$ – this is clear from Fig. 18.8. To find the amount by which income changes at other interest rates we take

$$\frac{d}{dr}\left(\frac{dY^*}{dh}\right) = -(1-b)^{-1} < 0, \tag{18.32}$$

which says that (dY^*/dh) falls as r increases, but since (dY^*/dh) is negative (18.32) says that (dY^*/dh) becomes *more negative* as r increases. Thus the fall in Y^* resulting from an increase in h is greater the higher is r. This too is apparent from Fig. 18.8. The effects of other parameter changes on the position of the IS curve are left as an exercise.

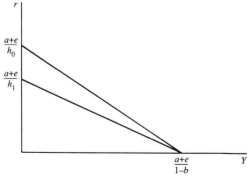

Fig. 18.8: The effect of an increase in h on the IS curve

The IS curve is a locus of equilibrium points. The IS curve shows combinations of interest rate and income which equate demand with supply in the goods market. Thus at any point on the IS curve, supply equals (planned) demand. What about

points off the IS curve? Consider the inequalities in (18.17) (excess demand) and
(18.18) (excess supply) but write these as

$$S(Y) < I(r_0) \qquad \text{for} \quad Y < Y^* \qquad\qquad (18.17)'$$
$$S(Y) > I(r_0) \qquad \text{for} \quad Y > Y^* \qquad\qquad (18.18)'$$

where we shall keep the interest rate fixed at r_0. Fig. 18.9 illustrates. At r_0,
equilibrium income is Y_0^*. Now at incomes less than Y_0^* we have a situation
similar to (18.17) where $S(Y) < I(r_0)$, so any point to the left of Y_0^* along r_0 is
generating excess demand. On the other hand at any point to the right of Y_0^* along
r_0 there is excess supply. Since we could have chosen *any* value of r to perform
this exercise it follows that all points to the south-west of IS are points of excess
demand (EDG) while all points to the north-east of IS are point of excess supply
in the goods market (ESG).

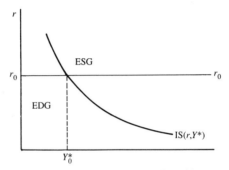

Fig. 18.9: Disequilibrium points off the IS curve

The IS curve is the first half of our IS-LM story. Clearly with the interest rate as
yet undetermined we must look elsewhere to complete the picture.

Exercise 18.2

1. Investment in this section is influenced by the cost of borrowing. If we regard savings
   as the supply of 'loanable funds' might not this supply also depend on the interest
   rate? (See Chapter 13.) How do we ensure that the general properties of the IS curve
   are preserved if S depends on both Y and r (I depending on r alone)? What further
   difficulties are encountered if part of investment is induced so that $S = S(Y, r)$ and
   $I = I(Y, r)$? Derive an expression for the slope of the IS curve in each of these cases
   providing your own assumption about the partial derivatives S_Y, S_r, I_Y, I_r.
2. Use the linear IS curve model of equation (18.29) and Fig. 18.8 to see how the curve
   shifts as a, e, and b change.
3. The IS curve mapped out for equation (18.29) in Fig. 18.8 is linear. Using the general
   form (18.25) and (18.27) show under what condition IS will be convex to the origin.
   [Hint: differentiate (18.27) once more with respect to r.] Is this condition reasonable?
4. Use the fact that I depends on r and Fig. 18.3 to derive the IS curve diagrammatically.
5. In identifying the zones ESG and EDG in Fig. 18.9, the interest rate was fixed and
   different incomes were considered to produce the inequalities in (18.17)' and (18.18)'.

Using a similar argument identify ESG and EDG points keeping *income* fixed and studying the implications of different interest rates.

18.3 Assets, money, and the LM relation

That the second part of the IS-LM story should be found in money and related markets is natural. This is because the interest rate, which figured in the determination of income in section 18.2, is strongly influenced by money market activity. The aim in this section is to present a brief account of the money market and the role of the interest rate. In this model, the interest rate will bring into equality the demand and supply of money at a given income level. We take a greatly simplified view of the world in which individuals may hold wealth in two forms: money and bonds. Both money and bonds are created by the government. For the present, this is the only role of government in th eeconomy. Money here is to be thought of as an amount held in a non-interest-bearing bank account, or alternatively in the form of cash in hand. The important point is that it is an asset which does not earn interest. 'Bonds' on the other hand represent an asset which bears interest. In this economy there is only one interest rate, r, and this is also the rate of return on the bond. In practice there are many interest-bearing assets, including easy access deposit accounts at banks, building society accounts and other financial institutions, but it is a useful simplification to collapse these into just one asset, a bond.

In view of this, the question to be answered is why, if wealth may be held in an interest-bearing form (bonds), should any money, which bears no interest, be held at all? We give two explanations for this behaviour. First, money is required to carry out *transactions*. In our simplified world we assume that goods and services may only be bought for cash or by cheque drawn on a current account. Thus individuals will wish to keep 'transactions balances'. Secondly, money will be held if individuals expect bond prices to fall in the near future. The idea here is that individuals may *speculate* as to future bond price movements and will want to hold cash to make a bond purchase when their price falls. Money deposits held for this purpose are often referred to as 'speculative balances'. We discuss each of these separately.

18.3.1 Transactions demand

Emphasis is often given to the role of income in influencing the transactions demand for money. Since consumption is positively related to income, the volume of transactions in goods and services will influence the volume of cash holdings. The tightness of the relationship between cash required and the volume of transactions depends on the stability or constancy of the *velocity of circulation*. This velocity is the average number of times a unit of currency (a pound or a dollar) is used in transactions. It is a measure of how hard each unit of currency is made to work. An increase in the volume of transactions may not produce an increased need for cash if each unit of currency changes hands more quickly so that the velocity of circulation increases. With a constant velocity, however, an increased volume of transactions will increase the cash requirement. We conclude then that the transactions demand for cash will depend positively on the level of income at a given velocity of circulation.

A related consideration is the role of the interest rate in the transactions demand. Individuals always have the option of selling bonds to provide themselves with cash for transactions. Naturally, they will be less inclined to cash in their bonds the higher is the interest rate. The prospect of losing interest on bonds will tend towards frequent encashment of a small number of bonds over a given period rather than one large encashment at the begining to finance transactions through a complete period. On the other hand, since the conversion of bonds into cash often involves a fixed brokerage fee, fewer encashments would be cheaper. Hence the rate at which cash is obtained from bond sales, and hence the average cash holding, will be determined by a trade-off between the two costs – the opportunity cost (the interest rate) and the brokerage cost. The higher is the interest rate the lower will be the desired average cash holding.

Transactions demand is therefore influenced both by income, determining the volume of transactions, and the interest rate, representing the opportunity cost of holding cash. The following formal model, due to William Baumol and James Tobin brings out these influences.

Suppose an individual has a given amount of income, Y deposited in a bank account at monthly intervals. During each month the individual consumes Y at a uniform rate, so that all of Y is consumed by the end of the month, half of Y is consumed by mid-month, and so on. Bank deposits earn interest at a rate r per dollar per month. To draw on the account, the individual makes personal trips to the bank at a cost of k per trip independent of the amount withdrawn from the account. If visits to the bank take place at regular intervals and the amount withdrawn each time is the same, then the *average* monthly balance is that which is in place exactly halfway through the month, i.e., the average balance is $Y/2$. The amount withdrawn on each visit is W and the number of visits is therefore $n = Y/W$. Each time W is withdrawn it is used uniformly until the next visit. In these intervals the average holding of cash is $W/2$ and so the interest lost over the month is therefore $rW/2$. Total costs of converting the account into cash are the sum of transactions costs of trips to the bank, and the interest opportunity cost of the average monthly cash-holding,

$$K = kn + r\frac{W}{2}$$

$$= k\frac{Y}{W} + r\frac{W}{2}$$

The individual chooses the size of withdrawal (and hence the *number of withdrawals*) so as to minimize K. So

$$\frac{dK}{dW} = -k\frac{Y}{W^2} + \frac{r}{2} = 0$$

and

$$W = \sqrt{\frac{2kY}{r}}$$

which is often referred to as the 'square root' formula. We may think of W as the demand for cash or simply the demand for money. It is easy to see that the square root formula implies that the demand for money is increasing in income and decreasing in the interest rate.

18.3.2 Speculative demand

The role of the interest rate is often emphasized through its effect on the speculative demand for cash. There are many ways of presenting the speculative demand, but we choose one of the simplest. We start by considering an individual speculator and then aggregate to derive the market speculative demand. As a preliminary, however, we should consider the market for bonds explicitly.

We assume that there is just one type of bond, though in practice the government issues many types of bonds of different maturities. Our bond, known as a *perpetuity*, is infinitely-lived and may be bought and sold in the market for bonds. Each bond is an obligation on the part of the government to pay a fixed amount, B (known as the *coupon*), in each period to the bearer of the bond.

The present value of a bond paying B in each future period must be equal to the price at which bonds exchange. Buyers of bonds are not going to offer a price which exceeds the total present value and suppliers are not going to accept a price below the total present value. Hence the bond price P_B must be equal to the total discounted value of the stream of Bs. To compute this total present value, consider first the value today of B received in one year's time. This is $B/(1 + r)$. Why? Consider the choice between receiving A now and an amount B in one year's time. Since A could be used for a one-year investment in a bond with an interest rate r per year, you should be indifferent between A now and B in one year's time if $A + rA = B$, or $A(1 + r) = B$. Hence $A = B/(1 + r)$ is the value *now* of receiving B in one year's time. Similarly, the present value of receiving B in two year's time is $B/(1 + r)^2$ and so on. Since a bond pays B in *each* future period (in our simple case) for ever the total present value is

$$B\left[\frac{1}{1 + r} + \frac{1}{(1 + r)^2} + \dots\right] = \frac{B}{(1 + r)}\left[1 + \frac{1}{(1 + r)} + \frac{1}{(1 + r)^2} + \dots\right]$$

$$= \left(\frac{1}{1 + r}\right) B \sum_{t=0}^{\infty}\left(\frac{1}{1 + r}\right)^t$$

$$= \frac{B}{r}$$

The price of the bond is therefore

$$P_B = \frac{B}{r}.$$

The bond price and the rate of interest on the bond (the yield) are inversely related. Remember the coupon, B, is fixed.

We postulate that an individual choosing between holding wealth in money or bonds will buy bonds when the price is low (the interest rate high) and hold cash when the price of bonds is high (the interest rate low), because by 'buying low' and 'selling high' individuals can make capital gains. The speculation enters into things because individuals must choose for themselves when is the best time to act. Bonds will be bought because an individual *believes* that the price cannot fall further and will be sold because an individual *believes* the price cannot rise further. Since different individuals hold different beliefs about whether to buy and to sell, the aggregate demand for bonds will be inversely related to bond price or positively related to the interest rate. The lower the interest rate (higher the bond price), the more people will be expecting it to rise (price to fall) and the more cash is held in anticipation.

The implication for the speculative demand for money is that speculative cash balances will be higher the lower is the interest rate on bonds and lower the higher the interest rate. However, underlying this relation is an assumed pattern of expectations in the market. We ignore the general equilibrium question of how these expectations are formed.

18.3.3 The LM relation

In general, then, the demand for money depends both on Y and on r. We may write, in view of our discussion,

$$L = L(Y, r) \qquad L_Y > 0, \ L_r < 0. \tag{18.33}$$

Fig. 18.10 illustrates the demand for money as a function of the interest rate for a *given* income level. The line L_T represents the least amount of cash society needs for transactions purposes, while the L-curve represents the total demand for money. The difference between L and L_T is the interest-sensitive demand for money, say L_S.

This completes our specification of the demand for money. The supply of money, M, is an exogenous constant fixed by the government. Equilibrium in the money market is given by the equality of the demand for and supply of money. This is illustrated in Fig. 18.10. More specifically, for a given income level Y, the equality of M and L is determined by the equilibrium interest rate r^*. Hence, the equilibrium condition is

$$M = L = L(Y, r^*). \tag{18.34}$$

Furthermore, for a given money supply, variations in income must be accompanied by variations in r^* if the equality of (18.34) is to be maintained. Equation (18.34) therefore implicitly gives r^* as a function of Y, just as the equation (18.25) implicitly gave Y^* as a function of r. Whereas (18.25) gave us the IS relation, (18.34) gives us the LM relation. The IS shows combinations of interest rate and income which give equilibrium in the goods market, while LM shows combinations of interest rate and income which give equilibrium in the money market. The LM relation has a positive slope, since total differentiation of (18.34) holding M constant gives,

$$0 = L_Y(Y, r^*)dY + L_{r^*}(Y, r^*)dr^* \tag{18.35}$$

or

$$\frac{dr^*}{dY} = \frac{-L_Y(Y, r^*)}{L_{r^*}(Y, r^*)} > 0. \tag{18.36}$$

A typical LM curve is illustrated in Fig. 18.11.

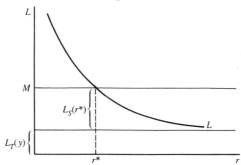

Fig. 18.10: Equilibrium in the money market

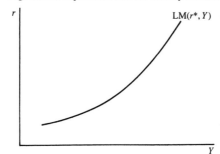

Fig. 18.11: An LM curve

As with the IS curve, some manipulations of LM are possible using a special (linear) case. These are left as an exercise. However, it is instructive to see what effect an increase in the money supply has on the LM curve. Holding *income* constant, total differentiation of (18.34) now gives

$$dM = L_{r^*}(Y, r^*)dr^*, \tag{18.37}$$

so that

$$\frac{dr^*}{dM} = \frac{1}{L_{r^*}(Y, r^*)} < 0. \tag{18.38}$$

An increase in the money supply, income constant, lowers the interest rate. As illustrated in Fig. 18.12, this is generally true at any given income level, hence

LM shifts out to the right from LM_0 to LM_1 as the money supply increases.

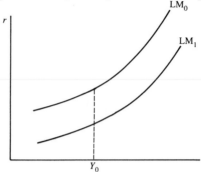

Fig. 18.12: The effect of an increase in the money supply on the LM curve

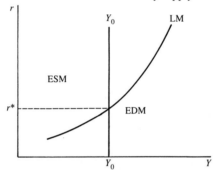

Fig. 18.13: Disequilibrium points off the LM curve

We are also able to identify disequilibrium zones around LM as we did with IS in Fig. 18.9. Given that $L_r < 0$ we have, for a given income level,

$$M < L(Y,r) \quad \text{for } r > r^* \tag{18.39}$$

and

$$M > L(Y,r) \quad \text{for } r < r^*, \tag{18.40}$$

where obviously (18.39) is the condition for excess supply of money and (18.40) that for excess demand. In terms of Fig. 18.13, if income is held fixed at Y_0, points along Y_0 at interest rates above r^* are points of excess supply of money, ESM, whereas points along Y_0 at interest rates below r^* are points of excess demand for money, EDM. Since this analysis may be conducted at any income level, we conclude that the area to the 'north-west' of LM is a zone of excess supply and the area to the 'south-east' a zone of excess demand for money.

This completes our discussion of the LM relation. In the next section we combine the IS and LM relations into a single model of overall equilibrium.

Exercise 18.3

1. Use Fig. 18.10 to study the effect on r^* of (a) an increase in Y with a fixed money supply, and (b) an increase in M at a fixed income.
2. Suppose $L(Y, r) = kY - lr$ where k and l are positive constants. Use (18.34) to find an expression for the equilibrium interest rate, r^*. Find the slope of the LM curve and examine the effects of changes in M, k, and l on both the position and slope of LM.
3. Under what circumstances will the LM curve be convex as drawn in Fig. 18.11. [Hint: differentiate (18.36) once more with respect to Y.]

18.4 Overall equilibrium

Our IS and LM relations give a system of two equations with two unknowns, Y^* and r^*. These two equilibrium values simultaneously satisfy the equilibrium conditions in the goods and money markets. As long as our IS and LM equations are sufficiently 'well-behaved' there will be one combination of Y^* and r^* which give overall equilibrium. This position is illustrated in Fig. 18.14.

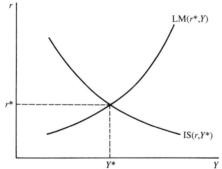

Fig. 18.14: Overall IS-LM equilibrium

Mathematically the overall equilibrium is given as the joint solution to the system (from (18.25) and (18.34))

$$S(Y^*) = I(r^*),$$
$$M = L(Y^*, r^*). \tag{18.41}$$

Explicit solutions for the unique pair Y^* and r^* are available if we employ the specific linear equation used earlier. For example, on the IS side we have (18.29)

$$Y^* = \frac{a + e - hr^*}{1 - b} \tag{18.42}$$

while using $L(Y, r) = kY - lr$ (question 2 of Exercise 18.3) gives

$$r^* = \frac{kY^*}{l} - \frac{M}{l}. \tag{18.43}$$

This problem could not be simpler; we have two independent linear equations in two unknowns. Substituting (18.43) into (18.42) and solving for Y^* gives

$$Y^* = \frac{a + e + hM/l}{1 - b + kh/l},$$

(18.44)

whilst substitution of (18.42) into (18.43) and solving for r^* gives

$$r^* = \frac{k(a + e) - M(1 - b)}{l(1 - b) + kh}.$$

(18.45)

Some comparative statics are obviously available for the whole equilibrium. For example, an increase in the autonomous component of investment e has the following effects:

$$\frac{dY^*}{de} = \frac{1}{1 - b + kh/l} > 0,$$

(18.46)

$$\frac{dr^*}{de} = \frac{k/l}{1 - b + kh/l} > 0.$$

(18.47)

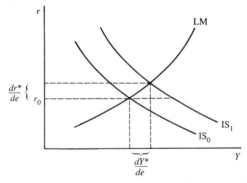

Fig. 18.15: The effect of an increase in e on IS-LM equilibrium

The intuition behind these effects takes account both of the goods *and* money market consequences of an increase in exogenous investment. The immediate effect is to increase the demand for goods and hence income. However, since the money supply is unchanged the increase in income puts increased pressure on the demand for money, and equilibrium in the money market can only be maintained if the interest rate rises. As the interest rate rises, speculative balances fall as cash is released for the purchase of bonds. The money released may then be used in the new higher volume of transactions created by the higher income. But the increase in r 'chokes off' some investment and therefore income is raised by less than in the simple income–expenditure model in which the interest rate does not change and is fixed at say r_0. The net result of all this activity is an increase in Y^* and in r^*. Fig. 18.15 illustrates this. It is clear by inspecting Fig. 18.15 and equations (18.46) and (18.47) that the size of these effects depends not only on how far IS shifts to the right as *e* increases but also on the slope of the LM curve. Further comparative-static effects are left as exercises. The further use of the IS-LM model to consider more comparative statics is the subjecte of the following chapter, and there our analysis will not be confined to special cases where explicit solutions are available.

Exercises 18.4

1. Figs 18.9 and 18.13 show zones of disequilibrium around the IS and the LM curves respectively. Use these two diagrams to complete the four zones of disequilibrium in Fig. 18.14.
2. Use (18.44) and (18.45) to find the effects of changes in a, b, h, k, and l on the equilibrium (Y^*, r^*). Illustrate each in an IS-LM diagram and provide economic reasons for your answers.

18.5 Summary

This chapter has introduced one of the basic tools of macroeconomic analysis, the IS-LM model. Used properly this framework is both powerful and insightful. The use of comparative-static analysis gives a good understanding of both the mechanics of the model and the economic interpretation of changes in equilibrium solutions.

The model has assumed fixed prices with the result that output (income) is demand-determined.

The two-way dependence of income on expenditure and expenditure on income results in a multiplier process in which a given increase in expenditure leads to a larger increase in income.

The IS-LM model uses the interest rate as a key link between money markets and goods markets, so the change in the demand for goods has the effect of increasing the transactions demand for money, leading, with a fixed money supply, to an increase in interest rates.

19 Government Expenditure and Taxation

The government plays an important part in macroeconomic analysis. Its effects on the income–expenditure model are two-fold. First, the government buys part of the economy's output. So far, we have been concerned with the expenditures of private agents; consumers and investors, but to this list we must add expenditures by the government. Secondly, the government affects the spending behaviour of private agents by altering their purchasing power. Direct taxes and transfers affect the amount of income available for spending – *disposable income*. In this chapter we look at government spending and taxation in the context of the income expenditure model. We conclude by showing the implications for the IS curve of including government for the IS curve. This serves as a preliminary to our discussion of fiscal and monetary policy in the IS-LM model in the following chapter.

19.1 Taxation and expenditure

Denote the total value of government expenditures by G. Ignoring transfers for the moment, we will denote the total value of direct taxes by T. Disposable income is then defined as income less taxes, or

$$Y_d \equiv Y - T. \tag{19.1}$$

Notice that T is the value of direct taxes such as the income tax. Indirect taxes, such as sales taxes or value-added tax which are incurred only if items are purchased, are easily subsumed in the definitions of C, I, and G. [Briefly, this is achieved by defining expenditure either in terms of 'factor cost' or 'market prices'. If we measure expenditure at market prices we will be including the indirect tax (or subsidy) element in the selling price. To convert expenditure into a factor cost valuation we simply subtract the value of net taxes, as we saw in Chapter 16. These alternative methods of valuing expenditure do not affect the theory of income determination but do require alternative interpretations of Y as being GDP at factor cost or GDP at market prices as appropriate.] For the moment we will treat T as an exogenous, known, and fixed amount.

One important change to the income–expenditure model is in the specification of consumption. This now depends on disposable income, so that (18.3) is replaced by

$$C = C(Y_d) \qquad 0 < C'(Y_d) < 1, \tag{19.2}$$
$$= C(Y - T). \tag{19.3}$$

The interpretation of $C'(Y_d)$ is now the marginal propensity to consume out of *disposable income*. With T and G exogenous, equilibrium income is now given not by (18.8) but by

$$Y^* = C(Y^* - T) + I + G. \tag{19.4}$$

Furthermore, income less consumption is now taken up by savings *and* tax payments to the government, so that in place of (18.15) we have

$$S(Y^* - T) + T = I + G. \tag{19.5}$$

Notice, it is no longer necessary that $S = I$, though this will be the case if government expenditure G equals tax revenue, T; that is if the government runs a *balanced budget*. In general it is only necessary that the sum of savings and taxation be equal to the sum of investments and government expenditure. Notice furthermore that (19.5) takes the 'withdrawals *equals* injections' form, since tax revenue is money not spent unless taken up by government expenditure or advanced to investors.

As before, we may subject the equilibrium income Y^* to a comparative-static analysis. We do this first using a linear functional form for $C(Y - T)$ and then consider the more general case. Suppose we specify the consumption function as

$$C = a + b(Y - T), \tag{19.6}$$

where b is a constant fraction, then equilibrium income is (substituting (19.6) into (19.4) and solving for Y^*)

$$Y^* = \frac{a + I + G - bT}{(1 - b)}. \tag{19.7}$$

Hence, the comparative-static effects on Y^* of changes in G and T are

$$\frac{dY^*}{dG} = (1 - b)^{-1} > 1 \tag{19.8}$$

and

$$\frac{dY^*}{dT} = \frac{-b}{(1 - b)} < 0. \tag{19.9}$$

Hence the autonomous consumption and exogenous investment multiplier, $(1 - b)^{-1}$, of Chapter 18 is also the government expenditure multiplier. An increase in direct taxation (a withdrawal from the system), however lowers equilibrium income. The effects of changes in G and T on equilibrium consumption are left as an exercise.

Returning to the general case, we may totally differentiate (19.4) to give

$$dY^*[1 - C'(Y^* - T)] + C'(Y^* - T)dT = dG. \tag{19.10}$$

Setting $dT = 0$ and solving for dY^*/dG gives

$$\frac{dY^*}{dG} = \frac{1}{[1 - C'(Y^* - T)]} > 0 \qquad (19.11)$$

which is analogous to (19.8), while setting $dG = 0$ and solving for dY^*/dT gives

$$\frac{dY^*}{dT} = \frac{-C'(Y^* - T)}{1 - C'(Y^* - T)} < 0 \qquad (19.12)$$

which is analogous to (19.9).

   This model is easily extended to allow for the fact that total tax revenue may itself depend on income so that higher incomes generate more tax. That is,

$$T = T(Y) \qquad 0 < T'(Y) < 1. \qquad (19.13)$$

The restriction on $T'(Y)$ simply means that the marginal tax rate is less than 100 per cent. In practice the system of tax rates is rather complicated with different income bands having different tax rates, and there is also an involved system of personal allowances depending on marital status and so on. However, (19.13) will suffice as a characterization of the income tax relationship for our purposes. The effect on (19.4) is as follows,

$$Y^* = C(Y^* - T(Y^*)) + I + G, \qquad (19.14)$$

and on (19.5)

$$S(Y^* - T(Y^*)) + T(Y^*) = I + G. \qquad (19.15)$$

   Before considering the comparative-statics of this equilibrium and how it affects the IS curve it may be useful to look at a special (linear) case. Suppose we make taxes a simple proportion of income,

$$T(Y) = tY \qquad 0 < t < 1, \qquad (19.16)$$

then using (19.16) in (19.6) and substituting in (19.4) gives the solution of equilibrium income at a particular interest rate as

$$Y^* = \frac{a + I + G}{1 - b(1 - t)}, \qquad (19.17)$$

and the comparative statics for a change in G are now

$$\frac{dY^*}{dG} = [1 - b(1 - t)]^{-1} > 0. \qquad (19.18)$$

   The effects of changes in a, b, and t are left as an exercise, as are the comparative statics on equilibrium consumption C^* in this case.

Returning to the general case, the total differential of (19.14) is

$$dY^*\{1 - C'(Y^* - T(Y^*)) + C'(Y^* - T(Y^*))T'(Y^*)\} = dG$$

or

$$dY^*\{1 - C'(Y^* - T(Y^*))[1 - T'(Y^*)]\} = dG,$$

so that

$$\frac{dY^*}{dG} = \{1 - C'(Y^* - T(Y^*))[1 - T'(Y^*)]\}^{-1} > 0 \qquad (19.19)$$

which is readily compared with the special case (19.18).

Results such as (19.19) provide a justification of government intervention in the economy. It appears that the economy can generate an equilibrium income level that is too 'low'. Since output is determined by demand, a low level of demand implies a low level of output. By implication, when output is below the economy's capacity, there are unused resources – particularly labour – and income (hence welfare) is lower than it need be. All that is lacking in this case is more demand. The apparent inability of private investors and consumers to spend enough on output to guarantee full employment then suggests a role for governments in compensating for the shortfall in demand.

There is one further modification of the income–expenditure model concerning the effect of government transfers, such as pensions, supplementary and unemployment benefits, family allowances (and so on), on disposable income. These payments from the government to private households add to disposable income. If we denote transfers by R we change (19.1) into

$$Y_d \equiv Y - T + R \qquad (19.20)$$

where in the simplest specification, R is a constant, known amount. Since the inclusion of R affects the analysis of equilibrium and its comparative statics in a straightforward way we leave the further analysis to the exercise. We will have more to say about G, T, and R in the following chapter.

Exercise 19.1

1. Using the model of equilibrium income in equation (19.7) and the consumption function in (19.6), show the effect of an increase in G and in T on equilibrium consumption, C^*.
2. Use (19.4) or (19.5) to show that the IS curve is negatively sloped, assuming that the investment function depends on the interest rate as in (18.23).
3. Use (19.17) to show the effect of an increase in a, b, and t on Y^*. What are the effects of changes in G, a, b and t on C^* in this case? What are the effects of changes in these parameters on T^* $(\equiv T(Y^*))$?
4. Use the consumption function (19.2) and the definition of Y_d in (19.20) to derive an expression for equilibrium income to replace (19.4) (i.e., where T is a constant). Using the definition of Y_d in (19.20) find an expression to replace (19.14) (i.e., where T depends

on income). In each case show the effect of an increase in R and Y^*. Hence, discuss (by comparing with earlier effects of changes in G and Y^*) whether an increase in direct government expenditure on goods might be preferred to an equivalent increase in state-provided assistance to families, as a means of increasing income.

19.2 The government budget

You may have the impression from the previous section that the government is unconstrained in its ability to generate additional demand. But we know from discussion by policy-makers that government expenditures and the way these expenditures are financed are the source of much discussion and debate.

A central concept in more general discussions of government policy is that of the *budget deficit* (or *surplus*). In the model without transfers, the budget deficit D is given by the excess of government expenditures over receipts,

$$D = G - T. \tag{19.21}$$

In the model with transfers this becomes

$$D = G + R - T. \tag{19.22}$$

Like households, governments may operate with deficits by borrowing at least for a short time. Unlike households, however, the government may use its powers to expand the economy's money supply in order to pay for its purchases of goods and services. In either case, as we will see, the financing of fiscal policy has monetary implications so that fiscal and monetary policy are intimately linked. The following chapter explores some of these linkages.

19.2.1 The government budget constraint

Suppose that, starting with a balanced budget in which government expenditure and tax revenue are equal, the government increases its expenditure. Other things being equal, this will create a budget deficit – the government is spending in excess of its (tax) income. Clearly, since bills must be paid, the government must find the shortfall of money from somewhere. There are four main ways in which governments may fill this gap:

(i)   borrowing from abroad (for example from the IMF or other countries),
(ii) borrowing from domestic residents and banks,
(iii) increasing taxation,
(iv) increasing the money supply.

The first of these is ruled out in this chapter because of the closed economy assumption. The three remaining ways of financing the deficit are accounted for in the following relationship, known as the *government budget constraint*;

$$G - T = \Delta M + \frac{\Delta B}{r}. \tag{19.23}$$

This says that the excess of expenditure over tax revenue must be equal to the increase in the money supply (cash) and the increased revenue from the sale of bonds. Clearly, the extent to which taxes are raised to reduce the deficit $G - T$, reduces the need to increase the money supply or to issue bonds. Increasing taxes is an obvious way of reducing the deficit and creating a balanced budget. Thus we have a pure fiscal policy in which $\Delta G = \Delta T$ with $\Delta M = \Delta B/r = 0$. We consider the consequences for national income of this type of policy in the next section.

The government may decide, however, to run a deficit and pay for the extra goods and services by increasing the money supply only. Thus, $G - T = \Delta M$ with $\Delta B/r = 0$, and the deficit is said to be *money-financed*. In principle there is no difficulty for the government to pay for extra goods and services by a new issue of high-powered money. As we noted in Chapter 17, the government is in the unique position of being able to print money. We see the implications of a money-financed deficit in the following chapter.

Finally, the government budget constraint tells us that with the tax and money supply unchanged, a government deficit may be financed by a bond-issue so that $G - T = \Delta B/r$, $\Delta M = 0$. This is the way governments borrow from banks and domestic residents. We saw in Chapter 17 that the public see government bonds as an asset which earns interest, r. Recall from section 18.3.2 that B/r is the price paid for bonds of total (face) value B, so that an increase in bonds of ΔB generates an extra revenue for the government of $\Delta B/r$, representing the amount paid by the public for the extra bonds. This revenue from bond sales is then used by the government to purchase goods and services. The implications of a bond-financed deficit are discussed fully below.

19.3 Fiscal policy in the income–expenditure model

We start by studying the model of section 19.1 further. In particular we take the version in which direct taxes are lump sum and exogenous. We know that, given a constant interest rate the effects of changes in government expenditure and taxation on equilibrium income are given by equations (19.8) and (19.9) in the special (linear) cases and by equations (19.11) and (19.12) in the more general formulation. We will develop the ideas of this section using the latter, more general, version. Hence, equation (19.4) is the appropriate equilibrium condition,

$$Y^* = C(Y^* - T) + I + G \tag{19.24}$$

and for convenience we rewrite (19.11) and (19.12),

$$\frac{dY^*}{dG} = \frac{1}{1 - C'(Y^* - T)} > 0 \tag{19.25}$$

and

$$\frac{dY^*}{dT} = \frac{-C'(Y^* - T)}{1 - C'(Y^* - T)} < 0. \tag{19.26}$$

In the language of the previous section the result in (19.25) represents the effect of a bond-financed increase in government expenditure, since we have not simultaneously allowed added government expenditure to be accompanied by increased taxes or an increase in the money supply (which of course is absent from the income–expenditure model). It is a fairly straightforward matter to use the effects in (19.25) and (19.26) to study the effects of a balanced budget increase in government expenditure. We start by observing that, according to (19.25) and (19.26), an increase in income may be generated by an increase in G or by a reduction in T.

Now, does the government regard increases in its own expenditure and reductions in direct tax as perfect substitutes? Apparently not, since a small reduction in T increases Y^* by

$$C'(Y^* - T)/[1 - C'(Y^* - T)],$$

whilst an *increase* in G of the same magnitude *increases* Y^* by

$$1/[1 - C'(Y^* - T)]$$

and the latter is clearly greater than the former. Notice that both of these changes have equivalent budgetary implications so that *as far as the budget is concerned* they are perfect substitutes. As far as their effectiveness in increasing Y^* is concerned they are not. This is because an increase in G increases expenditure directly while a reduction in T of equal magnitude first increases disposable income some of which is saved and the remainder consumed. Hence not all of the reduction in T is passed on as an increase in expenditure. This observation leads to a famous result known as the *balanced budget multiplier theorem*. A budgetary neutral increase in expenditure requires $dG = dT$. Totally differentiating (19.24) and enforcing this restriction gives

$$dY^*[1 - C'(Y^* - T)] = dG[1 - C'(Y^* - T)], \qquad (19.27)$$

so that $dY^*/dG = 1$. Thus *in this model* the balanced budget multiplier is unity: a budget-neutral increase in G increases Y^* by the same amount. The government expenditure multiplier is therefore smaller than (19.25) but is still positive.

This precise result is also obtained if the tax function takes the form in (19.16) or more generally (19.13). If the budget is balanced and is always to remain so, it must be that $T = G$ always, or alternatively $T(Y^*) = G$ for all G. Using (19.14) with $T(Y^*) = G$ gives,

$$Y^* = C(Y^* - G) + I + G. \qquad (19.28)$$

Hence,

$$dY\{1 - C'(Y^* - G)\} = dG\{1 - C'(Y^* - G)\} \qquad (19.29)$$

or again $dY^* = dG$. However, in this case we may explore the relationship between the policy instruments under the government's control and the balanced budget multiplier. Suppose the tax function takes the form in (19.16), viz.,

$$T(Y) = tY \qquad 0 < t < 1. \qquad (19.30)$$

In the earlier case, with a lump-sum tax the policy-maker's rule for arriving at the balanced budget multiplier was very simple; ensure that $dT = dG$. In (19.30), however, it is the tax *rate* that is under direct control, and the rule for knowing how much to change t when G increases so as to leave the budget in balance is a little more involved. This is because tax revenue T now increases automatically when G increases, because more taxable income is generated by the government expenditure multiplier. Fortunately we know by how much income increases as government expenditure and tax revenue increase by the same amount, since from (19.29)

$$dY = dG \ (= dT).$$
$$(19.31)$$

Now with the government able to vary t, the total tax revenue change is

$$dT = \frac{\partial T}{\partial t} dt + \frac{\partial T}{\partial Y} dY$$

or using (19.30)

$$dT = Ydt + t\, dY,$$
$$(19.32)$$

but $dT = dY$ so using this we may solve (19.32) for dt to give

$$dt = \frac{dY}{Y}(1 - t).$$
$$(19.33)$$

Thus the required change in the tax rate is the product of the proportional change in income dY/Y and the proportion of untaxed income. Since we know the current income Y, the current tax rate t, and $dY = dG$, the required change in the tax rate, dt may be calculated. Notice that the induced increase in the tax base following an increase in G is insufficient to bring about a balanced budget increase in tax revenue – the government must change the tax rate also. The formula for this, given by (19.33), is of course only appropriate for the particular model under consideration. If the tax schedule were other than the proportional one in (19.30) our formula for the change in tax instrument would be different.

We know from the previous section that unless the government operates on budget balance rules, increases in government spending will create budget deficits which must be financed in some way. These matters are best considered in the context of the full IS-LM model since they involve changes in the supplies of bonds and money.

Exercise 19.3

1. Suppose the tax function in an economy is given by $T(Y) = T_0 + tY$ where T_0 is a lump sum tax, and t the marginal tax rate, $0 < t < 1$. Conduct the balanced budget analysis of this section with this new tax function replacing (19.30). Assume that the government finds it politically undesirable to change the tax rate t, and prefers to change T_0. Find

an equivalent formula to (19.33) for the required change in lump sum tax to produce a balanced budget increase in G.
2. Suppose the government does not feel constrained to balance its budget. Using (19.24), the tax function (19.30), and the equilibrium condition $S(Y^* - T(Y^*)) + T(Y^*) = I + G$, derive an expression for the effect of an increase in government expenditure on the budget deficit, given an unchanged tax rate.

19.4 The modified IS curve

As a preliminary to the following chapter, we show here that the general properties of the IS curve are preserved when government expenditure and taxation are introduced. To do this we rewrite the equilibrium conditions (19.24) to take explicit account of the dependence of investment on the interest rate,

$$Y^* = C(Y^* - T) + I(r) + G. \tag{19.34}$$

The slope of the IS curve is found as in Chapter 18, by taking the total differential of (19.34) holding all but Y^* and r constant. Thus

$$dY^*\{1 - C'(Y^* - T(Y^*))[1 - T'(Y^*)]\} = dr\{I'(r)\},$$

which gives

$$\frac{dY^*}{dr} = \frac{I'(r)}{1 - C'(Y^* - T(Y^*))[1 - T'(Y^*)]} < 0. \tag{19.35}$$

We conclude that the negative slope of the IS curve is quite robust to alternative specifications of the consumption and tax functions.

Exercise 19.4

1. Derive an equation for the IS curve given the following information:

$$Y^* = C^* + I(r) + G, \quad C = a + bY_d, \quad Y_d = Y - T + R, \quad T = tY, \quad I(r) = e - hr,$$

where R is the level of lump-sum transfers from the government to the public. Find an expression for the slope of the IS curve. What factors affect the slope and how?

19.5 Summary

The government influences income and expenditure by its use of taxes, expenditure on goods and services, and transfer payments.

An increase in government spending on goods and services increases expenditure and income by an amount that depends on the marginal propensity to consume out of disposable income – the government expenditure multiplier.

An increase in taxation lowers expenditure by lowering disposable income.

The *balanced budget multiplier theorem* says that a budget-balanced increase in both government spending and taxation increases equilibrium income.

The general properties of the IS curve are preserved by the introduction of government spending and taxation.

20 Monetary and Fiscal Policy

In the previous chapter we included government spending and taxation in the income–expenditure model. It appears on the basis of this analysis that the government is able to increase output by increasing its own expenditure or by increasing the disposable income of households, either by lowering taxation or increasing expenditure on social programmes. These methods of stimulating the economy are known collectively as fiscal policy. More specifically, expansionary fiscal policy (that which increases GNP) is generally associated with a larger budget deficit.

The way the government finances an increase in its deficit was mentioned briefly in the previous chapter but a more comprehensive analysis using the full IS-LM model is required, and is presented here. First we present the bond-financed budget change in the IS-LM model. We then see how pure monetary policy works in IS-LM. Section 20.2 extends the analysis to consider the wealth effects of bond issues. Section 20.3 then shows how monetary policy and fiscal policy are connected. Section 20.4 discusses the problem of the cyclically adjusted budget deficit.

20.1 Fiscal policy in IS-LM

We can extend the linear IS-LM model of Chapter 18 to include the effects of increased government expenditure on GNP and the interest rate. Assume that taxation is proportional to income so that it takes the form of (19.16). Let the investment function be given by the linear form in (18.28) and the LM curve by the linear form (18.43). Then the IS curve is

$$Y^* = C(Y - T) + I(r) + G$$
$$= a + b(1 - t)Y + e - hr + G$$

and so (18.42) becomes

$$Y^* = \frac{a + e + G - hr}{1 - b(1 - t)},$$

so that (18.44) and (18.45) become

$$Y^* = \frac{a + e + G + hM/l}{1 - b(1 - t) + kh/l}, \tag{20.1}$$

$$r^* = \frac{k\{a + e + G\} - (1 - b(1 - t))M}{l\{1 - b(1 - t) + kh/l\}}, \tag{20.2}$$

giving

$$\frac{dY^*}{dG} = [1 - b(1 - t) + kh/l]^{-1} > 0 \tag{20.3}$$

and

$$\frac{dr^*}{dG} = kl^{-1}\{[1 - b(1 - t)] + kh/l\}^{-1} > 0 \tag{20.4}$$

These effects are summarized in Fig. 20.1. The original IS curve (IS_0) shifts to the right (IS_1) as G increases and LM is unaffected. Consequently, both the equilibrium interest rate and income increase. The reason for this is that the increased government expenditure generates higher income and the increased transactions require more money. With the total supply of money available in the economy unchanged cash may be released by those holding cash balances. An asset substitution of bonds for cash will achieve this, but if the bond market is in equilibrium, people will be holding their desired amount of bonds and will be induced to buy bonds (exchange cash for bonds) only if the interest rate increases. Thus, the supply of bonds increases, lowering the bond price and raising interest rates. This is the bond-financed increase in government expenditure again, but now we observe that there are interest rate consequences.

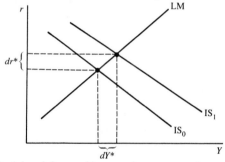

Fig. 20.1: A bond-financed increase in government expenditure

We may compare the effect on income in this case with that in the case with the interest rate fixed. Recall from (19.18) that if the interest rate is fixed the government expenditure multiplier is not (20.3) but

$$\frac{dY^*}{dG} = [1 - b(1 - t)]^{-1}. \tag{20.5}$$

For simplicity denote the expressions (20.3) and (20.5) by α and β respectively. That is,

$$\alpha = [1 - b(1 - t) + kh/l]^{-1} \tag{20.6}$$
$$\beta = [1 - b(1 - t)]^{-1} \tag{20.7}$$

It is easy to show that $\alpha < \beta$. If $\alpha < \beta$ it must be the case that $\beta^{-1} < \alpha^{-1}$, or

$$[1 - b(1 - t)] < [1 - b(1 - t) + kh/l],$$

or

$$0 < kh/l$$

Since k, h and l are all positive, this is clearly the case. It is easily demonstrated that $\alpha \geq \beta$ is not possible. Hence $\alpha < \beta$. This is evident from Fig. 20.2 which compares the two multipliers. We call β the 'constant interest rate multiplier' and α the 'constant money supply multiplier'. Notice that applying the constraint of unchanged r produces a disequilibrium in IS-LM unless compensatory changes in LM are permitted to accompany the shift in IS. We consider such changes presently. Clearly, if the interest rate could be held constant, the effectiveness of an increase in government expenditure in increasing income would be greater. The reason for this is straightforward. The multiplier α allows for the fact that when the interest rate rises to encourage the purchase of bonds, this discourages some investment and hence lowers income. Thus, whilst government expenditure has increased this has been partly offset by a reduction in investment expenditure. Some of the increased government expenditure has simply replaced some investment expenditure, or put another way some investment expenditure has been *crowded out* by government expenditure. In general in IS-LM equilibrium crowding out is unavoidable if the LM curve is unchanged but may not be a serious problem if the size of the crowding out is not too great, and if government expenditure and private investment are regarded as perfect substitutes.

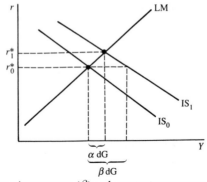

Fig. 20.2: Constant interest rate (β) and constant money supply (α) multipliers

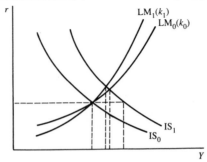

Fig. 20.3: The effect of an increase in k on crowding out ($k_1 > k_0$)

It is apparent from Fig. 20.2 that the extent of the crowding out will depend on the slopes of both IS and LM. This can be seen mathematically by constructing a measure of crowding out. One such measure might be

$$\Delta = \frac{\beta - \alpha}{\beta},$$

because for a given increase in government expenditure the difference in effects on income is accounted for by the relative magnitudes of α and β. A higher value of Δ indicates greater crowding out – that is a bigger fall in income for a given increase in the interest rate. It is easily established, by substituting from (20.6) and (20.7), that

$$\Delta = \frac{kh/l}{[1 - b(1 - t) + kh/l]}. \tag{20.8}$$

We know from question 2 in Exercise 18.3 that an increase in k increases the slope of the LM curve as drawn in Fig. 20.3, and we see from (20.8) that

$$\frac{d\Delta}{dk} = \frac{(h/l)[1 - b(1 - t)]}{\beta^2} > 0. \tag{20.9}$$

Hence an increase in k, by increasing the slope of LM, increases crowding out. The intuition here is simply that since a high value of k represents a strong transactions demand for cash, an individual will require a greater inducement to exchange cash for bonds. Hence the increase in the interest rate must be greater and consequently the reduction in investment is greater also. This is evident from Fig. 20.3. The effects of changes in the other parameters on Δ are left as an exercise.

We now turn to the more general case summarized in the equation system

$$S(Y - T(Y)) + T(Y) = I(r) + G, \tag{20.10}$$
$$M = L(Y, r), \tag{20.11}$$

where (20.10) is the IS curve including government expenditure and (20.11) the LM curve from (18.34). Henceforth, we omit the asterisks, but it is understood that all effects concern *equilibrium* income and interest rates.

To establish the effects of a bond-financed increase in G on the equilibrium pair Y and r we may take the total differential of (20.10) and (20.11) allowing only dY, dr and dG to be non-zero. This gives

$$dY\{S'(Y - T(Y))[1 - T'(Y)] + T'(Y)\} = drI'(r) + dG,$$
$$0 = dYL_Y(Y, r) + drL_r(Y, r),$$

or, rearranging in matrix form,

$$\begin{bmatrix} S'(Y - T(Y))[1 - T'(Y)] + T'(Y) & -I'(r) \\ \\ L_Y(Y, r) & L_r(Y, r) \end{bmatrix} \begin{bmatrix} dY \\ \\ dr \end{bmatrix} = \begin{bmatrix} dG \\ \\ 0 \end{bmatrix}. \tag{20.12}$$

Dividing both sides of (20.12) by dG, we may solve for dY/dG and dr/dG by Cramer's Rule. First, we have

$$\begin{bmatrix} S'(Y - T(Y))[1 - T'(T)] + T'(Y) & -I'(r) \\ \\ L_Y(Y,r) & L_r(Y,r) \end{bmatrix} \begin{bmatrix} dY/dG \\ \\ dr/dG \end{bmatrix} = \begin{bmatrix} 1 \\ \\ 0 \end{bmatrix}. \tag{20.13}$$

Hence, using Cramer's Rule (where $|A| < 0$ denotes the determinant of the square matrix in (20.13)),

$$\frac{dY}{dG} = |A|^{-1} \begin{bmatrix} 1 & -I'(r) \\ \\ 0 & L_r(Y,r) \end{bmatrix}$$

or

$$\frac{dY}{dG} = \frac{L_r(Y,r)}{\{S'(Y - T(Y))[1 - T'(Y)] + T'(Y)\}L_r(Y,r) + I'(r)L_Y(Y,r)}. \tag{20.14}$$

It is easily checked that the result for the special case in (20.3) is analogous to this and that $dY/dG > 0$. Similarly,

$$\frac{dr}{dG} = |A|^{-1} \begin{bmatrix} S'(Y - T(Y))[1 - T'(Y)] + T'(Y) & 1 \\ \\ L_Y(Y,r) & 0 \end{bmatrix}$$

or

$$\frac{dr}{dG} = \frac{-L_Y(Y,r)}{\{S'(Y - T(Y))[1 - T'(Y)] + T'(Y)\}L_r(Y,r) + I'(r)L_Y(Y,r)}, \tag{20.15}$$

which is analogous to (20.4) and positive.

The balanced budget multiplier for the IS-LM case may also be found. Recall that in this case we are looking at equilibria which are characterized by a balanced budget. That is we restrict the fiscal policy by insisting that $T(Y) = G$. In Chapter 19 we found that this was possible as long as the government was prepared to alter its tax instrument – the uniform tax rate in that case. Making use of the balanced budget condition in (20.10) gives

$$S(Y - G) + G = I(r) + G$$

or

$$S(Y - G) = I(r). \tag{20.16}$$

Using (20.16) and (20.11) we may derive the equivalent of (20.13) as

$$
\begin{bmatrix} S'(Y-G) & -I'(r) \\ -L_Y(Y,r) & -L_r(Y,r) \end{bmatrix} \begin{bmatrix} dY/dG \\ dr/dG \end{bmatrix} = \begin{bmatrix} S'(Y-G) \\ 0 \end{bmatrix}.
$$

(20.17)

Using Cramer's Rule to solve, and simplifying, gives

$$
\frac{dY}{dG} = \frac{-S'(Y-G)L_r(Y,r)}{-S'(Y-G)L_r(Y,r) - L_Y(Y,r)I'(r)}.
$$

(20.18)

Hence, dY/dG is positive but less than unity. This result is different to that of the income-expenditure model (which remember, behaves in a way similar to the IS-LM model but with the interest rate fixed) in which the balanced budget multiplier is unity. The difference between these two results is accounted for by the 'crowding out' effect. A balanced budget increase in government expenditure is expansionary but with an unchanged money supply, any expansion of income is accompanied by an increase in the interest rate which in turn causes a reduction in investment expenditure. The balanced budget change in the equilibrium interest rate is easily obtained from (20.17) using Cramer's Rule,

$$
\frac{dr}{dG} = \frac{S'(Y-G)L_Y(Y,r)}{-S'(Y-G)L_r(Y,r) - L_Y(Y,r)I'(r)}
$$

(20.19)

which, not surprisingly, is also positive.

In the models of this section increases in government expenditure have either been accompanied by increases in taxes (in the balanced budget case) or by an increase in bonds. Bond finance is implicit in the model of equations (20.12). However, although tax instruments remained unchanged as government expenditure increased, the increase in bonds was not required to finance the whole of the deficit. This is because there was some increase in the tax revenue *induced* through the increase in income. Recalling the definition of the deficit in (19.21) with $T = T(Y)$, the total increase in the deficit resulting from an increase in G is given by

$$
\frac{dD}{dG} = 1 - T'(Y)\frac{dY}{dG}
$$

(20.20)

(where use is made of the chain rule of differentiation). Substituting from (20.14) and simplifying gives

$$
\frac{dD}{dG} = \frac{S'(Y-T(Y))[1-T'(Y)]L_r(Y,r) + I'(r)L_Y(Y,r)}{\{S'(Y-T(Y))[1-T'(Y)] + T'(Y)\}L_r(Y,r) + I'(r)L_Y(Y,r)}.
$$

(20.21)

It is easily established that dD/dG is a positive fraction (Exercise 20.1, question 5), so that the increase in G does increase D but the increase in the deficit is less than the increase in government expenditure since some extra tax revenue is generated by the increase in income. This implies that the proportion of increased expenditure not financed by induced increases in tax revenue must be financed by a bond issue. The extra bonds must generate an immediate revenue for the government of an amount equal to the shortfall of the increased expenditure over induced tax increases.

In actual economies increases in tax revenue and increases in bonds are not mutually exclusive, and governments may change tax instruments and issue bonds at the same time. In closed economies, moreover, these two ways of financing an increased government expenditure are not the only ones. We now consider a complication which arises when consumers see an increase in government bonds as an increase in wealth.

Exercise 20.1

1. Use (20.8) to establish the effects of changes in $h, l, b,$ and t on the crowding-out measure. In each case illustrate the effects by appropriate movements in the IS or LM curves. Interpret the results.
2. Show that (20.14) in analogous to (20.3) and (20.15) is analogous to (20.4).
3. Show that (20.18) is a positive fraction.
4. Introduce lump-sum transfers R into the model of (20.10) and (20.11). Use Cramer's Rule to find dY/dR and dr/dR.
5. Show that (20.21) is a positive fraction.

20.2 Wealth effects

We now extend the analysis of the previous section by considering the effects of changes in private sector *wealth* on the process of adjustment to equilibrium. Both the previous section and the previous two chapters have mentioned *bonds* as an asset issued by the government and held by individuals. However, we have not persued the notion that total accumulated assets as well as current income may influence the level of private sector spending. This we do now.

We assume that accumulated savings are held either as money or bonds. The total wealth (or value of financial assets), A, of the economy is given by

$$A = M + \frac{B}{r} \tag{20.22}$$

The volume of money is M, B is the number of \$1 bonds, and $1/r$ the price of each bond. Therefore, B/r is the value of bonds. Notice that the government will pay interest on bonds of B each period to individuals. We also assert that private sector consumption is increasing in A, retaining the assumption from earlier that consumption is also increasing in disposable income. Specifically, we have

$$C = C(Y_d, A), \qquad 0 < C_{Y_D} < 1, \quad C_A > 0, \tag{20.23}$$

where $Y_d = (1 - t)(Y + B)$, with t the tax rate being the only tax instrument for the government. The private sector's total income includes interest payments on bonds, which are taxed at the rate t. For a given level of assets, equilibrium income is given by

$$Y = C((1 - t)(Y + B), A) + I(r) + G. \tag{20.24}$$

This is the modified IS curve.

For the modified LM curve, we observe that the demand for money will depend on the volume of transactions, which in turn now depends on A as well as on income. We have

$$L = L(Y, r, A), \qquad L_Y > 0, \quad L_r < 0, \quad L_A > 0. \tag{20.25}$$

Money market equilibrium is given by

$$M = L(Y, r, A) \tag{20.26}$$

which defines a *recursive* relationship between the money supply and the value of assets since A depends on M.

As a final ingredient to this analysis we use the (augmented) government budget constraint from (19.23)

$$G + B - T = \dot{M} + \frac{\dot{B}}{r} \tag{20.27}$$

where the time derivative has replaced the 'Δ' notation. The inclusion of B on the left hand side represents the interest payments on bonds which are an expenditure item for the government. A budget deficit at any instant requires an increase in the money supply or an increase in the value of bonds. In view of (20.22) we have

$$G + B - T = \dot{A}. \tag{20.28}$$

[We assume that neither the government nor individuals expect any change in the interest rate, so $\dot{r} = 0$.]

We now look at bond-financing of the deficit given that bonds affect wealth and therefore expenditure. In anticipation of some of the issues, consider that the government increases the number of bonds in the economy. There are three main effects.

(a) The increase in B increases expenditure by its effect on disposable income, and its effect on private sector wealth (via (20.23)).

(b) The increase in B increases tax revenue for the government.
    Notice that the net effect of (a) and (b) may be expansionary (i.e., to increase Y) or contractionary, since the demand for money will be increased (the LM-curve shifts to the left) by the increase in wealth.

(c) The increase in B increases the government's deficit by increasing interest payments. This tends to widen the budget deficit, calling for yet more bonds to be issued to finance the deficit.

These three effects point to a potential *instability* in this model, with bond financing creating the need for yet more bonds. The three effects also indicate that bond financing may be *contractionary*. The work of Alan Blinder and Robert Solow shows that the stability issue and the effectiveness issue are closely related.

First, we will *assume* stability, and make a comparative-static analysis of a change in bonds on output. Using (20.24) aand (20.26) we have

$$\begin{bmatrix} 1 - C_{Y_d}(1-t) & -I'(r) + C_A B/r^2 \\ \\ -L_Y & -L_r \end{bmatrix} \begin{bmatrix} dY/dB \\ \\ dr/dB \end{bmatrix} = \begin{bmatrix} C_{Y_d}(1-t) + C_A/r \\ \\ L_A/r \end{bmatrix}.$$

(20.29)

From which we find

$$\frac{dY}{dB} = |A|^{-1} \left\{ \left(C_{Y_d}(1-t) + \frac{C_A}{r} \right)(-L_r) + \left(I'(r) - C_A \frac{B}{r^2} \right) \left(\frac{L_A}{r} \right) \right\} \quad (20.30)$$

$$\frac{dr}{dB} = |A|^{-1} \left\{ (1 - C_{Y_d}(1-t)) \frac{L_A}{r} + \left(C_{Y_d}(1-t) + \frac{C_A}{r} \right) L_Y \right\}, \quad (20.31)$$

where $|A| > 0$ is the determinant of the square matrix in (20.29). It easy to establish that $dr/dB > 0$. However, the sign of dY/dB is ambiguous in general. This is because the IS curve shifts to the right as the number of bonds increases (the expenditure effect), while the LM curve shifts to the left (the effect of increased assets on the demand for money). [Draw these movements, and convince yourself that they imply that the interest rate must increase, but that output may increase or decrease.]

Now, the assumption of stability is crucial to this analysis since it guarantees that the movement from one static equilibrium to another is well-behaved. We have a direct way of studying this stability issue (at least in the neighbourhood of the new equilibrium) because (20.27) gives the dynamic relationship between the stock of bonds B and the budget deficit. (In (long-run) equilibrium, the budget must be balanced, with the stock of bonds and the money stock unchanged.) Ignoring monetray policy, so that $\dot{M} = 0$, and rearranging (20.27) gives

$$\dot{B} = r(G + B - t(Y + B)). \quad (20.32)$$

For stability, we require that increases in the number of bonds are *lower* the higher is the outstanding stock of bonds. That is, $d\dot{B}/dB < 0$. Differentiating (20.32) and noting that (from (20.30) and (20.31)) Y and r depend on B gives

$$\frac{d\dot{B}}{dB} = \frac{dr}{dB}(G + B - t(Y + B)) + r\left((1-t) - t\frac{dY}{dB} \right). \quad (20.33)$$

In the neighbourhood of the new steady-state, we have (from (20.32)) $\dot{B} = G + B - t(Y + B) = 0$ and so

$$\frac{d\dot{B}}{dB} = r\left((1-t) - t\frac{dY}{dB} \right).$$

This implies that a *necessary* condition for stability, $d\dot{B}/dB < 0$, is $dY/dB > 0$. A necessary and sufficient condition is

$$\frac{dY}{dB} > \frac{1-t}{t}$$

Hence, the case in which bond-financing is not effective in increasing output is precisely the case in which the system is unstable. A stable system will have output increased by a bond-financed increase in government expenditure.

The analysis of this section hangs crucially on the assumption that private agents perceive an increase in bonds as an increase in the *net* wealth of the economy. This view is often challenged by a proposition known as *Ricardian equivalence*. Bonds are an asset purchased by the private sector, but may be viewed equivalently as a liability of the government. Ultimately governments cannot create wealth, and bonds must eventually be replaced by higher taxation. Ricardian equivalence suggests that individuals will completely discount the increase in the number of bonds as representing an equivalent future tax burden. Bonds are not net wealth and so individuals will not change their spending patterns as bonds increase.

Ricardian equivalence is at the centre of an unresolved debate with opponents suggesting that individuals will not see new government debt as being completely offset, dollar for dollar, by future taxes since, with imperfect capital markets, the interest earned in the meanwhile will more than compensate for the future tax liability.

If the system is stable, then wealth effects may offset crowding out.

Exercise 20.2

1. Use (20.32) to find an expression for the required steady-state increase in bonds to finance an increase in government spending of dG.

20.3 Monetary control

As we saw in Chapter 17, the government has sole responsibility for issuing new notes and coin, and in principle the government can introduce new money into the economy simply by covering its expenditure by new issue. If the government prints extra notes to pay for goods and services, it is easy to see how the extra money finds its way into the economy – it initially goes into the bank accounts of the providers of goods and services. It remains to be seen how the government engages in a 'pure' monetary policy involving increases in the money supply for its own sake, not as a means of financing a deficit. In this section we also look at how the government attempts to ensure that the 'right' amount of money is available.

Suppose the budget is balanced and that the government seeks to increase the money supply for whatever reason. It is obvious from the government budget constraint (19.23) that with $G - T = 0$, $\Delta M > 0$ requires $\Delta B/r < 0$. This gives a clue as to how the increased money is put into circulation. The government uses an *open market operation* (discussed in Chapter 17) to buy back bonds from the public. The bonds are bought using the newly created money, which initially finds its way into the bank accounts of the members of the public who have sold their bonds. The amount of money in circulation has initially increased by the amount of the newly created notes and coin.

In practice governments face considerable difficulties in ensuring that the 'right' amount of money is in circulation. The constraining factors are inflation and interest rates. If the money supply grows too quickly, then inflation results, while if the money supply is inadequate for the value of transactions, the interest rate will rise. Neither inflation nor high interest rates are regarded as desirable. Inflation erodes the value of fixed nominal income and assets, while high interest rates lead to high mortgages and lower physical investment. First, consider the relationship between the money supply and prices. The following identity must be true,

$$MV \equiv PY$$

where M is the money supply (cash plus deposits), P the price level, Y real output (equals real income equals real expenditure) and V the velocity of circulation, or roughly how often each unit of currency on average changes hands. Given these definitions MV is the total volume of money changing hands, while PY is the total money value of expenditures. These two quantities are clearly equivalent. Now suppose that V and Y are fixed. Then an increase in M simply leads to an increase in P. We consider the relationship between the money supply and prices more fully in Chapter 22. However, it is clear that there is such a thing as too much money if undesirably high prices are the result.

In the IS-LM model, of course, it is P that is fixed and Y that is variable and so the interest rate constraint is of more immediate concern. Consider Fig. 20.4 which reproduces the money market equilibrium of Fig. 18.10. If the money supply is fixed at M_0 and the total demand for money curve is L_0 then the equilibrium interest rate is r_0. Now suppose income increases so that the demand for money curve shifts to L_1. With an unchanged money supply the interest rate increases to r_1, which will choke off investment expenditure, lowering output and income. If the interest rate r_0 was regarded as (politically) just right, then the interest rate r_1 will be too high, with the implication that M_0 is insufficient money given the new money demand conditions, L_1. Thus there is such a thing as too little money. One way of avoiding undesirable interest rate fluctuations when the demand for money changes is for the government to fix the interest rate, say at r_0, and to allow money demand to determine the appropriate quantity of money. In this case the increase in money demand to L_1 leads to an accommodating increase in the money supply. This guarantees interest rate stability but takes direct control of the money supply out of the government's hands, with possible inflationary consequences.

It appears from Fig. 20.4 that governments may opt to regulate the money market by 'price' (interest rate) controls or by quantity (money supply) controls. Many governments choose the latter as a broad policy, mainly because of their aversion to inflation and their belief that it is an ultimate responsibility of government to regulate the money supply. In practise the day-to-day operaton of monetary policy does not entirely disregard interest rate movements, and decisions are often taken to prevent sudden, large and destabilizing movements in the interest rate.

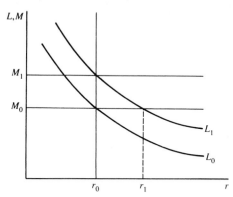

Fig. 20.4: Interest rate and money supply controls

Controlling the money supply is not an easy matter. In principle the government ought to be able to control the amount of cash, but we know from Chapter 17 that there is more to the money supply (say M1) than notes and coin. We can see some of the issues involved here by combining the credit-creating role of banks discussed in Chapter 17 and the government's budget constraint discussed earlier in this section.

First, we recall from Chapter 17 that the money supply (M1) is equal to notes and coin, C plus deposits D. Hence,

$$M = C + D$$

so that

$$\Delta M = \Delta C + \Delta D.$$

We further know that banks will use additional deposits to make additional loans or advances, A. Thus,

$$D = A$$

so that

$$\Delta M = \Delta C + \Delta A.$$

The government budget constraint discussed earlier in this section tells us that the size of the deficit is equal to the increase in the amount of government bonds ΔB_G plus the increase in the amount of cash ΔC. We may call the government deficit the public sector borrowing requirement (PSBR) so that

$$\text{PSBR} = \Delta B_G + \Delta C$$

or

$$\Delta C = \text{PSBR} - \Delta B_G$$

giving finally,

$$\Delta M = \Delta A + (\text{PSBR} - \Delta B_G).$$

Given our definition, an increase in bank loans to the private sector or a government deficit not financed by a bond issue both increase the money supply. For this reason governments take a great interest in the amount of advances made by banks and their own expenditure. It is clear from this that fiscal policy and monetary policy are rarely independent of each other in practice. For simplicity, however, we often study 'pure' forms of policy such as bond-financed increases in government expenditure or a budget-balanced increase in government expenditure and an open-market operation to increase the money supply. We look at the analytics of these policies now in the context of the IS-LM model.

20.4 Monetary policy in IS-LM

The money supply makes its appearance in IS-LM models as a fixed and entirely exogenous quantity. In this section we will use the term 'monetary policy' to mean any manipulation of the money supply, M1.

Consider the general IS-LM model used in section 20.1,

$$S(Y - T(Y)) + T(Y) = I(r) + G, \tag{20.34}$$

$$M = L(Y, r). \tag{20.35}$$

Taking total differentials of this system allowing only dY, dr, or dM to be non-zero gives, in matrix form,

$$\begin{bmatrix} S'(Y - T(Y))[1 - T'(Y)] + T'(Y) & -I'(r) \\[2mm] L_Y(Y, , r) & L_r(Y, r) \end{bmatrix} \begin{bmatrix} dY \\[2mm] dr \end{bmatrix} = \begin{bmatrix} 0 \\[2mm] dM \end{bmatrix}. \tag{20.36}$$

Dividing (20.24) by dM, we may solve for dY/dM and dr/dM by Cramer's Rule, to obtain

$$\frac{dY}{dM} = \frac{I'(r)}{\{S'(Y - T(Y))[1 - T'(Y)] + T'(Y)\}L_r(Y, r) + I'(r)L_Y(Y, r)} \tag{20.37}$$

and

$$\frac{dr}{dM} = \frac{S'(Y - T(Y))[1 - T'(Y)] + T'(Y)}{\{S'(Y - T(Y))[1 - T'(Y)] + T'(Y)\}L_r(Y, r) + I'(r)L_Y(Y, r)} \tag{20.38}$$

so that $dY/dM > 0$ and $dr/dM < 0$. These effects are illustrated in Fig. 20.5. This is the case of a pure monetary expansion by an open-market operation.

Monetary policy and fiscal policy, as we have defined them, are rarely used independently. In practice the interdependencies are complex, but we may illustrate how they may be made to work together. Recall from Fig. 20.2 that an increase in government expenditure forces up interest rates and induces some crowding out. If the money supply were to expand as government expenditure increases the effects of crowding out may be offset or eliminated. This is often known as an accommodating increase in the money supply. Diagrammatically, a completely accommodating monetary expansion is illustrated in Fig. 20.6.

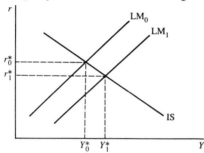

Fig. 20.5: Pure monetary expansion

Mathematically, the effect is obtained by noting that with the constraint that $dr = 0$, an increase in G must be accompanied by an increase in M. Thus M is no longer exogenous but *induced* by the necessity to keep interest rates unchanged. Suppose the unchanged interest rate is to be r, then total differentiation of (20.34) and (20.35) allowing only dY, dG, and dM to be non-zero gives, in matrix form,

$$
\begin{bmatrix}
S'(Y - T(Y))[1 - T'(Y)] + T'(Y) & 0 \\
\\
L_Y(Y, r) & -1
\end{bmatrix}
\begin{bmatrix}
dY \\
\\
dM
\end{bmatrix}
=
\begin{bmatrix}
dG \\
\\
0
\end{bmatrix}
\tag{20.39}
$$

where M is treated as endogenous – responding automatically to an increase in G. Dividing (20.39) by dG and using Cramer's Rule gives, after simplification,

$$
\frac{dY}{dG} = \{S'(Y - T(Y))[1 - T'(Y)] + T'(Y)\}^{-1}
\tag{20.40}
$$

$$
\frac{dM}{dG} = \frac{L_Y(Y, r)}{S'(Y - T(Y))[1 - T'(Y)] + T'(Y)}
\tag{20.41}
$$

It is apparent that (20.40) is positive and equivalent to (19.19) – the constant interest rate multiplier. The value for this multiplier for the by now familiar special case is (20.5). Hence (20.40) confirms the intuition behind Fig. 20.6.

Equation (20.41), however, is a new result and gives the government a rule for accommodating monetary expansion. The required increase in the money supply for a (small) change in government expenditure which will keep the interest rate constant at r is given by (20.41).

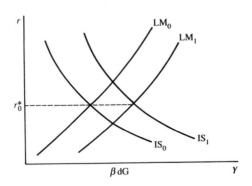

Fig. 20.6: An accommodating monetary expansion

Before leaving this introduction to monetary policy, it is worth pointing out that we have not explored any link there might be between the budget deficit, the money supply, and the price level. Hence the effects obtained here are strictly valid only within the fixed price IS-LM model. As we have seen, a monetary expansion will generally lead to an increase in real income at less than full employment. The effectiveness of an expansion of the money supply will depend at least in part on the slope of the IS curve. Verification of this is left as an exercise.

Exercise 20.4

1. Use the linear special case of IS-LM equilibrium income to show what factors influence the size of the money supply multiplier (20.37).
2. The result in (20.38) suggests that the interest rate falls as the money supply increases. Explain why.
3. The introduction of widespread use of cheque books and credit cards means that less cash needs to be held for day-to-day transactions. What effect would this have on IS-LM equilibrium?
4. We argued in this section that the 'constant interest rate' multiplier might come about if increases in government expenditure are accompanied by 'accommodating' increases in the money supply. The same effect would also be created by increases in government expenditure taking place along a perfectly elastic (i.e., horizontal) LM curve. How might such a slope for the LM curve come about?
5. Verify the equivalence between (20.40) and (19.19).

20.5 Cyclically adjusted budget deficits

In our specification of the budget deficit only taxes have been allowed to vary with income. We regarded this as plausible since in many economies 'income tax' is one of the more important sources of government revenue. Only slightly less obvious is the fact that government expenditure itself may vary over the cycle, but in the opposite direction to that of income – *countercyclically*. A moment's thought will reveal why this is so. At least a part of government expenditure takes the form of transfers, that is, state support for the poor, unemployed, sick, and aged. Unemployment and social security payments form a large part of this expenditure, and it is clear that these payments will tend to be higher when income and output are low, because of the associated high unemployment.

This has important implications both for how we interpret the budget deficit and for attempts to reduce the deficit by reducing government expenditure.

The interpretation of the deficit must change because, although we usually associate a budget deficit with attempts by the government to expand output (by injecting demand), a high government expenditure may not reflect high voluntary expenditures by the government on new employment-creating programmes, but merely the high support it is giving to the unemployed and their families. Whilst this type of support is easily justified on humanitarian grounds, it is not in the nature of employment-creating expenditure. This is reinforced by the fact that when income is low tax revenue is also low. In short, the size of the budget deficit does not give an accurate picture of the government's *fiscal stance*. The government deficit might be high not because the government is actively seeking expansion but because it is paying for the depression. One way around this difficulty is to evaluate the deficit not at the current level of income but at a 'benchmark' level of income. One such benchmark might be the full employment level of income, Y_f. At this level of income any excess of government expenditures over receipts arises not because of the costs of financing unemployment but because the government is *choosing* a relatively high level of discretionary expenditure to boost output.

These ideas may be put in mathematical terms quite simply. Since we are distinguishing between those government expenditures which automatically vary in a countercyclical way and those which are completely under government control, we will distinguish two types of government expenditure. This is achieved (using (19.22)) by making transfers depend on the level of income. Hence the budget deficit is

$$D = G - T(Y) + R(Y) \qquad (20.42)$$

where $0 < T'(Y) < 1$ and $R'(Y) < 0$. We will identify G with discretionary expenditure and $R(Y)$ with the exchequer cost of financing a low level of economic activity. If the current equilibrium level of income is low, then D will be high, not because the government is choosing to spend on job-creating activities, but because $T(Y)$ is low and $R(Y)$ is high. In a sense the low value of $T(Y)$ and high value of $R(Y)$ obscure the true position being chosen by the government in its discretionary expenditure. Fig. 20.7 illustrates this. Suppose the tax function takes the form $T(Y) = tY$ where t is a constant fraction. Discretionary government expenditure is fixed at G_0 and countercyclical expenditures are shown by the schedule $R(Y)$. At full employment income, Y_f, for simplicity we assume that there are no transfers. It is clear that, as drawn, there is a budget surplus at full employment. At Y_b there is a balanced budget and at Y_0 a budget deficit since $G + R(Y_0) > T(Y_0)$. But note that there is no difference in the government's chosen fiscal instruments between these three points – the tax rate and discretionary government expenditure are the same at Y_0, Y_b, and Y_f. Were it not for the fact that the government is financing high unemployment at Y_0, the budget would be in surplus if evaluated at full employment. Thus the government's fiscal instruments are not set to be expansionary but contractionary. So, as the budget deficit varies over the cycle we may judge to what extent these variations are due to changes in policy instruments and to what extent they are produced by cyclical effects by continually referring to the position which would prevail at the full-employment (benchmark) level of income. It is possible, of course, that the deficit would persist even at full employment if G itself is very high.

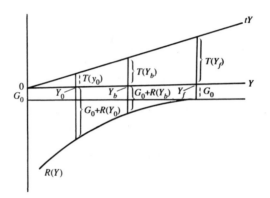

Fig. 20.7: Counter-cyclical induced government expenditure

A further difficulty concerns steps which may be taken by the government to reduce the size of its deficit. Such a move would be warranted for example by a government concerned to lower its costs by borrowing and so on. However, it is not obvious that an attempt to reduce government's discretionary expenditure will reduce the budget deficit. This is because the reduction in G will lower income, which will in turn lower taxes and raise transfers. It may be that the lower taxes and higher transfers will increase the deficit by more than the reduction in G. This is a potentially serious matter since it suggests that a policy change might be proposed which is entirely in the opposite direction to that required. We are able to study this problem and establish the conditions under which a reduction in G will have the desired effect of lowering the budget deficit using the income–expenditure model of the last chapter. The issue hinges on (*a*) how much equilibrium income falls as G falls and (*b*) how $T(Y)$ and $R(Y)$ respond to the fall in income.

Equilibrium income in the model with transfers (and a fixed interest rate) is given by

$$S(Y - T(Y) + R(Y)) + T(Y) = I + G + R(Y), \qquad (20.43)$$

from which we establish by total differentiation that

$$\frac{dY}{dG} = S'(Y - T(Y) + R(Y)) + [1 - S'(Y - T(Y) + R(Y))][T'(Y) - R'(Y)]^{-1} > 0. \qquad (20.44)$$

On the other hand the total change in the budget deficit as G and Y change is given by

$$dD = dG - dY[T'(Y) - R'(Y)]. \qquad (20.45)$$

If dG is negative (i.e., reducing discretionary expenditure) we will succeed in making dD negative (i.e., reducing the deficit) if

$$dG > dY[T'(Y) - R'(Y)] \qquad (20.46)$$

or if

$$\frac{dY}{dG} < [T'(Y) - R'(Y)]^{-1}. \qquad (20.47)$$

(Notice that dY is negative if dG is.) But dY/dG is given by (20.44) so that substituting from (20.44) into (20.47) and simplifying gives the condition that dD is reduced if

$$S'(Y - T(Y) + R(Y))[1 - T'(Y) + R'(Y)] > 0$$

or if

$$1 > T'(Y) - R'(Y). \qquad (20.48)$$

If condition (20.48) is met then a (small) reduction in G will have the desired effect of reducing the budget deficit. Since $R'(Y)$ is negative, there is of course no guarantee that (20.48) will be met, so that our earlier misgivings about the efficiency of such a policy may be justified. A similar approach may be adopted using the full IS-LM model, in which case whilst (20.47) will be the same we would have a different multiplier in (20.44). The condition (20.48) would consequently change also. This is left as an exercise.

Exercise 20.5

1. Replace (20.44) with a constant money supply government expenditure multiplier in the case of an IS-LM model with transfers $R(Y)$. Find an equivalent expression to (20.48) – a condition which will guarantee the effectiveness of a reduction in G reducing the budget deficit.
2. Using a diagram similar to Fig. 20.7, illustrate a situation in which a government has a full employment budget deficit. How would Figure 20.7 change to reflect a more realistic tax schedule with tax 'bands' of increasing marginal tax rates?

20.6 Summary

Fiscal policy multipliers in the IS-LM model depend on the way government expenditures are financed. Bond-financing may produce a 'crowding-out' effect. As the government issues bonds, interest rates increase to make bond-holding attractive to investors, but the increased interest rates reduce private sector spending on consumer durables and capital goods.

Crowding-out may be reduced when wealth effects are considered. If individuals see bonds as part of wealth, then bond financing might induce additional expenditure.

Wealth effects raise the issue of stability of the IS-LM system. The need for the government to pay interest on bonds raises the possibility that larger and larger bonds issues are required to finance interest payments on outstanding debt.

If the system is stable, then bond financing will have a net expansionary effect.

Increases in the money supply also increase GNP, but by lowering interest rates so as to stimulate private-sector spending.

An increase in government spending may be accompanied by an expansion of the money supply to take pressure off interest rates. The money-financed fiscal policy multiplier is therefore larger than the simple bond-financed fiscal policy multiplier. In fact, if the increase in the money supply is sufficient to maintain constant interest rates, then there will be no 'crowding out' and the multiplier will be the same as in the income-expenditure model.

When assessing whether current fiscal policy is expansionary or contractionary it is necessary to remove the cyclical components of government spending. An increase in the budget deficit may signal expansion, but if the increase in spending is on unemployment benefits and social programmes, it may signal the opposite.

This leads to the idea that reductions in government spending may not lead to a fall in the budget deficit. A fall in government spending on goods and services will lower GNP, increase unemployment, and hence increase the government expenditure on social programmes.

21   Aggregate Demand

The analysis of Chapters 19 and 20 was conducted with an important qualification in that prices were held fixed. As we have suggested, this assumption is technically equivalent to assuming a horizontal aggregate supply curve. Fig. 21.1 is a useful starting point for the analysis of this and the following chapters, because it not only depicts the type of aggregate supply curve implicit in the analysis thus far, but also shows a downward-sloping demand curve. Neither of these curves have yet been derived formally.

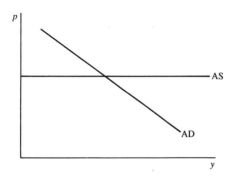

Fig. 21.1: Aggregate demand and supply

In this chapter we show that the demand curve, as illustrated in Fig. 21.1, does indeed represent the relationship between the economy's price level and output demanded. In fact the demand curve is derived from the now-familiar IS-LM model. This is unsurprising, since that model was put forward as representing the demand side of the economy – the goods and money markets jointly determine the level of demand for goods (which is equal to output or income in equilibrium) and the interest rate.

Once the demand curve is derived, and its properties studied, we turn to the important issue of what the economy's supply curve might look like. Generally we find that the slope of the aggregate supply curve is sensitive to the characteristics of the labour market. The labour market figures prominently in the study of aggregate supply since, in what is often referred to as the (Marshallian) short run (in which the aggregate capital stock is taken to be constant), the level of output produced depends on the level of employment, which in turn is determined by the demand for and supply of labour. Thus, whilst the goods and money markets determine demand it is the labour market (and more generally other markets for productive inputs) which determines aggregate supply.

The labour market and aggregate supply occupy us for the following two chapters. It will become apparent that our earlier conclusions on economic policy drawn from the IS-LM framework are changed substantially when conducted in the aggregate demand and supply framework.

21.1 Aggregate demand

With the price level fixed at some given value, the IS-LM model determined a resulting level of output and hence only a *point* on the aggregate demand curve. To find other points, and hence to trace out the whole demand curve, we need to vary the price level. As presented, however, the price level does not figure explicitly in the analysis. Consider, for example, the IS-LM equilibrium in (20.10)–(20.11), rewritten here for convenience as

$$S(Y - T(Y)) + T(Y) = I(r) + G, \tag{21.1}$$

$$M = L(Y, r). \tag{21.2}$$

The variables being determined here are the interest rate and *real* income. An assumption not emphasised since Chapter 16 is that all the quantities are being measured in *real* terms: real savings S, real tax revenue T, and investment I, real government expenditure G, real money balances L, and the real money supply M. The significance of the term *real* stems from the idea that we are interested in the determinants of physical increases in output or the volume of output. Increases in the value of output due to increased prices do not improve the real income positions of people since the increase in nominal income is absorbed into increases in nominal expenditure. In the IS-LM model with a fixed price level any increase in the value of output was a *real* increase in output volume. From now on in a world of potentially changing prices the difference between *real* and *nominal* quantities must be made explicit.

The simplest way to introduce prices in the IS-LM model summarized in (21.1) and (21.2) is the following. We suppose that all spending plans, including those of the government, are set in *real* terms. Thus if the price level increases by 5 per cent all nominal expenditure would also increase by 5 per cent to keep real expenditure unchanged. Thus the equation in (21.1) is invariant to changes in price. We assume also that the demand for cash balances L is expressed in real terms and so invariant to changes in price, i.e., a 5 per cent increase in price leads to a 5 per cent increase in nominal money demand. Thus the right-hand side of (21.2) does not depend on the price level. However, we assume that the left-hand side of (21.2) behaves in slightly different way. We assume that the government chooses the *nominal* value of the money supply, m. That is, a given change in the price level would not automatically cause an equal change in the nominal money supply. Hence, if the economy's price level is denoted by P, the real money supply M is given by

$$M = \frac{m}{P}. \tag{21.3}$$

Equilibrium in the money market just requires that *real* demand equals *real* supply, so that (21.2) becomes

$$\frac{m}{P} = L(Y, r). \tag{21.4}$$

The IS-LM system is then

$$S(Y - T(Y)) + T(Y) = I(r) + G,$$
$$\frac{m}{P} = L(Y, r). \tag{21.5}$$

With G and m predetermined we may derive the relationship between P and Y – the aggregate demand curve – by total implicit differentiation of (21.5) allowing only dY, dr, and dP to be non-zero. The total differential of (21.5) may be written in matrix form as

$$\begin{bmatrix} S'(Y - T(Y))[1 - T'(Y)] + T'(Y) & -I'(r) \\ \\ L_Y(Y, r) & L_r(Y, r) \end{bmatrix} \begin{bmatrix} dY \\ \\ dr \end{bmatrix} = \begin{bmatrix} 0 \\ \\ -(m/P^2)dP \end{bmatrix}.$$
$$\tag{21.6}$$

Dividing throughout by dP and using Cramer's Rule to find dY/dP gives

$$\frac{dY}{dP} = \frac{-I'(r)m/P^2}{|A|} \tag{21.7}$$

where $|A|$ is the determinant of the square matrix in (21.6),

$$|A| = \{S'(Y - T(Y))[1 - T'(Y)] + T'(Y)\}L_r(Y, r) + I'(r)L_Y(Y, r) < 0. \tag{21.8}$$

Hence $dY/dP < 0$ as we expect for a demand curve. This demand effect is brought about because while the price changes, for whatever reason, the monetary authorities are holding the nominal money stock constant. Thus for example as the economy's price level increases the effect is to *lower* the real money stock which puts upward pressure on the interest rate which in turn reduces investment demand. Diagrammatically the effect is shown in Fig. 21.2. In frame (*a*) we see the effect of an increased price on the IS-LM diagram whilst in frame (*b*) we graph the implied demand curve.

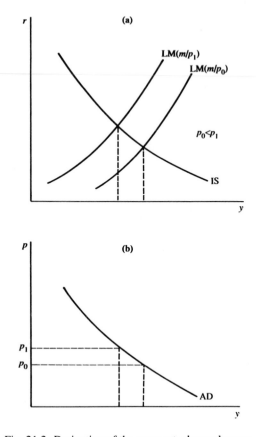

Fig. 21.2: Derivation of the aggregate demand curve

Exercise 21.1

1. The aggregate demand curve derived in this section assumes that while the government fixes *nominal* money supply, it fixes *real* government expenditure. It is common practice in several countries now for the government to set *nominal* government expenditure, say g, so that real government spending is g/P. Derive the AD curve under this alternative specification.
2. Derive an equation for the aggregate demand curve for the linear IS-LM model summarized in equations (20.1) and (20.2). Derive an expression for the slope of the AD curve and show what happens to this slope as b, t, k, h, and l change. Does the slope of AD depend on any other parameters in the linear case? Does the linear IS-LM model produce a linear, concave, or convex AD curve?

21.2 The position and slope of the aggregate demand curve

Recall that in the comparative-static analysis of Chapters 18 and 20, in which we studied the effects of various parameter changes on IS-LM equilibrium, a fixed price level was assumed. Hence, the effects on income we observed there translate into the aggregate demand framework as horizontal shifts of the demand curve at a given price level. An obvious example is the effect of a bond-financed increase in government expenditure G. To see this, fix the price level in (21.5) and totally differentiate with respect to G to obtain

$$\left.\frac{dY}{dG}\right|_P = \frac{L_r(Y, r)}{\{S'(Y - T(Y))[1 - T'(Y)] + T'(Y)\}L_r(Y, r) + I'(r)L_Y(Y, r)}, \quad (21.9)$$

which is identical to equation (20.14). The effects on both IS-LM and aggregate demand are shown in Fig. 21.3.

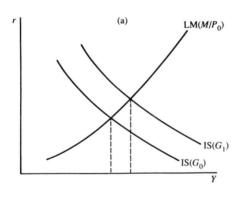

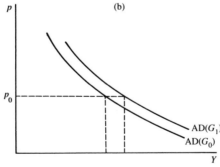

Fig. 21.3: The effect of an increase in government expenditure on aggregate demand

As we saw in question 2 of Exercise 21.1, parameter changes may also affect the slope as well as the position of the aggregate demand curve. Whilst this is not true of government expenditure in this model it is true of, for example, the marginal propensity to save – the term $S'(Y - T(Y))$ in $|A|$. By inspection it is apparent that an increase in the marginal propensity to save increases $|A|$, and moreover by inspection of (21.7) an increase in $|A|$ makes dY/dP smaller in absolute value, so that a given increase in P causes a smaller fall in Y when the marginal propensity to save is higher. Thus the demand curve is steeper the higher is the marginal propensity to save. This general result is easily illustrated using the linear example in question 2 of Exercise 21.1.

Finally, we note that as price increases and income falls as we move along the demand curve, the interest rate is also increasing. This can be seen by applying Cramer's Rule to (21.6) to find dr/dP. This gives

$$\frac{dr}{dP} = |A|^{-1} \left\{ \frac{-m}{P^2} \left[S'(Y_d)[1 - T'(Y)] + T'(Y) \right] \right\} > 0. \qquad (21.10)$$

This is quite intuitive given the mechanism by which price affects demand in this model. An increase in price lowers the real money supply and the interest rate increases to reduce money demand. The increase in the interest rate leads to a fall in demand for goods and hence to a fall in GNP.

Exercise 21.2

1. Establish whether, in the general case of equation (21.7), the aggregate demand curve is linear, concave, or convex.
2. According to the text, an analysis of (21.7) revealed that the aggregate demand curve is steeper the higher is the marginal propensity to save. Similarly, using (21.7) and (21.8), establish for the general case the effects on the slope of the aggregate demand curve of:

   (i)   an increase in the marginal tax rate,
   (ii)   an increased sensitivity of investment to the interest rate,
   (iii) an increased sensitivity of money demand to the interest rate,
   (iv)   an increased sensitivity of money demand to income.

21.3 Summary

Explicit introduction of the price level into the IS-LM model permits the derivation of a downward-sloping aggregate demand curve.

Earlier results on the effects of parameter changes on Y in the IS-LM model correspond to 'shifts' in the aggregate demand curve.

The aggregate demand curve is drawn for given money supply and exogenous expenditures, but movements along the curve, varying price and income, are also accompanied by changes in the interest rate.

22 Aggregate Supply and the Perfectly Competitive Labour Market

In the previous chapter, we claimed that the labour market plays a prominent role in the theory of aggregate supply. In the short run at least, the level of production in the economy will be determined by the level of employment, which in turn is determined by supply and demand conditions in the market for labour. The difficulty we encounter immediately is that of choosing the appropriate framework for studying the factors which determine the level of employment. There are a variety of views on this and it could be argued that many of the debates amongst economists on how the economy works come down to differences of opinion on how labour markets work or how they *should* work.

In this chapter and the following one we study this problem. Although we are ultimately interested in the supply relation between price and output, our immediate concern is with the effect of prices on employment. A further step, the analysis of the aggregate *production function*, then links employment with output.

The simplest model of the labour market is the perfectly competitive model. This has been analysed in Chapter 12. At the heart of the competitive model is the idea that a labour market will work in a way similar to a competitive goods market. Labour services are traded competitively and the 'price' of labour, the wage rate, adjusts to keep the balance between demand and supply. Economists are generally agreed that labour markets rarely, if ever, work in this way but are divided on whether it is possible or desirable that they could. It seems, however, a useful starting point.

The demand for and supply of labour are both specified as functions of the *real wage*. The intuition here is straightforward, but we devote a little time to the derivation of the labour demand and supply relations, because, as we shall see, the theory of aggregate supply depends critically upon them.

22.1 Aggregate labour demand

It is convenient to think of an economy with a large potential labour force of identical individuals – identical, that is, in terms of their productive skills – and a single commodity. Individual firms and workers are unable to influence the money wage paid to labour or the price of output. In other words all agents are *price takers* and the economy is perfectly competitive. There are no dominant firms and no unions.

The demand for labour arises when firms calculate the profit-maximizing level of employment, denoted by N for any given real wage. In Chapter 8 we made extensive use of the production function in the theory of the firm and in Chapter 12 we studied its role in deriving the firm's demand for labour function in some detail. A review of the main results here will help our discussion. For a typical firm in the economy a production function relates the level of employment N to the level of output, Y. The production function $f(N)$ is assumed to take the form

$$Y = f(N), \qquad f'(N) > 0, \qquad f''(N) < 0. \tag{22.1}$$

The last condition, $f''(N) < 0$, simply states that the marginal product of labour $f'(N)$ is diminishing. Thus as more labour is employed output increases but at a decreasing rate because extra labour becomes less productive at the margin. The typical firm maximizes profit π given the product price P and the money wage rate W and hence will choose the level of employment, N, which maximizes the difference between revenue and wage costs (assumed to be the only costs). Hence the problem for the firm is

$$\max_{N} \pi(N) = PY - WN \tag{22.2}$$

subject to

$$Y = f(N). \tag{22.3}$$

Substitution of (22.3) into (22.2) gives

$$\max_{N} \pi(N) = Pf(N) - WN \tag{22.4}$$

with the solution

$$\frac{W}{P} = f'(N_d) \tag{22.5}$$

where N_d denotes the firm's optimal demand for labour. Total implicit differentiation of (22.5) shows that

$$\frac{dN_d}{d\,(W/P)} = \frac{1}{f''(N_d)} < 0 \tag{22.6}$$

so that an increase in the real wage lowers the firm's demand for labour. We assume that this relationship between the real wage and the demand for labour will hold in the *aggregate* also. In fact we will henceforth treat $f(.)$ as the aggregate production function and N_d the aggregate employment level. This is illustrated in Fig. 22.1. At the real wage of $(W/P)_0$ the economy's employment level is N_0 since $(W/P)_0 = f'(N_0)$ satisfies (22.5). At the higher wage $(W/P)_1$, (22.5) is satisfied at a lower employment level $N_1, N_1 < N_0$.

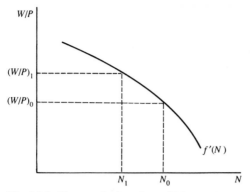

Fig. 22.1: The marginal product of labour curve

While the dependence of the economy's desired labour demand and the real wage may be obtained from (22.5), it is more conventional to write

$$N_d = N_d\left(\frac{W}{P}\right), \qquad N_d'\left(\frac{W}{P}\right) < 0, \qquad (22.7)$$

where $N_d(.)$ is clearly the inverse function of $f'(.)$. For example, assume that the aggregate production function is

$$f(N) = AN^{1/2} \qquad (22.8)$$

where A is a constant scale factor reflecting the number of firms. Clearly,

$$f'(N) = \frac{1}{2}AN^{-1/2} > 0$$

so that (22.5) implies

$$\frac{W}{P} = \frac{A}{2}N_d^{-1/2}$$

from which we solve for N_d to obtain the N_d function,

$$N_d = N_d\left(\frac{W}{P}\right) = \frac{A^2}{4\,(W/P)^2}. \qquad (22.9)$$

Clearly,

$$N_d'\left(\frac{W}{P}\right) = \frac{-A^2}{2\,(W/P)^3} < 0, \qquad (22.10)$$

thus confirming (22.7). This concludes the example. In general, we hypothesize that the economy's demand for labour function is given by (22.7).

Exercise 22.1

1. An economy's aggregate short-run production function is given by $Y = AN^\alpha$ where $0 < \alpha < 1$. Derive the marginal production of labour curve and hence the demand for labour curve. Does this economy have diminishing returns to labour?
2. Continuing with the previous question, suppose N is measured in millions of workers. Suppose $\alpha = 0.5$ and $A = 10$. How much labour would be demanded at a real wage equal to $1? What would labour demand be at a real wage of $2?
3. What factors might cause a shift to the right of the labour demand schedule? Consider the effects on the aggregate supply curve.

22.2 Aggregate labour supply

The theory of an individual's labour supply decision is given a full treatment in Chapter 12, to which readers are referred. The model of the individual's decision problem requires adaptation before we may use it as a basis for a theory of *aggregate* labour supply. First, there is the issue of aggregation, encountered also in the theory of labour demand. We assume that this issue is resolved and that there is no difficulty in aggregating individual labour supply functions. Secondly, it is apparent that 'labour supply' itself is an ambiguous phrase which may mean either the number of hours of effort an individual or fixed group of individuals are prepared to supply, or the number of individuals who are prepared to offer their services for a standard number of hours per day. This latter, all-or-nothing decision, is referred to as the *participation decision*. Our interest in this chapter is primarily that of employment determination and the model of aggregate labour demand developed in the previous section took the number of workers as being the important variable. Although, more generally, we may think of the level of employment as being measured in 'man-hours' this is an unnecessary complication and we will match the number of workers required for production by the *number of workers* prepared to supply labour.

Aggregate labour supply is the number of people of working age (the working population) who are offering themselves for work. People who do not offer themselves for work choose the alternative of non-market activity which we may call *non-participation*. Now, non-participation brings its own rewards, not least that non-participants are spared the disutility generally associated with work. In this simple competitive model it will turn out that all those who offer themselves for work will be given a job, and hence incur the disutility associated with work. As a consequence workers need to be compensated for this disutility by a wage payment. It is the *real wage* that matters, however, because workers are concerned about the spending power their labour commands. An increase of 5 per cent in both wages and prices will leave the real position of workers unchanged and hence will not affect labour supply, because the relative attractiveness of work over non-participation is unchanged. Finally we assume that individuals differ in their preferences between work and non-participation – some will have a strong preference for non-participation while others will find the disutility associated with work relatively low. The former group will need a high compensation (real wage) to tempt them into labour supply whilst the latter group will require a lower compensation. It seems plausible to assert that the lower the real wage the smaller will be the number of people who will be persuaded to offer themselves for work whilst at a high real wage relatively few people will choose non-participation and most will be supplying labour. (See question 4 of Exercise 22.2.)

This discussion clearly points to a labour supply function of the form

$$N_s = N_s \left(\frac{W}{P} \right), \qquad N_s' \left(\frac{W}{P} \right) > 0. \qquad (22.11)$$

Exercise 22.2

1. Consider how an individual's participation decision might be affected by:
   (a) the participation decision of others in the same household;
   (b) non-labour income;
   (c) dependents (children, aged parents);
   (d) unemployment benefits.
2. What factors cause the labour supply curve to shift to the left?
3. What would happen to labour supply if all money wages in the economy increased (say by 5 per cent) other things being equal?
4. In the model of the individual labour supply decision in section 12.1, assume the individual can work *either* for 8 hours per day *or* zero. Derive the individual's 'reservation wage' (i.e., the wage which she will be indifferent between working and not working). Sketch her labour supply curve in this case. [Hint: begin by drawing her 'reservation indifference curve', showing combinations of income and work which are indifferent to zero work. Then take the wage line, as before, and ask over what range its slope must be to induce choice of 8 hours per day rather than zero.]

22.3 The level of employment and output

The level of employment in the competitive model is determined by the equality of demand for and supply of labour – the equilibrating mechanisms being the real wage. We first show explicitly how this equilibrium level of employment and real wage are determined and then how output is determined.

Equilibrium in the competitive labour market is given by

$$N_d = N_s$$

or

$$N_d \left(\frac{W}{P} \right) = N_s \left(\frac{W}{P} \right). \qquad (22.12)$$

Diagrammatically the equilibrium is as shown in Fig. 22.2.

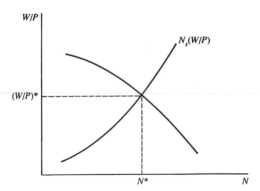

Fig. 22.2: Labour market equilibrium

We assume that the functions $N_d(W/P)$ and $N_s(W/P)$ are such that the equilibrium exists, and is unique. A simple argument establishes that the equilibrium illustrated in Fig. 22.2 is stable. A real wage above the equilibrium $(W/P)^*$ generates an excess supply. For a given price level, P, the money wage W is bid down as workers compete for jobs so that labour demand increases and labour supply falls until equilibrium is reached. This corresponds to the market mechanism discussed in detail in Chapter 2, and it is important to note that it implies direct competition between workers as well as a willingness to tolerate wage reductions. Shortly we will present some models which make some alternative, perhaps more realistic, assumptions about workers' behaviour. At a real wage below $(W/P)^*$ there is an excess demand for labour and it is firms who compete for workers, raising the money wage W and simultaneously reducing labour demand and increasing labour supply.

Notice that in this stability analysis the required adjustments in the real wage occur through adjustments in the *money wage W* for a *given* price level *P*. Labour markets do not determine the price level, only the money wage, but obviously changes in the price level affect labour markets because they affect the real wage. This price level is determined by aggregate demand and supply, as we shall see, and for the moment (that is, as far as the labour market is concerned) it is assumed to be exogenous. That the labour market brings about adjustments in the real wage by changes in the nominal wage is important. In fact it is this feature which is in part responsible for the competitive model producing the following remarkable result. *A change in the price level P does not affect the equilibrium real wage or level of employment.* This statement is easily proven.

Consider the equilibrium condition (22.12). It is apparent that an increase in price *P* for a *given* money wage *W* increases labour demand and reduces labour supply. The effects are

$$\frac{dN_d}{dP} = -\frac{W}{P^2}N_d'\left(\frac{W}{P}\right) > 0 \qquad (22.13)$$

and

$$\frac{dN_s}{dP} = -\frac{W}{P^2}N'_s\left(\frac{W}{P}\right) < 0. \tag{22.14}$$

This says that an increase in P given unchanged W increases the left-hand side of (22.12) and lowers the right-hand side, resulting in

$$N_d\left(\frac{W}{P}\right) > N_s\left(\frac{W}{P}\right). \tag{22.15}$$

We have just seen in our stability analysis, however, that the competitive labour market will eliminate excess demand by increasing the money wage. The money wage will continue to rise until (22.12) is restored. With unchanged N_d and N_s functions this must be at the original real wage. We may see this more formally using the labour market equilibrium condition (22.12). Total differentiation of this condition, allowing both W and P to vary, gives

$$\frac{dW}{P}\left\{N'_d\left(\frac{W}{P}\right) - N'_s\left(\frac{W}{P}\right)\right\} = \frac{W}{P^2}dP\left\{N'_d\left(\frac{W}{P}\right) - N'_s\left(\frac{W}{P}\right)\right\}$$

or

$$\frac{dW}{W} = \frac{dP}{P}. \tag{22.16}$$

Thus a change in P induces a change in W of equal proportion so that the real wage W/P is unchanged. No change in the real wage implies no change in equilibrium employment in this case, since the N_d and N_s schedules illustrated in Fig. 22.2 are unchanged. Any change in P is accompanied by an instantaneous change in W so that (22.12) always holds.

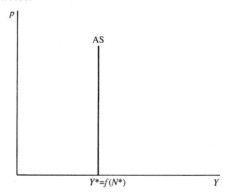

Fig. 22.3: The 'competitive' aggregate supply curve

When we study how employment influences output we see that this implication of the competitive labour market has important implications for aggregate supply.

To establish the form of the aggregate supply curve implied by this labour market behaviour we need to establish how price affects output supplied. First we know that price does not affect employment – this was established above. But if employment is unchanged, output is unchanged because $Y = f(N)$. Hence, *a change in price has no effect on output*. Output is independent of price. The aggregate supply curve is therefore *vertical* at a particular level of output Y^* determined by the equilibrium level of employment N^*. We shall call this level of output the 'natural' level of output. This aggregate supply curve is shown in Fig. 22.3.

This result might appear startling. For one thing, the derived aggregate supply function has a property at the other extreme to that assumed in Chapters 18 to 21 – it is vertical rather than horizontal. This clearly has serious consequences for our earlier analysis based on the IS-LM model, and so we now study output and price determination in the context of aggregate demand and supply. The supply curve we will use is that implied by the model of the labour market developed here. In Chapter 23 we will go on to change the labour market assumptions to reflect some commonly observed departures from the competitive model.

Exercise 22.3

1. Consider the effect on aggregate supply of:
    (a) an increase in the participation rate;
    (b) a reduction in the school-leaving age;
    (c) the abolition of mandatory retirement.
2. Show that the aggregate supply curve developed here is independent of labour demand and supply elasticities.

22.4 Comparative statics of the 'simple' aggregate demand and supply model

The aggregate demand side is given by the simultaneous system (21.5), rewritten here as
$$S(Y - T(Y)) + T(Y) = I(r) + G,$$
$$\frac{m}{P} = L(Y, r), \tag{22.17}$$
where the asterisks have been omitted for convenience.

The mathematical representation of the aggregate supply side is derived from the labour market equilibrium and the production function, viz.,
$$N_s = N_s\left(\frac{W}{P}\right), \tag{22.18}$$
$$N_d = N_d\left(\frac{W}{P}\right), \tag{22.19}$$
$$N = N_d\left(\frac{W}{P}\right) = N_s\left(\frac{W}{P}\right), \tag{22.20}$$
$$Y = f(N). \tag{22.21}$$

To make the model more interesting we may extend the aggregate supply system by introducing exogenous variables which determine the positions of the labour demand and labour supply schedules. First, suppose an increase in a parameter γ increases labour supply at any real wage. In terms of Fig. 22.2 an increase in γ shifts the $N_s(W/P)$ curve to the right. Mathematically we have

$$N_s = N_s(W/P, \gamma), \qquad \frac{\partial N_s}{\partial (W/P)} > 0, \qquad \frac{\partial N_s}{\partial \gamma} > 0. \qquad (22.22)$$

The obvious way to interpret γ is as a 'taste' parameter which reflects a change in labour's preferences in favour of work and away from leisure or non-participation.

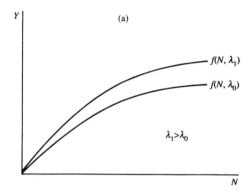

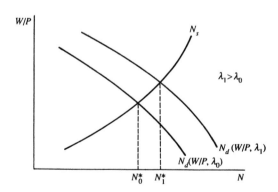

Fig. 22.4: The effect of a productivity improvement on the labour market

Secondly, we may introduce a parameter λ which affects the demand for labour. For example an increase in λ may reflect a technological improvement which makes a given workforce more productive. We assume that λ affects the production function *and* the demand for labour curve. Fig. 22.4 illustrates the effect of an increase in the 'technological' parameter λ. In frame (*a*) we see that λ increases Y at each level of employment. In frame (*a*) we see that it also increases the marginal product of labour, and in frame (*b*) we see that this is reflected in the shift to the right of the N_d curve (this being the marginal product of labour curve as we showed earlier).

Mathematically we replace (22.19) and (22.21) by

$$N_d = N_d(W/P, \lambda), \qquad \frac{\partial N_d}{\partial (W/P)} < 0, \qquad \frac{\partial N_d}{\partial \lambda} > 0 \qquad (22.23)$$

and

$$Y = f(N, \lambda) \qquad \frac{\partial Y}{\partial N} > 0, \qquad \frac{\partial Y}{\partial \lambda} > 0. \qquad (22.24)$$

By introducing the parameters γ and λ we are able to supplement 'demand side' comparative statics (the effects of monetary and fiscal policy) with 'supply-side' effects. The supply-side subsystem (22.18)–(22.21) becomes

$$N_s = N_s(W/P, \gamma), \qquad (22.25)$$
$$N_d = N_d(W/P, \lambda), \qquad (22.26)$$
$$N = N_d = N_s, \qquad (22.27)$$
$$Y = f(N, \lambda). \qquad (22.28)$$

For the purposes of the section, it is convenient to simplify this subsystem. What is of interest is how the level of employment N depends on γ and λ. Our analysis is simplified by deriving a *reduced-form* expression for the level of employment N. This involves nothing more than establishing how the labour market equilibrium is affected by changes in γ and λ. The labour market equilibrium may be written as

$$N_d\left(W/P, \lambda\right) - N_s\left(W/P, \gamma\right) = 0. \qquad (22.29)$$

Taking total differentials gives

$$d(W/P)\left\{ \frac{\partial N_d}{\partial (W/P)} - \frac{\partial N_s}{\partial (W/P)} \right\} + d\lambda\left\{ \frac{\partial N_d}{\partial \lambda} \right\} - d\gamma\left\{ \frac{\partial N_s}{\partial \gamma} \right\} = 0, \qquad (22.30)$$

from which we find that

$$\frac{d(W/P)}{d\lambda} = - \frac{\partial N_d/\partial \lambda}{\left\{ \partial N_d/\partial (W/P) - \partial N_s/\partial (W/P) \right\}} > 0 \qquad (22.31)$$

and

$$\frac{d(W/P)}{d\gamma} = - \frac{\partial N_s/\partial \gamma}{\left\{ \partial N_d/\partial (W/P) - \partial N_s/\partial (W/P) \right\}} < 0. \qquad (22.32)$$

These effects are easily confirmed diagrammatically. To find the effects of λ and γ on employment we simply use the fact that in equilibrium $N = N_s(W/P, \gamma)$ and differentiate with respect to λ and γ, bearing in mind the dependence of the real wage on λ and γ as given by (22.31) and (22.32). Thus

$$\frac{dN}{d\lambda} = \frac{\partial N_s}{\partial(W/P)} \left(\frac{d(W/P)}{d\lambda} \right) > 0, \qquad (22.33)$$

$$\frac{dN}{d\gamma} = \frac{\partial N_s}{\partial(W/P)} \left(\frac{d(W/P)}{d\gamma} \right) + \frac{\partial N_s}{\partial\gamma} > 0. \qquad (22.34)$$

We may therefore write the (reduced-form) employment function as

$$N = N(\gamma, \lambda), \qquad \frac{\partial N}{\partial\gamma} > 0, \qquad \frac{\partial N}{\partial\lambda} > 0. \qquad (22.35)$$

These effects are also confirmed by a diagrammatic analysis. The aggregate supply subsystem may now be written

$$N = N(\gamma, \lambda), \qquad (22.36)$$
$$Y = f(N, \lambda). \qquad (22.37)$$

The combined aggregate demand and supply equilibrium is given by

$$S(Y - T(Y)) + T(Y) = I(r) + G, \qquad (22.38)$$
$$\frac{m}{P} = L(Y, r), \qquad (22.39)$$
$$N = N(\gamma, \lambda), \qquad (22.40)$$
$$Y = f(N, \lambda). \qquad (22.41)$$

The aggregate demand and supply equilibrium is illustrated in Fig. 22.5.

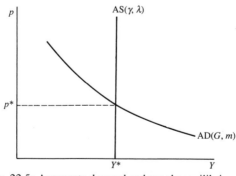

Fig. 22.5: Aggregate demand and supply equilibrium

It is apparent both from (22.38)–(22.41) and from Fig. 22.5 that equilibrium output is now determined solely by the aggregate supply subsystem (22.40) and (22.41). Given this equilibrium level of output (22.38) and (22.39) determine the interest rate and the price level. We have already established that the level of employment is invariant to changes in price. The system (22.38)–(22.41) may be simplified further by substituting (22.40) into (22.41). This gives a system of three independent equations determining the three endogenous variables Y, r, and P in terms of the four exogenous variables G, m, γ, and λ.

$$S(Y - T(Y)) + T(Y) = I(r) + G, \tag{22.42}$$

$$\frac{m}{P} = L(Y, r), \tag{22.43}$$

$$Y = f(N(\gamma, \lambda), \lambda). \tag{22.44}$$

This equilibrium is illustrated in Fig. 22.5.

Differentiating (22.42)-(22.44) totally allowing Y, r, P, G, m, γ, and λ to vary gives

$$
\begin{bmatrix}
S'(Y - T(Y))[1 - T'(Y)] + T'(Y) & -I'(r) & 0 \\
L_Y(Y, r) & L_r(Y, r) & m/P^2 \\
1 & 0 & 0
\end{bmatrix}
\begin{bmatrix}
dY \\
dr \\
dP
\end{bmatrix}
=
$$

$$
\begin{bmatrix}
dG \\
-dm/P \\
d\gamma(f_N N_\gamma) + d\lambda(f_N N_\lambda + f_\lambda)
\end{bmatrix}. \tag{22.45}
$$

The following effects are easily derived by application of Cramer's Rule to (22.45):

$$\frac{dY}{dG} = 0, \qquad \frac{dY}{dm} = 0$$

$$\frac{dY}{d\gamma} = |A|^{-1}\{-f_N N_\gamma I'[m/P^2]\} > 0, \qquad \frac{dY}{d\lambda} = |A|^{-1}\{-f_N N_\lambda - f_\lambda\}I'[m/P^2] > 0,$$

$$\tag{22.46}$$

where $|A| > 0$ is the determinant of the square matrix in (22.45). Thus both pure fiscal and pure monetary policy are entirely ineffective in increasing real output. This implies that the multiplier is completely neutralized by price adjustment. The only way of increasing output in this model is by 'supply-side' changes such as productivity improvements (via λ) or changes in workers' willingness to supply labour at any given real wage (via γ). It is often argued, for example, that reductions in unemployment benefit will make work a relatively more attractive proposition compared to non-participation, labour supply therefore expands, lowering the real wage and increasing employment. (See questio 5 of Exercise 22.4.) That the real wage does indeed fall as part of this process (represented by an increase in γ) is confirmed by (22.32).

The effects of changes in the exogenous variables on r and on P are left as exercises. The exercises also invite discussion of what happens to the money wage and hence the real wage in each case.

Exercise 22.4

1. Confirm the effects (22.31) and (22.32) diagrammatically. Hence confirm the effects (22.33) and (22.34).
2. What effect do the 'technological' and 'taste' parameters γ and λ have on the position of the aggregate supply curve?
3. Use the system (22.45) to find the effects of G, m, γ, and λ on r and P. Illustrate using IS-LM diagrams and AD-AS diagrams.
4. Consider a 'linear' version of the system (22.42)–(22.44):

$$(1 - b(1 - t))Y + tY = e - hr + G, \qquad (22.42)'$$

$$\frac{m}{P} = kY - lr, \qquad (22.43)'$$

$$Y = a_1\gamma^2 + a_2\lambda, \qquad (22.44)'$$

where the parameters in (22.42)$'$ and (22.43)$'$ are defined in Chapter 18, and a_1 and a_2 are positive constants. Find the effects on Y, P, and r of changes in b, t, e, h, k, l, γ, and λ and interpret the results. Illustrate each effect using IS-LM and AD-AS diagrams.
5. In the diagram constructed to answer question 4 of Exercise 22.2, assume the worker receives a payment $b > 0$ only if she chooses zero work. Show the effects of variations in b on her choice of whether to work 8 hours per day or zero.

22.5 Competitive labour markets and policy debates

The analysis of this chapter has a startling aggregate implication, summarized in (22.46). Expansions of aggregate demand have no impact on output if labour markets are frictionless and competitive. Price adjustment eliminates any 'multiplier' effects. Hence, the results of previous chapters have been dramatically changed.

This makes the issues on policy-making very clear. If labour markets behave in a competitive way, then government macroeconomic policy can only be directed towards price-stability and not towards employment and income expansion. An attractive and relatively simple strategy for a government may therefore be to:

(*a*) remove institutional inflexibilities from labour markets, or at least reduce them,

(*b*) reduce government spending, so reducing the deficit without any adverse employment or output consequences, (*c*) keep a check on price increases (inflation) by monetary and fiscal restraint, again without adverse employment or output consequences.

Strategies similar to this have been popular with many recent governments including those of the US, UK, and Canada. Before we pass judgement on the willingness of governments to embrace a 'frictionless' model of the economy, the subject of inflexibilities and rigidities in labour markets needs to be discussed. This we do in the next chapter.

22.6 Summary

A perfectly competitive labour market is one in which workers and firms act as price takers, and demand and supply decisions are based on the real wage which adjusts freely to eliminate excess demands and supplies of labour.

In this case, the aggregate supply curve for output is vertical, independent of the price level. The equilibrium level of output or 'natural rate' of output depends on the level of employment, which in turn depends on the underlying determinants of labour demand and labour supply: the production function and consumer preferences given non-market opportunities.

In fact the analysis of this chapter is a re-statement, in macroeconomic terms, of the model of perfectly competitive general equilibrium set out in Chapter 14.

23 Aggregate Supply and Labour Market Imperfections

In this chapter we look at some commonly suggested explanations as to why labour markets may not work in the way described in the previous chapter. Since the price mechanism there was able to function perfectly we are on safe ground by calling those market features which inhibit the mechanism collectively 'market imperfections'. The three types of imperfection considered all result in the same effect. That is, the money wage does not adjust freely to eliminate instantaneously excess demand for or excess supply of labour.

We first consider the effect of institutional rigidities in the money wage. A simple version of this argument is that trade unions successfully resist any attempts by workers to undercut each other in money wages and resist attempts by firms to encourage them. Secondly we study the effects of misperceptions of what is happening to wages and prices in the form of 'money illusion'. Finally we look at the difficulties caused by uncertainty and the need for agents to form *expectations* about prevailing prices.

This area has been, and continues to be, one of intense academic research. This is largely because policy implications are particularly sensitive to assumptions about the information agents have and the behaviour of labour markets in the presence of excess supply.

23.1 Downward money wage rigidity

The central idea here may be stated very simply. The economy inherits a labour market equilibrium from an earlier period. The equilibrium is characterized by a money wage–employment combination (W_0, N_0) at a price level P_0. This is illustrated in Fig. 23.1. The money wage W_0 comes to be regarded by workers collectively (that is, by a union) as representing a minimum standard of living which must be protected. The wage W_0 is therefore safeguarded. Naturally upward pressures on money wages (caused by excess demand for labour) will not be resisted, but downward pressures will. The consequences for aggregate supply will be as follows. Suppose we start in equilibrium at (W_0, N_0, P_0) in Fig. 23.1. The aggregate supply combination is $(P_0, f(N_0))$ or (P_0, Y_0) as shown in Fig. 23.2. We consider two types of price changes as movements away from P_0. First, price increases to $P_1 (> P_0)$. This reduces the real wage and generates a temporary excess demand for labour at real wage $W_0/P_1 (< W_0/P_0)$. Workers do not resist the bidding up of W_0 and the equilibrium is restored when the money wage is W_1 so that $W_1/P_1 = W_0/P_0$. Employment therefore remains at N_0 following the increase in price and output at Y_0. Now consider a price fall away from P_0 to P_2. This increases the real wage to W_0/P_2 creating an excess supply. Downward movements of the money wage are resisted so that disequilibrium is maintained.

Thus with excess supply $N_d < N_s$ and $N = N_d = N_2$ in Fig. 23.1. This situation will be maintained indefinitely as long as reductions in W are resisted. In terms of Fig. 23.2 the fall in price to P_2 results in a fall in output to $f(N_2) = Y_2$. Notice that the increase in real wage to W_0/P_2 in Fig. 23.1 has not only reduced labour demand compared with the initial equilibrium, but has also increased labour supply by making participation more worth while. Actual unemployment at W_0/P_2 is therefore $N_3 - N_2$.

The arrows in Fig. 23.2 remind us of the fact that the moves are one-directional. A fall in price from P_1 to P_0 does not leave output unchanged, since at price P_1 the wage floor is W_1 not W_0. Once at (P_1, Y_0) a fall in price back towards P_0 results in a fall in employment and output as will the fall in price from P_0 to P_2. The aggregate supply curve is vertical only for price increases and positively sloped only for price reductions.

Figs. 23.1 and 23.2 may be translated into a mathematical model quite simply. We know that at the initial equilibrium, (22.12) holds. However, when totally differentiating this condition we note that W is free to vary only for *increases* in P. Reductions in P are *not* accompanied by reductions in W in this model. Whilst the analysis of the previous section is appropriate for a price increase, a price fall is studied as follows. In general the level of employment N is given by the smaller of labour demand and supply:

$$N = \min\left[N_d\left(\frac{W}{P}\right), N_s\left(\frac{W}{P}\right)\right]. \tag{23.1}$$

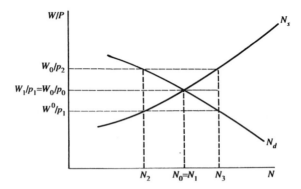

Fig. 23.1: The labour market with money wage rigidity

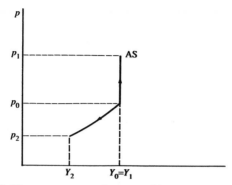

Fig. 23.2: The aggregate supply curve with money wage rigidity

Starting at the initial equilibrium and considering $dP < 0$ with $dW = 0$ we may study the effects on N_d and N_s by total differentials. We have

$$dN_d = dP\left\{-\frac{W}{P^2}N_d'\left(\frac{W}{P}\right)\right\}, \tag{23.2}$$

$$dN_s = dP\left\{-\frac{W}{P^2}N_s'\left(\frac{W}{P}\right)\right\}. \tag{23.3}$$

Since $dP < 0$ we have $dN_d < 0$ and $dN_s > 0$ so that employment is

$$N = N_d\left(\frac{W}{P}\right) < N_s\left(\frac{W}{P}\right).$$

That this level of employment is lower than initially is obvious, since employment was equal to labour demand before the price fall and is so following the price fall and $dN_d < 0$. Labour and employment have fallen.

In summary we have,

$$\frac{dP}{P} = \frac{dW}{W} \quad \text{if } dP > 0$$

and

$$\frac{dP}{P} < \frac{dW}{W} = 0 \quad \text{if } dP < 0$$

(implying an increased real wage and fall in employment).

There is now a large literature which attempts to explain why money wages may not adjust to eliminate an excess supply of labour. The challenge is not to rely on *ad hoc* explanations or irrationality, but to attempt to explain why optimizing, rational workers and firms may be reluctant to see downward wage adjustment.

The 'union' explanation has been mentioned already, but a full explanation here requires a specification of union objectives. In the absence of such a specification it is not clear that unions always prefer to resist wage changes with consequent adverse effects on employment and, possibly, membership. (See Chapter 12.)

The 'efficiency wage' theory is another explanation of wage rigidity but one which suggests that firms themselves might prefer to set wages above the market-clearing wage. A simple version of this theory relates worker productivity to the wage rate – a higher wage leading to higher productivity. With wages above the market level, employees earn rents with their current employer which they strive to keep by maintaining a high productivity so as to reduce the chances of layoff or dismissal. Thus a high wage buys high productivity for employers. However, if all firms find it profitable to offer wages above the competitive level, the level of employment will be *below* the competitive level, thus creating more unemployment. High unemployment lowers the probability that an unemployed worker will find a job, and so the threat of unemployment induces higher effort.

Yet another view of money wage rigidity arises when firms are isolated from labour market competition because the unemployed ('outsiders') are not regarded as perfect substitutes for current employees ('insiders'). Training in firm-specific skills is one example of an insider attribute not shared by outsiders. To the extent that firms seek to avoid turnover and training costs of replacing current workers with unemployed workers, current insiders have added bargaining power to secure a higher wage. The central theme is that 'outsiders' cannot bid down the 'insiders' wage to secure employment.

Whilst many models along these lines seek to explain why optimizing agents may prefer to isolate wages from competitive forces, other models attach significance to uncertainty and misperceptions about the real wage as a cause of apparently irrational behaviour by workers. We look at some of these issues now. We return to a model of efficiency wages in Chapter 28.

23.2 Perceptions and expectations

The competitive model of the labour market presumes that agents base their decisions on the appropriate criteria: a worker's labour supply depends on the real wage; firms demand labour on the basis of the real wage relative to marginal product. On the other hand, agents may attach greater significance to the nominal value of compensation rather than its purchasing power. This practice may be understandable following a long period of product-price stability. There may also be a difficulty in being able to calculate the real wage because of uncertainty about the price level. These issues are studied here.

23.2.1 Money illusion

In its extreme form money illusion on the part of workers involves considering the money wage as the all-important indicator of well-being. *All* price movements (in the extreme form) are ignored. There may be several reasons for this apparently irrational behaviour at least as a temporary phenomenon. If the economy has had a very long period of price stability workers will have become accustomed to judging real wage changes solely in terms of money wage changes. As long as prices are fixed, this is of course rational behaviour. When prices start to fluctuate workers persist in their habit of ignoring these, possibly because they regard any price changes they encounter day to day as relative price changes (increases in some prices being presumed to be offset elsewhere by price reductions) or as temporary. For a while at least workers see no reason to respond to price movements in terms of adjustments of the money wage. After some time, perhaps of continually rising prices (which become generally recognized as such), this money illusion becomes irrational and workers must recognize the effect the price increase is having on their real wage. Money illusion is at best therefore a temporary phenomenon.

To keep things simple in modelling this we assume that labour demand depends, as before, on the real wage – firms do not suffer from money illusion. Labour supply on the other hand depends solely on the money wage if workers suffer from money illusion. We may combine these two facts in a model of labour market equilibrium with the money wage on the vertical axis. The supply of labour is an upward sloping function of the money wage independent of price whilst the demand for labour is a downward sloping function of the money wage for a given price level, initially chosen to be P_0. At a price P_0, the labour market equilibrium is given by (W_0, N_0) in Fig. 23.3. The corresponding price output combination is (P_0, Y_0), where $Y_0 = f(N_0)$ is given in Fig. 23.4. Changes in price affect the demand for labour in Fig. 23.3 but do not affect labour supply. An increase in price shifts out N_d since at any given money wage a higher price level indicates a lower real wage and hence increased demand for labour. The increase in price from P_0 to P_1 shifts out N_d and, with the position of N_s unaffected, increases employment to N_1. Output correspondingly increases to $Y_1 = f(N_1)$ as shown in Fig. 23.4. On the other hand a fall in price from P_0 to P_2 shifts N_d to the left, representing a lower labour demand at a higher real wage. Employment falls to N_2 and output to $Y_2 = f(N_2)$.

To see the precise mechanics here we may model these effects as follows. The labour demand function is as before, but with money illusion in its extreme form labour supply depends only on the money wage. Equilibrium is therefore given by

$$N_d\left(\frac{W}{P}\right) = N_s(W) \tag{23.4}$$

with $N_d'(W/P) < 0$ as before and $N_s'(W) > 0$. Both P and W are free to vary, so that total differentiation of (23.4) gives

$$dP\left\{\frac{W}{P^2}N_d'\left(\frac{W}{P}\right)\right\} = \frac{dW}{P}\left\{N_d'\left(\frac{W}{P}\right) - PN_s'(W)\right\}. \tag{23.5}$$

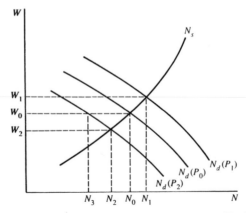

Fig. 23.3: The labour market with worker money illusion

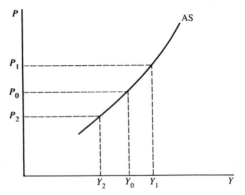

Fig. 23.4: The aggregate supply curve under worker money illusion

Inspection of (23.5) reveals that dP and dW always take the same sign so that $dW > 0$ if $dP > 0$ and $dW < 0$ if $dP < 0$. To see this, note that the sign of the braced terms on both sides of (23.5) is negative, so the equality in (23.5) can only hold if 'sign $[dP]$' = 'sign $[dW]$'. We will consider $dP > 0$, the case of $dP < 0$ being symmetrical. We may rearrange (23.5) to give

$$N_d'(W/P)\left\{ \frac{W}{P^2}dP - \frac{dW}{P} \right\} = -dW\{N_s'(W)\}, \qquad (23.6)$$

and dividing throughout by W gives

$$\frac{N_d'(W/P)}{P}\left\{ \frac{dP}{P} - \frac{dW}{W} \right\} = -\frac{dW}{W}\left\{ N_s'(W) \right\}. \qquad (23.7)$$

Now $dP > 0$ by assumption and so $dW > 0$. The right-hand side of (23.7) is therefore negative and so the left-hand side is also negative, implying

$$\frac{dP}{P} > \frac{dW}{W}. \qquad (23.8)$$

Thus both price and money wage increase but the price increase is proportionately larger than the money wage increase so that the real wage falls. The fall in the real wage stimulates labour demand and employment. Employment is always given by

$$N = N_d \left(\frac{W}{P} \right) = N_s(W).$$

Notice that the money wage is not inflexible in either direction in this version of the model. Workers do not appreciate the cause of variations in demands for their services (a variation in price) but do allow money wages to adjust to eliminate excess demands and supplies. Obviously the money wage adjustments are not complete enough to leave the real wage unchanged. The variations in employment imply that increases and decreases in price are accompanied by less than proportionate variations in the money wage. This is because the variations in the money wage are only required to remove excess demands or supplies, not to compensate workers for variation in the price level.

The model of downward money wage rigidity may be superimposed on this model easily. If workers will not allow the money wage to fall below W_0 as price falls to P_2 in Fig. 23.3 there will be an excess supply of labour with employment again determined by (23.2). Thus employment falls to N_3 – a lower level than if the money wage were flexible downwards.

23.2.2 Expectations

As with the previous model, the analysis of employment variations in response to price changes may or may not include the assumption of downward money wage rigidity. We will conduct the analysis on the assumption that W is free to vary to eliminate (perceived) excess demands and supplies in both directions. The analysis incorporating downward money wage rigidity is left as an exercise.

At the heart of the analysis of expectations is the problem of exactly how expectations are formed. In fact a large number of assumptions may be made about expectations formation. The theory of how expectations are formed is dealt with in Chapter 24. In this section we simply study the effects of expectations on the labour market. Furthermore we assume that the incomplete information which generates the need for expectations or forecasts to be made exists only on the workers' side. The uncertainty surrounds the price currently prevailing in the economy. Firms know the price for certain, whereas workers make a (hopefully informed) guess, known as the *expected price level*, P^e.

Our starting-point is that the economy has been operating at the same equilibrium for some time, so that workers' *expectations* about prices are correct. Initially, then, we assume $P = P^e$. Equilibrium is

$$N_d \left(\frac{W}{P} \right) = N_s \left(\frac{W}{P^e} \right) \tag{23.9}$$

with $P = P^e$ initially. Now suppose the actual price increases by $dP > 0$, how will expectations respond? This depends on how good workers are at forming price expectations. For the moment we will assume that the change in price expectations is a fraction α of the actual price change so that

$$dP^e = \alpha \, dP \tag{23.10}$$

with $0 \leq \alpha \leq 1$. The two extreme cases will be considered presently once we have studied the general properties of this assumption. Clearly, (23.10) is an arbitrary assumption and does not represent a *theory* of expectations formation. We consider the expectation process more fully presently. To consider the effects of a change in actual price, P on this model, we totally differentiate (23.9) allowing P, W and P^e to vary in accordance with (23.10). This gives

$$dW\left\{\frac{N'_s(W/P)}{P^e} - \frac{N'_d(W/P)}{P}\right\} = dP^e\left\{N'_s(W/P)\frac{W}{P^{e2}}\right\} - dP\left\{N'_d(W/P)\frac{W}{P^2}\right\}. \tag{23.11}$$

Recall that initially $P^e = P$. Using this and (23.10) gives, after rearranging,

$$\frac{dW}{W}\left\{N'_s\left(\frac{W}{P}\right) - N'_d\left(\frac{W}{P}\right)\right\} = \frac{dP}{P}\left\{\alpha N'_s\left(\frac{W}{P}\right) - N'_d\left(\frac{W}{P}\right)\right\}. \tag{23.12}$$

Both sides of the expression are positive but the braced term on the right-hand side is smaller than the braced term on the left-hand side (if α is a fraction), implying

$$\frac{dP}{P} > \frac{dW}{W}$$

which, as we now know, further implies a fall in the real wage and an increase in employment. This analysis is symmetric for price falls, $dP < 0$. The extreme case also emerges from (23.12). If $\alpha = 1$, workers accurately predict the price change and so $P^e = P$ always. This gives results equivalent to those in Chapter 22. On the other hand, if $\alpha = 0$ expectations are static and we observe effects similar to those of the money illusion model discussed earlier.

This expectations model therefore includes several cases, and the shape of the aggregate supply curve therefore depends on the assumed value of α which in turn reflects how correct expectations are, or rather, how good workers are at forming expectations. If expectations are correct ($\alpha = 1$) so that $P^e = P$ (i.e., there is perfect foresight) always, then the aggregate supply curve is vertical. If $0 < \alpha < 1$ then the aggregate supply curve will be positively-sloped. These cases are illustrated in Fig. 23.5.

Notice that this does not represent a *theory* of expectations and only tells us what happens to labour market activity if expectations lag or if they are correct. The way the analysis is presented is as if we associate the misalignment of expected and actual prices to be a feature of the 'short run' whilst the equality of expected and actual prices is associated with the 'long run'. We make this distinction clearer in the next chapter by considering how price expectations may be formed.

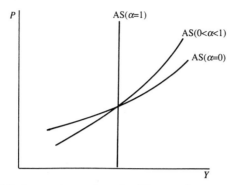

Fig. 23.5: Aggregate supply under different price expectations

Exercise 23.2

1. Conduct the analysis following (23.5) for the case in which $dP < 0$. Confirm mathematically that this analysis generates an aggregate supply curve as in Fig. 23.4.
2. Use a diagrammatic analysis to study a labour market in which workers suffer from money illusion but in which there is also downward money wage flexibility. What slope of aggregate supply curve is implied by this behaviour? Do asymmetries in the model require directional indicators as in Fig. 23.2?
3. Suppose *both* workers and firms face uncertainty about the actual price level. Firms form an estimate P_f^e and workers P_w^e. Suppose further that for price increases $P_f^e < P_w^e$ and for price falls $P_f^e > P_w^e$. Finally labour supply is $N_s(W/P_w^e)$ while labour demand is $N_d(W/P_f^e)$. What sort of behaviour do you expect to result from actual price variations and what properties do you expect the aggregate supply curve to have if (i) money wages are flexible in both directions, (ii) money wages are rigid downwards?

23.3 Summary

Several theories attempt to explain why money wages may not adjust to eliminate excess supply in the labour market. Some of these assume 'institutional' rigidities such as wage pressure from unions, pay comparability, and custom, while others rely on seemingly irrational behaviour by agents such as 'money illusion'. Other theories explain why it may be optimal or individually rational for firms and workers to agree on a level of compensation above the market-clearing level, with or without perfect information about price movements. We return to consider some of these theories in Chapter 28.

Whatever the source of the rigidity, the failure of the real wage to return to its equilibrium following a price change causes employment fluctuations and hence output fluctuations.

If there is perfect foresight about price movements and no restraints on the wage level reaching its competitive level, then we obtain the results of Chapter 22.

24 Aggregate Supply in the Long Run and the Short Run

The discussion at the close of the last chapter suggests that we may associate the equality of expected and actual prices with the 'long run', by which is meant, in these models, that period over which agents learn by their mistakes and adjust their expectations. Hence workers accumulate information on the actual price level over time. In the long run, therefore, with expected real wages in line with actual real wages, employment and output are invariant with respect to price – as in the competitive model of Chapter 22. This 'long-run' aggregate supply curve is shown as AS in Fig. 24.1.

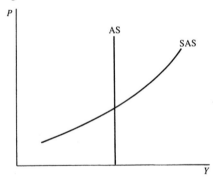

Fig. 24.1: Long-run and short-run aggregate supply

In the 'short run' with full money wage flexibility but expected price out of line with actual price, the aggregate supply curve has a shape given by SAS. For price increases the expected real wage is higher than the actual real wage, thus increasing employment and output. For price falls the expected real wage is lower than the actual, resulting in less labour being offered, a fall in employment, and a fall in output.

Two theories of expectations formation bring out this distinction between the short run and the long run. They are the *adaptive expectations* theory and the *rational expectations* theory. We consider each in turn. These concepts have also been used in the market analysis of Chapter 5.

We assume, as in the last chapter, that labour market equilibrium is given by

$$N_d\left(\frac{W}{P}\right) = N_s\left(\frac{W}{P^e}\right) \tag{24.1}$$

For convenience we will denote the actual and expected real wages by w and w^e respectively. Hence (24.1) becomes

$$N_d(w) = N_s(w^e). \tag{24.2}$$

Moreover, the analysis is simplified if we choose linear functional forms for N_d and N_s, specifically,

$$N_d = a - bw \qquad a, b > 0, \tag{24.3}$$

$$N_s = cw^e \qquad c > 0. \tag{24.4}$$

We conduct the analysis in terms of expectations of the *real wage*. Implicit in this, of course, is the price expectation since the money wage is known. The reason for adapting this strategy is simply a technical one. We will be studying labour market equilibrium through time using first-order difference equations and we avoid non-linearities by working in the expectation w^e rather than P^e.

24.1 Adaptive expectations

One mathematical formulation of the adaptive expectation is

$$w^e = w^e_{-1} + \lambda[w_{-1} - w^e_{-1}] \qquad 0 < \lambda < 1, \tag{24.5}$$

where w^e and w are current values of the expected and actual real wage, the -1 subscripts indicate the one-period lagged value – the value taken by the variable in the previous period, and λ is a positive constant. Equation (24.5) says that the expectation of the current value of the real wage is the expectation of the previous period real wage and a fraction of the previous period prediction error. The general form of (24.5) is often referred to as partial adjustment. Rearranging (24.5) we see that partial adjustment implies that w^e is nothing more than a weighted average of w_{-1} and w^e_{-1}, viz.,

$$w^e = \lambda w_{-1} + (1 - \lambda)w^e_{-1}. \tag{24.6}$$

Using (24.3) and (24.4) in (24.2) gives

$$w^e = \frac{a - bw}{c} \tag{24.7}$$

and lagging by one period gives

$$w^e_{-1} = \frac{a - bw_{-1}}{c}. \tag{24.8}$$

Substituting (24.7) and (24.8) into (24.5) (or into (24.6)) gives a linear first-order difference equation

$$w = \frac{\lambda a}{b} + \frac{[(1 - \lambda)b - \lambda c]}{b} w_{-1} \tag{24.9}$$

or

$$w = \frac{[(1 - \lambda)b - \lambda c]}{b} w_{-1} + \frac{\lambda a}{b} \tag{24.10}$$

which takes the form of the difference equations discussed in the appendix,

$$w - \alpha w_{-1} = \beta. \tag{24.11}$$

Obviously $\beta > 0$. Without further restrictions, however, α may be positive or negative but it is certainly a fraction because $b > [(1 - \lambda)b - \lambda c]$ implies $0 > -\lambda(b + c)$. Thus $0 < |\alpha| < 1$ so that the equilibrium in the labour market with adaptive expectations is stable. Thus, convergence to equilibrium may be monotonic ($\alpha > 0$) or oscillatory ($\alpha < 0$). The solution to (24.10) is

$$w_t = \left[w_0 - \frac{\beta}{1 - \alpha} \right] \alpha^t + \frac{\beta}{1 - \alpha}$$

or,

$$w_t = \left[w_0 - \frac{a}{b + c} \right] \alpha^t + \frac{a}{b + c} \tag{24.12}$$

where w_0 is an arbitrary starting point. Henceforth we will assume $0 < \alpha < 1$ for monotonic convergence. The implications of $-1 < \alpha < 0$ are left as an exercise. We immediately see that in the long run

$$\frac{W}{P} \equiv w = \frac{a}{b + c} \tag{24.13}$$

which is a constant, so that variations in P are accompanied by proportional variations in W and the long run real wage is constant. From (24.3) this implies constant employment and hence constant output. We conclude that the *long-run* aggregate supply curve under adaptive expectations is vertical.

24.1.1 The short run

To see the short-run properties we must be more specific about w_0. Suppose that a sudden, unanticipated fall in P increases the real wage so that $w_0 > a/(b + c)$. Then in period 1 we have, from (24.12)

$$w_1 = \left[w_0 - \frac{a}{b + c} \right] \alpha + \frac{a}{b + c}$$

or

$$w_1 = \alpha w_0 + (1 - \alpha)\frac{a}{b + c}$$

so that w_1 lies below the previous period 'surprise' real wage w_0 but above the long-run equilibrium.

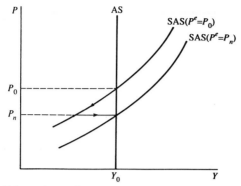

Fig. 24.2: Price and quantity movements under adaptive expectations

In period 2 we have

$$w_2 = \alpha^2 w_0 + (1 - \alpha^2)\frac{a}{b + c}$$

and with more weight attached to $a/(b + c)$ than to w_0, $w_2 < w_1$. In fact it is clear from (24.12) that as t increases w_t falls away from w_0, towards w_1. With no further reduction in P apart from the initial one, it is clear that a fall in w is being achieved by successive reductions in the money wage, W. From (24.3) it is clear that the initial jump in the real wage to w_0 lowers employment and hence output – we move down a short-term aggregate supply curve. Gradually, as w falls, employment increases and we move towards the long-run equilibrium level of output. The arrows in Fig. 24.2 show these movements. Because the initial move down SAS is caused by an unanticipated or 'surprise' fall in P, the short-run aggregate supply curve is often called the 'surprise' aggregate supply curve.

The analysis of a price increase is symmetric and since it is the magnitude of α which determines the spread of adjustment, the time taken to return to the long-run equilibrium depends on α, b, and c. Moreover each 'period' is associated with a different price (real wage) expectation and a different money wage.

Finally, the analysis is in the nature of a comparative-static exercise to discover the effects, or sequence of effects, of a *given* fall in price on aggregate supply. The analysis of aggregate demand and supply equilibrium and hence of endogenous price determination is taken up now in the following section.

Exercise 24.1

1. Conduct the adaptive expectation analysis on the assumption that $-1 < \alpha < 0$. Draw the equivalent diagram to Fig. 24.2 for this case.
2. Do you regard *adaptive expectations* as specified in this section as a plausible theory of expectation formation? Consider what would happen to real wages if workers persistently used the adaptive expectation mechanism outlined here in times of continually *increasing* inflation.

24.2 Rational expectations

Under adaptive expectations observations on previous values of the variable concerned are used to improve the forecast. Under rational expectations, however, *all* available information is used. Thus in the analysis of the previous section the cause of the initial fall in price and any information relating to it would be used by agents to predict the consequent effects. Moreover, in the context of a full model, agents are assumed to know the structural parameters and therefore to predict what equilibrium will be generated by any sudden change. Consider, for example, the full aggregate demand and supply model illustrated in Fig. 24.3. The initial position is at point A and there is an unexpected increase in demand to AD_1. Since this shift is unexpected, the economy moves along the static-expectations short-run aggregate supply curve, $SAS(P^e = P_0)$. Only if, and for as long as, the shift in aggregate demand is not perceived by workers, the actual price level in the short run will rise to P_1 and output to Y_1.

Although workers may not initially perceive the demand shift, they and all other agents including firms are assumed to know how the economy works. In particular they believe the economy operates in accordance with the analysis summarized in Fig. 24.3. Now, once the shift in demand is discovered workers must see, along with everybody else, that the only sustainable long-run equilibrium price under the new demand conditions is P_2, hence the rational expectation of the price level is $P^e = P_2$. The short-run aggregate supply curve shifts to $SAS(P^e = P_2)$ and expectations are fulfilled.

Whilst we do not have space to give a full consideration of the relative merits of different models of expectation formation, some remarks are in order. In the most general sense the rational expectations framework represents an improvement over the adaptive expectations approach. The rational expectation is often said to be 'model-based' and makes best use of *all* available information – not just past prices. By 'model-based' we mean that the demand and supply functions themselves are solved, using all relevant information, for the expected price. The implication is that if the shift in aggregate demand is the result of an announced policy and therefore not a surprise, then (as long as the policy-makers are believed) the move to the new long-run equilibrium takes place 'immediately'. It might appear to be stretching things to argue that all agents fully understand and agree on the way the model works. This is possible, but there are a large number of sources of information and expertise which may be tapped to improve the accuracy of the forecasts. Rational expectations do not imply the absence of all errors, only the absence of systematic (and hence avoidable) errors. In adaptive expectations there are systematic errors because people apparently never learn that they are systematically *underpredicting* the price rise.

We can use this discussion to summarize the main ingredients of aggregate supply in the long run and the short run.

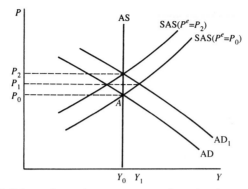

Fig. 24.3: Price and quantity movements under rational expectations

The long-run aggregate supply is given (regardless of how expectations are formed, but assuming money wage flexibility) by

$$Y = f(N),$$ (24.14)

$$N = N_s\left(\frac{W}{P^e}\right) = N_d\left(\frac{W}{P}\right),$$ (24.15)

$$P^e = P.$$ (24.16)

In the short run, aggregate supply is given by

$$Y = f(N),$$ (24.17)

$$N = N_s\left(\frac{W}{P^e}\right) = N_d\left(\frac{W}{P}\right),$$ (24.18)

$$P^e = P^e_{-1} + \lambda[P_{-1} - P^e_{-1}]$$ (adaptive expectations),   (24.19)

$$P^e = P_0.$$ (rational expectations)   (24.20)

Equations (24.19) and (24.20) are alternatives, with (24.20) representing the 'surprise' short-run expectations effect. In the absence of any surprises the short-run rational expectations formulation coincides with the long-run system (24.14)–(24.16).

We are now in a position to conduct a combined mathematical examination of aggregate demand and supply. We do this in the context of monetary and fiscal policy, and we will make exclusive use of the rational expectations formulation (allowing for 'surprises').

Exercise 24.2

1. In discussing rational expectations models it is important to distinguish 'surprises' or unanticipated shocks and anticipated or announced changes. Why?
2. Use Fig. 24.2 to show what happens if the change in demand is 'announced'.

24.3 Fiscal and monetary policy

As a preliminary we may replace the employment–wage subsystem in (23.9) with an *employment function* which allows us to eliminate the money wage. Recall from our discussion in Chapter 23 that with full money wage flexibility in both directions the money wage always adjusts to keep the labour market in equilibrium. Variations in employment only occur when expected prices diverge from actual prices. More specifically we may represent the employment relation as a function

$$N = N\left(\frac{P}{P^e}\right) \qquad N'(.) > 0 \tag{24.21}$$

If prices and expected prices vary by the same amount, the variation in the money wage ensures that the level of employment is unchanged at the 'natural' level of employment. As we have seen, an increase in actual price relative to expected price increases employment, whilst an increase in expected price relative to actual price lowers employment by symmetry. Equation (24.21) captures these effects. The aggregate supply side is therefore represented by

$$Y = f(N), \tag{24.22}$$

$$N = N\left(\frac{P}{P^e}\right). \tag{24.23}$$

The aggregate demand side will be represented by (22.17). The complete system determining the equilibrium level of income, Y, price P, interest rate r, and employment N is given by

$$S(Y - T(Y)) + T(Y) = I(r) + G, \tag{24.24}$$

$$\frac{m}{P} = L(Y, r), \tag{24.25}$$

$$Y = f(N), \tag{24.26}$$

$$N = N\left(\frac{P}{P^e}\right). \tag{24.27}$$

Note that in long-run equilibrium $P^e = P$, whilst in the short run variations in P are not accompanied by variations in P^e which we take as fixed. Taking total differentials of (24.24)–(24.27), allowing G, m, and P^e to vary gives

$$dY\{S'(Y - T(Y))[1 - T'(Y)] + T'(Y)\} = drI'(r) + dG, \tag{24.28}$$

$$\frac{dm}{P} - \frac{m}{P^2}dP = dYL_Y(Y, r) + drL_r(Y, r), \tag{24.29}$$

$$dY = dNf'(N), \tag{24.30}$$

$$dN = \left[\frac{dP}{P^e} - \frac{P}{P^{e2}}dP^e\right]N'\left(\frac{P}{P^e}\right). \tag{24.31}$$

Consider, first, the 'surprise' effects, so that $dP^e = 0$. In matrix form we have (after substituting (24.31) into (24.30) and rearranging)

$$
\begin{bmatrix}
S'(Y - T(Y))[1 - T'(Y)] + T'(Y) & -I'(r) & 0 \\
-L_Y(Y, r) & -L_r(Y, r) & -m/P^2 \\
1 & 0 & -N'(P/P^e)f'(N)/P^e
\end{bmatrix}
\begin{bmatrix}
dY \\
dr \\
dP
\end{bmatrix}
$$
$$
=
\begin{bmatrix}
dG \\
-dm/P \\
0
\end{bmatrix}. \quad (24.32)
$$

The effects of a 'surprise' increase in government expenditure or in the nominal money supply may now be studied. We will examine fiscal policy, the effects of monetary policy being left as an exercise. Setting $dm = 0$ in (24.32) and dividing through by dG we may establish the effects as usual using Cramer's Rule. First, the determinant of the square matrix in (24.32) is found to be

$$
|A| = \{S'(Y - T(Y))[1 - T'(Y)] + T'(Y)\}L_r(Y, r)f'(N)N'\left(\frac{P}{P^e}\right)/P^e + I'(r)m/P^2
$$
$$
+ I'(r)L_Y(Y, r)f'(N)N'\left(\frac{P}{P^e}\right)/P^e < 0
$$

(check this).

Application of Cramer's Rule gives the following comparative statics:

$$
\frac{dY}{dG} = \frac{L_r(Y, r)f'(N)N'(P/P^e)}{|A|P^e} > 0, \quad (24.33)
$$

$$
\frac{dr}{dG} = \frac{-L_Y(Y, r)f'(N)N'(P/P^e) - mP^e/P^2}{|A|P^e} > 0, \quad (24.34)
$$

$$
\frac{dP}{dG} = \frac{L_r(Y, r)}{|A|} > 0. \quad (24.35)
$$

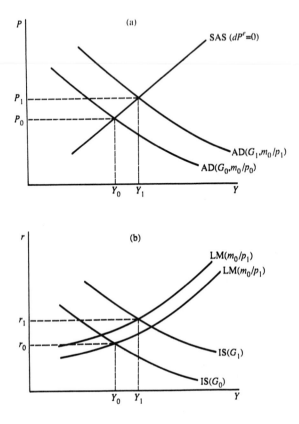

Fig. 24.4: Short-run 'surprise' effects of a bond-financed
increase in government expenditure

Thus a 'surprise' increase (i.e., P^e fixed) in government expenditure increases income, the interest rate and the price level. The effects are illustrated in Fig. 24.4. Frame (a) presents the main analysis in terms of (short-run) aggregate supply and aggregate demand while frame (b) shows the implications for the IS–LM model. Notice that since the increase in G increases P, the LM curve shifts to the right as the real money supply falls with an unchanged nominal money stock.

We know that under rational expectations the 'surprise' result is at best a short-run phenomenon and that in the long run actual and expected price changes must be in line. Thus, starting from a long-run equilibrium, a change in government expenditure must bring about another long-run equilibrium with the property that (by definition) $dP = dP^e$. Thus, returning to (24.31) and substituting $dP = dP^e$, we may derive the long-run comparative statics. With $dP = dP^e$, (24.31) and (24.30) imply that $dN = 0$ and $dY = 0$. The system (24.28)–(24.31) therefore reduces to

$$-drI'(r) = dG,$$

$$-drL_r(Y, r) - \frac{m}{P^2}dP = -\frac{dm}{P},$$

or, in matrix form,

$$\begin{bmatrix} -I'(r) & 0 \\ \\ -L_r(Y, r) & -m/P^2 \end{bmatrix} \begin{bmatrix} dr \\ \\ dP \end{bmatrix} = \begin{bmatrix} dG \\ \\ -dm/P \end{bmatrix}. \qquad (24.36)$$

Again concentrating on fiscal policy so that $dm = 0$, we have, using Cramer's Rule, the following comparative-static effects:

$$\frac{dr}{dG} = -\frac{1}{I'(r)} > 0, \qquad (24.37)$$

$$\frac{dP}{dG} = \frac{L_r(Y, r)}{I'(r)m/P^2} > 0, \qquad (24.38)$$

$$\frac{dY}{dG} = 0. \qquad (24.39)$$

These effects are illustrated in Fig. 24.5. Frame (*a*) shows the effects in terms of (long-run) aggregate supply and aggregate demand, while frame (*b*) follows through the implications for IS–LM. Notice that this time the increase in price reduces the real money supply so as to eliminate all income gains.

Figs 24.4 and 24.5 combined tell the complete story of a 'surprise' fiscal expansion. Note that if the fiscal expansion were to be 'announced', Fig. 24.5 would tell the short-run *and* the long-run story, since an announced policy change enables the new long-run equilibrium price to be calculated immediately by all agents, including workers.

The effects here are symmetrical for *reductions* in government expenditure.

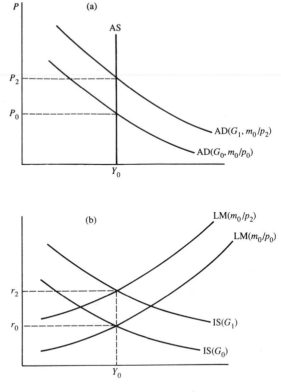

Fig. 24.5: Long-run effects of a bond-financed increase in government expenditure

Exercise 24.3

1. Use the system (24.24)–(24.27) to study the effect on output, the interest rate and the price level of (i) an unanticipated increase in the nominal money stock and (ii) a correctly predicted increase in the money stock. Illustrate your answers using the AD and AS diagram *and* the IS–LM diagram.
2. Given a choice, do you think that a government would prefer to make a 'surprise' reduction in the nominal money supply or an announced reduction?
3. Expansionary fiscal policy is effective in increasing real income under our formulation of rational expectations only if it takes the form of a 'surprise'. What scope is there for the government to continually affect output by continued 'surprises'?
4. In the system (24.24)–(24.27) output depends only on the labour input. Suppose in addition that there is a 'technological' parameter A, an increase in which increases both output at any given level of employment *and* the marginal product of labour at any given level of output. Hence the production function becomes $Y = f(N, A)$, $\partial Y/\partial N > 0$, $\partial Y/\partial A > 0$. Show diagrammatically that we are justified in making the (reduced form)

level of employment depend on A, so that (24.27) becomes $N = N(P/P^e, A)$, $\partial N/\partial A > 0$. Use these new relationships to establish the effect of an increase in A on long-run output, price, and the interest rate.

5. How would you introduce changes in labour's 'tastes' for work into the analysis? Adapt the model of this section to study the effect of a change in workers' tastes which makes work relatively more attractive.

24.4 Summary

Expectations may be modelled in a variety of ways. Two particularly useful models are the 'adaptive' and 'rational' expectations and approaches.

The long-run implications of these expectations mechanisms for wage and employment determinations are similar and imply (in the absence of wage frictions) a vertical aggregate supply curve.

'Rational' expectations appear to offer an optimizing and consistent model in which agents avoid making systematic errors. The learning process under adaptive expectations, however, is more evolutionary. This implies that the short-run period adjustment under adaptive expectations is likely to take longer than that under rational expectations. However, both approaches allow for short-run output variations when demand changes.

The rational expectations approach forces a distinction between 'announced' and 'unannounced' or 'surprise' changes in government policy. Only unannounced changes can produce the (temporary) misperceptions required for output to deviate from its equilibrium level.

25   The Open Economy

All national economies engage in trade to a greater or a lesser extent. For some, for example the UK, the volume of trade is particularly important since imports satisfy a large proportion of demands which cannot be met from domestic production. Moreover, export goods industries may account for a substantial proportion of total employment.

The trade and payments relationships between countries provide many opportunities for enhancing national output and employment, but they also place constraints on policy-makers. Stabilization policy, and our study of it, must take into account the consequences for the balance of payments and the exchange rate. Some of the links are implicit in many discussions about policy in the press. Why is a 'strong' dollar often linked to high US interest rates? Why is a 'strong' yen often linked to high Japanese exports? Why is inflation associated with exchange rate depreciation? Why does a balance of trade deficit mean a fall in revenue?

These and other issues are addressed through simple mathematical models in this and the following two chapters. As with the closed economy models, we build up from a strong initial set of assumptions to more 'realistic' models.

However, we are able to give only a sketch of open economy macroeconomics. Whole courses are devoted to international economics and balance of payments theory, and we do not have space to give the subject a full exposition. The following section summarizes some of the more important concepts to be used in this chapter including the balance of payments, capital mobility, role of reserves, and exchange rate regimes. Section 25.2 looks at how IS–LM is affected by the inclusion of trade. In section 25.3 the role of capital movements is considered and the IS–LM analysis extended appropriately to include balance of payments equilibrium.

25.1 The balance of payments and exchange rate regimes

We present here definitions of some useful concepts used in this chapter.

(i) *The balance of trade*

If we denote the total value of exports by X and the total value of imports by Q, the balance of trade *surplus* or value of net exports is given by

$$NX \equiv X - Q. \tag{25.1}$$

A balance of trade surplus ($NX > 0$) indicates that the income received from sale of output abroad exceeds the expenditure by domestic residents on foreign goods, while a balance of trade deficit (NX < 0) indicates an excess of expenditure on foreign goods over income received from the sale of goods abroad.

(ii) *Capital flows and the balance of payments*

In addition to the inter-country flow of goods (visible trade) there is also a flow of services such as insurance and tourism (invisible trade) and a flow of capital in the form of investment funds and so on. The balance of the flows of goods and services is known as the current account, while the balance of capital movements is known as the capital account. The sum of the two accounts representing the net overall external balance is known as the *balance of payments*, discussed in Chapter 16. If net capital inflows are denoted by K, then the overall balance of payments surplus is at its simplest given by

$$BS \equiv NX + K \tag{25.2}$$

$$BS = X - Q + K. \tag{25.3}$$

The balance of payments is in equilibrium when (25.3) is zero (note that K may be negative indicating a net capital *outflow*). Thus, a current account surplus (deficit) must be offset by a capital account deficit (surplus).

From the purely accounting point of view the balance of payments deficit or surplus must be offset by compensating flows of funds. Taken over a period of time a balance of payments deficit, for example (so that (25.3) is negative), must be financed from the official reserves. Put crudely, since those selling abroad are depositing in banks and those buying from abroad are drawing on banks, a balance of payments deficit will be associated with a net drawing on banks. Since the government is ultimately responsible for ensuring that the banking system is prepared for this, the net drawing on funds is entered into the balance of payments accounts as net funding from official reserves. We studied this in some detail in Chapter 17, but recall that because of this funding from reserves, balance of payments deficits can affect domestic monetary policy.

(iii) *Exchange rate and the terms of trade*

The *exchange rate*, ε, represents the *foreign price of domestic currency*. In many countries this corresponds to the everyday usage of the term. For example, in the UK we speak of one pound sterling as being worth x dollars, whilst in the US one dollar is said to be worth y pounds sterling. Each currency has many exchange rates, each associated with another currency with which it may be exchanged. In this chapter, however, we will assume that the country under consideration has just one exchange rate with 'the' foreign currency. We discuss the determinants of the exchange rate and exchange rate regimes presently.

The *terms of trade* are the ratio of prices of the goods traded between countries measured in a common currency. If the domestic price level is P, and P_f is the price of foreign output in units of the foreign currency, the home price expressed in terms of the foreign currency is εP. Similarly, the foreign price expressed in terms of the home currence is P_f/ε. The terms of trade, θ, are then

$$\theta = \frac{P_f/\varepsilon}{P}. \tag{25.4}$$

An *increase* in θ is described as the country experiencing a *deterioration* in the terms of trade, because it indicates that, measured in a common currency, the home economy is required to pay more per unit of the foreign good. Finally, on this, we mention a proposition which in its simplest form states that with a completely free flow of goods between countries and with no transport costs, identical goods should sell for the same price in each country measured in a common currency. That is,

$$P = P_f/\varepsilon. \tag{25.5}$$

Equation (25.5) is referred to as *purchasing power parity*, and it clearly implies, from (25.4), that $\theta = 1$.

(iv) *Exchange rate determination*

Purchases of goods and services, and capital transactions, take place in units of the local currency, for example purchases of US goods, services, and shares must be made in dollars. Many currencies, including the dollar and the pound sterling, are traded in an international 'market' for currencies. The relative amounts of currencies coming on to the foreign exchange market determine the structure of exchange rates. Consider sterling. Sterling will be 'supplied' to the market by those demanding foreign currencies in exchange – importers of goods and exporters of capital. Moreover, since more imports will be demanded as their (international) price falls, more sterling will be supplied and there will be a higher exchange rate. The cost of imports in sterling falls as ε increases. Similarly sterling will be 'demanded' on the foreign exchange market by those seeking to buy British goods and invest in British companies. British goods are cheaper (internationally) the lower is ε and thus export demand and demand for sterling will be higher. The picture is summarized in Fig. 25.1. The equilibrium exchange rate equates the demand and supply of sterling in the currency market. A sudden boost in imports (caused perhaps by increased domestic income) shifts out the S_s curve and reduces the exchange rate, reducing the supply of sterling and increasing demand until a new equilibrium exchange rate is reached. It should be apparent that an excess supply of (demand for) sterling on the foreign exchanges reflects an excess of imports (exports) over exports (imports). An excess supply of sterling reflects a deficit on the balance of payments whilst an excess demand for sterling reflects a surplus.

This adjustment in the exchange rate to bring about equilibrium in the market for sterling (in our example) assumes that exchange rate movements are unhindered by government intervention. Thus only in a purely flexible or 'floating' exchange rate regime would the exchange rate be free to adjust in this way.

In a world of fixed exchange rates, i.e., one in which the exchange rate is 'managed' to some degree by the government, the market mechanism is replaced by a system of official intervention designed to counteract changes in currency market conditions. Fig. 25.2 gives an example. Suppose that the government considers it appropriate to maintain the exchange rate at a particular level ε_0, and suppose further that the chosen fixed exchange rate is above the equilibrium exchange rate. The market consequently generates an excess supply of sterling as domestic residents buy relatively cheaper foreign goods and overseas buyers demand fewer of our relatively more expensive goods. However, ε_0 cannot be maintained by edict, since no single government is able to control events on international markets in this way. The government maintains ε_0 by entering the market and creating extra demand for sterling. The extent of the required 'mopping-up' exercise in Fig. 25.2 is x, and governments finance this type of activity from the central banks' foreign currency reserves. Thus x represents the extent to which foreign currency reserves are run down by the government intervening to maintain a 'high' exchange rate. It is apparent from Chapter 17 that the fall in reserves, R_f, leads to a fall in the total money supply. We conclude that an excess supply of sterling, associated with a balance of payments deficit, results in a fall in reserves and a fall in total money supply under a fixed exchange rate. On the other hand an exchange rate fixed below the equilibrium rate is associated with a balance of payments surplus and so reserves and the money supply increase. The endogeneity of the total money supply under a fixed exchange rate regime has important policy implications, as we shall see.

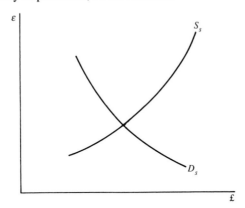

Fig. 25.1: The market for sterling

In practice we rarely encounter currencies which have entirely flexible exchange rates or totally fixed exchange rates. The extent of reserve movements in an economy with a predominantly floating exchange rate merely reflects the extent of intervention during the period. A government which believes its floating currency to be in the grip of a destabilizing speculation influence may wish to intervene to counter these, and hopefully bring the 'underlying' market forces to the fore.

In what follows we will treat only the extreme cases of a purely floating currency and a completely fixed currency. At any point in time we must specify clearly the exchange rate regime we assume to be in operation.

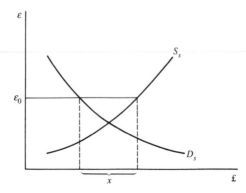

Fig. 25.2: Overvalued currency and excess currency supply

Exercise 25.1

1. What factors will influence the *position* of the supply and demand curves for sterling, S_s and D_s, in Fig. 25.1? Hence what changes might bring about a reduced need for official financing in order to maintain the fixed exchange rate as in Fig. 25.2?
2. The version of the purchasing power parity theorem introduced in this section states that the free flow of all goods across borders will eliminate price differentials when measured in common currency units. What factors might prevent the full elimination of price differentials?

25.2 IS-LM in the open economy

The IS-LM model of Chapter 19 is easily extended to incorporate trade and the exchange rate. To take the goods market first, we must make two changes to the income–expenditure equation (19.4). First, in the open economy some domestically produced output will be demanded by overseas buyers in the form of exports. Thus *total final expenditure* is $C+I+G+X$, where X is the value of exports. Secondly, some of these expenditures recorded as consumption, investment, and government spending will be on goods produced overseas, and so we must subtract the total value of imports, Q from total final expenditure to give expenditure on domestic output, $C + I + G + X - Q$. The term $(X - Q)$ is known as net exports or the balance of trade surplus. The behaviour of these two quantities is specified as follows. Exports are assumed to be a function only of the exchange rate ε, as long as domestic prices relative to foreign prices are fixed. (More generally, when domestic prices vary, the dependence is on the terms of trade θ. This is taken up in Chapter 27.) Given our earlier discussion about the exchange rate, we suggest that the following behaviour is reasonable:

$$X = X(\varepsilon) \qquad X'(\varepsilon) < 0. \tag{25.6}$$

Imports will be a function of the exchange rate for given prices *and* domestic income, since generally imports will increase with increases in consumption which, in turn, are generated by increases in income. Moreover, a higher exchange rate makes foreign goods relatively less expensive and therefore stimulates imports. Our import function therefore takes the form

$$Q = Q(Y, \varepsilon) \qquad Q_Y > 0, \quad Q_\varepsilon > 0. \tag{25.7}$$

Goods market equilibrium, from which we derive the open economy IS curve for a given exchange rate, is given by

$$Y = C(Y - T(Y)) + I(r) + G + X(\varepsilon) - Q(Y, \varepsilon) \tag{25.8}$$

or equivalently

$$S(Y - T(Y)) + T(Y) + Q(Y, \varepsilon) = I(r) + G + X(\varepsilon), \tag{25.9}$$

so that exports join the 'injections' side and a new 'withdrawal' is the value of imports. It is apparent from (25.9) that, for a given exchange rate, the open-economy IS curve has similar properties to those in the closed economy. Differentiation of (25.9) for a given exchange rate reveals

$$\frac{dY}{dr} = \frac{I'(r)}{S'(Y - T(Y))[1 - T'(Y)] + T'(Y) + Q_Y(Y, \varepsilon)} < 0. \tag{25.10}$$

It is easily established that an increase in the exchange rate reduces income for a given interest rate, so that IS shifts to the left, or

$$\frac{dY}{d\varepsilon} = \frac{X'(\varepsilon) - Q_\varepsilon(Y, \varepsilon)}{S'(Y - T(Y))[1 - T'(Y)] + T'(Y) + Q_Y(Y, \varepsilon)} < 0. \tag{25.11}$$

It is clear that in a fixed exchange rate world this last property of the open-economy IS curve will not play a role. However, (25.11) will be an essential part of the analysis under floating exchange rates. Denoting the balance of trade or net exports by NX we have

$$\text{NX} = X - Q \tag{25.12}$$

or

$$\text{NX}(Y, \varepsilon) = X(\varepsilon) - Q(Y, \varepsilon), \tag{25.13}$$

so that

$$\text{NX}_Y(Y, \varepsilon) = -Q_Y(Y, \varepsilon) < 0 \tag{25.14}$$

and

$$NX_\varepsilon(Y, \varepsilon) = X'(\varepsilon) - Q_\varepsilon(Y, \varepsilon) < 0. \qquad (25.15)$$

Notice from equation (25.9) that the equilibrium level of income need not produce equilibrium in the balance of trade. This latter equilibrium occurs when $NX = 0$. Equation (25.9) only requires that total withdrawals and total injections are equal and it does not imply that $Q(Y, \varepsilon) = X(\varepsilon)$. Thus suppose that the level of income which makes $NX = 0$ is given by $\tilde{Y}$. Then

$$0 = X(\varepsilon) - Q(\tilde{Y}, \varepsilon). \qquad (25.16)$$

If the actual equilibrium level of income Y given by (25.9) exceeds $\tilde{Y}$ then $NX < 0$ (balance of trade deficit), while $Y < \tilde{Y}$ implies $NX > 0$ (balance of trade surplus). The case of the balance of trade deficit is illustrated in Fig. 25.3. It is easily established that an increase in ε shifts the $NX = 0$ curve to the left, lowering $\tilde{Y}$, since from (25.16)

$$\frac{d\tilde{Y}}{d\varepsilon} = \frac{X'(\varepsilon) - Q_\varepsilon(\tilde{Y}, \varepsilon)}{Q_{\tilde{Y}}(\tilde{Y}, \varepsilon)} < 0. \qquad (25.17)$$

Before using this model of the open-economy IS curve and the balance of trade to present some tentative conclusions about the effects of trade deficits and exchange rate changes, we notice that in Fig. 25.3 the LM curve is drawn as being upward-sloping as in the closed economy. In fact, subject to one caveat, the LM curve has the same specification as in the closed economy. Specifically LM does not depend on the exchange rate, ε. We must take care, however, when interpreting the money market equilibrium conditions under fixed exchange rates. In view of our discussion surrounding Fig. 25.2 and equation (25.4) it is clear that under a fixed exchange rate the total money stock M is not fixed exogenously by the government. In the case of a balance of payment deficit M falls as the component R_f falls, whilst in the case of a payments surplus M increases as the component R_f increases. Thus balance of payments disequilibria induce movements in LM. A balance of payments deficit lowers M and so shifts LM to the left. A balance of payments surplus increases M and so shifts LM to the right. With a flexible exchange rate M is entirely under the control of the monetary authority and hence the LM curve is univariant to balance of payments disequilibria.

Some of the open economy considerations may be seen by a further analysis of the position illustrated in Fig. 25.3. We have already seen that since $Y > \tilde{Y}$ (equilibrium income exceeds the level of income required for balance of trade equilibrium) the IS–LM equilibrium represents a balance of trade deficit. This situation may be dealt with in a number of ways. First, it may be tolerated and the consequences for reserves accepted. Reserves fall as they are used to support the exchange rate shifting LM to the left, increasing domestic interest rates, and lowering income. The lower income reduces imports and so raises net exports until $Y = \tilde{Y}$. The balance of trade deficit is thus eliminated by *induced* monetary contraction. This is illustrated in Fig. 25.4.

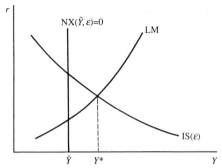

Fig. 25.3: IS-LM equilibrium and balance of trade deficit

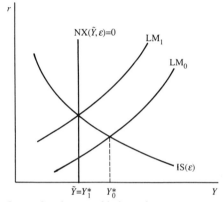

Fig. 25.4: Internal and external balance by monetary contraction

Secondly, it is clear that a similar result could have been obtained if the government chose to engage in monetary contraction by reducing the domestic credit component of money supply. In this simple model, the only difference in these two cases would be the difference in the final composition of the total money stock. Similarly a contractionary fiscal policy, shifting IS to the left, would also increase net exports, as would some combination of contracting monetary and fiscal policy.

Reducing domestic income to eliminate a balance of trade deficit is often not a government's preferred method. The reason is that, apart from creating political unpopularity, the reduced income may run counter to other policy objectives, notably growth and full employment. In Fig. 25.4, for example, if the target or full-employment income level were above Y the government would face a genuine policy dilemma.

A third way of handling the deficit is again to tolerate it and hope that underlying factors, such as low world demand or poor export performance, change. Instead of using reserves to maintain the value of the currency, the government may prefer to borrow from foreign governments either directly or via an international bank or fund such as the International Monetary Fund (IMF). In this case the situation remains as in Fig. 25.3, with the deficit financed by borrowing. Naturally, there is a limit to how much borrowing may be made.

Finally, the authorities might permit a change in the exchange rate. This may be done either by annoucing a new (more appropriate) *fixed rate* or by allowing the exchange rate to 'float' to find its new equilibrium level. To study the effect of a change in ε on both Y (the equilibrium level of income generated by IS–LM) and the balance of payments equilibrium level of $\tilde{Y}$, we write out the system as

$$S(Y - T(Y)) + T(Y) + Q(Y, \varepsilon) = I(r) + G + X(\varepsilon), \quad (25.18)$$

$$M = L(Y, r), \quad (25.19)$$

$$0 = X(\varepsilon) - Q(\tilde{Y}, \varepsilon). \quad (25.20)$$

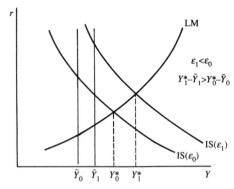

Fig. 25.5: Failure of devaluation to produce internal and external balance

We already know from equation (25.17) that $d\tilde{Y}/d\varepsilon < 0$ so that a devaluation (*fall in* ε) will increase $\tilde{Y}$. Total differentiation of the subsystem (25.18) and (25.19) allowing only ε, Y, and r to vary gives

$$\begin{bmatrix} S'(Y - T(Y))[1 - T'(Y)] + T'(Y) + Q_Y(Y, \varepsilon) & -I'(r) \\ \\ -L_Y(Y, r) & -L_r(Y, r) \end{bmatrix} \begin{bmatrix} dY \\ \\ dr \end{bmatrix}$$

$$= \begin{bmatrix} d\varepsilon[\mathrm{NX}_\varepsilon(Y, \varepsilon)] \\ \\ 0 \end{bmatrix}. \quad (25.21)$$

Using Cramer's Rule, this gives

$$\frac{dY}{d\varepsilon} = \frac{-NX_\varepsilon(.)L_r(.)}{-\{S'(.)[1 - T'(.)] + T'(.) + Q_Y(.)\}L_r(.) - I'(.)L_Y(.)} < 0 \quad (25.22)$$

(where the arguments of functions have been suppressed for simplicity).

Hence a devaluation (fall in ε) will also raise Y. If a devaluation increases both $\tilde{Y}$ and Y we must attempt to establish whether this will indeed bring $\tilde{Y}$ and Y closer. If there is a tendency for $\tilde{Y}$ to approach Y then there may be a case for successive devaluations. In fact, a priori we are unable to say whether a devaluation will reduce the balance of trade deficit. In Fig. 25.5 we illustrate a case in which the effect of a devaluation on Y is greater than on $\tilde{Y}$ so that the fall in ε *worsens* the balance of trade deficit.

In practice we would not expect this perverse result and it is not difficult to find conditions under which the expected result holds so that a devaluation does improve the balance of trade. The required condition to ensure that $dY/d\varepsilon$ is 'less negative' than $d\tilde{Y}/d\varepsilon$ is

$$\frac{dY}{d\varepsilon} > \frac{d\tilde{Y}}{d\varepsilon}. \tag{25.23}$$

Thus if we graph the (implicit) functions $\tilde{Y}(\varepsilon)$ and $Y(\varepsilon)$, where the latter incorporates the effect of interest rate changes, then an exchange rate fall will improve the balance of trade if the relative slopes of the functions are as illustrated in Fig. 25.6. The slope of $Y(\varepsilon)$ is given by (25.22) and the slope of $\tilde{Y}(\varepsilon)$ by (25.17). At the exchange rate ε_1, $Y_1 > \tilde{Y}_1$ and so there is a balance of payments deficit. A fall in ε towards ε_2 reduces the deficit and at ε_2 there is a trade balance, with income Y generating just enough imports to offset export earnings.

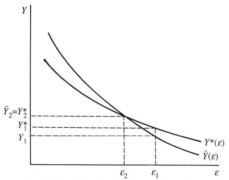

Fig. 25.6: Conditions required for a devaluation to produce internal and external balance

Further explorations of this problem are left as an exercise.

Our model of the open economy developed so far is incomplete. In particular, we have considered only trade flows thus far. Capital flows, many of which respond to interest rate differentials, are also very important and we must extend the IS–LM framework to account for these.

Exercise 25.2

1. Suppose an economy's IS-LM equilibrium is determined by $Y = b(Y - T) + h - er + G + X - qY$, $M = kY - lr$, where $b(Y - T)$ is consumption, T the lump-sum tax, $h - er$ investment (h and e constants), G government expenditure, X exports, and qY imports,

where q is the propensity to import. The balance of trade is achieved at an income level $Y = X/q$. Derive the condition under which $\tilde{Y} < Y$ as in Fig. 25.3.

2. Using the model of question 1 and the following information, obtain values for $\tilde{Y}$ and Y. $b = 0.8$, $e = 1$, $k = 0.6$, $l = 0.5$, $q = 0.4$, $T = 100$, $G = 200$, $M = 60$, $h = 100$, $X = 200$. Test the plausibility of your answer by using the conditions developed in your answer to question 1. Calculate the balance of trade surplus and deficit. Suppose the government seeks to bring Y into equality with $\tilde{Y}$ by pure fiscal policy. What change in government expenditure G is required to bring this about?

3. Under what circumstances will the conditions (25.23), illustrated in Fig. 25.6, be satisfied?

4. In each country in a two country world income is determined by $Y_i = C_i + I_i + X_i - Q_i$ $(i = 1, 2)$. In neither country does the interest rate play a role in determining aggregate demand and investment is a lump sum in each country. Consumption in each country is given by $C_i = b_i Y_i$ $(i = 1, 2)$, and imports by, $Q_i = q_i Y_i$ $(i = 1, 2)$. Given that there are only two countries, find a set of conditions jointly determining equilibrium income in each country. What is the effect on each country's income of an increase in investment in country 1, I_1? Assuming that each country initially had a balance of trade, what is the effect of the increase in I_1 on the balance of trade in each country?

25.3 IS–LM–BP analysis

Consider the definition of the balance of payments surplus in equation (25.2). We have specified the balance of trade component (net exports) as depending on income and the exchange rate. The final step, naturally, is to specify the capital account surplus. International capital is quite mobile between countries and will tend to flow into countries with high relative interest rates. In the so-called IS–LM–BP analysis the domestic interest rate will play a role, but it is worth remembering that it is variations in domestic rates *relative* to interest rates in other countries which induce capital movements. In what follows we treat the 'world' interest rate as fixed and exogenous, so that a variation in the domestic interest rate is a variation in the 'relative' interest rate. The higher is the interest rate, the higher are capital inflows. In addition to the interest rate, capital flows will also be affected by the exchange rate. The higher is the exchange rate, the more expensive it is to convert foreign currency into domestic currency and the lower are capital inflows. This suggests

$$K = K(r, \varepsilon) \qquad K_r(r, \varepsilon) > 0, \; K_\varepsilon(r, \varepsilon) < 0. \qquad (25.24)$$

The overall balance of payments surplus is therefore

$$BS(Y, r, \varepsilon) = NX(Y, \varepsilon) + K(r, \varepsilon) \qquad (25.25)$$

where $BS_Y(Y, r, \varepsilon) < 0$, $BS_r(Y, r, \varepsilon) > 0$, and $BS_\varepsilon(Y, r, \varepsilon) < 0$. The last result follows from (25.15) and (25.24). Balance of payments equilibrium is where the sum of the trade and capital account surpluses are zero. Notice now that a balance of trade deficit may be offset (without the need for government action or exchange rate movements) by a capital account surplus. Setting (25.25) equal to zero gives the condition for equilibrium in the balance of payments:

$$0 = NX(Y, \varepsilon) + K(r, \varepsilon). \tag{25.26}$$

For a given exchange rate, we may derive an equilibrium locus of combinations for Y and r which satisfy (25.26). Such a locus, known as the BP curve, is positively sloped. Total differentiation of (25.26) for an unchanged exchange rate reveals that

$$\frac{dY}{dr} = \frac{-K_r(r, \varepsilon)}{NX_Y(Y, \varepsilon)} > 0. \tag{25.27}$$

Because BP, like LM, is positively sloped we must make some further assumption about the relative steepness of LM and BP. The convention we adopt is that, because capital movements are very sensitive to changes in the interest rate, the BP curve is relatively flat. The intuition behind (25.27) is that, starting from one balance of payments equilibrium, an increase in income will increase imports and worsen the balance of trade. This will be offset by an improvement in the capital account brought about by an increase in the interest rate. If capital inflows are very interest sensitive then only a relatively small increase in the interest rate is required — hence BP in (r, Y)-space is relatively flat. Comparing the slope of LM and BP, we are assuming that

$$\left.\frac{dr}{dY}\right|_{BP} < \left.\frac{dr}{dY}\right|_{LM}.$$

or

$$\frac{NX_Y(Y, \varepsilon)}{-K_r(r, \varepsilon)} < \frac{-L_Y(Y, r)}{L_r(Y, r)}.$$

Multiplying through by $-K_r(r, \varepsilon)$, which is *negative*, gives

$$NX_Y(Y, \varepsilon) > \frac{L_Y(Y, r)K_r(r, \varepsilon)}{L_r(Y, r)},$$

and multiplying through by $L_r(Y, r)$ (negative) gives

$$L_r(Y, r)NX_Y(Y, \varepsilon) < K_r(r, \varepsilon)L_Y(Y, r).$$

As with IS and LM we may locate zones of disequilibrium in the balance of payments around BP. Fig. 25.7 shows a BP curve and its associated disequilibrium zones.

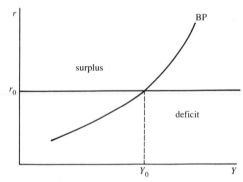

Fig. 25.7: The BP curve and zones of balance of payments disequilibrium

Consider an arbitrarily chosen interest rate r_0. By definition of the BP curve, the income level Y_0 is just right to produce balance of payments equilibrium. At the interest rate r_0, any level of income lower than Y_0 implies an associated lower level of imports and hence a balance of payments surplus. Similarly at incomes greater than Y_0 imports are higher and the balance of payments moves into deficit. In this way we identify the zone to the 'north-west' of BP as one of balance of payments surplus and the zone to the 'south-east' of BP as one of balance of payments deficit. The slope of the BP curve depends in particular on the sensitivity of investment flows to domestic interest rate changes. If, for example, capital moves very quickly between countries then only a slight increase in the interest rate is required to stimulate large inflows of capital which requires a large increase in income to reduce the surplus of stimulating imports. The slope of BP also depends on the propensity to import.

Finally, note that the BP curve is drawn for a given exchange rate. To see how BP shifts as the exchange rate appreciates (ε increases), totally differentiate (25.26) but hold the interest rate constant, say at r_0; then we find

$$\frac{dY}{d\varepsilon} = \frac{\mathrm{NX}_\varepsilon(Y, \varepsilon) + K_\varepsilon(r, \varepsilon)}{-\mathrm{NX}_Y(Y, \varepsilon)} < 0. \tag{25.28}$$

Hence at any interest rate an increase in ε lowers the level of income required to produce balance of payments equilibrium. Diagrammatically this implies that BP shifts to the left as ε increases and shifts to the right as the exchange rate falls.

In the following section we study the IS–LM–BP equilibrium and how it is affected by policy changes.

Exercise 25.3

1. The LM curve and BP curve may be written in general as $M = L(Y, r)$, $L_Y > 0$, $L_r < 0$, and $0 = \mathrm{BS}(Y, r, \varepsilon)$ for given M and ε, where $\mathrm{BS}(Y, r, \varepsilon) = \mathrm{NX}(Y, \varepsilon) + K(r, \varepsilon)$. Derive conditions under which the LM curve is steeper than the BP curve, using examples of linear forms for L, K and NX.

2. How would you model exchange controls which prohibited all capital movements in and out of a country?

25.4 Summary

The IS–LM model of income determination is readily extended to account for trade and capital flows between countries.

The *balance of payments* includes the *balance of trade* and *net capital flows*. The balance of trade or *net exports* depends on domestic income and the exchange rate. Net capital movements depend on domestic interest rates relative to overseas rates and the exchange rate.

The level of equilibrium income and the interest rate in an economy do not guarantee a balance on the balance of payments. Internal and external balance may be achieved simultaneously if domestic income can be manipulated so that net exports and net capital outflows are equal (i.e., by monetary and fiscal policy), or by changes in the exchange rate. Neither mechanism on its own guarantees success in bringing about balance of payments equilibrium.

Countries may (temporarily) sustain a balance of payments imbalance in the absence of corrective action by borrowing.

26 Macroeconomic Policy in an Open Economy

The IS-LM-BP model can be used to look at fixed-price policy effects for an open economy. Thus we can extend the results of Chapter 20, now taking account of balance of payments constraints and of exchange rate changes. Note that even with domestic and foreign prices fixed in their respective currencies, the *relative* price of goods between countries is affected by the exchange rate.

26.1 Overall equilibrium

The complete IS-LM-BP system is in overall equilibrium when the levels of equilibrium income and interest rate also generate equilibrium in the balance of payments. Overall equilibrium at a given exchange rate ε requires therefore

$$S(Y - T(Y)) + T(Y) + Q(Y, \varepsilon) = I(r) + G + X(\varepsilon), \tag{26.1}$$

$$M = L(Y, r), \tag{26.2}$$

$$0 = NX(Y, \varepsilon) + K(r, \varepsilon). \tag{26.3}$$

It is important to note that Y and r are determined by the subsystem (26.1) and (26.2). Condition (26.3) merely informs us that Y and r are such as to give balance of payments equilibrium for a given exchange rate. Fig. 26.1 illustrates this overall equilibrium on our adopted assumption that BP is less steep than LM.

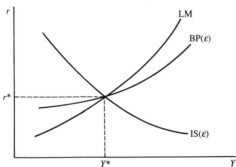

Fig. 26.1: Internal and balance of payments equilibrium

When considering the effects of policy changes, it is convenient to assume that the economy starts from a position of overall equilibrium. We are interested in establishing how the balance of payments acts as an aid or a constraint when a government attempts to increase output by fiscal or monetary expansion. The effects of these policies in an open economy depend critically on the underlying exchange rate regime. We consider the two extremes of fixed and floating exchange rates separately.

26.2 Policy under a fixed exchange rate

Under a fixed exchange rate, ε does not change in the system (26.1)–(26.3). Consider first an increase in M in the form of an increase in domestic credit. With a fixed exchange rate, income increases and the interest rate falls so that the economy moves into a balance of payments deficit. This is easily seen by considering the usual comparative statics on the subsystem (26.1) and (26.2), allowing M, Y, and r to change.

$$
\begin{bmatrix} S'(Y - T(Y))[1 - T'(Y)] + T'(Y) + Q_Y(Y, \varepsilon) & -I'(r) \\ \\ L_Y(Y, r) & L_r(Y, r) \end{bmatrix} \begin{bmatrix} dY \\ \\ dr \end{bmatrix} = \begin{bmatrix} 0 \\ \\ dM \end{bmatrix}.
$$

(26.4)

Ignoring the balance of payments effect for the moment, the change in M leads to changes in Y and r given by

$$
\frac{dY}{dM} = \frac{I'(r)}{\{S'(Y - T(Y))[1 - T'(Y)] + T'(Y) + Q_Y(Y, \varepsilon)\}L_r(Y, r) + I'(r)L_Y(Y, r)}, \quad (26.5)
$$

$$
\frac{dr}{dM} = \frac{S'(Y - T(Y))[1 - T'(Y)] + T'(Y) + Q_Y(Y, \varepsilon)}{\{S'(Y - T(Y))[1 - T'(Y)] + T'(Y) + Q_Y(Y, \varepsilon)\}L_r(Y, r) + I'(r)L_Y(Y, r)}. \quad (26.6)
$$

So $dY/dM > 0$ and $dr/dM < 0$. Moreover, those are only *temporary* effects, since the effect on (26.3) is to induce a balance of payments deficit. To see this, denote the right-hand side of (26.3) as $BP(Y, r, \varepsilon)$ (the balance of payments),

$$
BP(Y, r, \varepsilon) = NX(Y, \varepsilon) + K(r, \varepsilon), \quad (26.7)
$$

and so the effect of an increase in M on the balance of payments is

$$
\frac{dBP}{dM} = NX_Y \frac{dY}{dM} + K_r \frac{dr}{dM}. \quad (26.8)
$$

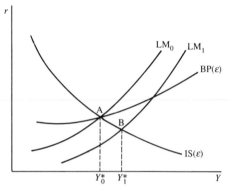

Fig. 26.2: Effect of an expansion of domestic credit (fixed exchange rate)

Substituting from (26.5) and (26.6) gives

$$\frac{d\text{BP}}{dM} = \{\text{NX}_Y I'(r) + K_r[S'(Y - T(Y))[1 - T'(Y)] + T'(Y) + Q_Y(Y, \varepsilon)]\}D^{-1} \quad (26.9)$$

where

$$D \equiv \{S'(Y - T(Y))[1 - T'(Y)] + T'(Y) + Q_Y(Y, \varepsilon)\}L_r(Y, r) + I'(r)L_Y(Y, r) < 0, \quad (26.10)$$

so we conclude that $d\text{BP}/dM < 0$. Since we started with an equilibrium on the balance of payments, an increase in the money supply moves the economy into a balance of payments deficit. Now the adjustment process under a balance of payments with a fixed exchange rate kicks in and reserves fall. In the absence of a *sterilization* policy the money supply falls and continues to do so until the balance of payments is back in equilibrium. That is, we require an increase in domestic credit to tbe matched by a reduction in foreign reserves, so that the total change in the money supply is zero. Since reserves fall until the total money supply is at its original level, the final position is one in which $dM = 0$, and hence $dY = 0$ and $dr = 0$. Diagrammatically these effects are shown in Fig. 26.2. The initial and final equilibrium is given by point A. The expansion of domestic credit shifts LM to LM_1 and there is a new (temporary) equilibrium at point B. At the lower interest rate and higher income level there is a balance of payments deficit, which prompts a reduction of reserves until total money stock is at its original level, at which income interest rate and the balance of payments are all at their original levels.

Under a fixed exchange rate an increase in G increases Y and r, as is apparent from the subsystem (26.1) and (26.2). However, with BP flatter than LM the new equilibrium is in a zone of balance of payments surplus so that reserves expand, the money supply expands and the new equilibrium is established at a higher income and interest rate than previously. These effects are illustrated diagrammatically in Fig. 26.3.

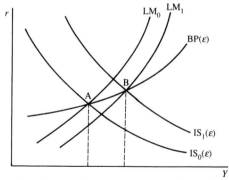

Fig. 26.3: Effect of an increase in government expenditure (fixed exchange rate)

It is apparent from this informal discussion that although (temporary) equilibria are determined by (26.1) and (26.2), the final equilibrium, after accommodating adjustments in the money supply caused by changes in reserves, is determined by (26.1) and (26.3). Thus using (26.1) and (26.3) we find that

$$dY/dM = 0 \tag{26.11}$$

and

$$
\begin{bmatrix}
S'(Y - T(Y))[1 - T'(Y)] + T'(Y) + Q_Y(Y, \varepsilon) & -I'(r) \\
NX_Y(Y, \varepsilon) & K_r(r, \varepsilon)
\end{bmatrix}
\begin{bmatrix}
dY \\
dr
\end{bmatrix}
=
\begin{bmatrix}
dG \\
0
\end{bmatrix}, \tag{26.12}
$$

so that

$$
\frac{dY}{dG} = \frac{K_r(r, \varepsilon)}{K_r(r, \varepsilon)\{S'(Y - T(Y))[1 - T'(Y)] + T'(Y) + Q_Y(Y, \varepsilon)\} + I'(r)NX_Y(Y, \varepsilon)} > 0. \tag{26.13}
$$

Notice that the government expenditure multiplier in this case (the right-hand side of (26.13)) depends not on the slope of the LM curve, but on the slope of the BP curve. To see this, rewrite the right-hand side of (26.13) as

$$
\{S'(Y - T(Y))[1 - T'(Y)] + T'(Y) + Q_Y(Y, \varepsilon) - I'(r)(-NX_Y(Y, \varepsilon)/K_r(r, \varepsilon))\}^{-1}, \tag{26.14}
$$

where $-NX_Y(Y, \varepsilon)/K_r(r, \varepsilon)$ is the slope of the BP curve in (r, Y)-space. Thus the extent of crowding out is determined by the balance of payments rather than by domestic money markets. As the slope of the BP curve increases, the final term in braces in (26.14) increases and so (26.14) and (26.13) fall. Conversely, a flatter BP curve leads to an increase in the government expenditure multiplier.

The intuition here is as follows. A balance of payments imbalance cannot be sustained because of adjustments in reserves. Hence a surplus caused by the (temporary) high interest rates is reduced as interest rates fall due to the expansion of the money supply through increased reserves. This process ends when the interest rate and income have adjusted to eliminate the balance of payments surplus. The extent of the fall in the interest rate and hence the size of the government expenditure multiplier depend on the slope of the BP curve. Since the BP curve is less steep than the LM curve, the government expenditure multiplier is larger than in the absence of a balance of payments constraint. We conclude that the induced monetary expansion following an increase in government spending makes the fixed exchange rate world favourable to fiscal policy.

The result in (26.11) says, on the contrary, that a fixed exchange rate makes an expansion of domestic credit completely ineffective in stimulating output. The balance of payments constraint simply leads to a fall in reserves equal to the initial expansion of the money supply.

Exercise 26.2

1. Rework the analysis of monetary and fiscal policy under fixed exchange rates on the assumption that the BP curve is steeper than the LM curve.
2. Use a linear model of (26.1)–(26.3) to confirm the statement immediately following the result in (26.13) that the extent of crowding out under fixed exchange rates depends on the slope of BP and is independent of the slope of LM.
3. The government expenditure multiplier under a fixed exchange rate is given by (26.13). Using the lump-sum tax schedule $T(Y) = T_0$ for all Y, derive an expression for the balanced budget multiplier for this case.
4. Although not expressed explicitly in this section, net exports depend on the level of 'world income', Y_f, so that

$$NX = NX(Y, \varepsilon, Y_f) \qquad NX_{Y_f} > 0.$$

Establish what effect an increase in world income will have on domestic income and interest rate in the case of a fixed exchange rate. What effect will Y_f have on reserves?

5. Capital inflows K depend negatively on the foreign interest rate r_f so that

$$K = K(r, \varepsilon, r_f) \qquad K_{r_f} < 0.$$

Establish the effect of an increase in r_f on Y, r and on reserves in the fixed exchange case. Illustrate using an IS–LM–BP diagram.

26.3 Policy under a flexible exchange rate

As a preliminary we note, from equations (26.1)–(26.3), that a change in the exchange rate effects the position both of IS and of BP. It is useful to know precisely how IS and BP are affected by ε. We will be considering comparative statics, which take us from one overall equilibrium to another, so that after the exchange rate adjustment we will still be looking for an equilibrium where all three curves, LM, IS, and BP, intersect. Identifying exchange-rate-induced shifts in IS and BP will help. More specifically we are interested in a locus of (r, Y) points which satisfy (26.1) and (26.3) at different exchange rates. Hence we are concerned for the moment with the behaviour of the subsystem (26.1) and (26.3) at different exchange rates.

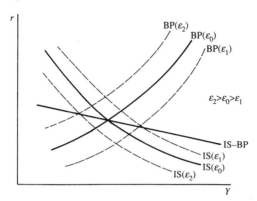

Fig. 26.4: The effect of variations in the exchange rate on IS and BP

Totally differentiating (26.1) and (26.3) allowing only Y, r, and ε to vary gives

$$
\begin{bmatrix}
S'(Y - T(Y))[1 - T'(Y)] + T'(Y) - NX_Y(Y, \varepsilon) & -I'(r) \\
\\
NX_Y(Y, \varepsilon) & K_r(r, \varepsilon)
\end{bmatrix}
\begin{bmatrix}
dY \\
\\
dr
\end{bmatrix}
=
$$

$$
\begin{bmatrix}
d\varepsilon\{NX_\varepsilon(Y, \varepsilon)\} \\
\\
d\varepsilon\{-BS_\varepsilon(Y, r, \varepsilon)\}
\end{bmatrix}.
\tag{26.15}
$$

After dividing through by $d\varepsilon$ and using Cramer's Rule we find

$$
\frac{dY}{d\varepsilon} = \frac{NX_\varepsilon(Y, \varepsilon)K_r(r, \varepsilon) - I'(r)\{NX_\varepsilon(Y, \varepsilon) + K_\varepsilon(r, \varepsilon)\}}{\lambda K_r(r, \varepsilon) + I'(r)NX_Y(Y, \varepsilon)} < 0
\tag{26.16}
$$

and

$$
\frac{dr}{d\varepsilon} = \frac{-\lambda\{NX_\varepsilon(Y, \varepsilon) + K_\varepsilon(r, \varepsilon)\} - NX_\varepsilon(Y, \varepsilon)NX_Y(Y, \varepsilon)}{\lambda K_r(r, \varepsilon) + I'(r)NX_Y(Y, \varepsilon)}
\tag{26.17}
$$

where $\lambda = S'(Y - T(Y))[1 - T'(Y)] + T'(Y) - NX_Y(Y, \varepsilon)$. The sign of (26.17) is ambiguous in general, but we assume it to be *positive* in what follows. Thus the locus of intersections of IS and BP generated by falls in the exchange rate is characterized by a lower income and higher interest rate as the exchange rate appreciates. What we will call the IS–BP locus is illustrated in Fig. 26.4. A fall in the exchange rate is characterized by a move down the IS–BP locus. This device will prove to be useful in interpreting the following analysis.

Consider first an increase in government expenditure. The effect is illustrated in Fig. 26.5, and we will go through the diagrammatic approach before specifying the mathematical model.

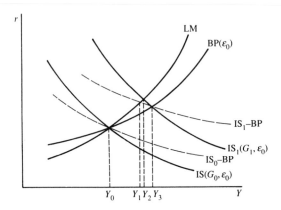

Fig. 26.5: Effect of an increase in government expenditure (flexible exchange rate)

The initial equilibrium income is Y_0 and the initial IS and BP curves are given by IS_0 and BP. Following the increase in government expenditure IS shifts out to IS_1. The economy's (temporary) equilibrium is at income Y_2 where IS_1 intersects with LM, which is in a zone of balance of payments surplus. The point at which IS_1 and BP intersect (income level Y_2) is known to be on an IS–BP locus. This locus is shown by the broken line IS_1–BP. The new overall equilibrium must occur where this locus intersects with LM, at income Y_1. Intuitively, the increase in government expenditure raises interest rates sufficiently (given the relative slopes of BP and LM) to create a balance of payments surplus ($Y = Y_2$). Exchange rate appreciation lowers this surplus, and lowers the interest rate and income. Although Y_1 is higher than Y_0 it is not as high as it might have been, because of crowding out. In the closed economy case the crowding out which takes income only to Y_2 instead of beyond is caused by the effect of increased interest rates in choking off investment expenditure. In the open economy under floating exchange rates there is another source of crowding out which reduces income even below Y_2. The movement from Y_2 to Y_1 is caused solely by exchange rate appreciation. Hence exchange rate appreciation by reducing exports and stimulating imports generates its own crowding out.

Whilst the IS–BP locus helps in the diagrammatic solution to the adjustment process, it is not necessary in the mathematical solution. Taking total differentials of (26.1)–(26.3), allowing changes in Y, r, ε, and G, and treating ε as endogenous gives

$$\begin{bmatrix} \lambda & -I'(r) & -NX_\varepsilon(Y,\varepsilon) \\ L_Y(Y,r) & L_r(Y,r) & 0 \\ NX_Y(Y,\varepsilon) & K_r(r,\varepsilon) & BS_\varepsilon(Y,r,\varepsilon) \end{bmatrix} \begin{bmatrix} dY \\ dr \\ d\varepsilon \end{bmatrix} = \begin{bmatrix} dG \\ 0 \\ 0 \end{bmatrix}, \quad (26.18)$$

where $\lambda = S'(Y - T(Y))[1 - T'(Y)] + T'(Y) - NX_Y(Y,\varepsilon)$. The determinant of the square matrix in (26.18) is given by,

$$\begin{aligned} \Delta = &\{S'(Y - T(Y))[1 - T'(Y)] + T'(Y)\}L_r(Y,r)BS_\varepsilon(Y,r,\varepsilon) \\ &+ I'(r)L_Y(Y,r)BS_\varepsilon(Y,r,\varepsilon) - NX_Y(Y,\varepsilon)L_r(Y,r)K_\varepsilon(r,\varepsilon) \\ &- NX_\varepsilon(Y,\varepsilon)L_Y(Y,r)K_r(r,\varepsilon) \end{aligned}$$

. Since the braced term is positive, $L_r < 0$, and $BS_\varepsilon < 0$, the first term in this expression is positive. With $I'(r) < 0$, $L_Y > 0$, and $BS_\varepsilon < 0$ the second term is also positive. Since $NX_Y < 0$, $L_r < 0$, and $K_\varepsilon < 0$ the third term is negative but is being subtracted. Finally, $NX_\varepsilon < 0$, $L_Y > 0$ and $K_r > 0$ so the last term is negative but again is being subtracted. We conclude that $\Delta > 0$.

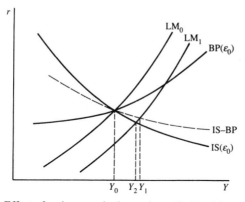

Fig. 26.6: Effect of an increase in domestic credit (flexible exchange rate)

Using Cramer's Rule we establish that,

$$\frac{dY}{dG} = L_r(Y,r)BS_\varepsilon(Y,r,\varepsilon)/\Delta > 0, \quad (26.19)$$

$$\frac{dr}{dG} = -L_Y(Y,r)BS_\varepsilon(Y,r,\varepsilon)/\Delta > 0, \quad (26.20)$$

$$\frac{d\varepsilon}{dG} = \{L_Y(Y,r)K_r(r,\varepsilon) - L_r(Y,r)NX_Y(Y,\varepsilon)\}/\Delta > 0. \quad (26.21)$$

The last result arises because LM steeper than BP implies that, as we showed in Chapter 25,

$$L_Y(Y, r)K_r(r, \varepsilon) > L_r(Y, r)NX_Y(Y, \varepsilon)$$

This confirms the diagrammatic analysis of Fig. 26.5. The increase in government expenditure increases income (from Y_0 to Y_1), increases the interest rate, and increases the exchange rate.

The effect of monetary policy may also be derived from (26.18) with the final vector $[dG \quad 0 \quad 0]'$ replaced by $[0 \quad dM \quad 0]'$. The mathematics of this is left as an exercise. The diagrammatic analysis of monetary policy is shown in Fig. 26.6. The increase in the money supply increases income from Y_0 to Y_2 but lowers the interest rate and creates a balance of payments deficit. The exchange rate depreciates and the new equilibrium occurs where the new LM curve intersects with IS–BP and Y_1. The depreciation reinforces the expansionary effect of an increased money supply.

Exercise 26.3

1. Use the system in (26.18) with $[dG \quad 0 \quad 0]'$ replaced by $[0 \quad dM \quad 0]'$ to study the effect of a monetary expansion in a flexible exchange rate model. Thus confirm the results of the diagrammatic exercise in Fig. 26.6.
2. Rework the analysis of this section on the assumption that the BP curve is steeper than LM.

26.4 A special case: perfect capital mobility

A useful special case of the analysis of this chapter will figure prominently in the following chapter. Specifically, we introduce the assumption of *perfect capital mobility*. If we think of international capital movements as being in terms of financial assets rather than in terms of physical quantities, it is clear that investors can easily relocate their investments in areas where interest rates are highest. For large investors and for financial institutions it is a relatively costless matter to move financial investments around, and so even quite small interest rate differentials between countries can be exploited at little cost. In the limiting case of perfect capital mobility *all* interest rate differentials between countries will be eliminated by capital movements.

The implication for our model of a small open economy is that its interest rate will be equal to the 'world' rate or the foreign rate r_f. We can rewrite (26.1)–(26.3) as

$$S(Y - T(Y)) + T(Y) - NX(Y, \varepsilon) = I(r_f) + G, \qquad (26.22)$$
$$M = L(Y, r_f), \qquad (26.23)$$
$$0 = NX(Y, \varepsilon) + K(r_f, \varepsilon). \qquad (26.24)$$

Essentially, the interest rate under perfect capital mobility is no longer *endogenous* but *exogenous*. This simplifies the analysis considerably.

Consider first the fixed exchange rate case and look at equation (26.22). With both ε and r_f fixed this gives one equation in the one unknown Y. So does equation (26.24). Either of these equations can be used to determine Y. Notice that (26.23) appears also to offer a formula for Y in terms of the exogenous variables M and r_f, but beware! Under a fixed exchange rate M is not exogenous because the reserve component of M is the equilibrating mechanism. So (26.23) is actually a formula for the equilibrium level of the money supply given the equilibrium level of income and the world interest rate.

Taking total differentials of (26.22) yields the following comparative statics:

$$\frac{dY}{dG} = \{S'(Y - T(Y))[1 - T'(Y)] + T'(Y) - NX_Y(Y, \varepsilon)\}^{-1} > 0, \quad (26.25)$$

$$\frac{dY}{dr_f} = \frac{I'(r_f)}{\{S'(Y - T(Y))[1 - T'(Y)] + T'(Y) - NX_Y(Y, \varepsilon)\}} < 0, \quad (26.26)$$

$$\frac{dY}{d\varepsilon} = \frac{NX_\varepsilon(Y, \varepsilon)}{\{S'(Y - T(Y))[1 - T'(Y)] + T'(Y) - NX_Y(Y, \varepsilon)\}} < 0. \quad (26.27)$$

You are asked to show as an exercise that (26.25) is the same as the government expenditure multiplier in an open economy version of the income–expenditure model. An increase in the world interest rate reduces output due to its impact on real investment at home, while an appreciation of the exchange rate reduces income because of its stimulus to imports. Notice that, as in the general case of this chapter, an increase in the money supply has no effect on output under a fixed exchange rate.

Now consider the flexible exchange rate case. Now, the money supply is fully exogenous and so (26.23) does represent a formula for equilibrium income for a given money supply and world interest rate. Either equation (26.22) or equation (26.24) will determine the equilibrium exchange rate. Using (26.23) and (26.24), for example, gives the comparative-static effect on Y of a change in the world interest as

$$\frac{dY}{dr_f} = \Delta^{-1}\{-L_{r_f}(Y, r_f)BS_\varepsilon(Y, r_f, \varepsilon)\} > 0$$

where $\Delta = L_Y(Y, r_f)BS_\varepsilon(Y, r_f, \varepsilon) < 0$.

Interestingly, the result for the general case that the flexible exchange rate regime is unfavourable to expansionary fiscal policy takes an extreme form in this special case. Notice, that since we can determine income *without* (26.22), G does not enter into the picture. In a flexible exchange rate world with perfect capital mobility an increase in government expenditure is completely 'crowded out'. You can confirm this in an exercise. The intuition is that, since a bond-financed increase in government spending creates upward pressure on domestic interest rates, there are large inflows of capital, which causes exchange rate appreciation. Exchange rate appreciation stabilizes the balance of payments surplus until interest rates are at their original level. That is when total expenditure is at its original level.

Exercise 26.4

1. Set up a simple income–expenditure model including the balance of trade. Confirm that (26.25) is equivalent to the government expenditure multiplier in this case. Explain. Sketch the effects derived in (26.25)–(26.27).
2. Use (26.22) and (26.23) to show that an increase in government expenditure has no impact on output in a flexible exchange rate regime with perfect capital mobility.
3. What is the effect of an increase in r_f on the exchange rate ε in the model of flexible exchange rates and perfect capital mobility?

26.5 Summary

The effectiveness of monetary and fiscal policy in increasing output in an open economy depends critically on the exchange rate regime in operation. The government expenditure multiplier and the money supply multiplier are both sensitive to exchange rate adjustment.

The specific results of this chapter depend on the relative sensitivity of balance of payments and domestic money markets to the interest rate (we have assumed LM has a steeper slope than BP), and on the relative sensitivity of domestic spending and the balance of payments to the exchange rate (we have assumed the IS–BP locus is downward-sloping).

In a fixed exchange rate regime, fiscal policy is effective in promoting increases in output because increases in government spending produces 'accommodating' increases in the money supply to reduce crowding out. The increase in the money supply is *induced* by increases in reserves.

A flexible exchange rate provides a more favourable environment for an increase in the money supply to make an impact on output. The reduction in interest rate produces increased investment, as usual, and a balance of payments deficit leading to a depreciation of the exchange rate. This depreciation leads to further increases in net demand from increased exports and reduced imports.

The relative effectiveness of fiscal and monetary expansions, under alternative exchange rate regimes, are exaggerated when there is perfect capital mobility. Fiscal policy becomes completely ineffective under flexible exchange rates, and is not offset by crowding out under fixed exchange rates.

27 Aggregate Demand and Supply in the Open Economy

It is extremely difficult to undertake a comprehensive study of demand and supply in the open economy. In particular, the analysis under flexible exchange rates is considerably more involved when we allow for endogenous prices. Our analysis is therefore selective.

We continue to give separate consideration to the two extreme types of exchange rate regime, and we show that the general shape of the aggregate supply in Chapter 24, in which a distinction is made between short-run (static expectations) and long run (fulfilled expectations), is also appropriate here although the information requirements of agents are greater. This is because in forming price expectations, agents must consider the influence of the balance of payments on the way the economy adjusts and on the economy's equilibrium.

Before we study aggregate demand and supply in the open economy context we will adapt the IS-LM-BP framework slightly so as to simplify the analysis. We do this by assuming the special case of section 26.4: that of perfect capital mobility. This implies that the BP curve is horizontal at a particular interest rate. It also implies that the IS–BP locus is horizontal at this interest rate. This is a considerable simplification.

To build up the open economy AD–AS model we first establish the properties of the aggregate demand function under alternative exchange rate regimes. We will subsequently discuss the aggregate supply curves and then subject the entire system to comparative-static analysis under fixed and floating exchange rates.

27.1 Aggregate demand

Under the new assumption of perfect capital mobility and hence fixed interest rate we rewrite the system (26.1)–(26.3) as follows:

$$S(Y - T(Y)) + T(Y) - NX(Y, \theta) = I(r_f) + G, \tag{27.1}$$

$$\frac{m}{P} = L(Y, r_f), \tag{27.2}$$

$$0 = BS(Y, r_f, \theta), \tag{27.3}$$

where r_f is the 'world' interest rate which the domestic rate must equal. Our concern in this chapter is with the determination of prices in the open economy. Hence, trade flows and the balance of payments now depend not just on the exchange rate, ε, but on the terms of trade, θ, defined in Chapter 25 as the international relative price of domestic output, or

$$\theta = \frac{P_f/\varepsilon}{P}. \tag{27.4}$$

In what follows the foreign price P_f will remain fixed, but even so the balance of payments will be influenced by both ε and P through θ. In the fixed-price world of IS–LM–BP it was only necessary to make explicit the dependence of trade and the balance of payments on ε. Clearly, when the domestic price level is allowed to change it is important to make explicit the dependence of NX and BS ($\equiv$ NX + K) on θ, since it is the relative price of domestic and foreign goods which, when measured in a common currency, determines the strength of export and import demand. The higher is θ the higher will be net exports and the higher the balance of payments surplus. So,

$$NX_\theta(Y, \theta) > 0,$$
$$\tag{27.5}$$
$$BS_\theta(Y, r_f, \theta) > 0.$$

A fall in either P or ε, moreover, will increase θ; hence

$$\frac{\partial\theta}{\partial P} = \frac{-\theta}{P} < 0,$$

$$\tag{27.6}$$

$$\frac{\partial\theta}{\partial\varepsilon} = \frac{-\theta}{\varepsilon} < 0,$$

and an increase in θ is referred to as a *deterioration* in the terms of trade.

In the case of the *fixed exchange rate*, we know from section 26.2 that equilibrium is determined by the subsystem (26.1) and (26.3). Similarly, equilibrium is determined here by equations (27.1) and (27.3). Moreover, (27.3) seems only to fix the interest rate, so that any variation in Y as P varies is obtained solely from (27.1). We know that under the fixed exchange rate regime the money supply will always accommodate such a change in the absence of sterilization. Fig. 27.1 illustrates this. An increase in price from P_0 to P_1 lowers income from Y_0 to Y_1. Mathematically, the effect of a change in price on equilibrium income as obtained from (27.1) by total differentiation is simply

$$\frac{dY}{dP} = \frac{-NX_\theta(Y, \theta)\theta/P}{S'(Y - T(Y))[1 - T'(Y)] + T'(Y) - NX_Y(Y, \theta)} < 0, \tag{27.7}$$

since $NX_\theta(Y, \theta) > 0$. This confirms that the aggregate demand curve for an open economy with a fixed exchange rate is downward-sloping. It is easily established, again from (27.1), that the *position* of the aggregate demand curve depends on tax parameters and exogenous expenditures (as in the closed economy) and also on the exchange rate ε, the foreign price level P_f, and the interest rate. For example, from (27.1),

$$\frac{dY}{d\varepsilon} = \frac{-NX_\theta(Y, \theta)\theta/\varepsilon}{S'(Y - T(Y))[1 - T'(Y)] + T'(Y) - NX_Y(Y, \theta)} < 0, \tag{27.8}$$

so that a revaluation of the exchange rate (increase in ε) will shift the aggregate demand curve to the left, lowering income at each price level.

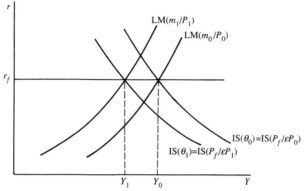

Fig. 27.1: Effect of an increase in price on output
(fixed exchange rate and perfect capital mobility)

In the case of the *flexible exchange rate* we saw in section 26.3 the entire system (26.1)–(26.3) was necessary to determine demand-side equilibrium. With the assumption of a fixed interest rate, however, one of the equations (27.1)–(27.2) may be eliminated. Since exchange rate adjustment will always remove an imbalance in the balance of payments by shifting the IS curve, it is equation (27.1) which can be eliminated. Note that with a 'horizontal' BP curve at r_f, we have removed the necessity to monitor the relative effects of ε on IS and BP. Hence total implicit differentiation of the subsystem (27.2) and (27.3), treating ε as endogenous gives

$$
\begin{bmatrix} L_Y(Y, r_f) & 0 \\ \\ -NX_Y(Y, \theta) & -BS_\varepsilon(Y, r_f, \theta) \end{bmatrix} \begin{bmatrix} dY \\ \\ d\varepsilon \end{bmatrix} = -dP \begin{bmatrix} m/P^2 \\ \\ BS_\theta(Y, r_f, \theta)\theta/P \end{bmatrix}, \qquad (27.9)
$$

so that, using Cramer's Rule,

$$
\frac{dY}{dP} = \frac{m/P^2}{-L_Y(Y, r_f)} < 0. \qquad (27.10)
$$

This is illustrated in Fig. 27.2.

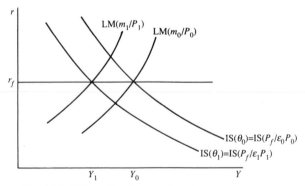

Fig. 27.2: Effect of an increase in price on output
(flexible exchange rate and perfect capital mobility)

This confirms that the aggregate demand curve for an open economy with a flexible exchange rate is downward sloping. The *position* of this demand curve depends on m and r_f, as may be seen from (27.2) and (27.3). Notice that the terms of trade are endogenous. An increase in the nominal money stock m shifts the flexible exchange rate aggregate demand curve to the right, since total differentiation of (27.2) and (27.3) holding price constant is

$$
\begin{bmatrix}
L_Y(Y, r_f) & 0 \\
-NX_Y(Y, \theta) & -BS_\varepsilon(Y, r_f, \theta)
\end{bmatrix}
\begin{bmatrix}
dY \\
d\varepsilon
\end{bmatrix}
=
\begin{bmatrix}
dm/P \\
0
\end{bmatrix}.
\tag{27.11}
$$

Exercise 27.1

1. Which economic variables are constant and which vary as we move down along an open economy aggregate demand curve under perfect capital mobility and (i) a fixed exchange rate, (ii) a flexible exchange rate?
2. Establish the main properties of an open economy aggregate demand curve when capital is mobile, but not perfectly mobile (i.e., the general case of Chapter 26).

27.2 Aggregate supply

For simplicity the open economy can be assumed to have no particular implication for the aggregate supply functions developed in Chapter 24, although the open economy provides a different informational environment in which expectations are formed. In the following analysis we make use of the distinction first developed in Chapter 24 between the 'short-run' aggregate supply relation based on static expectations (the 'surprise' supply curve) and the 'long-run' aggregate supply based on correct expectations. In summary we have, from Chapter 24 equations

(24.22) and (24.23)

$$Y = f(N) \tag{27.12}$$

$$N = N\left(\frac{P}{P^e}\right) \qquad N'(.) > 0. \tag{27.13}$$

One interesting extension of the basic aggregate supply story is to assume that, in addition to labour, output also depends on an imported *intermediate* good. We consider an example along these lines in section 27.4 below.

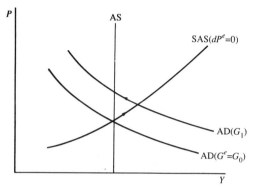

Fig. 27.3: Short-run and long-run effects of a 'surprise' increase in G ($G_1 > G_0$)

27.3 Fiscal and monetary policy in the open economy AD–AS model

Rather than conduct a full analysis of fiscal and monetary policy for each exchange rate regime, we take just two examples of policy changes. First we look at fiscal policy under a fixed exchange rate, and secondly we study monetary policy under flexible exchange rates. This choice is not arbitrary, since monetary policy (in the sense of increases in the money supply) will not affect the position of the AD function under fixed exchange rates, nor will fiscal policy (in the sense of increased government expenditure) affect aggregate demand under flexible rates, at least when capital is perfectly mobile. Each of these propositions follows immediately from the discussion of the previous section and it is unnecessary to pursue them formally.

Recall that under a *fixed exchange rate* the position of the aggregate demand curve depends on fiscal instruments, the exchange rate, the (fixed) interest rate and the foreign price level. Aggregate supply and demand combined under fixed exchange rates determine output and the price level in the system:

$$S(Y - T(Y)) + T(Y) - NX(Y, \theta) = I(r_f) + G, \tag{27.14}$$

$$Y = f\left(N\left(\frac{P}{P^e}\right)\right). \tag{27.15}$$

The discussion of section 26.4 suggests that (27.14) above determines aggregate demand under a fixed exchange rate with perfect capital mobility, whereas (27.15) simply combines (27.12) and (27.13). The 'surprise' aggregate supply curve makes $dP^e = 0$ whereas in the 'long run' $dP^e = dP$ and $P^e = P$.

Consider first a 'surprise' increase in government expenditure, G, then total differentiation of (27.14) and (27.15) with $dP^e = 0$ and remembering that θ depends on P gives

$$\begin{bmatrix} S'(Y + T(Y))[1 - T'(Y)] + T'(Y) - NX_Y(Y, \theta) & NX_\theta(Y, \theta)\,\theta/P \\ \\ 1 & -f'(N)N'(P/P^e)/P^e \end{bmatrix} \begin{bmatrix} dY \\ \\ dP \end{bmatrix} =$$

$$\begin{bmatrix} dG \\ \\ 0 \end{bmatrix}, \qquad\qquad (27.16)$$

which gives

$$\frac{dY}{dG} = \frac{-f'(.)N'(.)/P^e}{-f'(.)N'(.)/P^e\{S'(.)[1 - T'(.)] - NX_Y(Y, \theta)\} - NX_\theta(Y, \theta)\theta/P} > 0 \qquad (27.17)$$

and

$$\frac{dP}{dG} = \frac{-1}{-f'(.)N'(.)/P^e\{S'(.)[1 - T'(.)] - NX_Y(Y, \theta)\} - NX_\theta(Y, \theta)\theta/P} > 0, \qquad (27.18)$$

where the more obvious arguments of functions have been dropped.

These effects are illustrated in Fig. 27.3. The 'expected' aggregate demand curve is $AD(G^e)$, where the expected government expenditure is G^e. Actual government expenditure, however, is $G_1 > G_0$. This is not anticipated by workers, so that the price and output both increase, according to (27.17) and (27.18) above. When G_1 is correctly perceived of course we have the 'long-run' result that $dP^e = dP$, which implies, as in Chapter 24, that $dY = 0$. From (27.16) we may directly calculate the long-run impact on price of an increase in government expenditure as

$$\frac{dP}{dG} = \frac{1}{NX_\theta(Y, \theta)\theta/P}. \qquad (27.19)$$

It is easily established that (27.19) is greater than (27.18).

In the *flexible exchange rate* case we know from the previous chapter that equilibrium in the aggregate demand sector is determined not by the IS-function, (27.14), but by equations (27.2) and (27.3) – the LM and BP functions. In place of (27.14) and (27.15) the aggregate demand and supply system is now

$$\frac{m}{P} = L(Y, r_f),$$ (27.20)

$$0 = BS(Y, r_f, \theta)$$ (27.21)

$$Y = f\left(N\left(\frac{P}{P^e}\right)\right).$$ (27.22)

For the 'surprise' effect of an increase in the nominal money stock, m, we totally differentiate (27.20)–(27.22) allowing endogenous changes in ε but setting $dP^e = 0$. This gives the system

$$
\begin{bmatrix}
L_Y(Y, r_f) & m/P^2 & 0 \\
BS_Y(Y, r_f, \theta) & -BS_\theta(Y, r_f, \theta)\theta/P & BS_\varepsilon(Y, r_f, \theta) \\
1 & -f'(N)N'(P/P^e)/P^e & 0
\end{bmatrix}
\begin{bmatrix}
dY \\
dP \\
d\varepsilon
\end{bmatrix}
=
\begin{bmatrix}
dm \\
0 \\
0
\end{bmatrix}.
$$ (27.23)

The determinant of the matrix in (27.23) is negative and given by

$$\Delta = BS_\varepsilon(Y, r_f, \theta)\{L_Y(Y, r_f)f'(N)N'(P/P^e)/P^e + m/P^2\} < 0.$$

Hence, the 'surprise' effect may be summarized as

$$\frac{dY}{dm} = \{BS_\varepsilon(Y, r_f, \theta)f'(N)N'(P/P^e)/P^e\}\Delta^{-1} > 0,$$ (27.24)

$$\frac{dP}{dm} = BS_\varepsilon(Y, r_f, \theta)/\Delta > 0,$$ (27.25)

$$\frac{d\varepsilon}{dm} = \{-BS_Y(Y, r_f, \theta)f'(N)N'(P/P^e)/P^e + BS_\theta(Y, r_f, \theta)\theta/P^e\}\Delta^{-1} < 0.$$ (27.26)

Thus a 'surprise' increase in money supply increases output and the price level and induces a depreciation in the exchange rate. These effects are illustrated in Fig. 27.4.

The 'long-run' effects are easily obtained from the conditions $dP^e = dP$ and $dY = 0$, which eliminates (27.22) and gives us

$$
\begin{bmatrix}
m/P^2 & 0 \\
-BS_\theta(Y, r_f, \theta)\theta/P & BS_\varepsilon(Y, r_f, \theta)
\end{bmatrix}
\begin{bmatrix}
dP \\
d\varepsilon
\end{bmatrix}
=
\begin{bmatrix}
dm \\
0
\end{bmatrix}.
$$ (27.27)

The fully anticipated effects are, therefore, $dY/dm = 0$ and

$$\frac{dP}{dm} = \frac{P^2}{m} > 0, \tag{27.28}$$

$$\frac{d\varepsilon}{dm} = \frac{BS_\theta(Y, r_f, \theta)\theta}{mBS_\varepsilon(Y, r_f, \theta)/P} < 0, \tag{27.29}$$

so that a fully anticipated increase in the money supply increases the price level, causes exchange rate depreciation, but leaves output unchanged. It is easily established that the net effect of (27.28) and (27.29) is to leave the terms of trade, θ, unaffected. The terms of trade in the long run are independent of monetary policy.

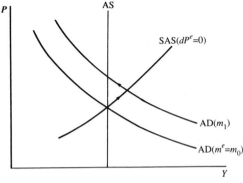

Fig. 27.4: Short-run and long-run effects of a 'surprise' increase in m ($m_1 > m_0$)

It is important to remember that these effects depend on the simplifying assumption of perfect capital mobility.

Exercise 27.3

1. Use equations (27.14) and (27.15) to determine the effect on output and price in the 'short run' of an increase in (a) the 'world' interest rate r_f and (b) the foreign price level P_f under fixed exchange rate.
2. Use equations (27.20) to (27.22) to determine the effect on output and price in the 'short run' and in the 'long run' of an increase in (a) the world interest rate r_f and (b) the foreign price level P_f under flexible exchange rates. Compare with your answers to question 1.
3. Suppose (27.22) is replaced by

$$Y = f\left(N\left(\frac{P}{P^e}\right), \alpha\right) \qquad f_\alpha > 0,$$

where α is a parameter representing technological improvement. Demonstrate the effect on income, price, and the exchange rate in both the 'short run' and the 'long run' of an increase in α.

4. The results in the text on the effect of an increase in m on AD–AS equilibrium assume that the foreign price is fixed. Suppose, on the contrary that changes in P and ε are accompanied by variations in P_f by purchasing power parity so that θ does not change (and always equals unity). How are the results (27.24)–(27.26), and (27.28) and (27.29), affected by purchasing power parity?

27.4 An imported intermediate good

Suppose that this economy requires an additional input, say oil, which must be imported from abroad. The price of oil per barrel is fixed in terms of foreign currency at Π_f. Denote the quantity of oil demanded by domestic producers by V. Clearly, V is the inverse marginal product curve of oil and so

$$V = V\left(\frac{\Pi_f/\varepsilon}{P}\right) \qquad V'(.) < 0 \qquad (27.30)$$

so that the demand for oil depends inversely on the price of oil to domestic producers.

The production function is

$$Y = f(N, V) \qquad f_N > 0 \quad f_V > 0, \quad f_{NN} < 0, \quad f_{VV} < 0.$$

Finally, we assume, perhaps unrealistically, that total domestic aggregate demand is independent (directly) of oil imports.

As an illustration of this model, consider a fixed exchange rate and perfect capital mobility; let us examine the impact on output and price of a 'surprise' increase in the price of oil Π_f. Equations (27.14) and (27.15) are replaced by

$$S(Y - T(Y)) + T(Y) - NX(Y, \theta) = I(r_f) + G \qquad (27.31)$$

$$Y = f\left(N\left(\frac{P}{P^e}\right), V\left(\frac{\Pi_f/\varepsilon}{P}\right)\right). \qquad (27.32)$$

A comparative-static analysis of (27.31) and (27.32) reveals that an increase in Π_f reduces output and increases domestic prices. The derivation of this result is straightforward, and is left as an exercise.

This is a simple example of a *supply shock* to the economy which leads to increased prices and lower output and employment.

Exercise 27.4

1. Confirm the results of this section that an increase in Π_f increases price and reduces output.
2. How might aggregate demand depend on oil imports?

3. Conduct an analysis of the 'long-run' effect of an increase in oil prices, using the model of this section.

27.5 Summary

The effects on output and prices in the open economy are similar to those in the closed economy, in the sense that 'surprises' affect both output and prices while in the long run only prices are affected.

The added results are that an increase in the money supply, for example, also causes exchange rate depreciation in both the short run and the long run.

The basic model can be extended to account for the use of an imported intermediate good in the production process. In a simple version of this model, a 'surprise' increase in the foreign price of oil reduces output and increases domestic prices under a fixed exchange rate.

28 Inflation and Unemployment

Inflation and unemployment are often regarded as the two biggest threats to economic stability and economic welfare. Inflation appears to have distortionary effects on relative prices and erodes the purchasing power of those on fixed incomes, which fail to keep pace with higher prices. It reduces the real burden of outstanding debt but reduces the value of investors' portfolios. Unemployment is an indicator of unused resources and is associated with personal hardship.

This chapter extends the analysis of earlier chapters to consider a stylized inflation process and any associated implications for unemployment. Further developments are necessary because, while earlier chapters have considered the determination of output and the price *level*, an analysis of inflation requires a study of price changes. Inflation, after all, refers to a *continual* increase in the general price level, not merely relative price movements or once-for-all price changes. In addition to forging a link between price levels and price *changes* we will switch emphasis away from output to unemployment – both being indicators of economic activity.

For simplicity, we consider only a closed economy. In sections 28.3 and 28.4 we consider alternative specifications of the way labour markets and product markets work, which shed different light on inflation dynamics and on persistent unemployment. Thus we take up some issues raised in Chapter 23.

28.1 Dynamics of aggregate demand and supply

In this section we develop a dynamic formulation of the aggregate demand and supply model of Chapter 24. Specifically we seek a reformulation of the system in equations (24.24)–(24.27) which emphasizes the dynamics of inflation and output. Unfortunately an analysis of anything but the simplest dynamic system takes us outside the scope of this book, and so we are aiming for 'bare-bones' simplicity.

Consider first aggregate demand. Assume that we can solve the static LM equation for the interest rate in terms of output at the real money supply, and suppose further that we can substitute into the static IS curve and find a 'reduced form' solution for output, Y, in terms of the real money supply, m/P, and exogenous expenditure components, A. Now, suppose we can express this reduced form in natural logarithms so that

$$\ln Y = \alpha \ln A + \beta(\ln m - \ln P), \qquad (28.1)$$

where $\alpha > 0$ and $\beta > 0$ are 'multipliers'. Now lag (28.1) by one period to obtain

$$\ln Y_{-1} = \alpha \ln A_{-1} + \beta(\ln m_{-1} - \ln P_{-1}) \qquad (28.2)$$

and subtract (28.2) from (28.1), giving

$$\ln Y - \ln Y_{-1} = \alpha(\ln A - \ln A_{-1}) + \beta(\ln m - \ln m_{-1}) - \beta(\ln P - \ln P_{-1}). \quad (28.3)$$

Now, it is shown in section 1.6 of the mathematical appendix that the difference between the natural logs of two values of a variable is approximately equal to the proportional rate of change between these two values. Hence (28.3) is approximately equivalent to

$$\hat{Y} = \alpha\hat{a} + \beta(\mu - \pi), \quad (28.4)$$

where $\hat{Y}$ is the growth rate of output, $\hat{a}$ the growth rate of autonomous spending, μ the growth rate of the money supply, and π the inflation rate. For simplicity we will assume that $\hat{a} = 0$. That is, the autonomous component of private and public sector spending is constant. We have

$$\hat{Y} = \beta(\mu - \pi). \quad (28.5)$$

Notice that something similar to (28.5) could have been derived from the classical identity $MV = PY$ discussed in section 20.3, since taking logs gives

$$\ln M + \ln V = \ln P + \ln Y,$$

and then taking differences (as in (28.2) and (28.3)) gives (assuming V is constant)

$$\hat{Y} = \mu - \pi. \quad (28.6)$$

We refer to (28.5) as the dynamic aggregate demand curve, and it says simply that the growth of aggregate demand is equal to the growth rate of the *real* money supply (the growth rate of the nominal money supply *less* the inflation rate). We note in passing that the inverse aggregate demand curve is given by

$$\pi = \mu - \frac{\hat{Y}}{\beta}. \quad (28.7)$$

The dynamic aggregate supply relation is derived by applying a similar log-linear approach to the reduced form of the labour market equilibrium condition and the production function. In essence, the message of Chapter 24 as far as this reduced-form relationship is concerned is that deviations from the equilibrium level of output are due only to deviations between actual and expected prices and, in the dynamic context, between actual and expected inflation rates. Hence,

$$\ln Y - \ln Y_0 = \frac{1}{\gamma}(\pi - \pi^e) \quad (28.8)$$

where γ is a positive constant. We can rewrite (28.8) as

$$\pi = \gamma(\ln Y - \ln Y_0) + \pi^e. \tag{28.9}$$

In summary, we have a quite simple two-equation system (28.5) and (28.9) which relates output and inflation dynamics to the rate of growth of the money supply and expectations about inflation. We assume that the growth rate of the money supply, μ, is under the control of the government, but that still leaves inflationary expectations unspecified. As in Chapter 24 we will look at alternative specifications of how agents form expectations about inflation. Moreover, we can discuss the 'long-run' equilibrium by presuming that in the 'long run' the expected inflation rate is equal to the actual inflation rate.

28.1.1 The 'long run' and the 'steady state'

Let us now impose on the model summarized in (28.5) and (28.9) the 'long-run' requirement that expectations are correct. From (28.9) setting $\pi^e = \pi$ implies that $\ln Y = \ln Y_0$. That is, in the long-run equilibrium the level of output is constant at Y_0. This implies in turn that the growth rate of output, $\hat{Y}$, is zero, which is possible if and only if $\pi = \mu$ by (28.5). An equilibrium in which output growth is zero is, in this context, referred to as a steady state of the system (28.5) and (28.9). In steady state all dynamic effects are the same in all periods. In this steady state, actual and expected inflation rates are equal to each other and both are equal to the rate of growth of the money supply. Note that the 'steady-state' condition that $\hat{Y} = 0$ implies and is implied by the 'long-run' condition that $\pi = \pi^e$ as long as the rate of money growth is constant.

The main result, then, is that a once-for-all increase in the rate of monetary growth increases the steady-state inflation rate but has no effect on output in the long run.

As in Chapter 24, many of the most interesting aspects of the model are the adjustments that the economy makes between (steady-state) equilibria, and this requires us to specify the short-run behaviour of inflation expectation.

28.1.2 Short-run adjustment under adaptive expectations

An extreme version of adaptive expectations is to assert that the expected inflation rate is equal to the previous period's *actual* inflation rate. So,

$$\pi^e = \pi_{-1}. \tag{28.10}$$

Using this in (28.9) and rearranging gives

$$\pi - \pi_{-1} = \gamma(\ln Y - \ln Y_0), \tag{28.11}$$

while dynamic aggregate demand is

$$\hat{Y} = \beta(\mu - \pi). \tag{28.12}$$

To investigate the possible short-run dynamics of this system we will assume that we start in a steady state in which $\mu = \pi = \pi^e$ and $\hat{Y} = 0$. This situation is depicted in Fig. 28.1, in which we show the aggregate demand curve (conditional on μ and $\ln Y_{-1}$) and the aggregate supply curve (conditional on π_{-1}). The initial steady-state equilibrium is at point A.

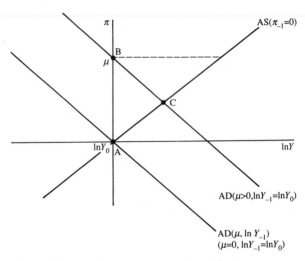

Fig. 28.1: Dynamic aggregate demand and supply curves

Now suppose that μ increases to some positive value which is maintained by the government. The only new steady-state equilibrium possible is at point B in Fig. 28.1 at which $\pi = \mu > 0$, $\pi = \pi^e$ and $\ln Y = \ln Y_0$. Moreover, the path taken by the economy between A and B may not be so direct, depending on values taken by the parameters γ and β. We can use (28.11) and (28.12) to study some of the possibilities. These are illustrated in Figs 28.2 and 28.3.

Equations (28.11) and (28.12) give equations of motion for the inflation rate and the (log of) real output respectively. Equation (28.11) indicates that inflation is increasing as long as $\ln Y > \ln Y_0$, and decreasing as long as $\ln Y < \ln Y_0$, while (28.12) indicates that output is increasing as long as $\pi < \mu$ and decreasing when $\pi > \mu$. Fig. 28.2 indicates direction arrows for π and $\ln Y$, and shows four distinct zones.

In zone I, $\ln Y > \ln Y_0$ and $\pi < \mu$ and so both π and $\ln Y$ are increasing. In zone II, $\ln Y > \ln Y_0$ but $\pi > \mu$ so that inflation increases as output falls. In zone III both inflation and output are falling while in zone IV output is increasing and inflation falling. The exact path taken by the economy through the zones I–IV from A to B depends on the relative sizes of the parameters β and γ. One possibility is shown in Fig. 28.3. Note that the economy initially moves to point C in Fig. 28.1 when μ increases and expectations are temporarily fixed, so we move immediately into zone I.

In Fig. 28.3, the economy passes through zone I, which is a typical 'boom' period with both increases in output and inflation. As the economy passes into zone II there appears to be 'stagflation' in which output is falling and inflation increasing. Zone III is a depression in which both inflation and output are falling, and zone IV is a recovery period in which inflation is falling and output rising.

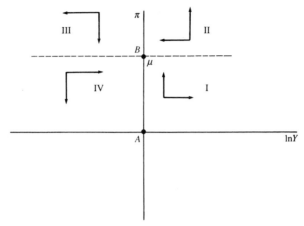

Fig. 28.2: Zones of disequilibrium adjustment

This analysis points to two main conclusions. First, if the rate of growth of the money supply is permanently increased to $\mu > 0$ then inflation increases until $\pi = \mu$ when it stabilizes at the new steady-state rate. In the new steady state, prices are increasing but at the same annual rate in each successive year. If, on the other hand, the monetary authorities continued to increase the rate of growth of the money supply then prices would increase at an accelerating rate, indicated by a continual increase in the rate of inflation. The second conclusion turns this argument around. A monetary policy which systematically reduces the rate of growth of the money supply seems an appropriate way of reducing the rate of inflation, even though reducing μ to zero may involve an adjustment path which takes the economy through periods of high unemployment. You should think about these possibilities as an exercise.

28.1.3 Adjustment with rational expectations

As in Chapter 24, the adjustment process following a change in policy depends on whether the policy was announced or not.

If the policy is announced then agents can solve for the new long-run steady-state equilibrium and the economy moves directly from A to B, in terms of Fig. 28.1.

A 'surprise' policy change, on the other hand, since expected inflation remains unchanged, moves the economy from A to C in terms of Fig. 28.1, since $\pi^e = \pi = 0$. However, point C is not a rational expectations equilibrium, and as soon as agents learn about the policy change the economy moves from C to B, the new long-run equilibrium.

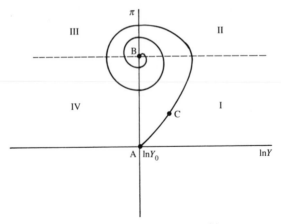

Fig. 28.3: Adjustment to a new steady state

Exercise 28.1

1. Use equations (28.11) and (28.12), and a diagram such as Fig. 28.3 to consider a path taken by the economy under an anti-inflation policy in which the rate of growth of the money supply, initially at $\mu > 0$, is reduced to $\mu = 0$.

28.2 Inflation and unemployment

Most policy debate takes place around the combined issues of inflation and *unemployment* rather than output. The link between output and unemployment is considered here and enables us to use the model of the previous section.

First, we propose that there is a stable relationship between output and *employment*. Theoretically, this is achieved by considering the inverse of the economy's production function. Hence if we define the (inverse) function as

$$g = f^{-1},$$

then employment, N is given by

$$N = g(Y). \tag{28.13}$$

The second step is to find a link between the level of employment and the level of unemployment. If the total working population is L, then, as seen in Chapter 17, we have unemployment given by

$$U = L - N. \tag{28.14}$$

So, as long as L is constant there is an inverse one-for-one relationship between employment and unemployment. An increase in employment by one thousand reduces unemployment by the same amount. In Chapter 17 we discussed why L may not be constant, but for now, we assume it is. Hence combining (28.13) and (28.14) we have

$$U = L - g(Y), \tag{28.15}$$

which gives the relationship between unemployment and output.

The *empirical* basis for a stable relationship between output and unemployment is summarized in a proposition known as 'Okun's Law' after the American economist Arthur Okun. Okun's Law points to a stable statistical (linear) relationship between output and unemployment such that, at least in the US, a sustained increase in output of 2.5 per cent is associated with a *reduction* in unemployment of around one per cent. Thus, if we consider deviations of output and employment from their equilibrium values, Y_0 and U_0 respectively, then we have

$$U - U_0 = g(Y_0) - g(Y) \tag{28.16}$$

or, using the log-linear approximations,

$$\ln U - \ln U_0 = a(\ln Y_0 - \ln Y), \tag{28.17}$$

where a is a positive constant. It is now straightforward to conduct the analysis of the previous section in terms of unemployment rather that output by substituting from (28.17).

The aggregate supply curve (28.9) becomes

$$\pi = \frac{\gamma}{a}(\ln U_0 - \ln U) + \pi^e. \tag{28.18}$$

The relationship between the unemployment level and the inflation rate in (28.18) is often known as the expectations-augmented Phillips curve. This is a prevalent vehicle for the study of inflation and unemployment, but it is clear that it only tells part of the story – the 'supply side'. In fact (28.18) is best interpreted not as an equilibrium condition for the economy but as representing the choices or opportunities for inflation and unemployment. Even so, its use requires great care since it leaves unspecified the expected inflation rate π^e. Some simple examples may be studied in the following exercises to illustrate.

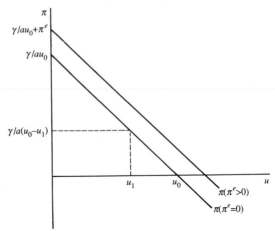

Fig. 28.4: Expectations-augmented Phillips curves

In Fig. 28.4 we illustrate the expectations-augmented Phillips curve (28.18). As long as expected inflation is zero, the trade-off between inflation and unemployment is given by the curve $\pi(\pi^e = 0)$. An expansion of demand, according to the analysis of the previous section, raises inflation and output if expectations are unchanged. In Fig. 28.4 this corresponds to a move up *along* the Phillips curve $\pi(\pi^e = 0)$. As inflation expectations adjust upwards, the available trade-offs between inflation and unemployment worsen.

Exercise 28.2

1. Suppose an economy's short-run Phillips curve, which we may treat as a 'supply curve', is given by

$$\pi = \pi^e + (\ln U_0 - \ln U) \tag{1}$$

and suppose expectations are formed by $\pi^e = \pi_{-1}$. Now suppose that the government maintains a constant rate of growth of money demand and that we may represent the 'demand' side of the economy in terms of unemployment by

$$\pi = \mu + (\ln U - \ln U_{-1}). \tag{2}$$

Finally, suppose that economy starts with $\ln U_{-1} = \ln U = \ln U_0 = 5$, $\pi = \mu = \pi^e = 0$.

   (a) What rate of growth of nominal demand, μ, must the government choose to reduce (log) unemployment immediately to 4? What unemployment level and inflation rate result in the first, second, third, and fourth periods?

   (b) Can the government reduce unemployment permanently to 4 by any constant rate of growth of demand? Illustrate using diagrams similar to Fig. 28.4.

2. Use the model of question 1 but now assume that $\ln U = \ln U_0 = 5$, and $\pi = \mu = \pi^e = 10$. Suppose μ is suddenly reduced to 5 for each future period. What unemployment level and inflation rate result in the first, second, and third periods, and in the long run?

3. Rework your answer to question 2 but with equation (1) in question 1 replaced by $\pi = 0.8 \pi^e + (\ln U_0 - \ln U)$. Illustrate using diagrams similar to Fig. 28.4.

28.3 Prices, wages, and unemployment – the non-competitive case

The analysis of the previous two sections has brought us to a model of price changes, giving the dynamics of price changes and unemployment, or output. One of the key issues which arises in discussion about inflation episodes and solutions to them concerns the behaviour of money wages. However, the model developed so far is silent on the issue, since it is based on reduced-form aggregate supply and aggregate demand functions. In particular aggregate supply is specified in terms of deviation between actual and expected price, and money wages adjust only if workers anticipate price changes. In this section we open up the issue of money wage dynamics. In so doing we build up our inflation–unemployment model from a different angle to allow for explicit wage changes and non-competitive pricing behaviour by firms.

To start, we suggest a relationship between wage changes and excess demand in the labour market. This relationship captures the idea that money wages do not adjust *instanteously* to excess labour demand. Hence we write

$$\hat{W} = \lambda(\ln U_0 - \ln U), \tag{28.19}$$

where $\hat{W} \equiv \dot{W}/W$ is the proportional change in wages or *wage inflation* and λ is the positive constant measuring the speed of wage adjustment. In a frictionless spot-auction market, $\lambda \to \infty$, indicating instantaneous adjustment of wages in response to excess demand. Excess demand in the labour market is measured by deviations of actual employment from the equilibrium or 'natural' level. If $U < U_0$ there is an excess demand for labour, and wages increase at a rate λ for every unit difference between $\ln U_0$ and $\ln U$. Among other things, this means that the rate of increase in the money wage is proportional to the unemployment gap (expressed in logs). If $U > U_0$ then there is an excess supply of labour and competition for jobs between workers lowers money wages so that $\hat{W} < 0$. However, in this model, the speed with which wages fall and hence the speed with which the excess supply of labour (unemployment) is eliminated depends on λ, which in turn depends on various institutional features of the labour market.

Why might λ be 'small' so that money wages are slow to adjust? There are many explanations, but two popular stories concern 'contracts' and labour 'unions'. A third popular story is considered in some detail in the following section. In the case of contracts, workers and firms will often agree on a nominal wage to prevail over a period of time, say one or two years. During that time, if labour market conditions change, there is little opportunity to revise wages. One side will always have an incentive to resist entering into renegotiations that are neither legally required nor entirely costless. Often, if the contract has been negotiated between a firm and a union, there may be considerable resistance to wage reductions in the presence of excess supply. Unions seek to protect living standards of members and the existing workforce, and whilst labour movements as a whole may be concerned about excess labour supply, the incentives for any single union to negotiate wage reductions are very weak.

Whilst (28.19) captures the response of wage changes to excess labour demand, it does not tell the whole story. Consider a case in which a contract is due for renegotiation at a time when there is no excess demand or excess supply of labour but workers expect a positive rate of price inflation, $\pi^e > 0$ during the next period before the contract can once more be renegotiated. In this case we can expect some adjustment in money wages to reflect the anticipated inflation. If both firms and workers collectively agree on the expected rate of inflation, then workers can expect full compensation. More generally, there will be part compensation in anticipation of future inflation. Hence (28.19) becomes

$$\hat{W} = \lambda(\ln U_0 - \ln U) + \pi^e \qquad (28.20)$$

where π^e is the expected inflation rate and we assume that wages increase one-for-one with expected inflation. Often, labour contracts will include automatic, though partial adjustment for loss of real income over the previous period. This retrospective compensation is often referred to as cost-of-living-adjustment or COLA, and suggests that (28.19) be written

$$\hat{W} = \lambda(\ln U_0 - \ln U) + c\pi_{-1}, \qquad (28.21)$$

where the positive constant c represents the proportion of 'catch-up' specified in the COLA clause.

The relationship between unemployment, expected inflation, and wage changes in (28.20) is sometimes called the Friedman-Phelps expectations-augmented Phillips curve.

To forge a link between (28.20) and price inflation we look at a form of non-competitive pricing in which prices are a constant mark-up on unit costs. If we treat the wage as the unit cost of output and the mark-up as some constant k, then we have prices given by

$$P = kW, \qquad (28.22)$$

so that

$$\dot{P} = k\dot{W} \qquad (28.23)$$

and

$$\frac{\dot{P}}{P} \equiv \pi = \frac{k\dot{W}}{kW} = \frac{\dot{W}}{W} \equiv \hat{W}. \qquad (28.24)$$

We conclude that with a constant mark-up this pricing behaviour makes the inflation rate equal to the proportional rate of change of wages.

Thus wage changes are always passed on in terms of higher prices. This has implications for the appropriate economic policy for the control of inflation, and points to the possibility that some form of wage and price controls may be warranted. Although currently out of vogue, such policies have been introduced from time to time. The preference amongst many policy makers more recently is to affect inflationary expectations directly through tight monetary policy and to let 'excessive' wage settlements have their consequences for unemployment.

28.4 Persistent unemployment

In many of the current inflation models, high levels of unemployment relative to the equilibrium level are 'temporary'. Moreover, the 'equilibrium' level of unemployment is identified as the market-clearing level, that is, the level of unemployment consistent with an equality between labour demand and labour supply. This level of unemployment includes those people between jobs or who are otherwise temporarily displaced. In Chapter 17 we referred to this as 'frictional' unemployment. One implication is that in this equilibrium only 'frictions' cause willing individuals to be jobless. This apart, there are no workers who are willing to work at the equilibrium wage rate who cannot find a job. Since the labour market works like a spot-auction market, individuals should not experience *persistent* unemployment.

However, this seems to be counter-factual. As we saw in Chapter 17, workers in many countries experience long (a year or more) spells of unemployment. Much effort has been put into trying to match the theory more closely to the facts. This issue has been around for a long time and was, for example, a major force behind the 'employment theory' of the English economist John Maynard Keynes. Since the traditional 'Keynesian' story lacked a sound microeconomic basis for labour market rigidities, many recent attempts have been made to explain the apparent 'market failure'. We consider some such explanations now.

28.4.1 Efficiency wages

Suppose each firm in the economy has an identical production function which relates labour input to output. The efficiency wage theory has two special features. One is that the labour input is not simply the number of workers but is the product of the number of workers and the amount of effort each worker supplies, e. So,

$$Y = f(eN). \qquad (28.25)$$

The idea behind this specification is that workers whose performances cannot be monitored accurately have an incentive to 'shirk' – that is, to provide less than their maximum effort. Thus output depends not only on the number of workers but on how much effort each worker supplies.

The second feature is that firms can influence effort supplied by monitoring, and by making it costly for workers to be caught shirking. One way of achieving this is by threatening dismissal if caught, and simultaneously raising the wage above the worker's *reservation wage* or alternative wage, which will be below the current wage for all but the 'marginal' worker. Thus we have effort supplied as a function of the wage rate:

$$e = e(W), \qquad e'(W) > 0. \qquad (28.26)$$

Each price-taking firm chooses the wage and the level of employment to maximize profit

$$\pi = pf(eN) - WN, \qquad (28.27)$$

giving the first-order conditions

$$pf'(eN)e'(W)N - N = 0, \qquad (28.28)$$

$$pf'(eN)e - W = 0. \qquad (28.29)$$

Together (28.28) and (28.29) show that the wage should be chosen so that the elasticity of the effort function with respect to the wage is one. That is,

$$\frac{We'(W)}{e(W)} = 1. \qquad (28.30)$$

This condition clearly shows that the choice of wage by the firm depends only on the properties of the effort function and not on labour market conditions generally. If the optimal wage, according to (28.30), is above the market-clearing level, then since all firms are identical, the economy's employment level will fall below the competitive equilibrium. The presence of unemployment among people with reservation wages *below* the going wage rate re-enforces the disciplining effect of the threat of dismissal. Notice that no firm has an incentive to lower the wage, because to do so would increase shirking. The picture is presented for a single firm in Fig. 28.5.

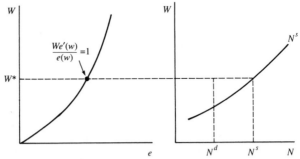

Fig. 28.5: The 'efficiency wage' and employment level in a typical firm

28.4.2 'Insider-outsider' theory

Assar Lindbeck and Dennis Snower have proposed a general set of models in which employment confers sufficient benefits on both employees and employers that firms are reluctant to exchange an incumbent workforce (insiders) for new workers (outsiders). Thus the unemployed, being a source of new workers, are not regarded as perfect substitutes for those already employed. Because existing filled jobs are generating rents and because these rents cannot be competed away by outsiders, wages are above the competitive level and unemployment results.

The formal analysis of these issues can take a variety of forms depending on exactly what aspect of the employer-employee relationship is responsible for outsiders being unable to compete for jobs held by insiders. One possibility is that replacing existing workers is costly for the firm in terms of hiring and firing costs. Workers can exploit this and share in the firm's cost savings by having a stable labour force. Another possibility is that existing workers are more productive than outsiders. This may arise through on-the-job training or simply 'learning-by-doing'. This increases the incumbent workers' marginal products relative to those of outsiders. On the other hand on-the-job training will be costly for the firm, giving it an incentive to reduce the level of employment.

A simple model will illustrate. Consider a firm engaged in an activity for just two periods. Employment is n and output is produced according to the production function, $f(n)$, with $f'(n) > 0$ and $f''(n) < 0$. When hiring on the open market, the firm pays the competitive wage W_1 which is fixed. The firm is a price taker in the product market and product price is $p > 0$. In the first period the firm incurs a cost $c > 0$ of training each worker. In the second period the firm is able to produce a times as much output as in the first period using the same level of employment. We ignore turnover aspects for simplicity. The firm chooses the level of employment for the two periods, and the second-period wage, subject to the (feasibility) constraints that neither the firm nor the worker can do better over the two periods by opting for two *single*-period contracts. Formally, the firm solves the following problem

$$\max_{n,W_2} \quad \pi = \underbrace{pf(n) - W_1 n - cn}_{\text{first period profit}} + \underbrace{paf(n) - W_2 n}_{\text{second period profit}}$$

$$\text{s.t.} \quad \pi \geq \sum_{i=1}^{2}(pf(n) - W_i n),$$

$$W_1 \leq W_2,$$

Where W_1 is fixed and $a > 1$. Assume for the moment that the first constraint is satisfied with equality. Then it is optimal for the firm to set $W_2 = W_1 = W$ and employment so that

$$pf'(n) = \frac{2}{1+a}W + \frac{c}{1+a}. \tag{28.31}$$

Now compare this with the normal condition for a single-period firm offering a competitive wage (as in condition (12.20)),

$$pf'(n) = W. \tag{28.32}$$

Because of concavity of $f(.)$ the optimal employment level in (28.31) will be lower than in (28.32) if the right-hand side of (28.31) is *greater* than W, that is, if

$$W(1 - a) + c > 0, \tag{28.33}$$

or if training costs c are high relative to the output premium a. We now need to check that this condition is consistent with the contract being feasible for the firm. It is easily checked that this requires

$$pf(n)(a - 1) \geq cn. \tag{28.34}$$

Since (28.33) is independent of price and output, whereas (28.34) does depend on price and output, it is not difficult to construct examples where both conditions are met simultaneously. Hence there are circumstances under which the two-period contract is preferred by firms but this implies lower employment than the competitive case. The implication is that if all firms behave in this way, aggregate employment will be lower than in a world of only single-period contracts. On-the-job training permits increased output, with training costs being covered by savings on the wage-bill.

28.4.3 A unionized economy

Many commentators claim that unions create unemployment. The basis for this claim can be seen in Fig. 12.4, in which a *monopoly union* sets the wage and the employer sets the level of employment. Any wage set above the 'competitive' wage reduces employment.

In general, we may think of the union as the institutional embodiment of the 'insider' theory. However, the analysis of Chapter 12 points to an important issue which may offset or eliminate any tendency for unions to restrict employment. Unions may seek to *bargain* over employment as well as over the wage. Depending on both the relative weight attached to employment in the union's utility function, and on the relative bargaining strengths of the firm and the union, an efficient contract may involve a higher level of employment than in th ecompetitive equilibrium. The complete answer to this issue therefore requires an analysis of the bargaining process and information on the preferences of unions.

Exercise 28.4

1. Show how an increase in the product price affects the wage and the employment level in the simple efficiency wage model.
2. Suppose that, in addition to wages, the firm spends an amount T per worker on training. Set up an efficiency wage model with training costs included, and if $T = t/W$ where t is a constant. Provide a rationale for why $T'(W) < 0$, and show what happens to W and N if t increases.
3. Construct a numerical example using the model of section 28.4.2 for which the conditions (28.33) and (28.24) are jointly satisfied. Can you produce a counter-example?

28.5 Summary

Much policy debate is concerned with inflation and unemployment or output. The main policy implication of the dynamic analysis of section 28.1 is that deflationary monetary policy may involve periods of temporarily high unemployment (output below the natural rate); see question 1 of Exercise 28.1.

If mark-up pricing behaviour is assumed, then the wage–inflation link is such that prices affect wage demands *and* wage settlements induce additional inflation one-for-one. This is sometimes referred to as the wage–price 'spiral'.

Persistent unemployment may be a feature of an economy even when the agents are profit-maximizing and utility-maximizing. The 'efficiency-wage' model is an example of such a market failure. Imperfect monitoring of effort gives workers an incentive to shirk. Higher wages and consequently lower employment result, and the unemployed are unable to offer to work at below the going wage. The efficiency wage model is just one of a series of relatively recent contributions which address the issue of labour market rigidities.

The insider-outsider model (which may feature unions as an example) explores the implications of turnover, training costs and other factors which make the unemployed and the employed imperfect substitutes. Thus the unemployed are unable to compete effectively for jobs. Again equilibrium may entail persistent unemployment.

29 A Simple Model of Economic Growth

This chapter is concerned with the mathematical modelling of the growth of productive potential of the economy. Hence the 'long-run equilibrium' level of output identified in Chapter 24 is regarded not as a constant but as variable over time and driven by *real* forces. In Chapter 24 we saw that permanent changes in the long-run equilibrium output might result from changes in the conditions underlying labour supply (worker preferences, for example) or those underlying labour demand (the production function, for example). These changes were in the nature of comparative-static effects, however, whereas our concern here is with the evolution of productive potential; we are looking primarily for *systematic* and sustained influences. Our attention is therefore drawn to the dynamics of the economy's production function. The dominant systematic and recurrent determinants of the economy's growth rate are the capital stock, the size of the labour force, natural resources, and technological change.

To model this apparently complex growth process requires, at least initially, some simplifying assumptions, and before we study the mathematics of the growth model itself we will develop the mathematics of the production function to be used. We will then derive some key relationships and study the properties of equilibrium growth which emerge.

29.1 Long-run production

Particular challenges have to be met when we consider technological change and the use of (ultimately finite) natural resources. Abstracting from these difficulties means specifying a production function which depends only on the amounts of capital and labour used as inputs. We write the general form of aggregate production function as

$$Y = F(K, N), \tag{29.1}$$

with

$$F_K(K, N) > 0, \quad F_N(K, N) > 0, \quad F_{KK}(K, N) < 0 \quad \text{and} \quad F_{NN}(K, N) < 0,$$

$$F_{KN}(K, N) = F_{NK}(K, N) > 0.$$

In other words, output depends on capital and labour, both of which have positive but diminishing marginal products. Production functions were discussed in some detail in Chapter 8, but for now one important feature about the production function in (29.1) is that since $F(.,.)$ is assumed to be continuous and differentiable, there is scope for substitution between capital and labour. In fact there will be many different ways of producing any given level of output depending on the relative amounts of K and N used. The *marginal rate of substitution* between capital and labour is defined as the (small) increase in labour required to replace a (small) reduction in capital so that output remains unchanged. Total differentiation of (29.1), setting $dY = 0$ and rearranging, gives the marginal rate of substitution as,

$$\frac{dK}{dN} = \frac{-F_N(K,N)}{F_K(K,N)} < 0. \tag{29.2}$$

Equation (29.2) is the slope of an *isoquant* (see Chapter 8) which allows limited substitution possibilities between capital and labour. This potential for variability in the capital–labour ratio, and also in the capital–output ratio, has important consequences for the growth model to be discussed presently.

Whilst (29.1) represents the general form of the production function, the particular growth model of this section uses a class of production function which exhibits constant returns to scale. That is, a λ-fold increase in both K and N produces a λ-fold increase in output Y if $F(K,N)$ exhibits constant returns to scale, viz.,

$$\lambda Y = F(\lambda K, \lambda N). \tag{29.3}$$

For example, a doubling of both inputs leads to a doubling of output. One production function which exhibits constant returns to scale, and one which we will use to illustrate the more important proposition, is the Cobb-Douglas form,

$$Y = K^\alpha N^{1-\alpha} \qquad 0 < \alpha < 1. \tag{29.4}$$

That (29.4) satisfies (29.3) is demonstrated as follows:

$$(\lambda K)^\alpha (\lambda N)^{1-\alpha} = \lambda^\alpha \lambda^{1-\alpha} K^\alpha N^{1-\alpha}$$
$$= \lambda K^\alpha N^{1-\alpha}$$
$$= \lambda Y$$

for any value of λ. One useful feature of a constant returns to scale production function, and one which follows from (29.3), is that output per worker only depends on the capital–labour ratio, because setting $\lambda = 1/N$ gives

$$\frac{Y}{N} = f\left(\frac{K}{N}, 1\right) \tag{29.5}$$

or

$$y = f(k), \tag{29.6}$$

where y and k are the output–labour and capital–labour ratios respectively, and (29.6) is often referred to as the *intensive* form of the production function. The mechanics of the intensive form may be illustrated using the Cobb-Douglas special case, from which we have

$$\frac{Y}{N} = \frac{K^\alpha N^{1-\alpha}}{N^\alpha N^{1-\alpha}}$$

$$= \left(\frac{K}{N}\right)^\alpha$$

or $y = k^\alpha.$ (29.7)

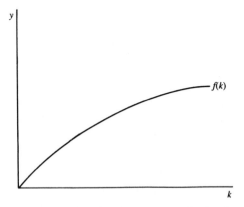

Fig. 29.1: The intensive-form aggregate production function

A further useful result is that the increment to output per worker from an increase in capital per worker is precisely equal to the marginal product of capital for a constant returns to scale production function. This is easily illustrated using the Cobb-Douglas example. From (29.4) the marginal production of capital is

$$\frac{\partial Y}{\partial K} = \alpha K^{\alpha-1} N^{1-\alpha}$$

$$= \alpha \left(\frac{K}{N}\right)^{\alpha-1},$$

whilst from (29.7) the increment to output per worker from an increase in capital per worker is

$$\frac{\partial y}{\partial k} = \alpha k^{\alpha-1}$$

$$= \alpha \left(\frac{K}{N}\right)^{\alpha-1}.$$

The production function is illustrated in Fig. 29.1. The concavity of the intensive form production function is a result of the diminishing marginal product of capital. In the Cobb-Douglas case we have

$$\frac{\partial}{\partial k}\left(\frac{\partial y}{\partial k}\right) = \alpha(\alpha - 1)\left(\frac{K}{N}\right)^{\alpha-2} < 0$$

$$= \alpha(\alpha - 1)K^{\alpha-2}N^{1-\alpha}N$$

$$= \frac{\partial}{\partial K}\left(\frac{\partial Y}{\partial K}\right)N.$$

If the marginal productivity of capital is diminishing then the marginal productivity of capital per head is diminishing also as long as the production function displays constant returns to scale. Notice also that $f(0) = 0$, which is satisfied by the Cobb-Douglas form in (29.7).

A final property of constant returns to scale production functions which will prove useful is the fact that the total output may be expressed as the sum of each factor's marginal product multiplied by the amounts of each factor used. Hence, returning to (29.1), we have

$$Y = F_K(K,N)K + F_N(K,N)N \tag{29.8}$$

if $F(K,N)$ is a constant returns to scale production function. To illustrate this result, which is an application of a mathematical theorem for homogeneous functions known as Euler's Theorem, we may use the Cobb-Douglas production function (29.4). From this,

$$F_K(K,N) = \alpha K^{\alpha-1}N^{1-\alpha}$$

and

$$F_N(K,N) = (1 - \alpha)K^{\alpha}N^{-\alpha}.$$

Now

$$F_K(K,N)K + F_N(K,N)N = \alpha K^{\alpha}N^{1-\alpha} + (1 - \alpha)K^{\alpha}N^{1-\alpha}$$

$$= K^{\alpha}N^{1-\alpha} = Y.$$

Moreover, from (29.7)

$$\frac{Y}{N} = F_K(K,N)\frac{K}{N} + F_N(K,N)$$

or

$$y = F_K(K,N)k + F_N(K,N). \tag{29.9}$$

Now consider Fig. 29.2, which reproduces the intensive form production function, and consider a particular capital–labour ratio, k_0. At k_0 the marginal product of k is given by the slope of the tangent to $f(k)$ at k_0. The economy under consideration here is assumed to be perfectly competitive so that each factor is paid its marginal product. Thus the slope of the tangent at k_0 is the rental on capital or the profit rate. Furthermore the output per head produced by the capital–labour ratio k_0 is y_0. It is apparent by inspection of Fig. 29.2 that we have

$$y_0 = w + rk_0, \tag{29.10}$$

where w is the intercept of the tangent on the y-axis. Rearranging (29.10) gives

$$w = y_0 - rk_0. \tag{29.11}$$

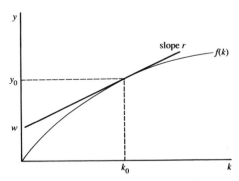

Fig. 29.2: The wage rate and the profit rate

Now, y_0 is output per worker and rk_0 total profit per worker when the capital–labour ratio is k_0. Using Euler's Theorem, summarized in intensive form by (29.9), means that w must equal the marginal product of labour – the wage rate in a competitive economy. Output per worker less total profit per worker must equal the wage per worker. Observe that as k increases beyond k_0, the rate of profit falls and the wage rate rises.

This discussion suggests that the production process is intimately linked with the distribution of income between capital and labour. As the capital–labour ratio increases the ratio of the rate of profit to the wage rate falls. In general, however, we are unable to say whether the ratio of the share of capital to the share of labour, rK/wL, increases, remains the same, or falls as the capital–labour ratio increases. In intensive form this share ratio is given by

$$\Omega(k) = \frac{rk}{w}$$

$$= \frac{rk}{f(k) - rk}$$

$$= \frac{f'(k)k}{f(k) - f'(k)k},$$

from which we find

$$\frac{d\Omega(k)}{dk} = \frac{f''(k)f(k)k + f'(k)[f(k) - f'(k)k]}{[f(k) - f'(k)k]^2},$$

and since $f'(k) > 0$, $f''(k) < 0$, and $[f(k) - f'(k)k] > 0$ the sign of this is ambiguous in general. In the special case of the Cobb-Douglas production function, however, we can easily establish that

$$\frac{d\Omega(k)}{dk} = 0,$$

so that the ratio of factor shares stays constant as k increases. That the ratio of factor shares is constant in the Cobb-Douglas case may also be inferred from Euler's Theorem. Moreover, not only is the *ratio* constant but each factor's *share* is also constant, because Euler's Theorem tells us that

$$rK + wN = F_K(K,N)K + F_N(K,N)N$$
$$= \alpha K^\alpha N^{1-\alpha} + (1-\alpha)K^\alpha N^{1-\alpha}$$
$$= \alpha Y + (1-\alpha)Y.$$

Hence capital's share of output is α and labour's share is $(1-\alpha)$. Also, in the Cobb-Douglas case,

$$\Omega = \frac{\alpha}{(1-\alpha)}.$$

This analysis of the constant returns to scale, aggregate production function provides the basis of the growth model of this chapter. It is apparent, from Fig. 29.2 for example, that given the production function and a theory explaining how k is determined we will be able to solve the model for y, r, and w.

Exercise 29.1

1. The intensive form of the production function makes output per head a function of capital per head simply by dividing the constant returns to scale production function through by the size of the labour force. Mathematically, it would be equally valid to divide the original production function through by the amount of capital, K, giving as the intensive form output per unit of capital as a function of labour per unit of capital, say $Y/K \equiv u$ as a function of $L/K \equiv l$. Construct the equivalent diagrams to Fig. 29.2 with u and l instead of y and k. Identify r and w. Find the steady-state value of l.
2. Using (29.1) and the associated restrictions, find an expression for the slope of the isoquant for the case of the Cobb-Douglas production function. Show that the Cobb-Douglas production function has convex isoquants.

29.2 Dynamics

So far we have said nothing about dynamics, and nor have we discussed how the production function relates to *growth* as such. Dynamics will enter the scene both because of increases in the capital stock (the result of net investment) and the growth of the labour force.

First, the growth of the capital stock. We know from the simple model of income determination that the change in the capital stock at any point in time is identically equal to net investment which in turn is funded from the economy's savings. Hence

$$I \equiv \dot{K} = sY, \tag{29.12}$$

where s is the economy's savings ratio. The growth in the capital stock is therefore given by

$$\frac{\dot{K}}{K} \equiv \hat{K} = \frac{sY}{K}$$

or

$$\hat{K} = \frac{s}{v},$$

where the capital-output ratio is v,

$$v \equiv \frac{K}{Y} = \frac{k}{y}. \tag{29.13}$$

Now, we are conducting the analysis all in terms of labour units in the intensive form and so we are interested in the change in k, $\dot{k}$, and its proportional rate of change $\hat{k} \equiv \dot{k}/k$. Clearly,

$$\dot{k} = \frac{\dot{K}N - \dot{N}K}{N^2}$$

and

$$\hat{k} \equiv \frac{\dot{k}}{k} = \frac{N}{K}\left[\frac{\dot{K}N - \dot{N}K}{N^2}\right]$$

$$= \frac{\dot{K}}{K} - \frac{\dot{N}}{N} = \hat{K} - \hat{N}$$

Thus the growth rate of the capital–labour ratio is equal to the growth of the capital stock *less* the growth of the labour force. In the growth model under consideration this latter growth rate is an exogenously determined parameter. We will denote by n the (exogenous) rate of growth of labour, so the growth rate of the capital–labour ratio can be written

$$\hat{k} = \frac{s}{v} - n \tag{29.14}$$

or, using (29.13),

$$\hat{k} = \frac{sy}{k} - n, \tag{29.15}$$

and using (29.7) gives the growth of GNP per head as

$$\hat{y} = \alpha\hat{k} \tag{29.16}$$

while the growth of GNP is found using (29.4). Finally,

$$\dot{Y} = \alpha\dot{K}K^{\alpha-1}N^{1-\alpha} + (1-\alpha)\dot{N}K^{\alpha}N^{-\alpha}$$

and dividing by Y,

$$\hat{Y} = \alpha \frac{\dot{K}}{K} + (1 - \alpha) \frac{\dot{N}}{N}$$

or

$$\hat{Y} = \alpha \hat{K} + (1 - \alpha)n. \tag{29.17}$$

We claim that a steady state will be achieved when $\dot{k} = 0$. Before we explain why this is so, consider the consequences of this for (29.15). Clearly the (equilibrium) steady state requires that k satisfies

$$sf(\bar{k}) = n\bar{k}. \tag{29.18}$$

Fig. 29.3 illustrates this. First we derive $sf(k)$, which is straightforward, since at any value of k the vertical distance to $f(k)$ may be split into two sections of length $sf(k)$ and $(1 - s)f(k)$, as, for example, at $\bar{k}$. If $sf(k)$ indicates total savings then $(1 - s)f(k)$ indicates total consumption. Secondly we plot nk, which again is straightforward, and since n is a constant, the graph of nk is linear. Equation (29.18) is satisfied only at $\bar{k}$, at which an output per head of $\bar{y} = f(\bar{k})$ is generated. With the steady-state capital–labour ratio thus determined, the steady-state rate of profit and wage rate may also be derived. Fig. 29.3 summarizes this.

We now justify our assertion that $\dot{k} = 0$ is a condition for the steady state in which output grows at a constant rate. The labour force is assumed to be growing at the constant rate n, so that, from (29.17), output will only grow at a constant rate if the capital stock grows at a constant rate. With both the growth in the capital stock and the growth of the labour force being constant, the ratio of capital to labour, k, must be constant, and hence $\dot{k} = 0$. Thus (29.18) is indeed the steady-state equilibrium condition. As usual with equilibria we ought to establish existence, uniqueness, and stability. First, existence. From (29.14) we can see that the steady state implies

$$s = vn, \tag{29.19}$$

where v is the capital–output ratio $K/Y = k/y$. Since s is a fraction and n *any* non-negative number, (29.19) is guaranteed to be satisfied if v is perfectly free to take any value in the range $[0, \infty)$. The value taken by v, of course, is precisely that required for steady-state equilibrium. In the Cobb-Douglas case it is easily established that

$$v = k^{\alpha - 1}, \tag{29.20}$$

so that for any given α, v may take any value in the interval $[0, \infty)$ if k is also free to vary in the range $[0, \infty)$. This is the case as evidenced by the structure of the (y, k) diagrams.

Secondly, there is the issue of uniqueness. Rearranging (29.19) gives

$$v = \frac{s}{n}. \tag{29.21}$$

Since the left-hand side of (29.21) is monotonically increasing in v, starting at zero for $v = 0$, and the right-hand side is a positive constant, it is apparent that one and only one value of v satisfies (29.21) and hence satisfies the steady state.

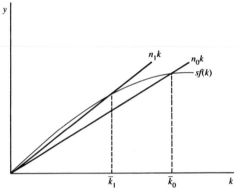

Fig. 29.3: Effect of an increase in the labour force growth rate

Finally, there is the issue of stability, which we study using equation (29.15) and Fig. 29.3. We have established that $\bar{k}$ is the (unique) steady-state capital–labour ratio. Now consider a value of $k > \bar{k}$. It is obvious from Fig. 29.3 that for any $k > \bar{k}$, $nk > sf(k)$, which implies from equation (29.15) that $\dot{k} < 0$. Hence for $k > \bar{k}$, k declines until $\bar{k}$ is reached. A symmetric argument reveals that $\dot{k} > 0$ for any $k < \bar{k}$. We conclude that the steady-state equilibrium is stable.

We may now make inferences about the growth rate of the economy's output. In the Cobb-Douglas case, the growth in output is given by (29.17), rewritten here as

$$\hat{Y} = \alpha\hat{K} + (1 - \alpha)n. \tag{29.22}$$

We know also that $\hat{K} = s/v$, thus

$$\hat{Y} = \alpha\frac{s}{v} + (1 - \alpha)n. \tag{29.23}$$

However, from (29.14) we also know that in steady-state equilibrium $s/v = n$. Hence (29.23) is simply

$$\hat{Y} = n. \tag{29.24}$$

The economy's equilibrium growth rate is equal to the rate of population growth. This is of course a consequence of the fact that since capital per head is constant in steady state, output per head is also constant. Mathematically, for any constant returns to scale production function,

$$y = f(k)$$

so that

$$\dot{y} = f'(k)\dot{k} \tag{29.25}$$

and so $\dot{k} = 0$ implies $\dot{y} = 0$.

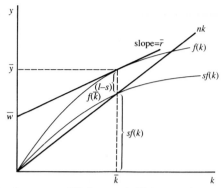

Fig. 29.4: Steady-state equilibrium capital/labour and output/labour ratios

Let us now summarize the steady-state equilibrium, firstly for any constant returns to scale production function and then in the Cobb-Douglas case.

$$s\bar{y} - n\bar{k} = 0 \tag{29.26}$$

or

$$\frac{s}{v} - n = 0,$$

$$\bar{y} = f(\bar{k}), \tag{29.27}$$

$$\bar{r} = f'(\bar{k}), \tag{29.28}$$

$$\bar{w} = \bar{y} - \bar{r}\bar{k}. \tag{29.29}$$

Condition (29.26) says that steady-state growth rate of capital should equal the rate of population growth. Condition (29.27) simply says that the steady-state level of output per head is given by the production function. Conditions (29.28) and (29.29) say that the interest rate should equal the marginal product of capital and the wage rate should equal the marginal product of labour in steady state.

Out of steady state, growth is given by

$$\hat{Y} = \frac{\dot{N}}{N} + \frac{f'(k)}{f(k)}\dot{k} \tag{29.30}$$

or

$$\hat{Y} = n + \frac{f'(k)}{f(k)}\dot{k}.$$

In steady-state equilibrium, $\dot{k} = 0$ and

$$\hat{Y} = n. \tag{29.31}$$

In the Cobb-Douglas case (29.27)–(29.31) become

$$\bar{y} = \bar{k}^\alpha, \tag{29.32}$$

$$\bar{r} = \alpha\bar{k}^{\alpha-1}, \tag{29.33}$$

$$\bar{w} = (1 - \alpha)\bar{k}^\alpha. \tag{29.34}$$

Out of steady state, growth is given by

$$\hat{Y} = \alpha \left[\frac{s}{v} - n \right] + n. \tag{29.35}$$

In steady state,

$$\hat{Y} = n. \tag{29.36}$$

Two parametric changes are generally considered in this simple growth model: an increase in the saving ratio, s, and an increase in population growth, n.

Consider first an increase in s. This is illustrated in Fig. 29.5. Mathematically we see the effect on $\bar{k}$, $\bar{y}$, $\bar{r}$, and $\bar{w}$ in the following way. First substitute (29.27) into (29.26) and totally differentiate, giving

$$\frac{\partial \bar{k}}{\partial s} = \frac{-f(\bar{k})}{sf'(\bar{k}) - n} > 0. \tag{29.37}$$

The earlier stability analysis argued that $sf(k) < nk$ for $k > \bar{k}$ and $sf(k) > nk$ for $k < \bar{k}$. Thus as we increase k from a value below $\bar{k}$ to one above $\bar{k}$, both $sf(k)$ and nk increase but $sf(k)$ increases by less than nk. Hence stability requires $sf'(k) < n$. This is obviously so in Fig. 29.3. This establishes that (29.37) is positive. This is confirmed in Fig. 29.5, as is

$$\frac{\partial \bar{y}}{\partial s} = f'(\bar{k}) \frac{\partial \bar{k}}{\partial s} > 0. \tag{29.38}$$

Thus both the steady-state capital–labour ratio and output per head increase. From (29.28) and (29.29) we establish that

$$\frac{\partial \bar{r}}{\partial s} = f''(\bar{k}) \frac{\partial \bar{k}}{\partial s} < 0, \tag{29.39}$$

$$\frac{\partial \bar{w}}{\partial s} = \frac{\partial \bar{y}}{\partial s} - \frac{\partial \bar{r}}{\partial s} \bar{k} - \frac{\partial \bar{k}}{\partial s} \bar{r}$$

$$= \left[f'(\bar{k}) - \bar{k}f''(\bar{k}) - \bar{r} \right] \frac{\partial \bar{k}}{\partial s}$$

$$= -\bar{k}f''(\bar{k}) \frac{\partial \bar{k}}{\partial s} > 0 \tag{29.40}$$

(because $\bar{r} = f'(\bar{k})$). Thus steady-state rate of profit falls and the wage rate increases. Steady-state growth is unaffected, however, because from (29.31), $\hat{Y}$ is independent of s. We consider this result a little further. Notice that it is the *steady-state* growth rate that is independent of s. The result in (29.38), on the other hand, says that output per head is increased by an increase in the savings ratio, which suggests that growth in output exceeds the growth of population.

This apparent contradiction is easily resolved, and is most easily seen using the Cobb-Douglas example.

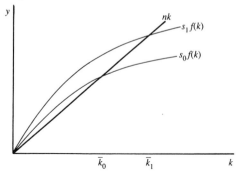

Fig. 29.5: Effect of an increase in the savings rate

Consult equation (29.35) – the general equation for growth rate – and let the initial savings ratio and steady-state capital output–ratio be s_0 and v_0 respectively. Now suppose that the savings ratio suddenly increases to $s_1 > s_0$, but suppose that the capital–output ratio does not change immediately. This results because capital equipment cannot be obtained instantly. We then have for a time a situation in which $s_1/\bar{v}_0 > n$ (recall $s_0/\bar{v}_0 = n$)

$$\hat{Y} = \alpha \left[\frac{s_1}{\bar{v}_0} - n \right] + n > n. \tag{29.41}$$

Thus out of steady state the adjustment to the new capital–output ratio requires a growth rate in excess of n, the steady-state growth rate. As v rises to the new required value, say $\bar{v}_1$, $\hat{Y}$ returns to its steady-state value. The increase in $\bar{k}$ (and $\bar{y}$) is achieved by temporary increase in the growth rate out of steady state. Without specifying the time taken for capital adjustments to be made we can say nothing about how long it takes for the economy to return to the steady-state growth path.

Fig. 29.5 shows the result of an increase in n. Mathematically we see from total differentiation of (29.26) that

$$\frac{\partial \bar{k}}{\partial n} = \frac{\bar{k}}{\{sf'(\bar{k}) - n\}} < 0 \tag{29.42}$$

and, moreover,

$$\frac{\partial \bar{y}}{\partial n} = f'(\bar{k}) \frac{\partial \bar{k}}{\partial n} < 0 \tag{29.43}$$

Thus an increase in population growth lowers both capital per head and output per head in steady state. Also, from (29.28) and (29.29),

$$\frac{\partial \bar{r}}{\partial n} = f''(\bar{k}) \frac{\partial \bar{k}}{\partial n} > 0 \tag{29.44}$$

and

$$\frac{\partial \bar{w}}{\partial n} = -\bar{k} f''(\bar{k}) \frac{\partial \bar{k}}{\partial n} < 0. \tag{29.45}$$

These effects are obvious. It is also obvious that the steady-state growth rate will also increase, since, from (29.31), $\partial \hat{Y}/\partial n = 1$. Notice that despite the increase in the growth rate of output, output per head falls. This is a direct consequence of the reduction in capital per head.

Exercise 29.2

1. Using the Cobb-Douglas example, show what happens to $\bar{k}$, $\bar{y}$, and $\hat{Y}$ if α increases. Interpret the result.
2. What, if anything, can be said about the effect of (*a*) a change in *s* and (*b*) a change in *n* on aggregate consumption in this simple growth model?
3. If the production function is Cobb-Douglas, it is possible to ascertain the effect on the share of output going to labour of an increase in the steady-state capital–labour ratio. Why is it not possible in general to establish this with a constant returns to scale production function?

29.3 Summary

The long-run growth of productive potential is determined by population growth, capital accumulation, natural resources, and technological change.

In the simple model of this chapter, with a constant saving rate, the steady-state growth rate is equal to the population growth rate.

An increase in the saving rate increases capital per head and increases the steady-state wage rate but reduces the rate of profit. An increase in the saving rate has no impact on the steady-state *growth rate*, but it will in general affect the path taken by the economy in adjusting from one steady state to another.

Appendix 1: Mathematics Revision

In this Appendix we review all the mathematical concepts and techniques – beyond the most basic algebra and geometry – that are used in this book. The treatment is brief, since the material is intended as a revision of topics the reader will already have studied.

Review 1. Topics in algebra

1.1 Exponents

If a number a is multiplied by itself n times, where n is a positive integer, we write the result as a^n, and n is called the exponent in the expression. It follows from this definition that, for any other positive integer m

$$a^n a^m = a^{n+m},$$
$$a^n/a^m = a^{n-m}.$$

This latter result leads naturally to the cases

$$a^0 = 1,$$
$$a^{-q} = 1/a^q.$$

If we set

$$a^n = b,$$

then $a = b^{1/n}$, since $b^{1/n}$ multiplied by itself n times gives b. For a similar reason,

$$(a^p)^q = a^{pq}, \qquad (ab)^p = a^p b^p.$$

1.2 Logarithms

If $a^n = b$, where a and b are positive numbers, then n is called the logarithm of b to the base a, and is written $n = \log_a b$. It follows that $\log_a a = 1$, $\log_a 1 = 0$, and that $\log_a(bc) = \log_a b + \log_a c$ and $\log_a(b/c) = \log_a b - \log_a c$.

If two numbers x and y are related as

$$y = x^m,$$

then since $a^{\log_a x} = x$, we have

$$y = a^{m \log_a x} \Rightarrow \log_a y = m \log_a x$$

for any base a.

Note that we use the notation 'ln' to denote '$\log_e$', the 'natural' logarithm, where e is a number defined in section 1.7 below.

1.3 Absolute values

Given any number x, its absolute value is written $|x|$ and satisfies the rules

$$|x| = x \quad \text{if } x \geq 0,$$
$$|x| = -x \quad \text{if } x < 0.$$

For example, $|3| = |-3| = 3$. Loosely, the absolute value can be thought of as the numerical part of the number, disregarding its sign.

It follows that $|x| \, |y| = xy$ if both numbers have the same sign, but $|x| \, |y| = |xy| > x.y$ if they have opposite signs.

1.4 Subscripts and $\sum$-notation

When we have a list of n numbers, it is useful to distinguish elements of the list by introducing a subscript, say i, and denoting a general element by x_i, with i taking on possible values $1, 2, \ldots, n$. Statements about x_i are then thought of as being true of any element of the list. For example there may be a list of n goods, and x_i, $i = 1, \ldots, n$ is the quantity of the ith good, with $x_i \geq 0$.

We may have a two-dimensional list of numbers. For example the n goods may be divided up among and consumed by m individuals. We then introduce a second subscript, say $j = 1, \ldots, m$, and denote by x_{ij} the quantity of the ith good consumed by the jth individual. There are clearly $n.m$ such quantities or elements in this two-dimensional list.

We could continue in this way and introduce a subscript for each dimension along which it is necessary to distinguish the values taken by the variable. For example, if the individuals consume goods over a number of time periods, we could introduce the subscript $t = 1, \ldots, T$, where T is the number of time periods, and x_{ijt} would be the quantity of the ith good individual j consumes in the tth period. Superscripts rather than subscripts may sometimes be used simply to make presentation clearer, e.g., x_{ij}^t may be considered neater than x_{ijt}.

Given a list of numbers $x_1, x_2, \ldots, x_n$, we can denote their sum by

$$\sum_{i=1}^{n} x_i = x_1 + x_2 + \ldots + x_n.$$

Where we have two subscripts, care must be taken, since clearly

$$\sum_{i=1}^{n} x_{ij} = x_{1j} + x_{2j} + \ldots + x_{nj} \qquad j = 1, \ldots, m$$

has a different meaning to

$$\sum_{j=1}^{m} x_{ij} = x_{i1} + x_{i2} + \ldots + x_{im} \qquad i = 1, \ldots, n.$$

In our goods example, the former would be the sum of quantities consumed by individual j (if these quantities can be meaningfully added, which may well not be the case), the latter the total consumption of good i across all individuals.

If all elements of the list x_{ij} can be summed we can write

$$\sum_{i=1}^{n} \sum_{j=1}^{m} x_{ij} = x_{11} + x_{12} + \ldots + x_{1m} + x_{21} + \ldots + x_{2m} + \ldots + x_{nm}.$$

Products can also obviously be summed:

$$\sum_{i=1}^{n} p_i x_i = p_1 x_1 + p_2 x_2 + \ldots + p_n x_n;$$

$$\sum_{i=1}^{n} p_i x_{ij} = p_1 x_{1j} + \ldots + p_n x_{nj} \qquad j = 1, \ldots, m;$$

$$\sum_{i=1}^{n} \sum_{j=1}^{m} p_i x_{ij} = p_1 x_{11} + \ldots + p_1 x_{1m} + p_2 x_{21} + \ldots + p_2 x_{2m} + \ldots + p_n x_{nm}.$$

In the case where the p_i are prices of goods and the x_{ij} are quantities consumed by individuals, you should explain what the last two of the above sums are.

Note also the following:

$$\sum_{i=1}^{n} a x_i = a \sum_{i=1}^{n} x_i \quad \text{for any constant } a;$$

$$\sum_{i=1}^{n} x_i^2 \neq \left(\sum_{i=1}^{n} x_i \right)^2 \quad \text{for } n > 1, \ x_i \neq 0;$$

$$\sum_{i=1}^{n} (a + x_i) = na + \sum_{i=1}^{n} x_i.$$

1.5 Sequences and series

A *sequence* is an ordered list of numbers $\{x_1, x_2, \ldots, x_n\} = \{x_k\}_{k=1}^n$. If n is a finite number then we have a finite sequence of length n, while if n increases indefinitely then we have an infinite sequence, which may be denoted by $\{x_k\}_{k=1}^\infty$. Sequences are often defined by a formula for calculating each term, for example

$$x_k = 1/k \qquad k = 1, \ldots, n,$$

$$x_k = a + (k-1)d \qquad k = 1, \ldots, n,$$

$$x_k = ar^{k-1} \qquad k = 1, \ldots, n,$$

for numbers a, d, and r.

A *series* is the sum of the terms of a sequence, for example $\sum_{k=1}^n 1/k$, $\sum_{k=1}^n [a+(k-1)d]$, and $\sum_{k=1}^n ar^{k-1}$ are the series defined by the above three sequences. The second and third of these series are called, respectively, the arithmetic series with common difference d, and the geometric series with common ratio r.

It is frequently useful to define a series, denoted by S_n, by a formula. For example

$$S_n = \sum_{k=1}^n [a + (k-1)d] = na + d \sum_{k=1}^n (k-1) = n[2a + (n-1)d]/2$$

$$S_n = \sum_{k=1}^n ar^{k-1} = a + ar + ar^2 + \ldots + ar^{n-1},$$

therefore

$$rS_n = ar + ar^2 + \ldots + ar^n,$$

therefore

$$(1-r)S_n = a - ar^n,$$

therefore

$$S_n = a(1 - r^n)/(1 - r) \qquad \text{(assuming } r \neq 1\text{)}.$$

Note that in this last expression, for the sum of n terms of the geometric series, if $|r| < 1$ then as $n \to \infty$, $S_n \to a/(1 - r)$, and the series is said to be *convergent*. If $|r| > 1$, then $S_n \to \pm\infty$ as $n \to \infty$ and the series is *divergent*. If $r = 1$, $S_n = na$ which is divergent. What if $r = -1$?

1.6 Compounding and discounting in discrete time

Time is divided into discrete periods of equal length, with $t = 0, 1, \ldots, T$ denoting the periods. i is the per period interest rate expressed as a decimal. \$1 lent at *compound interest* from periods 0 to t accumulates to $\$1(1+i)^t$. It follows that if $\$1$ is to be received at period t, its *present value* is $\$1(1+i)^{-t}$, since this sum lent at $t=0$ will accumulate to exactly $\$1$ at period t. We say that the future $\$1$ is *discounted* to its present value. Then the present value of a stream of payments R_t, $t=0,1,\ldots,T$ is

$$V_0 = R_0 + R_1(1+i)^{-1} + \ldots + R_T(1+i)^{-T} = \sum_{t=0}^{T} R_t(1+i)^{-t}.$$

If $R_t = R$, a constant, for $t = 0, \ldots, T$, then, using the expression for S_n of a geometric series with $r = (1+i)^{-1}$, $a = R$, and $n = T + 1$:

$$V_0 = R(1 + (1+i)^{-1} + \ldots + (1+i)^{-T}) = R[(1+i) - (1+i)^{-T}]/i.$$

Alternatively, if $R_0 = 0$, $R_t = R$, $t = 1, \ldots, T$, then

$$V_0 = R((1+i)^{-1} + \ldots + (1+i)^{-T}) = R(1 - (1+i)^{-T})/i.$$

In the former case, as $T \to \infty$, $V_0 \to R(1+i)/i$; in the latter case, as $T \to \infty$, $V_0 \to R/i$.

The following result will prove useful. We know that compounding gives the value of an amount R_0 invested at a per period rate of interest i for t periods as

$$Z_t = R_0(1+i)^t.$$

Taking natural logs gives

$$\ln Z_t = \ln R_0 + t \ln(1+i).$$

Now consider one further period of compounding:

$$Z_{t+1} = R_0(1+i)^{t+1},$$

or

$$\ln Z_{t+1} = \ln R_0 + (t+1)\ln(1+i),$$

giving

$$\ln Z_{t+1} - \ln Z_t = \ln(1+i).$$

Moreover, it can be shown that $\ln(1+i)$ is approximately equal to i, with the approximation being closer for smaller i [test this with $i = 0.05$ and $i = 0.1$].

$$\ln(1+i) \approx i.$$

We conclude that the rate of growth of an amount between periods is approximately equal to the difference in the natural logs of the amounts between periods.

1.7 The number e, continuous compounding and discounting

The mathematical constant $e = 2.71828\ldots$ can be defined as

$$\lim_{x \to \infty} (1 + 1/x)^x.$$

It plays an important role in processes in which growth or contraction can be thought of as taking place continuously. Suppose, as in the previous section, i is an interest rate per period, but now suppose that interest at the rate i/n is to be calculated n times per period. Thus, after t periods (where t can now be thought of as any real number and not just an integer), \$1 will have compounded to

$$V_t = \$1(1 + i/n)^{nt} = \$1(1 + 1/x)^{xit} \qquad \text{where } x \equiv n/i.$$

Then, as interest is calculated more and more often, $n \to \infty$, therefore $x \to \infty$, therefore

$$\lim_{n \to \infty} V_t = \lim_{x \to \infty} \$1[(1 + 1/x)^x]^{it} = \$1\,e^{it}.$$

It follows that if \$1 is to be received at t, its present value is $V_0 = \$1\,e^{-it}$.

Instead of thinking of i as an interest rate, we could think of it more generally as a growth rate, so a magnitude with value x_0 at $t = 0$ and growing continuously at a rate g per period takes the value $x_t = x_0 e^{gt}$ at $t \geq 0$.

1.8 Matrices, determinants, and Cramer's Rule

Suppose that we wish to solve two simultaneous linear equations in two unknowns, which can be written

$$a_{11}x_1 + a_{12}x_2 = b_1,$$
$$a_{21}x_1 + a_{22}x_2 = b_2,$$

where a_{ij}, b_i $(i, j = 1, 2)$ are any real numbers and the x_j are the unknowns. These can be put into *matrix notation* by defining:

the 2×2 matrix

$$\mathbf{A} = \begin{bmatrix} a_{11} & a_{12} \\ a_{21} & a_{22} \end{bmatrix};$$

the 2×1 (column) vector

$$\mathbf{x} = \begin{bmatrix} x_1 \\ x_2 \end{bmatrix};$$

the 2×1 (column) vector

$$\mathbf{b} = \begin{bmatrix} b_1 \\ b_2 \end{bmatrix};$$

and then writing the matrix equation

$$\mathbf{A}\mathbf{x} = \mathbf{b}.$$

This implicitly defines a rule for multiplication of matrices and vectors: the product $\mathbf{A}\mathbf{x}$ is the 2×1 vector

$$\begin{bmatrix} a_{11}x_1 + a_{12}x_2 \\ a_{21}x_1 + a_{22}x_2 \end{bmatrix},$$

and equality of two vectors means equality of their corresponding components.

We define the *inverse* matrix of $\mathbf{A}$ as the matrix $\mathbf{A}^{-1}$ which satisfies $\mathbf{A}^{-1}\mathbf{A} = \mathbf{A}\mathbf{A}^{-1} = \mathbf{I}$, where

$$\mathbf{I} = \begin{bmatrix} 1 & 0 \\ 0 & 1 \end{bmatrix}$$

is the *unit matrix* or the *identity matrix*. This is by analogy with ordinary numbers, where the inverse of a is a^{-1} and $aa^{-1} = a^{-1}a = 1$. $\mathbf{I}$ is the matrix counterpart of the number 1. Then we can solve the above matrix equation by multiplying through by $\mathbf{A}^{-1}$ to obtain

$$\mathbf{A}^{-1}\mathbf{A}\mathbf{x} = \mathbf{A}^{-1}\mathbf{b} \implies \mathbf{I}\mathbf{x} = \mathbf{A}^{-1}\mathbf{b} \implies \mathbf{x} = \mathbf{A}^{-1}\mathbf{b}.$$

Since $\mathbf{A}^{-1}$ and $\mathbf{b}$ contain given numbers, the right-hand side of the last equation gives a solution for the $\mathbf{x}$ vector.

The only *number* which does not have a reciprocal is 0, i.e. 0^{-1} is not defined. However, it turns out that many matrices may not possess an inverse, and so the above procedure can only be applied when the matrix $\mathbf{A}$ does in fact possess an inverse $\mathbf{A}^{-1}$. The problem is to compute this matrix.

There is a procedure called Cramer's Rule, which is often computationally easier than the above method of using an inverse matrix, and is equivalent to it. Moreover, when, as in most applications in this book, the system of linear equations we wish to solve contains only general mathematical expressions rather than numbers, Cramer's Rule provides the only sensible method of solution. We therefore now consider Cramer's Rule and the main concept it uses, that of the *determinant*. We define the determinant of the above matrix $\mathbf{A}$ as

$$|\mathbf{A}| \equiv \begin{vmatrix} a_{11} & a_{12} \\ a_{21} & a_{22} \end{vmatrix} = a_{11}a_{22} - a_{21}a_{12}.$$

That is, the determinant $|\mathbf{A}|$ is a number derived from the coefficients of the system by taking the product of the diagonal terms a_{11} and a_{22}, and subtracting the product of the other two diagonal terms a_{21} and a_{12}.

If $|\mathbf{A}| = 0$, the determinant is said to be *singular* and it can be shown that the equations possess no solution.

If $|\mathbf{A}| \neq 0$, the determinant is *non-singular* and we can always find unique solutions for x_1 and x_2. Cramer's Rule tells us that these are given by

$$x_1^* = \begin{vmatrix} b_1 & a_{12} \\ b_2 & a_{22} \end{vmatrix} / |\mathbf{A}| = \frac{b_1 a_{22} - b_2 a_{12}}{a_{11} a_{22} - a_{21} a_{12}},$$

$$x_2^* = \begin{vmatrix} a_{11} & b_1 \\ a_{21} & b_2 \end{vmatrix} / |\mathbf{A}| = \frac{b_2 a_{11} - b_1 a_{21}}{a_{11} a_{22} - a_{21} a_{12}}$$

(clearly if $|\mathbf{A}| = 0$ neither of these expressions is defined). To solve for x_j^* we replace the jth column in the determinant $|\mathbf{A}|$ with the column of constants $[b_1, b_2]$, and then take the ratio of this new determinant to the determinant $|\mathbf{A}|$. [Note that the notation $|\ |$ to denote a determinant has a completely different meaning to the use of $|\ |$ to denote absolute values; an unfortunate double use of notation but the context should usually make clear which meaning is intended.]

Suppose now that we have *three* simultaneous linear equations in three unknowns:

$$a_{11}x_1 + a_{12}x_2 + a_{13}x_3 = b_1,$$
$$a_{21}x_1 + a_{22}x_2 + a_{23}x_3 = b_2,$$
$$a_{31}x_1 + a_{32}x_2 + a_{33}x_3 = b_3.$$

The determinant of the system is now written

$$|\mathbf{A}| = \begin{vmatrix} a_{11} & a_{12} & a_{13} \\ a_{21} & a_{22} & a_{23} \\ a_{31} & a_{32} & a_{33} \end{vmatrix} = ?$$

and is again a single number. We have to show how it is derived.

Step 1: Choose any row or column (preferably the one with the most zeroes in it). Say we choose column 1. Then

Step 2:

$$|\mathbf{A}| = a_{11} \begin{vmatrix} a_{22} & a_{23} \\ a_{32} & a_{33} \end{vmatrix} - a_{21} \begin{vmatrix} a_{12} & a_{13} \\ a_{32} & a_{33} \end{vmatrix} + a_{31} \begin{vmatrix} a_{12} & a_{13} \\ a_{22} & a_{33} \end{vmatrix}$$

$$= a_{11}(a_{22}a_{33} - a_{32}a_{23}) - a_{21}(a_{12}a_{33} - a_{32}a_{13}) + a_{31}(a_{12}a_{33} - a_{22}a_{13}).$$

Careful comparison will show that this is an application of the general formula for expansion of a 3×3 determinant along column j:

$$|\mathbf{A}| = (-1)^{1+j}a_{1j}A_{1j} + (-1)^{2+j}a_{2j}A_{2j} + (-1)^{3+j}a_{3j}A_{3j} \qquad j = 1, 2, 3.$$

Expansion along row i gives

$$|\mathbf{A}| = (-1)^{i+1}a_{i1}A_{i1} + (-1)^{i+2}a_{i2}A_{i2} + (-1)^{i+3}a_{i3}A_{i3} \qquad i = 1, 2, 3.$$

The terms denoted A_{ij} are the 2×2 subdeterminants formed by deleting row i and column j from $|\mathbf{A}|$.

Whichever form of expansion is chosen results in exactly the same value for $|\mathbf{A}|$.

Again, if $|\mathbf{A}| = 0$ the equations have no solution, if $|\mathbf{A}| \neq 0$ they have a unique solution which, by Cramer's Rule, is given by

$$x_j^* = |\mathbf{B}_j|/|\mathbf{A}| \qquad j = 1, 2, 3$$

where $\mathbf{B}_j$ is the 3×3 determinant formed by replacing the jth column $[a_{1j}, a_{2j}, a_{3j}]$ by the column of constants $[b_1, b_2, b_3]$.

Review 2. Calculus

2.1 Functions

A function is a rule which, for every element in some set D, gives one and only one element in a set Y. If x is an element of D and y an element of Y then we may write the function as $y = f(x)$, where $f(.)$ denotes the rule for associating elements of the sets. D is called the *domain* of the function. The set of y-values in Y given by $y = f(x)$ is called the *range* of the function. The domain of a function should always be specified: a given rule $f(.)$ defined on different domains corresponds to different functions.

In this book, the domains of all functions are sets of real numbers, or sets of n-tuples of real numbers. The functions themselves are usually thought of as the standard functions of mathematical analysis, where the rules of association $f(.)$ take a straightforward algebraic form. Examples are:

rule	*domain*	*form*
$y = bx$	x a real number	linear
$y = a + bx$	x a real number	linear affine
$y = ax^b$	x a positive real number	power
$y = e^{f(x)}$	x a real number	exponential
$y = \log x$	x a positive real number	logarithmic
$y = a + bx + cx^2$	x a real number	quadratic
$y = a + bx + cx^2 + gx^3$	x a real number	cubic
$y = a + bx_1 + cx_2$	x_1, x_2 real numbers	linear affine
$y = ax_1^b x_2^c$	x_1, x_2 positive real numbers	'Cobb-Douglas'

where the parameters a, b, c, g are always real numbers.

The reader should feel confident about being able to sketch the graphs of these functions for particular parameter values.

In the case of functions of two (or more) variables, 3-dimensional graphs are cumbersome and the device of the *contour* or *level* is often used. Given the function $y = f(x_1, x_2)$, the value of the function is set equal to some constant, $\bar{y}$, and this implicitly defines a set of (x_1, x_2)-values which satisfy the *equation* $\bar{y} = f(x_1, x_2)$. This set, called the *contour set* or *level set* can then be graphed in 2 dimensions (in the case of functions of 3 or more variables all except two of the variables also have to be set equal to fixed values). For example,

(a) $\bar{y} = a + bx_1 + cx_2 \Rightarrow x_2 = (\bar{y} - a - bx_1)/c$, which defines a line in (x_1, x_2)-space.

(b) $\bar{y} = ax_1^b x_2^c \Rightarrow x_2 = (\bar{y}/a)^{1/c} x_1^{-b/c}$ which, for $\bar{y}$, a, b, c all positive defines a smooth, negatively sloped, convex-to-the-origin curve in (x_1, x_2)-space.

Changing the value of $\bar{y}$ in a specified way causes the contour to shift in a way which can be deduced from the parameter values. This way of obtaining an idea of the shape of a function has been of considerable historical importance in economics.

Of course, in economics, variables and functions are often given letters to suggest their economic meaning. For example prices are denoted by p, consumption c, income y; demand functions $D(.)$, supply functions $S(.)$, consumption functions $C(.)$, and so on.

2.2 Limits

2.2.1 Limit of a sequence

Given some infinite sequence $\{x_k\}_{k=1}^{\infty}$, this sequence has a limit L if, for any $\varepsilon > 0$, there exists a K (often depending on ε) such that for all $k > K$, $|x_k - L| < \varepsilon$. L, if it exists, may or may not be a term of the sequence. For example the sequence $x_k = 1/k$ has limit 0, which is not a term of the sequence. However, for suitably large k, the limit L, if it exists, will be a good approximation for the values of terms of the sequence.

Intuitively, the terms of the sequence converge more and more closely to the limit as k increases, when the limit exists. We write this formally as: $\lim_{k \to \infty} x_k = L$. A sequence which possesses a limit is called *convergent*. The sequence $x_k = ar^{k-1}$ is convergent if $|r| < 1$, since then $\lim_{k \to \infty} x_k = 0$. However, if $|r| > 1$ then $|r^{k-1}|$ becomes indefinitely larger as $k \to \infty$, and so we write $\lim_{k \to \infty} x_k = \pm \infty$ in this case (under what conditions would we have plus or minus infinity respectively?). Such a sequence is clearly *divergent*.

2.2.2 Limit of a function

Consider a function $f(x)$, where the domain is a set of real numbers. Choose some number a in this set and choose a sequence of x values in the domain which converge to a. For each term in the sequence there will be a value $f(x)$, and by definition of the function there is a value $f(a)$. Then, the question is: as the sequence of x-values tends to a, does the sequence of $f(x)$ values tend to $f(a)$? If the answer is yes, for *all* such x-sequences, then the function is said to be *continuous* at $x = a$. If for some x-sequences the $f(x)$ sequence does not have a limit, or tends to a limit other than $f(a)$, the function is discontinuous at a. For example:

(i) $y = bx$, x a real number, is continuous at all points in its domain.

(ii)

$$y = \begin{cases} 1/(1-x) & \text{for } x \neq 1 \\ 1 & \text{for } x = 1 \end{cases} \quad x \text{ a real number}$$

Since, for $x > 1$, $\lim_{x \to 1} 1/(1-x) = -\infty$, while for $x < 1$, $\lim_{x \to 1} 1/(1-x) = \infty$, the limit of the $f(x)$ sequence does not exist as $x \to 1$ and so the function has a discontinuity at this point.

(iii)

$$y = \begin{cases} 1, & \text{for } 0 \leq x \leq 1. \\ 2, & \text{for } x > 1. \end{cases}$$

Since, for $x > 1$, $\lim_{x \to 1} f(x) = 2 \neq f(1) = 1$, this function has a discontinuity at $x = 1$.

Why could we assert that the function in (i) is everywhere continuous? Let the sequence of x-values be defined by $a+h_k$, $k = 1, \ldots$ where h_k is any sequence such that $\lim_{k \to \infty} h_k = 0$. Then $f(x) = b(a+h_k) = b.a+bh_k = f(a)+bh_k$. Then $\lim_{x \to a} f(x) = \lim_{k \to \infty} [f(a)+bh_k] = f(a) + b \lim_{k \to \infty} h_k = f(a)$.

2.3 The derivative

Given a function $y = f(x)$ where x is a real number, by changing x from a particular value x_0 to $x_0 + h$ we change y to $f(x_0 + h)$, and we denote this difference $f(x_0 + h) - f(x_0)$ by Δy. The *average rate of change* of y per unit change in x is then

$$\frac{\Delta y}{\Delta x} = \frac{f(x_0 + h) - f(x_0)}{h}.$$

The problem with this expression as a measure of the rate at which y changes when x changes is that, for almost all functions, its value depends on the value of $h = \Delta x$.

Think now of a sequence of h values tending to zero. This will define a corresponding sequence of Δy values, also tending to zero, and a sequence of $\Delta y/\Delta x$ values. Consider then the limit

$$\lim_{\Delta x \to 0} \left[\frac{\Delta y}{\Delta x} \right] = \lim_{h \to 0} \frac{f(x_0 + h) - f(x_0)}{h}.$$

If this limit exists and is unique for any sequence of h-values it is called the *derivative* of $f(x)$ at x_0, and the function is said to be *differentiable* at x_0. We denote the derivative most commonly by dy/dx, $df(x)/dx$, or $f'(x)$.

An important geometric interpretation of the derivative at x_0 is that it is the slope of the tangent line drawn to the graph of the function at the point x_0. That is, the derivative measures the slope of a function at a point.

We can always find the general expression for the derivative of a function from the above definition. For example, given the function $y = x^2$, we have

$$\frac{\Delta y}{\Delta x} = \frac{(x+h)^2 - x^2}{h} = \frac{x^2 + 2hx + h^2 - x^2}{h} = 2x + h.$$

Then $\lim_{h \to 0}(2x + h) = 2x$ and so $d(x^2)/dx = 2x$.

However, it is far simpler to refer to a standard table. For this book, all we require are:

the *power function* rule: $d[ax^b]/dx = abx^{b-1}$, for any real numbers a, b (note the cases $b = 0$, and $b = 1$).

the *exponential function* rule: $d[e^{f(x)}]/dx = f'(x)e^{f(x)}$ (so, for $f(x) = ax$, this implies that $d[e^{ax}]/dx = ae^{ax}$, and for $a = 1$, that $d[e^x]/dx = e^x$, a very useful fact).

the *logarithmic function* rule: $d[\log x]/dx = 1/x$.

Also of great importance, especially in economics where often we deal with general functions, are the rules for differentiating combinations of functions. Given two differentiable functions $f(x)$ and $g(x)$ defined on the same domains,

$$d[f(x) \pm g(x)]/dx = f'(x) \pm g'(x);$$

$$d[f(x)g(x)]/dx = g(x)f'(x) + f(x)g'(x);$$

$$d[f(x)/g(x)]/dx = \frac{1}{[g(x)]^2}(g(x)f'(x) - f(x)g'(x)),$$

and also the very important 'function of a function' or 'chain' rule

$$\frac{df[g(x)]}{dx} = f'[g(x)]g'(x) = \frac{df}{dg}\frac{dg}{dx},$$

where f is a function of some variable g which is itself a function of x.

In general, the derivative of a function of x is also a function of x, and so may be differentiated to yield another function of x, which may be differentiated ... and so on. Thus we have

$$f(x); \quad f'(x) \quad \text{or} \quad df/dx; \quad f''(x) \quad \text{or} \quad d^2f/dx^2; \quad f'''(x) \quad \text{or} \quad d^3f/dx^3 \dots,$$

yielded by successive differentiation. In economics we often need the second derivative $f''(x)$, sometimes the third derivative $f'''(x)$, but very rarely anything of higher order than that.

The geometric interpretation of the second derivative is also important. Since it is the derivative of the first derivative, it measures the slope of the graph of the first derivative function at a point. If this slope is negative, this tells us that the first derivative – the slope of the function itself – is decreasing, and so the function must be concave in shape at that point. If the second derivative is positive, the first derivative, the slope of the function, is increasing, and so the function must be convex at that point. If the second derivative is zero, the slope of the function is neither increasing or decreasing. Fig. A1 illustrates these cases.

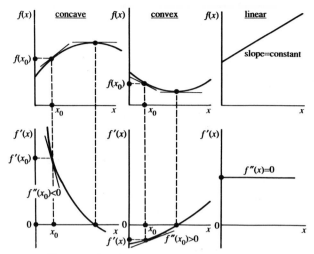

Fig. A1: First- and second-order derivatives

2.4 The integral

Given some function $f(x)$, let $F(x)$ denote a function such that $F'(x) = f(x)$. That is, when $F(x)$ is differentiated it yields $f(x)$. Then $F(x)$ is called the *anti-derivative* or the *indefinite integral* of $f(x)$. Since $d[F(x)]/dx = d[F(x) + C]/dx$ for any constant C, we write

$$\int f(x)dx = I(x) + C = F(x),$$

where $I(x)$ can be thought of as the specific function we find by reversing the process of differentiation and C is an arbitrary *constant of integration*. For example if $f(x) = ax^b$, then

$$\int f(x)dx = \frac{ax^{b+1}}{b+1} + C = F(x).$$

If $f(x) = e^x$, then

$$\int e^x dx = e^x + C = F(x).$$

The only rule we need in this book for integrating combinations of functions is

$$\int [f(x) \pm g(x)]dx = \int f(x)dx \pm \int g(x)dx,$$

while it is also useful to remember that

$$\int \frac{d}{dx}[f(x)]dx = f(x) + C.$$

The *definite integral* is

$$\int_a^b f(x)dx = [F(x)]_a^b = F(b) - F(a) = I(b) - I(a),$$

where a and b are the *limits of integration*, $f(x)$ must be continuous on the interval $[a, b]$, and $F(x)$ is any indefinite integral of $f(x)$. Thus

$$\int_a^b \alpha x^\beta \, dx = \left[\frac{\alpha x^{\beta+1}}{\beta+1} + C\right]_a^b = \frac{\alpha b^{\beta+1}}{\beta+1} + C - \left[\frac{\alpha a^{\beta+1}}{\beta+1} + C\right] = \frac{\alpha}{\beta+1}(b^{\beta+1} - a^{\beta+1}).$$

An important geometric interpretation of the definite integral is that it is the area under the graph of the function $f(x)$ between the values $x = a$ and $x = b$.

We can clearly think of the value of the definite integral $\int_a^b f(x)dx$ as a function of the limits of integration a, b. A very useful rule is then:

$$\frac{d}{da}\left[\int_a^b f(x)dx\right] = -f(a); \qquad \frac{d}{db}\left[\int_a^b f(x)dx\right] = f(b).$$

If $a < b < c$ then we can always write

$$\int_a^c f(x)dx = \int_a^b f(x)dx + \int_b^c f(x)dx.$$

Finally a small matter of notation. A limit of integration may sometimes be a variable in a problem. In that case it is notationally clearer to use a different symbol for the 'dummy variable of integration', as in

$$V(x) = \int_0^x f(\chi)d\chi, \qquad V'(x) = f(x);$$

$$B(q) = \int_0^q f(x)dx, \qquad B'(q) = f(q);$$

$$S(p) = \int_p^{p^0} f(x)dx, \qquad S'(p) = -f(p).$$

2.5 Partial differentiation

We wish to extend the concept of the derivative to cases in which y is a function of more than one variable. We begin with the case $y = f(x_1, x_2)$. If we take x_2 as fixed at some value x_2^0, then we can regard y as a function of x_1 alone. Then we can define, for some given x_1-value x_1^0,

$$\frac{\Delta y}{\Delta x_1} = \frac{f(x_1^0 + h_1, x_2^0) - f(x_1^0, x_2^0)}{h_1}$$

as the average rate of change of y from the point (x_1^0, x_2^0) with respect to changes in x_1, and the *partial derivative* with respect to x_1 is

$$\lim_{h_1 \to 0} \frac{\Delta y}{\Delta x_1} = \lim_{h_1 \to 0} \frac{f(x_1^0 + h_1, x_2^0) - f(x_1^0, x_2^0)}{h_1}.$$

This limit is commonly denoted by $\partial y / \partial x_1$, $\partial f / \partial x_1$, or $f_1(x_1, x_2)$. In a similar way we can define the partial derivative with respect to x_2.

For example let $y = x_1^2 + 2x_1 x_2 + x_2^2$. Then

$$\frac{\Delta y}{\Delta x_1} = \frac{(x_1 + h_1)^2 + 2(x_1 + h_1)x_2 + x_2^2 - (x_1^2 + 2x_1 x_2 + x_2^2)}{h_1}$$

$$= \frac{x_1^2 + 2h_1 x_1 + h_1^2 + 2x_1 x_2 + 2h_1 x_2 + x_2^2 - x_1^2 - 2x_1 x_2 - x_2^2}{h_1}$$

$$= 2x_1 + h_1 + 2x_2;$$

therefore

$$\lim_{h_1 \to 0} \frac{\Delta y}{\Delta x_1} = 2(x_1 + x_2) = \partial y / \partial x_1 = f_1(x_1, x_2).$$

Now show that $\partial y / \partial x_2 = 2(x_1 + x_2)$ also. This example makes clear the trick: we simply treat all variables but one as constants and differentiate the function as if it were a function of just that variable alone. We do this with respect to each variable in turn, so that we obtain as many first-order partial derivatives as there are variables, in this case two.

An important example which occurs repeatedly in this book is the function $y = ax_1^b x_2^c$. Applying the above rule gives

$$\frac{\partial y}{\partial x_1} = abx_1^{b-1} x_2^c, \qquad \frac{\partial y}{\partial x_2} = acx_1^b x_2^{c-1}$$

as a simple application of the rule for differentiating power functions.

The geometric interpretation, for a function of two variables, is as follows. The graph of the function is a surface in 3 dimensions. When we fix a value of x_2, we are taking a cross-sectional slice through this surface, perpendicular to the x_2-axis, parallel to the x_1-axis. This defines a graph which is in two dimensions, a curve relating y to x_1 alone. The partial derivative at (x_1^0, x_2^0) is then the slope of the tangent to this curve at the point (x_1^0, x_2^0). It is therefore taking the slope of the surface at a particular point and *in a particular direction*, viz. perpendicular to the x_2-axis and parallel to the x_1-axis. The partial derivative at (x_1^0, x_2^0) with respect to x_2 is taking the slope of the surface in a direction perpendicular to the x_1-axis and parallel to the x_2-axis.

Differentiating $f(x_1, x_2)$ with respect to x_1 and x_2 will yield two functions of x_1 and x_2 which can again be partially differentiated to give four second-order partial derivatives:

$$\frac{\partial^2 y}{\partial x_1^2}, \ \frac{\partial^2 y}{\partial x_1 \partial x_2}, \ \frac{\partial^2 y}{\partial x_2 \partial x_1}, \ \frac{\partial^2 y}{\partial x_2^2} \quad \text{or} \quad f_{11}(x_1, x_2), \ f_{12}(x_1, x_2), \ f_{21}(x_1, x_2), \ f_{22}(x_1, x_2).$$

For example, the Cobb-Douglas function above gives:

$$\frac{\partial^2 y}{\partial x_1^2} = ab(b-1)x_1^{b-2}x_2^c, \qquad \frac{\partial^2 y}{\partial x_2 \partial x_1} = abcx_1^{b-1}x_2^{c-1},$$

$$\frac{\partial^2 y}{\partial x_1 \partial x_2} = abcx_1^{b-1}x_2^{c-1}, \qquad \frac{\partial^2 y}{\partial x_2^2} = ac(c-1)x_1^b x_2^{c-2}.$$

The fact that $\partial^2 y/\partial x_2 \partial x_1 = \partial^2 y/\partial x_1 \partial x_2$ is not peculiar to this example. An important theorem in calculus, Young's Theorem, tells us that this is always true for a function with continuous second-order partial derivatives.

The second order partials $\partial^2 y/\partial x_i^2$, $i = 1, 2$, give the slope of the corresponding first-order partial with respect to x_i. For example, if $\partial^2 y/\partial x_1^2 < 0$, this tells us that, with x_2 fixed, the function $\partial y/\partial x_1$ is falling as x_1 increases. The second-order cross-partials $\partial^2 y/\partial x_i \partial x_j$, $i, j = 1, 2$, $i \neq j$, tell us the effect on the partial $\partial y/\partial x_j$ of a change in the other variable x_i, with x_j held constant. For example, if $f(x_1, x_2)$ is a production function with x_1 labour and x_2 capital, then $\partial^2 f/\partial x_2 \partial x_1$ gives the effect on the marginal product of labour, $\partial f/\partial x_1$, of a change in the amount of capital that labour has available to work with.

2.6 Total differentiation

Given a function $f(x)$, where x is a real number and the function is differentiable at a point x_0, the *differential* of the function at x_0 is defined as the linear equation

$$dy = f'(x_0)dx,$$

where dy and dx are thought of as variables. As we vary dx, so dy varies in such a way that $dy/dx = f'(x_0)$. The graph of the differential is the tangent line to the graph of the function at x_0.

An important application of the differential is to obtain an approximation to the change in y which follows from changing x. If, as before, we define $\Delta x = (x_0 + h) - x_0$, $\Delta y = f(x_0 + h) - f(x_0)$, then in general it is *not* true that $\Delta y = f'(x_0)\Delta x$. But if Δx is taken to be suitably small and denoted by dx, then $dy = f'(x_0)dx$ may be accepted as a good enough approximation for Δy.

Given a function of two variables, $y = f(x_1, x_2)$, the differential of the function at the point (x_1^0, x_2^0) is the linear equation

$$dy = \frac{\partial y}{\partial x_1}dx_1 + \frac{\partial y}{\partial x_2}dx_2 = f_1(x_1^0, x_2^0)dx_1 + f_2(x_1^0, x_2^0)dx_2.$$

Again, dy, dx_1, and dx_2 are thought of as variables, while $f_1(x_1^0, x_2^0)$ and $f_2(x_1^0, x_2^0)$ are numbers, given by the values taken by the partial derivatives at (x_1^0, x_2^0). Again, an important application of the differential in this case is that it is a close approximation to the value

$$\Delta y = f(x_1^0 + h_1, x_2^0 + h_2) - f(x_1^0, x_2^0)$$

when the changes in x_1 and x_2, h_1 and h_2, are suitably small.

A second important application of the differential is to *contour differentiation*. Suppose that we have defined a contour of the function $y = f(x_1, x_2)$ by setting $y = \bar{y}$. We now want to change x_1 and x_2 in such a way as to leave the value of y unchanged at $\bar{y}$, that is, we set

$$dy = f_1(x_1, x_2)dx_1 + f_2(x_1, x_2)dx_2 = 0.$$

This tells us that the values of the variables dx_1 and dx_2 must satisfy

$$\frac{dx_2}{dx_1} = -\frac{f_1(x_1, x_2)}{f_2(x_1, x_2)},$$

where again f_1 and f_2 are thought of as being evaluated at a specific point. Recall that we could regard the contour as defining x_2 as a function of x_1, say $x_2 = \Phi(x_1)$. Then the derivative of this function at (x_1^0, x_2^0) is $\Phi'(x_1^0) = dx_2/dx_1 = -f_1/f_2$. Thus the slope of a tangent to the graph of the contour at (x_1^0, x_2^0) is given in terms of the partial derivatives of the function f at that point. This idea has received very wide application in economics.

Also important is the idea of *implicit differentiation*. Suppose we have some function $f(x_1, x_2)$, where x_1 and x_2 must be chosen to satisfy

$$f(x_1, x_2) = 0.$$

This is known as an implicit function, since implicitly it gives x_2 as a function of x_1. Alternatively we could think of it as defining a contour with $\bar{y} = 0$. Now suppose we wish to change x_1 and x_2 in such a way that the value of the function continues to equal zero. Then, just as before, dx_1 and dx_2 must satisfy

$$f_1 dx_1 + f_2 dx_2 = 0 \implies dx_2/dx_1 = -f_1/f_2.$$

This is known as implicit differentiation, and contour differentiation clearly generalizes it.

The extension of differentials to the case of functions of more than two variables is immediate. If $y = f(x_1, \ldots, x_n)$, then $dy = \sum_{i=1}^{n} f_i(x_1^0, \ldots, x_n^0)dx_i$ at the point $(x_1^0, \ldots, x_n^0)$. Contour and implicit differentiation, as well as the interpretation of dy as an approximation to Δy for suitably small changes in the x_i, are then straightforward extensions of the two-variable case.

2.7 The 'dot' and the 'hat' calculus

Consider a continuous and differentiable function of time $x(t)$. Many economic variables may be written as depending on time – prices change over time, as does investment, money supply, and so on. The derivative of x with respect to time is usually written

$$\dot{x}(t) \equiv \frac{dx}{dt}.$$

We can also introduce the idea of the *proportional change* in x over time or the *growth rate* of x as

$$\hat{x}(t) \equiv \frac{1}{x} \frac{dx}{dt}.$$

The following rules are easily derived:

$$z(t) = x(t) + y(t), \qquad \hat{z}(t) = \frac{\dot{x}(t) + \dot{y}(t)}{x(t) + y(t)};$$

$$z(t) = x(t)y(t), \qquad \hat{z}(t) = \hat{x}(t) + \hat{y}(t).$$

Review 3. Maxima and minima

3.1 Unconstrained maxima and minima of functions of one variable

A differentiable function $f(x)$, where x is a real number, achieves a *local maximum* at a point x_0 if

$$f'(x_0) = 0; \quad f''(x_0) < 0.$$

It achieves a *local minimum* at x' if

$$f'(x') = 0; \quad f''(x') > 0.$$

To see this in the case of a maximum, take the differential to the function at x_0,

$$dy = f'(x_0)dx.$$

If $f'(x) \neq 0$, then a (positive or negative) dx could always be found to make $dy > 0$, i.e. to increase the value of the function. If $f(x_0)$ is locally a maximum this cannot be done, implying $f'(x_0) = 0$. A similar argument applies in the case of a minimum.

We also need a condition on the second derivative of the function because the condition on the first derivative does not distinguish between maxima, minima, and a third class of points, points of inflexion, at which a function changes shape from concave to convex (or conversely) and so also has a zero derivative. The condition $f''(x_0) < 0$ says that the slope of the function is decreasing through x_0, which implies it is positive to the left of x_0 and negative to the right. In other words the function describes a 'hill' around x_0 with its peak at x_0. Similarly the condition $f''(x') > 0$ says that the slope of the function is increasing through x', from negative to positive, and so the function describes a 'valley' through x' with its lowest point at x'.

To find a maximum or minimum of a function requires us to differentiate the function, set the derivative equal to zero, and solve the resulting equation, and then, by plugging the solution value into the expression for the second derivative, check that we really do have a maximum or a minimum. There may in fact be: no solution to the first step, e.g. the function $y = a + bx$, $b \neq 0$, has neither a maximum nor minimum; or several solutions, in which case these must be compared and that one of them found which yields the highest or lowest function value.

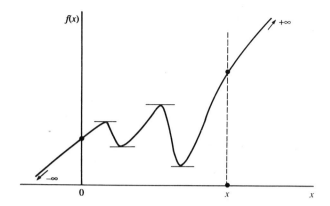

Fig. A2: Local and global maxima and minima

However, it must be emphasized that this calculus-based method finds only *local* maxima or minima, that is, points which yield highest or lowest values of a function relative to small intervals of *x*-values around themselves. Fig. A2 illustrates this. Values of $f(x)$ exist which are higher than its values at the two local maxima, and lower than its values at the two local minima. So, since we are really interested in *global* maxima and minima, i.e. points which are optimal over the *entire domain* of the function, the procedure we have just described does not solve the problem in this case. Indeed nothing can: the function illustrated in the figure has neither a global maximum nor a global minimum.

In economics this type of problem is solved in two ways. We may make assumptions which restrict the form the function can take, so that any local maximum or minimum is also global. Also in many economic problems it is natural to place constraints on the values of *x* over which the function can be maximized or minimized, and these will usually have the effect of ensuring that a global maximum or minimum exists (for example, suppose in the figure that the values of *x* we can consider must lie between 0 and $\bar{x}$). We consider the question of constraints further in section 3.3 below.

3.2 Unconstrained maxima and minima of functions of more than one variable

A differentiable function $f(x_1, \ldots, x_n)$ achieves a local maximum at a point $(x_1^0, \ldots, x_n^0)$ only if

$$f_1(x_1^0, \ldots, x_n^0) = 0,$$

$$f_2(x_1^0, \ldots, x_n^0) = 0,$$

$$\cdots\cdots\cdots\cdots\cdots$$

$$f_n(x_1^0, \ldots, x_n^0) = 0,$$

that is, only if all its partial derivatives are simultaneously zero at that point. To see this, take the differential of the function at $(x_1^0, \ldots, x_n^0)$,

$$dy = f_1(x_1^0, \ldots, x_n^0)dx_1 + \ldots + f_n(x_1^0, \ldots, x_n^0)dx_n.$$

If any partial $f_i \neq 0$, then it will be possible to find dx_i such that $dy > 0$, and so the function cannot be at a maximum at that point. If dy cannot be made positive for any dx_i whatever, this implies that each $f_i = 0$.

A similar argument leads to the conclusion that the function is minimized at $(x_1', \ldots, x_n')$ only if each partial $f_i(x_1', \ldots, x_n') = 0$.

So, as in the case of functions of a single variable, the 'first-order conditions' do not distinguish between maxima and minima. It is possible to give conditions in terms of the second-order partial derivatives of the functions, which allow us to distinguish between the two cases, but in this book we make no use of them and so shall not take the space to develop them. Essentially, whenever we solve a maximization or minimization problem in two or more variables, we simply assume that the relevant second-order conditions are satisfied.

Note that the conditions give n equations in n unknowns. In general, depending on the shape of the function f, we may have no solutions, a unique solution, or several solutions to these equations. The discussion of local vs. global maxima and minima given in the previous section applies directly here. When we solve an unconstrained maximization problem in economics, we usually build in assumptions to ensure that a local maximum will exist and will also be a global maximum.

3.3 Constrained maxima and minima

Refer again to Fig. A2, and assume we wish to solve the problem of maximizing the function $f(x)$ *subject to the constraint* $0 \leq x \leq \bar{x}$. Then clearly the solution is at $x = \bar{x}$. But at $\bar{x}$, $f'(\bar{x}) > 0$. So, in this constrained problem, it is clearly no longer a necessary condition for a solution that $f'(x) = 0$. If it were possible to choose $dx > 0$, then at $\bar{x}$ we could have $dy > 0$, but the constraint rules out such a dx. For all feasible dx, i.e. $dx \leq 0$, the value of the function cannot be increased at $x = \bar{x}$. Thus this point is a local maximum and, by inspection of the figure, also a global maximum.

This example then tells us that the method we used for unconstrained problems cannot be expected to carry over to constrained problems – we have to start from scratch. We now do this in the context of a 2-variable 1-constraint maximization problem:

$$\max f(x_1, x_2) \quad \text{s.t.} \ g(x_1, x_2) = b,$$

where g is the *constraint function* and b is the *constraint constant* (and s.t. is read 'subject to'). Essentially, the problem restricts the set of (x_1, x_2)-pairs over which we are maximizing the function f to the points lying along a contour of the function g.

Throughout this book we use the *Lagrange multiplier* procedure to solve this type of problem. This procedure, applied to the above problem, takes the following steps:

1. Form a *Lagrange function*, by introducing a *Lagrange multiplier* λ, and adding $\lambda[b - g(x_1, x_2)]$ to $f(x_1, x_2)$, i.e.

$$L(x_1, x_2, \lambda) = f(x_1, x_2) + \lambda[b - g(x_1, x_2)].$$

2. Solve the problem: $\max_{x_1, x_2, \lambda} L(x_1, x_2, \lambda)$, which, note, is an *unconstrained* maximization problem, to obtain the first-order conditions:

$$\frac{\partial L}{\partial x_1} = f_1(x_1, x_2) - \lambda g_1(x_1, x_2) = 0,$$

$$\frac{\partial L}{\partial x_2} = f_2(x_1, x_2) - \lambda g_2(x_1, x_2) = 0,$$

$$\frac{\partial L}{\partial \lambda} = b - g(x_1, x_2) = 0.$$

Then it can be shown that the values x_1 and x_2 which satisfy these equations are, on certain conditions (which we always assume hold in economic problems), precisely the values which solve the original constrained problem. A proof of this proposition requires mathematical ideas well in advance of the level of this book, but we shall try to give an intuitive idea of the procedure with the help of our example.

We assume it is possible to solve the constraint $g(x_1, x_2) = b$ for x_2 in terms of x_1, and we write

$$x_2 = \gamma(x_1), \quad \frac{dx_2}{dx_1} = \gamma'(x_1) = -\frac{g_1(x_1, x_2)}{g_2(x_1, x_2)}.$$

For the explanation of the derivative γ', recall the discussion of contour differentiation in section 2.6 of this appendix. Substituting for x_2 in the function f now gives an unconstrained problem in x_1 alone,

$$\max_{x_1} f(x_1, \gamma(x_1)).$$

In effect, we have ensured that the solution to the problem *must* lie along the constraint contour. Then the first-order condition for a solution is

$$\frac{df}{dx_1} = f_1 + f_2 \gamma'(x_1) = 0 \quad \text{(using the function of a function rule)}.$$

Rearranging this condition and using the above expression for γ' gives

$$-f_1(x_1, x_2)/f_2(x_1, x_2) = -g_1(x_1, x_2)/g_2(x_1, x_2) \quad (= \gamma').$$

This is, however, one equation in the two unknowns x_1 and x_2. To solve for the optimal values of these variables we need a further equation, which is given by the constraint $g(x_1, x_2) = b$. Using this and the condition $f_1/f_2 = g_1/g_2$ we can obtain the solution to the problem.

How does this translate into the Lagrange procedure? Note that if we take the first two conditions derived under the Lagrange multiplier procedure, take the terms in λ to the right-hand side, and then eliminate λ by taking ratios we obtain precisely the condition $f_1(x_1, x_2)/f_2(x_1, x_2) = g_1(x_1, x_2)/g_2(x_1, x_2)$. Thus the three conditions derived from the Lagrange procedure reduce to the two conditions which, we saw by substitution, solve the problem! The two approaches are completely equivalent.

If that is the case, why then use the Lagrange procedure? The answer is that when we take problems with any number n of variables and m ($< n$, to avoid possible problems of over-determinacy) constraints, the Lagrange procedure generalizes neatly and simply. We just form the Lagrange function

$$L(x_1, \ldots, x_n, \lambda_1, \ldots, \lambda_m) = f(x_1, \ldots, x_n) + \sum_{j=1}^{m} \lambda_j(b_j - g^j(x_1, \ldots, x_n)),$$

where g^j is the jth constraint function, and write the necessary conditions for the solution as

$$\frac{\partial L}{\partial x_i} = f_i - \sum_{j=1}^{m} \lambda_j g_i^j = 0 \qquad i = 1, \ldots, n,$$

$$\frac{\partial L}{\partial \lambda_j} = b_j - g^j = 0 \qquad j = 1, \ldots, m.$$

This usually allows a clearer and more perceptive analysis than if we used the m constraints to solve for m variables in terms of the remaining $n - m$ variables, substituted into the function f and solved as an unconstrained problem.

Two further reasons. The Lagrange multipliers, the λ_j, turn out to have very interesting interpretations, as shadow prices or marginal valuations of the b_j in the constraints with which they are associated. Secondly, the Lagrange procedure generalizes readily to cases in which the constraints take other forms, namely direct restrictions on the variables such as $x_i \geq 0$, or inequality constraints such as $g^j(x_1, \ldots, x_n) \leq b_j$. However, we do not develop these points in this book and this is left as a subject for further study.

Finally, note that nothing in the above discussion would have changed if the problem had been to *minimize f* subject to the constraint. First-order conditions are the same. We finesse the question of second-order conditions by assuming they always are what we need them to be.

Review 4. First-order linear difference and differential equations

4.1 First-order linear difference equations

Let time be divided into equal, discrete periods denoted by $t = 0, \ldots, T$, where T is possibly infinite. A variable x_t takes values at each t, and this sequence of values is related according to the function

$$x_t = a + b x_{t-1} \qquad t = 1, \ldots, T,$$

where a and b are real numbers. This is clearly a linear equation, and is called first-order because the value x_t does not depend on a value of x more than one period away. A *solution* of this equation is an expression which gives x_t in terms of parameters alone. To solve the equation we need one *initial condition*, which is often a given value of x_0; we take that case here. Then the solution can easily be derived as follows:

$$x_1 = a + bx_0,$$
$$x_2 = a + bx_1 = a + ab + b^2 x_0 \quad \text{(by substitution)},$$
$$x_3 = a + bx_2 = a + ab + ab^2 + b^3 x_0,$$

and so by induction

$$x_t = a + ab + ab^2 + \ldots + ab^{t-1} + b^t x_0 = a(1 - b^t)/(1 - b) + b^t x_0,$$

using the expression for the sum of a geometric series with $b \neq 1$. If $b = 1$, then

$$x_t = at + x_0.$$

Assume $b \neq 1$, and rearrange the solution for x_t to obtain

$$x_t = a/(1 - b) + (x_0 - a/(1 - b))b^t.$$

If T is infinite, then we have three cases.
(i) $|b| < 1$. Then $\lim_{t \to \infty} x_t = a/(1 - b)$, and x_t converges to this 'long-run equilibrium' or 'steady-state' value.
(ii) $|b| > 1$. Then $\lim_{t \to \infty} x_t = \pm\infty$, and no long-run equilibrium exists.
(iii) $b = -1$. Then x_t oscillates indefinitely between the values x_0 and $a - x_0$, and again no long-run equilibrium exists (to see this, set t alternately even and odd in the above expression for x_t, with $b = -1$).

Finally, if $b = 1$ then clearly $\lim_{t \to \infty} x_t = \infty$.

4.2 First-order linear differential equations

Now let t be a real variable, with $t \geq 0$. We now regard the value of the variable x as a differentiable function of t, $x(t)$. A linear first-order differential equation (with constant coefficients) gives a relationship between $x(t)$ and its derivative $x'(t)$ of the form

$$x'(t) + ax(t) = b,$$

where a and b are real numbers, and we assume $a \neq 0$. A *general solution* for this equation is a function $x(t)$ which satisfies the equation as an identity, for all t. The general solution will, as we shall see, contain an undetermined constant. A *particular solution* is found as the function which also satisfies an initial condition, often on the value $x(0)$. We shall take that case here.

To find the general solution, multiply through the above equation by the *integrating factor* e^{at}:

$$e^{at}(x'(t) + ax(t)) = e^{at}b.$$

We then note that the left-hand side of this equation is simply the derivative $d[e^{at}x(t)]/dt$, using the product rule of differentiation and recalling that $de^{at}/dt = ae^{at}$. Therefore integrating through the above equation gives

$$\int e^{at}(x'(t) + ax(t))dt = \int \frac{d}{dt}[e^{at}x(t)]dt = e^{at}x(t) = \int e^{at}b\,dt = c + e^{at}b/a,$$

where c is a constant of integration. This gives the general solution

$$x(t) = ce^{-at} + b/a.$$

To check, note that $x'(t) = -ace^{-at}$, and plugging this and $x(t)$ back into the original differential equation satisfies it identically.

To determine c, let $x(0) = x_0$, some given value, so that

$$x(0) = b/a + c = x_0 \implies c = x_0 - b/a.$$

Thus the particular solution with initial condition $x(0) = x_0$ is

$$x(t) = b/a + [x_0 - b/a]e^{-at}$$

As $t \to \infty$, there are two possibilities.

(i) $a > 0$. Then, since $\lim_{t\to\infty} e^{-at} = 0$ in that case, $\lim_{t\to\infty} x(t) = b/a$, and this is the long-run equilibrium or steady-state value of x.

(ii) $a < 0$. Then $\lim_{t\to\infty} e^{-at} = \infty$, and $\lim_{t\to\infty} x(t) = \pm\infty$, depending on the sign of $[x_0 - b/a]$, so there is no long-run equilibrium.

If $a = 0$, then the differential equation is $x'(t) = b$, which can be integrated directly to give $x(t) = bt + c$ with $c = x_0$, and again there is no long-run equilibrium.

Appendix 2: Answers to Selected Questions

Chapter 1

Exercise 1.1

1. The simplest 'neo-classical' model of the problem is as follows:

   endogenous variables: x_1, x_2 are the amounts of land devoted to each crop; y_1 and y_2 are outputs of the crops.

   exogenous variables: x, the total amount of land owned by the farmer; p_1 and p_2, price per unit of each crop.

   functions and equations: $y_1 = f_1(x_1)$, $y_2 = f_2(x_2)$, with $f_i' > 0$, $f_i'' < 0$, $i = 1, 2$, are the functions giving outputs as functions of the land devoted to each crop;

   $R_1 = p_1 y_1$, $R_2 = p_2 y_2$, give the farmer's revenues as functions of the outputs he sells;

   $\pi = R_1 + R_2 - F = p_1 y_1 + p_2 y_2 - F = p_1 f_1(x_1) + p_2 f_2(x_2) - F$ is the farmer's profit, with F the rent he pays for the land, which is a fixed cost independent of how the land is cultivated;

   $x_1 + x_2 = x$ expresses the condition that land used must equal what is available. (Why is this an equation and not an identity?)

   solution principle: assume that the farmer chooses x_1 and x_2 to maximize profit, i.e., he solves the problem

   $$\max_{x_1, x_2} \; p_1 f_1(x_1) + p_2 f_2(x_2) - F \qquad \text{s.t. } x_1 + x_2 = x.$$

   the solution: optimal x_1^*, x_2^* must satisfy the conditions

   $$p_1 f_1'(x_1^*) = \lambda,$$
   $$p_2 f_2'(x_2^*) = \lambda,$$
   $$x_1^* + x_2^* = x,$$

   where λ is the Lagrange multiplier associated with the constraint (note that F is irrelevant to the solution).

   interpretation: the essential feature of the solution is that land is optimally allocated so as to equalize the *marginal revenue products* of the crops, $p_i f_i'$, since each is equal to λ. (Give an intuitive explanation of this result in terms a farmer would understand. Try to work out a diagrammatic illustration of this solution to the problem.)

Chapter 2

Exercise 2.1

1. *minimum* of q_S and q_D.

2.
    (i)   out of equilibrium;
    (ii)   both in and out of equilibrium;
    (iii) out of equilibrium;
    (iv) only in equilibrium.

4.
    (*a*) an equilibrium may not exist;
    (*b*) if the market is out of equilibrium it may not move back to an equilibrium position.

5.
    (*a*) $p^* = 33.\dot{3}$, $q^* = 133.\dot{3}$;
    (*b*) $p^* = 2.08 = q^*$;
    (*c*) $p^* = 20$, $q^* = 30$;
    (*d*) $p^* = 2.372$, $q^* = 14.372$.

6.

There is a discontinuity in the supply function, through which the demand curve passes, so there is no intersection and no equilibrium. At any price below \$5 each seller finds it unprofitable to produce and so shuts down production. At a price of \$5 and over every seller wants to sell a positive quantity.

Exercise 2.2

1.
$$dp^*/dy = f/(b + \beta); \quad dp^*/dp_1 = -c/(b + \beta); \quad dp^*/dp_2 = e/(b + \beta);$$
$$dp^*/dw_1 = \gamma/(b + \beta); \quad dq_D^*/dy = \beta f/(b + \beta); \quad dq_D^*/dp_1 = -\beta c/(b + \beta);$$
$$dq_D^*/dp_2 = \beta e/(b + \beta); \quad dq_S^*/dw_1 = -b\gamma/(b + \beta).$$

2. $dp^*/dy = c/(b + \beta) > 0; \quad dq^*/dy = \beta c/(b + \beta) > 0.$

3.
    (i)   Coal and oil are substitutes, if p^*, q^* are equilibrium price and quantity of coal and p_2 is the price of oil then $dp^*/dp_2 > 0$, $dq^*/dp_2 = (dq_S^*/dp^*)(dp^*/dp_2) > 0$.
    (ii)   If convenience foods are normal goods $dp^*/dy > 0$, $dq^*/dy > 0$.

(iii) Computers and computer games are complements, so if p^*, q^* are equilibrium price and quantity of computer games and p_1 is the price of computers, $dp^*/dp_1 < 0$, $dq^*/dp_1 < 0$.

(iv) θ denotes weather conditions and let supply increase with θ. Then

$$dp^*/d\theta = -(\partial S/\partial\theta)/(\partial S/\partial p - \partial D/\partial p) < 0$$

and

$$dq^*/d\theta = (dq_D^*/dp)(dp^*/d\theta) > 0.$$

Exercise 2.3

3. This rule says that the quantity traded adjusts at a speed proportional to the difference between the price buyers want to pay and the price sellers wish to receive. Substitution gives

$$\frac{dq}{dt} = \gamma\left(\frac{a\beta + \alpha b}{b\beta}\right) - \gamma\left(\frac{b+\beta}{b\beta}\right)q(t) = A - Bq(t),$$

therefore, $q(t) = \frac{A}{B} + Ce^{-Bt} = q^* + Ce^{-Bt}$ where $q^* = (a\beta + \alpha b)/(b+\beta)$ is equilibrium quantity traded. Therefore, we have stability if $Ce^{-Bt} \to 0$, i.e. if $B > 0$. Since $B \equiv \gamma(b+\beta)/b\beta$:

(i) $b > 0$, $\beta > 0$ then $B > 0$, stability.
(ii) $b < 0$, $\beta > 0$, then if $|\beta| < |b|$, stability, otherwise not.
(iii) $b < 0$, $\beta < 0$, then $B < 0$, instability.
(iv) $b > 0$, $\beta < 0$, then if $|\beta| > |b|$, stability, otherwise not.

Chapter 3

Exercise 3.1

1. If $q_D = a - bp$, $e_p = bp/D(p)$. Therefore $p = 0 \Rightarrow e_p = 0$. As $p \to a/b$, $D(p) \to 0$ and so $e_p \to \infty$ (note: at $p = a/b$, e_p is undefined).

2. Engel curve slopes positively for normal good, negatively for inferior good. $e_y > 0$ is first case. $e_y < 0$ is second. Case in which Engel curve has zero slope, $e_y = 0$, would usually be included as 'normal'.

3. $\log q_D = \log a - b_1 \log p + b_2 \log y - b_3 \log p_1 + b_4 \log p_2$, therefore

$$e_p = b_1, e_y = b_2, e_{p_1} = -b_3, e_{p_2} = b_4.$$

For total expenditure to be constant we require $\partial E/\partial p = D(p)(1 - e_p) = 0$, i.e., $e_p = 1$. Thus we must have $b_1 = 1$. In this case the demand curve is a rectangular hyperbola.

4. $E = p.q_D = ap - bp^2 = \alpha q_D - \beta q_D^2$. Therefore $dE/dp = a - 2bp$ and $dE/dq_D = \alpha - 2\beta q_D$. At a maximum of E, $dE/dp = 0$ and this $\Rightarrow p = a/2b$. In that case:

$$e_p = bp/D(p) = b(a/2b)/(a - ab/2b) = 1.$$

5.

    (a) $e_p = 1$, then $dE/dp = dE/dq_D = 0$;
    (b) $e_p > 1$, then $dE/dp < 0$, $dE/dq > 0$;
    (c) $e_p < 1$, $dE/dp > 0$, $dE/dq < 0$.

Exercise 3.2

1.

$$T = tq = \left(\frac{a\beta + \alpha b}{b + \beta}\right)t - \frac{b\beta t^2}{b + \beta},$$

therefore

$$\frac{dT}{dt} = \frac{a\beta + \alpha b}{b + \beta} - \frac{2b\beta t}{b + \beta} = 0 \Rightarrow t^* = \frac{a\beta + \alpha b}{2b\beta}.$$

2.

$$D(p) = a - bp, \quad S(x) = \alpha + \beta p/(1 + v),$$

therefore

$$a - bp = \alpha + \beta p/(1 + v) \Rightarrow$$

  (a) $\hat{p} = \dfrac{(a - \alpha)}{b + \beta/(1 + v)} > \dfrac{a - \alpha}{b + \beta} = p^*$ since $v > 0$.

  (b) $\hat{x} = \dfrac{\hat{p}}{1 + v} = \dfrac{(a - \alpha)}{b(1 + v) + \beta} < x^* = p^*$.

  (e) $\hat{q} = \alpha + \dfrac{\beta(a - \alpha)}{b(1 + v) + \beta} < \alpha + \dfrac{\beta(a - \alpha)}{b + \beta} = q^*$.

  (f) $T = v\hat{x}\hat{q} = v\left(\dfrac{\alpha - a}{b(1 + v) + \beta}\right)\left(\alpha + \dfrac{\beta(a - \alpha)}{b(1 + v) + \beta}\right)$.

  (g) from (b) and (e) we have $\hat{x}\hat{q} < x^*q^*$.

4. Since $\hat{p} = p^* + \beta t/(b + \beta)$, we require $\beta/(b + \beta) < 0$. This is the case if: $\beta < 0, b > 0$ and $b > |\beta|$; or $b < 0, \beta > 0$ and $|b| > \beta$.

5. $E = p^*q^*$. Therefore, using 'function of a function' rule:

$$\frac{dE}{dt} = q^*\frac{dp^*}{dt} + p^*\frac{dq^*}{dp^*}\frac{dp^*}{dt} = q^*\left(1 + \frac{p^*}{q^*}\frac{dq^*}{dp^*}\right)\frac{dp^*}{dt} = q^*(1 - e_p)\frac{dp^*}{dt}.$$

Since $q^* > 0, dt > 0$, and $dp^*/dt > 0$, E increases if $e_p < 1$, falls if $e_p > 1$, and remains constant if $e_p = 1$.

Chapter 4

Exercise 4.1

2.

$$D_1(p_1, p_2, y) = S_1(p_1, p_2, \theta),$$

$$D_2(p_1, p_2, y) = S_2(p_1, p_2),$$

$$|A| = \left(\frac{\partial D_1}{\partial p_1} - \frac{\partial S_1}{\partial p_1} \right) \left(\frac{\partial D_2}{\partial p_2} - \frac{\partial S_2}{\partial p_2} \right) - \left(\frac{\partial D_2}{\partial p_1} - \frac{\partial S_2}{\partial p_1} \right) \left(\frac{\partial D_1}{\partial p_2} - \frac{\partial S_1}{\partial p_2} \right).$$

$|A| > 0$ if demand slopes downward and supply slopes upward and if the product of own price effects dominates the product of cross-price effects. Remaining analysis is exactly as in the text.

5.

$$dp_1^* = \frac{1}{|A|} \left[\frac{\partial D_1}{\partial p_2} \frac{\partial D_2}{\partial y} - \frac{\partial D_1}{\partial y} \left(\frac{\partial D_2}{\partial p_2} - \frac{\partial S_2}{\partial p_2} \right) \right] dy.$$

Assume goods are substitutes and good 2 is inferior. If

$$\left| \frac{\partial D_1}{\partial p_2} \frac{\partial D_2}{\partial y} \right| < \left| \frac{\partial D_1}{\partial y} \left(\frac{\partial D_2}{\partial p_2} - \frac{\partial S_2}{\partial p_2} \right) \right|$$

then $dp_1^* > 0$. If

$$\left| \frac{\partial D_1}{\partial p_2} \frac{\partial D_2}{\partial y} \right| > \left| \frac{\partial D_1}{\partial y} \left(\frac{\partial D_2}{\partial p_2} - \frac{\partial S_2}{\partial p_2} \right) \right|$$

then $dp_1^* < 0$.

$$dp_2^* = \frac{1}{|A|} \left[\frac{\partial D_2}{\partial p_1} \frac{\partial D_1}{\partial y} - \frac{\partial D_2}{\partial y} \left(\frac{\partial D_1}{\partial p_1} - \frac{\partial S_1}{\partial p_1} \right) \right] dy.$$

If

$$\left| \frac{\partial D_2}{\partial p_1} \frac{\partial D_1}{\partial y} \right| > \left| \frac{\partial D_2}{\partial y} \left(\frac{\partial D_1}{\partial p_1} - \frac{\partial S_1}{\partial p_1} \right) \right|$$

then $dp_2^* > 0$. If

$$\left| \frac{\partial D_2}{\partial p_1} \frac{\partial D_1}{\partial y} \right| < \left| \frac{\partial D_2}{\partial y} \left(\frac{\partial D_1}{\partial p_1} - \frac{\partial S_1}{\partial p_1} \right) \right|$$

then $dp_2^* < 0$.

Exercise 4.2

1.

$$D_1(p_1, p_2) \equiv a_1 - b_{11}p_1 - b_{12}p_2$$

$$= \alpha_1 + \beta_{11}p_1 + \beta_{12}p_2 \equiv S_1(p_1, p_2),$$

$$D_2(p_1, p_2) \equiv a_2 - b_{21}p_1 - b_{22}p_2$$
$$= \alpha_2 + \beta_{21}p_1 + \beta_{22}p_2 \equiv S_2(p_1, p_2),$$

$$p_1^* = \frac{(a_1 - \alpha_1)(b_{22} + \beta_{22}) - (a_2 - \alpha_2)(b_{12} + \beta_{12})}{(b_{11} + \beta_{11})(b_{22} + \beta_{22}) - (b_{21} + \beta_{21})(b_{12} + \beta_{12})},$$

$$p_2^* = \frac{(a_2 - \alpha_2)(b_{11} + \beta_{11}) - (a_1 - \alpha_1)(b_{21} + \beta_{21})}{(b_{11} + \beta_{11})(b_{22} + \beta_{22}) - (b_{21} + \beta_{21})(b_{12} + \beta_{12})}.$$

Suppose $p_1^0 > p_1^*$ is imposed:

$$p_2^0 = \frac{(a_2 - \alpha_2) - (b_{21} + \beta_{21})p_1^0}{(b_{22} + \beta_{22})},$$

$$p_2^0 = p_2^* + \Delta p_2 = p_2^* - \frac{(b_{21} + \beta_{21})}{(b_{22} + \beta_{22})}\Delta p_1.$$

If

$$\frac{b_{21} + \beta_{21}}{b_{22} + \beta_{22}} > 0$$

then $p_2^0 < p_2^*$.

$$E_{S_1}(p_1, p_2) = S_1(p_1, p_2) - D_1(p_1, p_2),$$
$$E_{S_1}(p_1^*, p_2^*) = 0,$$
$$E_{S_1}(p_1^0, p_2^*) > 0,$$
$$E_{S_1}(p_1^0, p_2) = (\alpha_1 - a_1) + (b_{11} + \beta_{11})p_1^0 + (\beta_{12} + b_{12})p_2,$$
$$E_{S_1}(p_1^0, p_2^0) - E_{S_1}(p_1^0, p_2^*) = (\beta_{12} + b_{12})(p_2^0 - p_2^*) < 0.$$

2. Let $p_1^0 < p_1^*$ be imposed:

$$p_1^* = \frac{(a_1 - \alpha_1)(b_{22} + \beta_{22}) + (a_2 - \alpha_2)(b_{12} + \beta_{12})}{(b_{11} + \beta_{11})(b_{22} + \beta_{22}) - (b_{21} + \beta_{21})(b_{12} + \beta_{12})},$$

$$p_2^* = \frac{(a_2 - \alpha_2)(b_{11} + \beta_{11}) + (a_1 - \alpha_1)(b_{21} + \beta_{21})}{(b_{11} + \beta_{11})(b_{22} + \beta_{22}) - (b_{21} + \beta_{21})(b_{12} + \beta_{12})},$$

$$p_2^0 = \frac{(a_2 - \alpha_2) + (b_{21} + \beta_{21})p_1^0}{b_{22} + \beta_{22}},$$

$$p_2^* = \frac{(a_2 - \alpha_2) + (b_{21} + \beta_{21})p_1^*}{b_{22} + \beta_{22}}.$$

$$p_2^0 = p_2^* + \Delta p_2 = p_2^* + \frac{(b_{21} + \beta_{21})}{(b_{22} + \beta_{22})}\Delta p_1.$$

Since $\Delta p_1 < 0$, $p_2^0 < p_2^*$.

$$E_{S_1}(p_1^0, p_2^*) < 0,$$

$$E_{S_1}(p_1^0, p_2^0) - E_{S_1}(p_1^0, p_2^*) = -(b_{12} + \beta_{12})(p_2^0 - p_2^*) > 0,$$

therefore $|E_{S_1}(p_1^0, p_2^0)| < |E_{S_1}(p_1^0, p_2^*)|$. Excess demand is reduced, but is still positive.

Chapter 5

Exercise 5.1

2. From (5.6), $p^*(t) = k_1 e^{\delta_1 t}$. Replace p by $p_R = p(t)/P(t) = p(t)/e^{it}P_0$.

$$\frac{p^*(t)}{e^{it}P_0} = k_1 e^{\delta_1 t},$$

$$p^*(t) = k_1 P_0 e^{(\delta_1 + i)t}.$$

3. We only obtain a traditional downward-sloping demand curve in two cases:
   (i) $0 < \alpha g < r$,
   (ii) $\varepsilon r > -\eta \alpha g > 0$. Even in this case, however, if the demand curve is shifting then the points we observe will not lie on the same demand curve.

5. Consider

$$D(p(t), t) = \gamma p^{-\varepsilon} e^{\alpha g t},$$

$$S(p(t), t) = e^{-rt} \beta p^{\eta}.$$

$-r$ is the effect of increase in real wages. Therefore

$$p^*(t) = \left((\gamma/\beta) e^{(\alpha g + r)t} \right)^{\frac{1}{\varepsilon + \eta}} = k_1 e^{\hat{\delta}_1 t},$$

$$q^*(t) = k_2 e^{\hat{\delta}_2 t},$$

$$\hat{\delta}_1 = (\alpha g + r)/(\varepsilon + \eta),$$

$$\hat{\delta}_2 = (\eta \alpha g - \varepsilon r)/(\varepsilon + \eta),$$

$$\hat{p}^* = \hat{\delta}_1,$$

$$\hat{q}^* = \hat{\delta}_2.$$

Then the various cases can be analysed as before.

Exercise 5.2

1. $q_S^t = \alpha + \beta p_t$; $q_D^t = a - b p_{t-1}^e$.

(i)   *naive expectations:* $p_{t-1}^e = p_{t-1} \Rightarrow p_t = (-b/\beta)^t(p_0 - p^*) + p^*$, therefore the market is stable if $|b/\beta| < 1$, otherwise unstable.

(ii) *rational expectations:* N buyers with demands $q_i^t = \hat{a} - \gamma p_{t-1}^e$. If n are naive, r rational, we have

$$q_D^t = n(\hat{a} - \gamma p_{t-1}) + r(\hat{a} - \gamma p_t) = N\hat{a} - \gamma(np_{t-1} + rp_t).$$

Define $a \equiv N\hat{a}$, $b \equiv (n + r)\gamma$, and $s \equiv r/(n + r)$. Then for equilibrium

$$\alpha + \beta p_t = a - b[(1 - s)p_{t-1} + sp_t].$$

Therefore

$$p_t = \left(\frac{-(1 - s)b}{\beta + sb}\right)^t p_{t-1} + p^*,$$

where $p^* = (a - \alpha)/(b + \beta)$. Therefore stable if $|(1 - s)b/(\beta + sb)| < 1$, otherwise unstable.

3. Stability requires $3(1 - s)/(2 + 3s) < 1$, implying $s > 0.1\dot{6}$. In general we require $s > (\beta - b)/2\beta$.

4. Naive expectations model, speed of convergence greater the smaller is β/b. Rational expectations model, speed of convergence determined by $(1 - s)\beta/(b + s\beta) < \beta/b$, therefore convergence is faster for $0 < s \leq 1$.

Chapter 6

Exercise 6.1

1. Take three bundles A, B, C such that A, C lie on indifference curve u_2 and B, C lie on indifference curve u_1; that is, the indifference curves u_1, u_2 intersect at C. Suppose further that A is preferred to B. We now show that this contradicts the assumption of transitivity; for if A is indifferent to C and C is indifferent to B then transitivity requires A and B to be indifferent to each other; or, if A is preferred to B and B is indifferent to C then transitivity requires A to be preferred to C. Thus the assumption that preferences are transitive rules out the possibility that indifference curves (sets) intersect.

2. $-dx_i/dx_j = u_j/u_i = (b_j/b_i)(x_i/x_j)$. If $v = \alpha u^\beta$ then using the 'function of a function' rule:

$$-dx_i/dx_j = v_j/v_i = (\beta \alpha u^{\beta-1} u_j)/(\beta \alpha u^{\beta-1} u_i) = u_j/u_i \quad i, j = 1, ..., n.$$

In general, $-dx_i/dx_j = v_j/v_i = T'u_j/T'u_i = u_j/u_i$, where $v = T[u]$.

4. $u_i > 0$, but u_{ii} *cannot be signed* for an ordinal utility function. For example if $u = x_1^{1/2}x_2^{1/2}$ then $u_{11} = -(1/4)x_1^{-3/2}x_2^{1/2} < 0$, but if $v = u^3$, then $v_{11} = (3/4)x_1^{-1/2}x_2^{3/2} > 0$, and v is just as valid as u, since it is a positive monotonic transformation of u. Thus in ordinal utility theory it is not appropriate to speak of 'diminishing marginal utility'.

5.

$$\frac{d^2 x_2}{dx_1^2} = \frac{d}{dx_1} \left[\frac{-u_1(x_1, x_2)}{u_2(x_1, x_2)} \right]$$

$$= \frac{-1}{u_2^2} \left[\left(u_{11} + u_{12} \frac{dx_2}{dx_1} \right) u_2 - \left(u_{21} + u_{22} \frac{dx_2}{dx_1} \right) u_1 \right]$$

$$= \frac{-1}{u_2^2} \left[u_2 u_{11} - \frac{u_1 u_2 u_{12}}{u_2} - u_1 u_{21} + \frac{u_1^2 u_{22}}{u_2} \right]$$

$$= \frac{-1}{u_2^3} \left[u_2^2 u_{11} - 2 u_1 u_2 u_{12} + u_1^2 u_{22} \right]$$

where we make use of the fact that x_2 is implicitly a function of x_1 with $dx_2/dx_1 = -u_1/u_2$, and that $u_{12} = u_{21}$. For convexity of the indifference curve we require $d^2 x_2/dx_1^2 > 0$, hence, since $u_2^3 > 0$, we require

$$u_2^2 u_{11} - 2 u_1 u_2 u_{12} + u_1^2 u_{22} < 0.$$

6. The more x_1 and less x_2 the consumer has, the greater the increase in x_1 required to compensate for a marginal reduction in x_2.

Exercise 6.2

1. $x_1 = c_1 + s(y - p_1 c_1 - p_2 c_2)/p_1$; $x_2 = c_2 + (1 - s)(y - p_1 c_1 - p_2 c_2)/p_2$ where $s \equiv b_1/(b_1 + b_2)$. To derive this, define $\hat{x}_i = x_i - c_i, i = 1, 2$, and redefine the problem as:

$$\max_{\hat{x}_1, \hat{x}_2} \hat{x}_1^{b_1} \hat{x}_2^{b_2} \quad \text{s.t. } p_1(\hat{x}_1 + c_1) + p_2(\hat{x}_2 + c_2) = y.$$

The c_i can be interpreted as minimum levels of consumption required for subsistence. Note: the main effect of introducing $c_i > 0$ is that $\partial x_1/\partial p_2 < 0$, $\partial x_2/\partial p_1 < 0$, as compared to the 'pure' Cobb-Douglas case.

4. Nothing! Multiplying all prices and income by k leaves the budget constraint and hence the optimal solution unchanged. Thus consumers' demands are unaffected by expressing prices as kp_i and income as ky. We could therefore set $k = 1/p_j$, implying we choose good j as the 'numeraire' with $kp_j = 1$; or set $k = 1/y$, implying we set income at 1; or $k = (\sum_{i=1}^{n} p_i)^{-1}$, implying $0 < kp_i < 1$ for each i. This property of consumer demands is often called 'absence of money illusion'.

Exercise 6.3

1.

$p =$	\$5	\$10	\$15
$e_p =$	1.09	1.05	1.03

Note how market elasticity converges to consumer 1's elasticity of 1 as consumer 2's demand share falls.

2. From (6.29), $e_y = 2(2/3)\delta^1 + 1(1/3)\delta^2$ where $h = 1$ denotes the rich group.

$$\delta^1 = (4/3)(dy^1/dy); \qquad \delta^2 = (4/1)(dy^2/dy).$$

Hence:
(a) $e_y = (4/3)(4/3).1 + (1/3)(4/1).0 = 1.78$;
(b) $e_y = (4/3)(4/3).0 + (1/3)(4/1).1 = 1.33$;
(c) $e_y = (4/3)(4/3)(1/2) + (1/3)(4/1)(1/2) = 1.56$.

3. Summing through (6.27) gives

$$x = \sum_h x^h = \sum_h a^h(p) + b(p) \sum_h y^h = a(p) + b(p)y.$$

Chapter 7

Exercise 7.1

1.
(a) If $v = T[u]$ with $T' > 0$, the Hicksian demand functions become $H_i(p_1, \ldots, p_n, v)$ and the expenditure function becomes $y(p_1, \ldots, p_n, v)$, while the forms of these functions are unchanged.
(b) If prices all increase by a multiple k this leaves the solution of the expenditure minimization problem unchanged, but the expenditure required to achieve a particular utility level increases by a multiple k, i.e., $e(kp_1, \ldots, kp_n, u) = ke(p_1, \ldots, p_n, u)$.

2. *Marshallian*: $x_i = (1/2)y/p_i = 50/p_i$, $\qquad i = 1, 2$. *Hicksian*: $x_i = (p_j u/p_i)^{1/2}$, $i, j = 1, 2$, $i \neq j$. (Take $u = (50)(25) = 1250$ in graphing the Hicksian demand function.)

4. We must have $b_1/(b_1 + b_2) = 0.4$. Set $b_1 + b_2 = 1$. Then $b_1 = 0.4$, $b_2 = 0.6$. Then Hicksian demand and expenditure functions are

$$x_1 = (2/3)^{0.6}(p_2/p_1)^{0.6}u; \quad x_2 = (3/2)^{0.4}(p_1/p_2)^{0.4}u;$$
$$y = [(2/3)^{0.6} + (3/2)^{0.4}]p_1^{0.4}p_2^{0.6}u.$$

Exercise 7.2

1. Hicksian and Marshallian consumer surpluses for good 1 coincide. Take the utility function $u = u_1(x_1) + \alpha x_2$. The marginal rate of substitution is $-dx_2/dx_1 = u_1'(x_1)/\alpha$,

which is independent of x_2, i.e., at given x_1 the slopes of the indifference curves are the same for all x_2.

2.

(a) $C = \$29.3$, $E = \$41.4$; Marshallian CS = $34.7.

(b) $C = -\$41.4$, $E = -\$29.3$; Marshallian CS = $34.7.

3.   Maximum lump sum is $29.29 per consumer. Its total revenue will change from $50,000 to $64,645.

Chapter 8

Exercise 8.1

2.

(i) $f_l(l, k) = ab_1 l^{b_1-1}k^{b_2}$. Therefore, $f_l(\lambda l, \lambda k) = ab_1(\lambda l)^{b_1-1}(\lambda k)^{b_2}$
      $= ab_1 l^{b_1-1}k^{b_2}\lambda^{b_1-1+b_2} = \lambda^{b_1+b_2-1}f_l(l, k)$ and likewise for f_k.

(ii) Total payments to inputs are $f_1 x_1 + f_2 x_2$. This exceeds, equals, or falls short of $f(x_1, x_2)$ according as $n > 1, n = 1$, or $n < 1$. For the Cobb-Douglas case we have

$$f_l l + f_k k = [ab_1 l^{b_1-1}k^{b_2}]l + [ab_2 l^{b_1}k^{b_2-1}]k = (b_1 + b_2)al^{b_1}k^{b_2} = nal^{b_1}k^{b_2}$$

if $b_1 + b_2 = n$.

(iii) $\dfrac{dx_2}{dx_1} = \dfrac{-f_1}{f_2} = \dfrac{-ab_1 x_1^{b_1-1}x_2^{b_2}}{ab_2 x_1^{b_1}x_2^{b_2-1}} = \dfrac{-b_1}{b_2}\dfrac{x_2}{x_1}$ as required. Along a ray x_2/x_1 is constant.

3.

(i) $f_l = ab_1 l^{b_1-1}k^{b_2} > 0, f_{ll} = a(b_1 - 1)l^{b_1-2}k^{b_2} < 0$, if $b_1 < 1$. Similarly for k.

(ii) $f_{lk} = f_{kl} = ab_1 b_2 l^{b_1-1}k^{b_2-1} > 0$.

(iii) Since $f_{ll} < 0, f_{kk} < 0$, and $f_{lk} > 0$, as we have just seen, (8.6) is necessarily satisfied (the key assumption here is $0 < b_1, b_2 < 1$, and the value of $b_1 + b_2$ is irrelevant).

Exercise 8.2

2. $\partial C_1/\partial w = l_1$ (Shephard's lemma). Therefore $(\partial C_1/\partial w)(w/C_1) = wl_1/C_1$. Similarly for k_1.

3. $C_1 = 23.8\, q^{4/3}$, $C_1' = 31.7\, q^{1/3}$, $C_1/q_1 = 23.8\, q^{1/3}$.

Exercise 8.3

1. $C_0 = 3.2\, q^2 + 200$; AVC $= 3.2\, q$; ATC $= 3.2\, q + 200/q$.

Exercise 8.4

2. Long-run costs and short-run costs for production *next period* will both be higher, and the envelope relation continues to hold between them. However, the envelope relation will *not* hold between long-run costs and short-run costs of production *this* period. In case (*a*) short-run costs will be higher than they are now, in case (*b*), lower.

3. $C_1 = (C_1(q_1)/q_1)q_1 \Rightarrow C'_1 = C_1(q_1)/q_1 + q_1 d[C_1(q_1)/q_1]/dq_1$. Likewise $C'_0 = C_0(q_0)/q_0 + q_0 d[C_0(q_0)/q_0]/dq_0$. Tangency at $\bar{q}$ implies $C'_1(\bar{q}) = C'_0(\bar{q})$ and $C_0(\bar{q})/\bar{q} = C_1(\bar{q})/\bar{q}$. Hence

$$\bar{q} d[C_1(\bar{q})/\bar{q}]/dq_1 = \bar{q} d[C_0(\bar{q})/\bar{q}]/dq_0$$

as required.

Chapter 9

Exercise 9.1

1. If q_0^f is supply of firm $f = 1, ..., F$ and $S_0^f(p)$ its supply function, then market supply is simply $q_0 = \sum_{f=1}^{F} q_0^f = \sum_{f=1}^{F} S_0^f(p) = S_0(p)$. If each firm has the same value of $p^0 = $ AVC, then there is a discontinuity in market supply at p^0. It would disappear if there was a continuous distribution of firms' values of p^0 across some interval $[0, \bar{p}^0]$ (for which, strictly, we require an infinity of firms).

2. With n firms, the short-run market supply function is $q = (0.16)np$.

Exercise 9.2

1. In case (*a*), each firm's marginal cost is falling, the firm's profit function is *increasing* with output, so firms expand indefinitely – no perfect competitive equilibrium exists. In such a market, we would expect monopoly or oligopoly to develop. In case (*b*), each firm's marginal cost is constant. If price exceeds marginal cost we have the same result as case (*a*). If price is less than marginal (= average) cost, output is zero. If price equals marginal cost, firms' outputs and the number of firms in the market are indeterminate.

5. $p^* = \$0.67$, $n^* = 1499$, $q^* = 999.\dot{3}$.

Chapter 10

Exercise 10.1

1. $\pi(q^*) = \$58,523.80$ is the maximum bid.

2. If $p = a - bq$, then $R(q) = aq - bq^2$, $R'(q) = a - 2bq$. Hence at $p = 0$, $q = a/b$, while at $R'(q) = 0$, $q = a/2b$. $p - R'(q) = bq$ in this case, and this clearly increases with q.

$R'(q)$ becomes negative at $q > a/2b$, which is permissible since $p > 0$ for q such that $a/2b \leq q \leq a/b$.

3. If $p = aq^{-1}$, $e = 1$, therefore $R'(q) = 0$, i.e., $R(q) = pq = a = constant$ for $q > 0$. Thus profit increases as $q \to 0$, but falls to zero at $q = 0$, so there is a discontinuity in the profit function at this point. Intuitively, the firm should produce the smallest possible output at the highest possible price. The same is true if $p = aq^{-b}$ with $b > 1$. Only if $b < 1$ is there a determinate profit-maximizing output.

4. $R'(q) = p(1 - 1/e) = 0$ if $e = 1$. This is the first-order condition for a maximum of revenue. For profit maximization $R'(q) = C'(q) > 0$, implying $e > 1$ at this point.

5. If $p = \alpha - \beta q$, $C(q) = aq^2$, $\pi(q) = \alpha q - (a + \beta)q^2$. We then have: $q = (\alpha - t)/2(\beta + a)$; $p^* = [\alpha(\beta + 2a) + \beta t]/2(\beta + a)$; $dq^*/dt = -1/2(\beta + a) < 0$; $dp^*/dt = \beta/2(\beta + a) > 0$.

6. The second-order condition gives: $2p' + qp'' < C''$. Since $p' < 0$, this condition could be satisfied if $C'' < 0$. We require the slope of the MR curve to be 'more negative' – greater in absolute value – than that of the marginal cost curve.

7. Problem is: max $R(q)$ s.t. $R(q) - C(q) \geq 0$. If at the optimum q^* we have $R(q^*) > C(q^*)$, then the constraint can be ignored and we have $R'(q^*) = 0$. If on the other hand $R(q^*) = C(q^*)$, then $R(q^*)/q^* = C(q^*)/q^*$ and price equals average cost (sketch the diagrams).

8. Let r be the royalty rate. Author maximizes $rR(q)$ and so would set output q^* such that $R'(q^*) = 0$. Publisher maximizes $(1 - r)R(q) - C(q)$, and so sets output $\hat{q}$ such that $R'(\hat{q}) = C'(\hat{q})/(1 - r) > 0$. This must imply $\hat{q} < q^*$.

9. max $_l \pi(l) = R(l) - w(l)l \Rightarrow R'(l^*) - (w'(l^*)l^* + w(l^*)) = 0$. Rearranging and using $\eta = w/lw'$ gives the result, with $w^* = w(l^*)$.

Exercise 10.2

1. Assume costs are in dollars. The firm seeks to max $_{q_1,q_2} p_1q_1/r + p_2q_2 - (q_1 + q_2)^2$. Using the given demand functions we get:

$$q_1^* = 100(1 - r)/(r + 4);$$
$$p_1^* = (300r + 200)/(r + 4);$$
$$dq_1/dr = -500/(r + 4)^2 < 0;$$
$$q_2^* = 50(2r + 3)/(r + 4);$$
$$p_2^* = (100r + 650)/(r + 4);$$
$$dq_2/dr = 250/(r + 4)^2 > 0.$$

If a tax t is introduced, the problem is max $_{q_1,q_2} p_1q_1/r + p_2q_2 - (q_1 + q_2)^2 - tq_1/r$. Then profit-maximizing outputs are $\hat{q}_1 = q_1^* - t/(r + 4)$; $\hat{q}_2 = q^* + t/2(r + 4)$.

4. The problem is max $_{q_1,q_2}$ $R(q_1+q_2)-C_1(q_1)-C_2(q_2)$, implying the first-order condition

$$R'(q_1^* + q_2^*) = C_1'(q_1^*) = C_2'(q_2^*).$$

5. The problem is max $_{l_1,l_2}$ $R(l_1 + l_2) - w_1(l_1)l_1 - w_2(l_2)l_2$. The first-order conditions imply

$$w_1^*(1 + 1/\eta_1) = w_2^*(1 + 1/\eta_2),$$

and so $\eta_1 < \eta_2 \Rightarrow w_1^* < w_2^*$.

Exercise 10.3

1. $\pi(p) = \int_p^{p^0} (100x - 5x^2)dx - 5\int_p^{p^0}(100 - 5x)dx$. Therefore the first-order condition gives

$$\pi'(p^*) = 5p^{*2} - 125p^* + 500 = 0$$

from which we obtain $p^* = \$5$ (= marginal cost, as we expect). Then, $Q^* = \int_5^{20}(100 - 5x)dx = 562.5$, $\pi(p^*) = \$2812.50$.

2. Now $\pi(p) = (p - 5) \int_p^{p^0}(100 - 5x)dx$. The first-order condition gives

$$\pi'(p) = 7.5 p^{*2} - 225p^* + 1500 = 0$$

from which we obtain $p^* = \$10$, $Q^* = 250$, $\pi(p^*) = \$1250$.

Exercise 10.4

1.
   (a) $\hat{\theta} = 65$. $\pi(\hat{\theta} = 65) = \$8100$; $\pi(\hat{\theta} = 20) = \$7200$.
   (b) $T^* = \$301.20$. Resulting profit is $\$5067$.

Chapter 11

Exercise 11.1

2. The cartel's profit function is: $\pi = aq_1 - b_1q_1^2 + aq_2 - b_2q_2^2 - c_3q_3^2 - c_4q_4^2$ where q_3 and q_4 are the outputs of firms 1 and 2 respectively. The cartel maximizes this subject to the condition: $q_1 + q_2 = q_3 + q_4$. Using the Lagrange approach, we obtain 5 conditions: $a - 2b_kq_k = \lambda$, $k = 1, 2$; $2c_iq_i = \lambda$, $i = 1, 2$; and $q_1 + q_2 - q_3 - q_4 = 0$. Replacing λ with, say, $2c_3q_3$ we obtain the four conditions: $a - 2b_kq_k = 2c_3q_3$, $k = 1, 2$; $2c_4q_4 = 2c_3q_3$, and $q_1 + q_2 - q_3 - q_4 = 0$, to solve for the outputs. Thus we obtain

$$q_i^* = (a/2c_i)(1/b_1 + 1/b_2)/(1/b_1 + 1/b_2 + 1/c_3 + 1/c_4) \qquad i = 3, 4,$$

and

$$q_k^* = (a/2b_k)(1/b_1 + 1/b_2)/(1/b_1 + 1/b_2 + 1/c_3 + 1/c_4) \qquad k = 1, 2.$$

Exercise 11.2

2. The set of points achievable by lump sum redistributions in the numerical example must satisfy the equation $\pi_2 = 1500 - \pi_1$. The set required consists of the points on this line between the end points $(945.25, \ 554.75)$ and $(989.61, \ 510.39)$.

Exercise 11.3

1. Firm 2's best response function is $q_2 = (a - bq_1)/2(b + c_2)$. Inserting this into its profit function gives

$$\pi_2 = \left[a - b \left(q_1 + \frac{(a - bq_1)}{2(b + c_2)} \right) \right] \frac{(a - bq_1)}{2(b + c_2)} - c_2 \left[\frac{(a - bq_1)}{2(b + c_2)} \right]^2.$$

Minimizing this with respect to q_1 gives the first-order condition

$$\frac{d\pi_2}{dq_1} = \frac{b}{2(b + c_2)}(bq_1 - a) = 0 \ \Rightarrow \ q_1 = a/b,$$

as required. Intuitively, if firm 1 produces $q_1 < a/b$, there is some $q_2 > 0$ which yields positive profit. A similar result holds when 2 minimaxes 1.

2. Firm i minimaxes j by producing output $q_i = (a - c)/b$, i.e. by driving price down to c, each firm's marginal cost. In that case the best j can do is produce zero output and earn zero profit – any positive output would generate losses.

Chapter 12

Exercise 12.1

1. $\max_l (24 - l - \bar{x})^a (m + wl - \bar{y})^b \ \Rightarrow \ l = [bw(24 - \bar{x}) - a(m - \bar{y})]/w(a + b).$
   $\partial l/\partial w = a(a + b)(m - \bar{y})/[w(a + b)]^2; \quad \partial l/\partial m = -a/w(a + b).$

5. Using the Slutsky equation and 'function of a function' rule, $\partial S/\partial t = (\partial S/\partial w)\,(\partial w/\partial t) = (\partial H/\partial w + l\,\partial S/\partial m)\partial w/\partial t = -w_0(\partial H/\partial w + l\,\partial S/\partial m) \gtrless 0$ since $\partial S/\partial m < 0$.

6. The income target can be expressed by the condition $m + wl = y^0$, with $y^0 > m$ (why?). Therefore $l = (y^0 - m)/w \equiv S(w, m, y^0). \quad \partial S/\partial m = -1/w < 0. \quad \partial S/\partial w = -(y^0 - m)/w^2 < 0.$ The labour supply curve is a rectangular hyperbola.

7. The required sum is given by the difference in expenditure function values,

$$m(w_A, u_A) - m(w_B, u_A) = 24(w_B^{b/(a+b)} - w_A^{b/(a+b)}) + (\bar{y} + \bar{x})(w_B^{a/(a+b)} - w_A^{a/(a+b)})/$$
$$(w_A w_B)^{a/(a+b)}.$$

Exercise 12.2

1. Since $q = f(l(p, w, v), k(p, w, v))$, we have

$$\partial q/\partial w = (\partial f/\partial l)(\partial l/\partial w) + (\partial f/\partial k)(\partial k/\partial w) < 0,$$
$$\partial q/\partial v = (\partial f/\partial l)(\partial l/\partial v) + (\partial f/\partial k)(\partial k/\partial v) < 0,$$
$$\partial q/\partial p = (\partial f/\partial l)(\partial l/\partial p) + (\partial f/\partial k)(\partial k/\partial p) > 0,$$

from (12.17)–(12.19).

3. Given the first-order condition $pf_l(l, \bar{k}) = w$, in the Cobb-Douglas case we have

$$l = (pb_1 \bar{k}^{b_2}/w)^{1/(1-b_1)}.$$

Market equilibrium conditions are:

$$p^{-\alpha} = n_0 \bar{k}^{b_2} (w/b_1 p \bar{k}^{b_2})^{b_1/(b_1-1)} \qquad \text{(output market)},$$

$$w^\beta = n_0 (w/b_1 p \bar{k}^{b_2})/^{1/(b_1-1)} \qquad \text{(labour market)}.$$

This system is linear in logs. Using Cramer's Rule therefore gives:

$$\log p^* = [(1 + \beta)(b_1 - 1)\log n_0 - b_2(1 + \beta)\log \bar{k} - \beta b_1 \log b_1]/[\alpha(1 + \beta) + b_1\beta(1 - \alpha)],$$

$$\log w^* = [\alpha(1 - b_1)\log n_0 - b_1(1 - \alpha)\log \bar{k} + \alpha \log b_1]/[\alpha(1 + \beta) + b_1\beta(1 - \alpha)].$$

4. First-order condition becomes $p(1 - 1/e)f_l = w$. In linear case, marginal revenue product $= (a - 2bq)f_l$ while marginal value product is $(a - bq)f_l$. Equating each of these to given wage rate results in lower l in former case, since $f_{ll} < 0$ at equilibrium.

6. The condition for strict concavity of the Cobb-Douglas function becomes

$$(b_1(b_1 - 1)l^{b_1-2}k^{b_2})(b_2(b_2 - 1)l^{b_1}k^{b_2-2}) - [b_1 b_2 l^{b_1-1}k^{b_2-1}]^2 > 0.$$

The left-hand side is equal to $b_1 b_2(b_1 - 1)(b_2 - 1)(l^{b_1-1}k^{b_2-1})^2$, and so cancelling terms gives $(b_1 - 1)(b_2 - 1) > b_1 b_2$ implying $b_1 + b_2 < 1$. If $b_1 + b_2 \geq 1$, the firm's long-run equilibrium output and labour demand do not exist.

Exercise 12.3

1. The efficient wage and profit pair is: $w = \$374.70$, $L = 500.4$. The point on the labour demand curve at which profit $= \$100$ is $w = \$490$, $L = 20$.

2. Solving the problem: max $w^{3/4}L^{1/4}$ s.t. $L^{\alpha} - wL = 0$ ultimately gives the condition

$$L^{\alpha}(1 - 3\alpha/2) = 0.$$

But this has no solution with $L > 0$. The reason is easy to see. The profit constraint gives $w = L^{\alpha-1}$. Thus $d^2w/dL^2 = (\alpha - 1)(\alpha - 2)L^{\alpha-3} > 0$ if $\alpha < 1$. The profit contour is convex, not concave.

Chapter 13

Exercise 13.1

1. Demand functions: $y_1 = a(\bar{y}_1 + \bar{y}_2/(1+r))$; $y_2 = (1 - a)(\bar{y}_1(1+r) + \bar{y}_2)$. $b = y_1 - \bar{y}_1 = (a - 1)\bar{y}_1 + a\bar{y}_2/(1+r)$.

Exercise 13.2

1. $b = sk = \alpha(r+\delta)^{(b_1-1)/\beta}s^{-b_2/\beta}p^{1/\beta}w^{-b_1/\beta}$ where $\beta \equiv 1 - (b_1 + b_2)$, $\alpha \equiv (b_1^{b_1}b_2^{1-b_1})^{1/\beta}$.

2. $db = s\,dI + (k - k_0)ds$ from (13.24). From (13.22), $\partial k/\partial r = s/pf_{kk} < 0$. Hence with $ds = 0$, $\partial b/\partial r = s\,\partial I/\partial r = s\,\partial k/\partial r < 0$. Again from (13.22), $\partial k/\partial s = (r + \delta)/pf_{kk} < 0$. Hence $\partial I/\partial s < 0$ but we may have $(k - k_0) > 0$ and so $\partial b/\partial s \gtrless 0$.

Chapter 14

Exercise 14.1

1. There are n consumers, m goods, and F firms, with x_{ij} the jth consumer's consumption of the ith good, x_{if} the fth firm's output of the ith good, and l_{if} the fth firm's use of labour in producing good i. The wage rate is w and the given output prices p_i. Then the equilibrium condition is

$$\sum_{i=1}^{m}\sum_{f=1}^{F} l_{if}(w^*, p_i) = nL$$

where the firm's labour demands satisfy the conditions

$$p_i\partial h_i^f(l_{if}^*)/\partial l_{if} = w \qquad i = 1, \ldots, m, \quad f = 1, \ldots, F$$

with h_i^f firm f's production function for good i. Consumers' incomes are

$$y_j^* = w^* L + \sum_{f=1}^{F} \theta_{fj} \pi_f^* \qquad j = 1, \ldots, n$$

where π_f is firm f's profit, θ_{fj} is j's shareholding in firm f. Summing these conditions over consumers, recalling that $\sum_j \theta_{fj} = 1$, and assuming each consumer's budget constraint $\sum_i p_i x_{ij}^* = y_j$ is satisfied, we have

$$\sum_j \sum_i p_i x_{ij}^* = \sum_j y_j^* = \sum_i \sum_f p_i x_{if}^* \Rightarrow \sum_i p_i e_i^* = 0$$

where $e_i^* = (\sum_j x_{ij}^* - \sum_f x_{if}^*)$ is net imports or exports of the ith good.

2. $w^* = \$10.77$; $l_1^* = 86.2$; $l_2^* = 13.8$; $x_1^* = 92.85$; $x_2^* = 29.71$; $y_1^* = \$1467.04$; $y_2^* = \$687.07$; $e_1^* = -47.58$; $e_2^* = 95.17$.

4. Labour demands fall to $l_1 = 69.44$, $l_2 = 11.11$, thus there is unemployment in this economy. Outputs fall to $x_1 = 83.33$, $x_2 = 26.67$. The economy is below its production possibility curve. Incomes fall to $y_1 = 1316.62$, $y_2 = 616.68$. Consumer 1's utility falls from 51.87 to 46.55, consumer 2's from 32.92 to 29.53. Consumers are all made better off by reducing the wage to its equilibrium. Further improvements in utility (at given world prices) can only be made by increasing labour productivity, thus shifting outward the production possibility curve.

5.

$$dw^*/dp_1 = -\partial l_1/\partial p_1/(\partial l_1/\partial w^* + \partial l_2/\partial w^*) > 0,$$
$$dw^*/dL = 2/(\partial l_1/\partial w^* + \partial l_2/\partial w^*) < 0,$$
$$\partial l_1^*/\partial p_1 = -f_l^1/p_1 f_{ll}^1 > 0.$$

6. $dx_2/dx_1 = -\phi_x^1(x_1)/\phi_x^2(x_2)$. So, $d^2 x_2/dx_1^2 = -(\phi_x^2 \phi_{xx}^1 - (\phi_x^1 \phi_{xx}^2)(dx_2/dx_1))/[\phi_x^2]^2$. Since

$$\phi_{xx}^i = -f_{ll}^i/[f_l^i]^3 > 0, \quad i = 1, 2,$$

we have $d^2 x_2/dx_1^2 < 0$ as required.

Exercise 14.2

1. With m goods and n consumers, Walras's Law is $\sum_{i=1}^{m} p_i \sum_{j=1}^{n} (x_{ij} - \bar{x}_{ij}) = 0$. Choose good n as the numeraire, and define $r_i \equiv p_i/p_n$, $i = 1, \ldots, n - 1$. Then we take $n - 1$ demand = supply conditions to determine the $n - 1$ relative prices. If each good is traded sequentially *at equilibrium prices*, then (14.38) becomes: $p_n^* \sum_{i=1}^{n} r_i^* (\sum_j (x_{ij}^* - \bar{x}_{ij})) = nM^*$ where $r_n^* = 1$. This then determines the *absolute* level of equilibrium prices given M^* and the equilibrium *relative* prices r_i^*.

2.

    (*a*) $r^* = 3/7$.

    (*b*) $x_{11}^* = 65/3$, $x_{21}^* = 65/7$; $x_{12}^* = 25/3$, $x_{22}^* = 75/7$. Therefore MRS $= x_{21}^*/x_{11}^* = x_{22}^*/3x_{12}^* = 3/7$.

    (*c*) $T^* = (3/7)(65/3 - 60/3) + (75/7 - 70/7) = 1.43$. Therefore $p_2^* = \$1398.60$, $p_1^* = \$599.40$.

3. Because a Marshallian demand is unchanged when all prices are multiplied by λ, i.e.

$$D_i(p_1, p_2, v) = D_i(\lambda p_1, \lambda p_2, \lambda v) = \lambda^0 D_i.$$

5. Velocity $= 1$. We have $p_2^*[r^*(x_{11}^* - \bar{x}_{11}) + (x_{22}^* - \bar{x}_{22})] = M^*$.

Exercise 14.3

4.

    (*a*) $\theta_{11} = 1$, $\theta_{22} = 1$, $\theta_{21} = 0$, $\theta_{12} = 0$. $r_1^* = 8.16$, $r_2^* = 75.21$.

    (*b*) $\theta_{ij} = 1/2$, $i,j = 1, 2$. $r_1^* = 9.24$, $r_2^* = 60.62$.

    (*c*) $\theta_{11} = 1$, $\theta_{21} = 1$, $\theta_{12} = \theta_{22} = 0$. $r_1^* = 16.07$, $r_2^* = 60.62$.

6. The wage rate has dimension: dollars/hour of work; a price has dimension dollars/unit of the good. Therefore price/wage has dimension (dollars/unit of good)/(dollars/hour of work) = hour of work/unit of good, or the amount of working time it takes to buy a unit of the good.

Chapter 15

Exercise 15.1

1. The equation of the utility possibility frontier is $u_1 = 70.71 - u_2$.

2. Initial equilibrium price ratio is $p_1/p_2 = 1$. $u_1 = 2, u_2 = 8$. The desired utility distribution (6,4) can be achieved by transferring all 8 units of 2's initial endowment of good 2 to consumer 1, and announcing relative prices $p_1/p_2 = 1$.

Exercise 15.2

1. Along *the Marshallian* demand curve market demand increases to 77,000 when price falls by \$1, so the increase in Marshallian consumer surplus is \$73,500. Along their individual *Hicksian* demand curves, type 1 consumers increase demand to 10.98, type 2 to 23.9, and type 3 to 41.6. Thus their CVs are \$10.49, \$21.95, and \$40.80 respectively. Hence aggregate CV is \$73,240. Thus the Marshallian measure over-states this by 0.35%.

Exercise 15.3

1.

$$(1/2)[p(\hat{q}) - C'(\hat{q})](q^* - \hat{q}) = (1/2)[(a - \alpha) - (b + \beta)\hat{q}](q^* - \hat{q})$$
$$= (1/2)[(a - \alpha)(q^* - \hat{q})$$
$$+ (b + \beta)q^*(q^* - \hat{q}) - (b + \beta)(q^{*2} - \hat{q}^2)].$$

But $a - bq^* = \alpha + \beta q^*$, so the above becomes

$$(1/2)[(a - \alpha)(q^* - \hat{q}) + (a - \alpha)(q^* - \hat{q}) - (b + \beta)(q^{*2} - \hat{q}^2)]$$

which gives the required result.

2. $\partial LM/\partial b = \partial LM/\partial \beta = -(1/2)(q^{*2} - \hat{q}^2) < 0.$

3. Since monopoly output is undefined, we cannot define a welfare loss.

4. $LM = \int_{1.90}^{2.51}(10q^{-1/2} - q^2)dq = 1.13.$ $(1/2)(7.25 - 3.61)(2.51 - 1.90) = 1.11.$

Exercise 15.4

1. $EB(t) = \int_{q(t)}^{q^*}[a - bq - \beta q]dq = a(q^* - q(t)) - (1/2)(b + \beta)(q^{*2} - q^2(t)).$

Exercise 15.5

1. From section 3.2 we have: $q_i(t_i) = (a_i\beta_i - b_i\beta_i t_i)/(b_i + \beta_i)$. Using (15.56) we then obtain $t_1 = (a_1 b_2/a_2 b_1)t_2$. Then substitute for t_1 in (15.55), solve for t_2, and then solve for t_1.

2. If $T > 0$, then we require positive taxes, i.e., $p_i > C'_i$, and so (15.54) $\Rightarrow \lambda \neq 0$. We cannot have $\partial T/\partial t_i = q_i + t_i q'_i < 0$ at an optimum, since this would imply a reduction in t_i would increase T *and* increase welfare. Then (15.54) gives $\lambda = t_i q'_i/(q_i + t_i q'_i) < 0.$

3. Simply, T is greater than the sum of the maximum tax revenues which could be raised in these markets.

Exercise 15.6

1.
   (a) $q_1^* = 10$, $q_2^* = 25$; $\hat{q}_1 = 8$, $\hat{q}_2 = 25.$
   (b) $t^* = 2.$
   (c) $r^* = 2$, $G_1 = G_2 = 2$, $G_3 = 32$, $G_4 = 0.$
   (d) $q_1 = 6$, $q_2 = 25.$

Exercise 15.7

1.

   (*a*) $125.

   (*b*) $100 − $25 = $75.

   (*c*) $t' = \$7.75$. Welfare loss = $60.06.

Exercise 15.8

1.

   (*a*) $100 - 2q_t = 10 \Rightarrow q_t = 45$, $3 \times 45 = 135 > 80$.

   (*b*) $q_0^* = 28.38$, $q_1^* = 26.72$, $q_2^* = 24.9$, $p_0^* = \$43.23$, $p_1^* = \$46.56$, $p_2^* = $
     50.21.

   (*c*) $\rho_0^* = \$33.23$, $\rho_1^* = \$36.56$, $\rho_2^* = \$40.21$.

$$(36.56 - 33.23)/33.23 = 0.1; \quad (40.21 - 36.56)/36.56 = 0.1.$$

   (*d*) (ii) $q_t^* = 26.67$, ρ_t^* is constant.

2. $(T + 1)\hat{q} > R$.

Chapter 16

Exercise 16.3

1. Let v_i be the value added by firm i. Then we have

   (i)

$$v_A = 5000 - 3000 = 2000,$$
$$v_B = 500 - 300 = 300,$$
$$v_C = 6000 - 2000 = 4000,$$

   so total value added is $v_A + v_B + v_C = 6300$. Since there is complete depreciation,
   this is also *net* value added, or national income.

   (ii)

$$v_A = 5000 - 3000 = 2000,$$
$$v_B = 500 - 200 = 300$$
$$, v_C = 6000 - 500 = 5500,$$

   so total net value added, or national income, is 7800.

2.

   (i)

$$v_p = 1500,$$
$$v_i = 2500 - 1000 = 1500,$$
$$v_f = 5000 - 500 - 0.5(2500) = 3250,$$

   so total net value added is 6250.

3.

$$Y \equiv C + I + G + X - Q$$

but

$$Y - C \equiv S + T = I + G + X - Q$$

and so $S + T + Q \equiv I + G + X$.

5. For the GDP deflator, use equation (16.14):

$$GDPD = \frac{(6 \times 30) + (5 \times 20) + (2 \times 6)}{(5 \times 30) + (3 \times 20) + (4 \times 6)}$$

$$= \frac{292}{234} = 1.2478.$$

For the consumer price index, use equation (16.15):

$$CPI = \frac{(6 \times 14) + (5 \times 10) + (2 \times 5)}{(5 \times 14) + (3 \times 10) + (4 \times 5)}$$

$$= \frac{144}{120} = 1.2.$$

6. Use the data in Table 16.4.

Chapter 17

Exercise 17.3

3. Banks must satisfy *three* demands for money: customers' demands for loans, customers' demands for cash, and the banks' cash requirements. For a given nominal volume of money, the amount available for loans must go down if cash requirements increase. Since loans determine the *effective* volume of money, a reduction in loans reduces the effective volume of money.

Exercise 17.4

1.
   (a) Increased leisure preference affects the *equilibrium* by reducing labour supply. A larger proportion of workers will choose not to work. The natural rate will increase.
   (b) Labour-saving technology in one sector need not increase unemployment. However, if there is no job-creation of a suitable type in other sectors, there will be some structural unemployment and an increase in the natural rate, since aggregate labour demand has fallen. Alternatively, if labour becomes cheaper relative to capital, then again there is no need for the natural rate to increase.

(c) A reduction in exports will reduce the demand for labour in the export goods industries. If wages do not adjust downwards then there will be some demand-deficient unemployment. The natural rate need not change.

Chapter 18

Exercise 18.1

1. Simply differentiate the expression in (18.10) with respect to I, a, and b in turn. For example,

$$\frac{dC^*}{db} = \frac{I(1-b) + (a+bI)}{(1-b)^2} > 0.$$

An increase in the marginal propensity to consume increases equilibrium consumption for two reasons: more is consumed out of each unit of income and more income is generated in equilibrium (see (18.12)). In terms of Fig. 18.2 the slope of the consumption function increases at each level of income but the intercept remains unchanged.

3.

(i) At least one of a and α greater than zero and $b + \beta > 1$, or both a and α equal to zero and $b + \beta < 1$.

These are linear functions and so there is at most one equilibrium.

4.

$$Y^* = \frac{a + \alpha}{1 - b - \beta}, \quad \frac{dY^*}{da} = \frac{1}{1 - b - \beta}.$$

Exercise 18.2

1. In this case,

$$\frac{dY}{dr} = \frac{I'(r) - S_r(Y, r)}{S_Y(Y, r)},$$

so $S_r > 0$ is sufficient to make $dY/dr < 0$. If I also depends on Y, then

$$\frac{dY}{dr} = \frac{I_r(Y, r) - S_r(Y, r)}{S_Y(Y, r) - I_Y(Y, r)}$$

so that, if $I_Y > 0$, we require $S_Y > I_Y$ for $dY/dr < 0$.

Exercise 18.3

2.

$$r^* = \frac{kY - M}{l}; \quad \frac{dr^*}{dk}\bigg|_Y = \frac{Y}{l} > 0, \quad \frac{dr^*}{dM}\bigg|_Y = \frac{1}{l} > 0$$

and

$$\frac{dr^*}{dl}\bigg|_Y = \frac{-kY + M}{l^2} < 0.$$

An increase in M and k increases the interest rate at each income level, while an increase in l lowers the LM curve.

$$\frac{dr^*}{dY} = \frac{k}{l} > 0; \qquad \frac{d}{dk}\left(\frac{dr^*}{dY}\right) = \frac{1}{l} > 0$$

and

$$\frac{d}{dl}\left(\frac{dr^*}{dY}\right) = \frac{-k}{l^2} < 0.$$

So an increase in k increases the slope of LM and an increase in l reduces the slope. An increase in M has no effect on the slope of LM.

Exercise 18.4

2. Simply differentiate (18.44) and (18.45) with respect to a, b, h, k, and l. For example,

$$\frac{dY^*}{dh} = \frac{[M(1-b) - k(a+e)]/l}{[1 - b + kh/l]^2} < 0 \quad \text{and} \quad \frac{dr^*}{dh}$$

$$= \frac{-k[k(a+e) - M(1-b)]}{[l(1-b) + kh]^2} < 0.$$

[Note that $k(a+e) - M(1-b) > 0$ if the equilibrium interest rate is positive (by 18.45).] An increase in the interest sensitivity of the demand for money reduces equilibrium income and reduces the equilibrium interest rate.

Chapter 19

Exercise 19.1

1. $C^* = a + b(Y^* - T)$ and differentiate with respect to G and T, remembering that Y^* depends on G and T, as in (19.7). Thus

$$\frac{dC^*}{dG} = b\frac{dY^*}{dG} = \frac{b}{1-b} > 0,$$

$$\frac{dC^*}{dT} = b\frac{dY^*}{dT} - b = \frac{-b^2}{1-b} - b = \frac{-b}{1-b} < 0.$$

[Can you explain why $dC^*/dT = dY^*/dT$?]

3. Simply differentiate (19.17) with respect to a, b, and t. For example, ·

$$\frac{dY^*}{dt} = \frac{-b(a + I + G)}{[1 - b(1-t)]^2} < 0.$$

Again,

$$C^* = a + b(Y^* - T)$$

where Y^* depends on a, I, G, b, and t as in (19.17) and $T = tY$. So,

$$C^* = a + b(1 - t)Y^*$$

and, for example,

$$
\begin{aligned}
\frac{dC^*}{dt} &= -bY^* + b(1 - t)\frac{dY^*}{dt} \\
&= \frac{-bY^*[1 - b(1 - t)]}{[1 - b(1 - t)]} - \frac{b^2(1 - t)Y^*}{[1 - b(1 - t)]} \\
&= \frac{-bY^*}{[1 - b(1 - t)]} < 0.
\end{aligned}
$$

4.

$$C = C(Y - T + R)$$

and

$$Y^* = C(Y^* - T + R) + I + G$$

(note that R is a transfer and is not part of government expenditure on goods and services),

$$
dY^*\{1 - C'(Y^* - T + R)\} \\
= dR\{C'(Y^* - T + R)\}.
$$

So,

$$
\begin{aligned}
\frac{dY^*}{dR} &= \frac{C'(Y^* - T + R)}{1 - C'(Y^* - T + R)} \\
&< \frac{1}{1 - C'(Y^* - T + R)} = \frac{dY^*}{dG}.
\end{aligned}
$$

If $C = C(Y - T(Y) + R)$, then $Y^* = C(Y^* - T(Y^*) + R)$ and

$$
dY^*\{1 - C'(Y^* - T(Y^*) + R) + C'(Y^* - T(Y^*) + R)T'(Y^*)\} \\
= dR\{C'(Y^* - T(Y^*) + R)\}.
$$

Again, it is easily checked that $dY^*/dR < dY^*/dG$ in this case. So an increase in government direct expenditure on goods and services is more effective than an equivalent increase in family support in its effect on increasing income. This is because an increase in direct government spending has a full impact on output during the 'first round' of the multiplier process, whereas part of family income support is saved in the 'first round'.

Exercise 19.3

1.

$$dT = \frac{\partial T}{\partial T_0} dT_0 + \frac{\partial T}{\partial t} dt + \frac{\partial T}{\partial Y} dY$$

$$= \frac{\partial T}{\partial T_0} dT_0 + Y dt + t \, dY$$

Note that (19.31) still holds, and $\partial T / \partial T_0 = 1$, so,

$$dY = dT_0 + Y dt + t \, dY$$

and the required change in the tax rate is

$$dt = \frac{dY}{Y}(1 - t) - \frac{dT_0}{Y}$$

where $dY = dG$ is known. Now suppose $dt = 0$, then

$$dT_0 = dG(1 - t).$$

2. The government budget is given by (19.21):

$$D = G - T(Y^*),$$

so

$$\frac{dD}{dG} = 1 - T'(Y^*)\frac{dY^*}{dG}$$

where dY^* / dG is given by (19.25). Since both $T'(Y^*)$ and dY^*/dG are positive fractions,

$$dD/dG > 0$$

Exercise 19.4

1. Clearly,
$$Y^* = \frac{a + bR + e + G - hr}{1 - b(1 - t)},$$

$$\frac{dY^*}{dr} = \frac{-h}{1 - b(1 - t)}.$$

Thus, as before, the slope of the IS curve is affected by b and h. Exogenous transfers do not affect the slope of the IS curve, but an increase in the tax rate makes the IS curve steeper. [An increase in t makes dY^*/dr *less* negative, but dr/dY^* *more* negative.]

Chapter 20

Exercise 20.1

1. Simply differentiate (20.8) with respect to each of these parameters in turn. For example,

$$\frac{d\Delta}{dl} = \frac{-kh[1 - b(1 - t)]/l^2}{[1 - b(1 - t) + kh/l]^2} < 0.$$

3. Since $S'(.) > 0$ and $L_r(.,.) < 0$, the numerator is positive. Since $L_Y(.,.) > 0$ and $I'(.) < 0$, the denominator is positive. The denominator is smaller than the numerator since $-L_Y I'$ is positive.

4.

$$S(Y - T(Y) + R) + T(Y) - R = I(r) + G,$$

$$M = L(Y, r),$$

$$\begin{bmatrix} S'(.)[1 - T'(Y)] + T'(Y) & -I'(r) \\ L_Y(Y, r) & L_r(Y, r) \end{bmatrix} \begin{bmatrix} dY/dR \\ dr/dR \end{bmatrix} = \begin{bmatrix} (1 - S') \\ 0 \end{bmatrix},$$

$$\frac{dY}{dR} = |A|^{-1} \begin{vmatrix} (1 - S') & -I'(r) \\ 0 & L_r \end{vmatrix}$$

$$= \frac{L_r(1 - S')}{|A|} > 0,$$

$$\frac{dr}{dR} = |A|^{-1} \begin{vmatrix} S'[1 - T'] + T' & (1 - S') \\ L_Y & 0 \end{vmatrix}$$

$$= \frac{-L_Y(1 - S')}{|A|} > 0$$

(recall $|A| < 0$).

Exercise 20.2

1. In steady state $G + B = T$ so $dG + dB = t(dY + dB)$ and

$$dB = \left[t\frac{dY}{dG} - 1 \right] dG.$$

Exercise 20.4

1. Use (18.44), with b replaced by $b(1 - t)$ for the model with income tax,

$$\frac{dY}{dM} = \frac{h/l}{1 - b(1 - t) + kh/l} > 0,$$

so, for example,

$$\frac{d}{dh}\left(\frac{dY}{dM}\right) = \frac{-h/l^2[1 - b(1 - t)]}{[1 - b(1 - t) + kh/l]^2} < 0.$$

(Recall that an increase in h makes the IS curve less steep.)

2. The availability of more cash means that demand for money needs to be increased; this is achieved by asset substitution between bonds and money brought about by a fall in interest rates.

4. If the liquidity preference schedule was perfectly elastic at some interest rate. This is a case known as the 'liquidity trap'.

Exercise 20.5

1.

$$S(Y - T(Y) + R(Y)) + T(Y) = I(r) + G + R(Y),$$
$$M = L(Y, r);$$

$$\begin{bmatrix} S'(.)[1 - T'(Y) + R'(Y)] + T'(Y) - R'(Y) & -I'(r) \\ \\ -L_Y(Y, r) & -L_r(Y, r) \end{bmatrix} \begin{bmatrix} dY \\ \\ dr \end{bmatrix} = \begin{bmatrix} dG \\ \\ 0 \end{bmatrix}.$$

So,

$$\frac{dY}{dG} = \frac{-L_r(Y, r)}{-\{S'(.)[1 - T'(Y) + R'(Y)] + T'(Y) - R'(Y)\}L_r(Y, r) - I'(r)L_Y(Y, r)}$$

and substituting into (20.47) gives:

$$-L_r(Y, r)[T'(Y) - R'(Y)] < -\{S'(.)[1 - T'(Y) + R'(Y)] + T'(Y) - R'(Y)\}L_r(Y, r) - I'(r)L_Y(Y, r)$$

and dividing by $-L_r$ gives, in place of (20.48),

$$S'(Y - T(Y) + R(Y))[1 - T'(Y) + R'(Y)] + I'(r)\frac{L_Y(Y, r)}{L_r(Y, r)} > 0.$$

Chapter 21

Exercise 21.1

1. Replace (21.5) with

$$S(Y - T(Y)) + T(Y) = I(r) + \frac{g}{P},$$

$$\frac{m}{P} = L(Y, r);$$

$$\begin{bmatrix} S'(Y - T(Y))[1 + T'(Y)] + T'(Y) & -I'(r) \\ \\ L_Y(Y, r) & L_r(Y, r) \end{bmatrix} \begin{bmatrix} dY \\ \\ dr \end{bmatrix} = \begin{bmatrix} -(g/P^2) \\ \\ -(m/P^2) \end{bmatrix} dP,$$

$$\frac{dY}{dP} = \frac{1}{|A|} \left\{ -(g/P^2)L_r(Y, r) - I'(r)(m/P^2) \right\} < 0,$$

since $L_r < 0, I' < 0$, and $|A| < 0$.

2. Replace M in (20.1) with m/P to give

$$Y = \frac{a + e + G + hm/Pl}{1 - b(1 - t) + kh/l}.$$

So,

$$\frac{dY}{dP} = \frac{-hm/P^2 l}{1 - b(1 - t) + kh/l} < 0.$$

Now, differentiate this with respect to b, t, k, h, and l in turn. For example,

$$\frac{d}{db} \left(\frac{dY}{dP} \right) = \frac{-(1 - t)hm/P^2 l}{[1 - b(1 - t) + kh/l]^2} < 0.$$

Thus an increase in b makes the aggregate demand curve flatter. Demand is more responsive to a price fall the higher is the marginal propensity to consume. Slope of AD also depends on M.

Exercise 21.2

1.

$$\frac{d}{dP} \left(\frac{dY}{dP} \right) = \frac{2PI'(r)m/P^4}{|A|} > 0$$

so the demand curve is convex.

Chapter 22

Exercise 22.1

1.
$$\frac{dY}{dN} = \alpha AN^{\alpha-1} = \frac{W}{P},$$

so

$$N = \left[\frac{W}{P}\frac{1}{\alpha A}\right]^{\frac{1}{\alpha-1}}.$$

2. If $\alpha = 0.5$, $A = 10$,

$$N = \left[\frac{1}{5}\left(\frac{W}{P}\right)\right]^{-2}$$

$$= \left[\frac{5P}{W}\right]^2.$$

If $W/P = 1$, $N = 25$ million; if $W/P = 2$, $N = 6.25$ million.

Exercise 22.2

2. Increased preferences for leisure. Any factor which increases reservation wages in the economy.

3. According to (22.11) labour supply would increase by a proportion determined by the elasticity of labour supply

$$\eta_s = \frac{N'\left(\frac{W}{P}\right)}{N\left(\frac{W}{P}\right)}\left(\frac{W}{P}\right)$$

as long as prices remain fixed.

Exercise 22.3

1. (*a*) The labour supply curve shifts to the right and increases aggregate supply at each price level.

Exercise 22.4

3.
$$\frac{dr}{dG} = |A|^{-1}L_Y(Y,r)m/P^2 > 0, \qquad \frac{dr}{dm} = 0,$$

$$\frac{dr}{d\gamma} = |A|^{-1}\{-f_N N_\gamma\}\{S'(Y-T(Y))[1-T'(Y)]+T'(Y)\}m/P^2 < 0,$$

$$\frac{dr}{d\lambda} = |A|^{-1}\{-f_N N_\lambda - f_\lambda\}\{S'(Y - T(Y))[1 - T'(Y)] + T'(Y)\}m/P^2 < 0.$$

4. Straightforward application of Cramer's Rule.

Chapter 23

Exercise 23.2

3. Suppose prices have been unchanged for some time so that $P_f^e = P_W^e$ and

$$N_s\left(\frac{W_0}{P_w^e}\right) = N_d\left(\frac{W_0}{P_f^e}\right).$$

First, think of an actual price increase, so P_f^e and P_w^e increase but $\Delta P_f^e < \Delta P_w^e$, thus

$$N_s\left(\frac{W_0}{P_w^e}\right) < N_d\left(\frac{W_0}{P_f^e}\right).$$

(*a*) If money wages are flexible upwards, then W increases from W_0 but only until the excess demand perceived by firms is eliminated. Labour expects an even higher wage increase but this will not be forthcoming. In the new equilibrium

$$N_s\left(\frac{W_1}{P_w^e}\right) < N_d\left(\frac{W_1}{P_f^e}\right)$$

where W_1 is the wage which eliminates the excess demand as perceived by firms. Further wage increases are resisted by firms. An increase in the actual price level therefore *reduces* employment and output.

(*b*) If money wages are flexible downwards, then a fall in the actual price creates an excess supply of labour. The excess supply as perceived by workers is larger than that perceived by firms. The wage falls until the excess supply as seen by labour is eliminated. Again employment is determined by the 'short-side' (demand in this case) as labour resists further wage reductions.

(*c*) If money wages are rigid downward the effect in (*b*) is even stronger.

Chapter 24

Exercise 24.3

1.

(i) Using (24.32), setting $dG = 0$ and using Cramer's Rule, gives

$$\frac{dY}{dm} = |A|^{-1} I'(r) \left[\frac{1}{P} + \frac{N'(P/P^e)f'(N)}{P^e} \right] > 0,$$

$$\frac{dr}{dm} = |A|^{-1} \{S'(Y - T(Y))[1 - T'(Y)] + T'(Y)]\} \left[\frac{1}{P} + \frac{N'(P/P^e)f'(N)}{P^e} \right] < 0,$$

$$\frac{dP}{dm} = |A|^{-1} \frac{I'(r)}{P} > 0.$$

(ii) Using (24.30) and setting $dG = 0$ gives

$$\frac{dY}{dm} = 0,$$

$$\frac{dr}{dm} = 0,$$

$$\frac{dP}{dm} = \frac{P}{m} > 0.$$

2. If the objective of a reduction in the money supply is to reduce the price level, then the government would choose to announce the policy.

Chapter 25

Exercise 25.1

2. Imperfect information and transportation costs.

Exercise 25.2

1.

$$Y = \left\{ 1 - b + q + \frac{ek}{l} \right\}^{-1} \left\{ h + G + X - bT + \frac{eM}{l} \right\}$$

and

$$\tilde{Y} = \frac{X}{q},$$

so $\tilde{Y} < Y \Rightarrow$

$$X \left\{ 1 - b + \frac{ek}{l} \right\} < q \left\{ h + G - bT + \frac{eM}{l} \right\}.$$

2.

$$\tilde{Y} = \frac{X}{q} = \frac{200}{0.4} = 500,$$

$$Y = \frac{h + G + X - bT + eM/l}{1 - b + q + ek/l}$$

$$= \frac{100 + 200 + 200 - 0.8(100) + 60/0.5}{1 - 0.8 + 0.4 + 0.6/0.5}$$

$$= \frac{540}{1.8} = 300$$

There is a balance of trade surplus. Y needs to be increased by 200. Therefore

$$\Delta Y = \frac{\Delta G}{1.8}$$

or $\Delta G = 200 \times 1.8 = 360$.

4.

$$Y_i = C_i + I_i + Q_j - Q_i \qquad i \neq j.$$

So,

$$Y_1 = b_1 Y_1 + I_1 + q_2 Y_2 - q_1 Y_1,$$

$$= \frac{I_1 + q_2 Y_2}{1 - b_1 + q_1}.$$

Similarly,

$$Y_2 = \frac{I_2 + q_1 Y_1}{1 - b_2 + q_2}.$$

Solving for Y_1 and Y_2 gives

$$Y_1 = \left\{ 1 - b_1 + q_1 - \frac{q_1 q_2}{1 - b_2 + q_2} \right\}^{-1} \left\{ I_1 + \frac{q_2 I_2}{1 - b_2 + q_2} \right\},$$

$$Y_2 = \left\{ 1 - b_2 + q_2 - \frac{q_1 q_2}{1 - b_1 + q_1} \right\}^{-1} \left\{ I_2 + \frac{q_1 I_1}{1 - b_1 + q_1} \right\}.$$

From these we see that

$$\frac{dY_1}{dI_1} = \left\{ 1 - b_1 + q_1 - \frac{q_1 q_2}{1 - b_2 + q_2} \right\}^{-1},$$

$$\frac{dY_2}{dI_1} = \left\{ 1 - b_2 + q_2 - \frac{q_1 q_2}{1 - b_1 + q_1} \right\}^{-1} \left\{ \frac{q_1}{1 - b_1 + q_1} \right\}.$$

The signs of these are ambiguous in general. Initially, both countries have a balance on the balance of trade. So

$$Q_1 = Q_2,$$

$$q_1 Y_1 = q_2 Y_2,$$

and

$$\frac{dQ_i}{dI_1} = q_i \frac{dY_i}{dI_1} \quad \text{for } i = 1, 2.$$

Chapter 26

Exercise 26.2

2. Consider the following linear system:

$$(1 - b)(1 - t)Y + tY + qY = h - er + G + X$$

(savings + taxes + imports) = (investment + government spending + exports),

$$0 = X - qY + \kappa r$$

= (exports − imports + capital inflows).

We obtain

$$\begin{bmatrix} (1 - b)(1 - t) + t + q & e \\ \\ -q & \kappa \end{bmatrix} \begin{bmatrix} dY \\ dr \end{bmatrix} = \begin{bmatrix} dG \\ 0 \end{bmatrix}$$

and so

$$\frac{dY}{dG} = \frac{\kappa}{\kappa[(1 - b)(1 - t) + t + q] + eq} > 0.$$

4.

$$S(Y - T(Y)) + T(Y) + Q(Y, \varepsilon) = I(r) + G + X(\varepsilon),$$
$$0 = NX(Y, \varepsilon, Y_f) + K(r),$$

so

$$\frac{dY}{dY_f} = D^{-1} \begin{vmatrix} 0 & -I'(r) \\ \\ -NX_{Y_f}(Y, \varepsilon, Y_f) & K_r(r, \varepsilon) \end{vmatrix} < 0,$$

$$\frac{dr}{dY_f} = D^{-1} \begin{vmatrix} S'(.)[1 - T'(.)] + T'(.) + Q_Y & 0 \\ \\ NX_Y(Y, \varepsilon, Y_f) & -NX_{Y_f}(Y, \varepsilon, Y_f) \end{vmatrix}.$$

Reserves must fall. (Draw the diagram.)

Exercise 26.3

1.

$$\frac{dY}{dM} = \Delta^{-1}\left\{I'(.)BS_\varepsilon - NX_\varepsilon K_r\right\} > 0,$$

$$\frac{dr}{dM} = \Delta^{-1}\left\{\lambda BS_\varepsilon + NX_\varepsilon NX_Y\right\} < 0,$$

$$\frac{d\varepsilon}{dM} = \Delta^{-1}\left\{-\lambda K_r - I'(.)NX_Y\right\} < 0,$$

Exercise 26.4

3.

$$M = L(Y, r_f),$$

$$0 = NX(Y, \varepsilon) + K(r_f, \varepsilon),$$

so

$$\frac{d\varepsilon}{dr_f} = \Delta^{-1}\begin{vmatrix} L_Y & L_{r_f} \\ \\ NX_Y & BS_{r_f} \end{vmatrix}$$

$$> 0,$$

where $\Delta = L_Y BS_\varepsilon < 0$.

Chapter 27

Exercise 27.1

1.
  (i) Constant: exchange rate, interest rate, domestic credit. Changing: terms of trade (deteriorating), price (falling), reserves (falling), demand (increasing).
  (ii) Constant: domestic credit, reserves, interest rate. Changing: price (falling), demand (increasing), terms of trade;

$$\frac{d\theta}{dP} = -\frac{\theta}{P} - \frac{\theta}{\varepsilon}\frac{d\varepsilon}{dP}$$

where

$$\frac{d\varepsilon}{dP} = \frac{-L_Y BS_\theta \theta/P - NX_Y(m/P^2)}{-L_Y BS_\varepsilon}$$

$$= \frac{\varepsilon}{P} + \frac{NX_Y(m/P^2)}{L_Y BS_\varepsilon}$$

$$> 0,$$

so $d\theta/dP < 0$ and so as P falls we observe a deterioration in the terms of trade.

Exercise 27.3

1. (*a*) (short run)

$$\frac{dY}{dr_f} = \{-f'(.)N'(.)/P^e I'(.)\} D^{-1} < 0,$$

$$\frac{dP}{dr_f} = I'(.)D^{-1},$$

where $D = -f'(.)N'(.)/P^e \{S'(.)[1 - T'(.)] - NX_Y(Y, \theta)\} - NX_\theta(Y, \theta)\theta/P$.

3.

$$\frac{dY}{d\alpha} = \Delta^{-1}f_\alpha(m/P^2)\, BS_\varepsilon > 0.$$

Exercise 27.4

1.

$$\frac{dY}{d\Pi_f} = D^{-1}\{-f_V V_{\Pi_f} NX_\theta \theta/P\} < 0,$$

$$\frac{dP}{d\Pi_f} = D^{-1}\{S'(.)[1 - T'(.)] - NX_Y\}f_V V_{\Pi_f} > 0,$$

where D is given in the answer to question 1 of Exercise 27.3.

Chapter 28

Exercise 28.1

1. Straightforward. Notice that the adjustment path is again in a counter-clockwise path, with output initially depressed below Y_0.

Exercise 28.2

1. (*a*) to find μ, solve (1) and (2) to give

$$\mu = \pi^e + \ln U_0 + \ln U_{-1} - 2 \ln U$$
$$= 0 + 5 + 5 - 2(4) = 2.$$

So in period 1: $\mu = 2$, $\ln U = 4$ and, using (2) $\pi = 1$. Again using (1) and (2), we can solve for $\ln U$ and π given $\mu = 2$, $\pi^e = 0$, $\ln U_0 = 5$, and the appropriate value of $\ln U_{-1}$:

$$\ln U = 0.5[(\pi_{-1} - \mu) + \ln U_0 + \ln U_{-1}]$$
$$= 0.5[(1 - 2) + 5 + 4] = 4 \quad \text{(period 2)},$$
$$\pi = 2 + (4 - 4) = 2 \quad \text{(period 2)};$$
$$\ln U = 0.5[(2 - 2) + 5 + 4] = 4.5 \quad \text{(period 3)},$$
$$\pi = 2 + (4.5 - 5) = 1.5 \quad \text{(period 3)};$$
$$\ln U = 0.5[(1.5 - 2) + 5 + 4.5] = 4.5 \quad \text{(period 4)},$$
$$\pi = 2 + (4.5 - 4.5) = 2 \quad \text{(period 4)}.$$

(*b*) The new equilibrium is reached some time after period 4 when $\ln U = \ln U_{-1} = 4$, and $\pi = \pi^e = \mu = 2$.

2. and 3. Use similar methods to question 1. Note that 'long-run' means, in this case, $\ln U = \ln U_{-1}$.

Exercise 28.4

1. Since (28.30) is always satisfied at the optimum, W is clearly independent of p. Differentiating (28.29) gives

$$\frac{dN}{dp} = -\frac{f'(eN)e}{pe^2 f''(eN)} > 0.$$

2. Profit is now given by

$$\pi = pf(eN) - N(W + T(W))$$

where $W + T(W)$ are total labour costs, and the analysis proceeds as before. Now, assuming that $T'(W) < 0$ raises the possibility that an increase in the wage lowers the average labour cost by reducing turnover.

Chapter 29

Exercise 29.1

1.

$$Y = k^\alpha N^{1-\alpha} \qquad 0 < \alpha < 1,$$
$$F_k = \alpha k^{\alpha-1} N^{1-\alpha} > 0, \quad F_N = (1 - \alpha)k^\alpha N^{-\alpha} > 0,$$
$$F_{kk} = \alpha(\alpha - 1)k^{\alpha-2}N^{1-\alpha} < 0, \quad F_{NN} = -\alpha(1 - \alpha)k^\alpha N^{-\alpha-1} < 0,$$
$$F_{Nk} = F_{kN} = \alpha(1 - \alpha)k^{\alpha-1}N^{-\alpha} > 0.$$

Along an isoquant Y is constant, so

$$dY = dk\, F_k + dN F_N = 0$$

or, the slope of the isoquant is

$$\frac{dk}{dN} = -\frac{F_N}{F_k} = \frac{(1-\alpha)}{\alpha}\frac{k}{N}$$

and the change in the slope is

$$\frac{d}{dN}\left(\frac{dk}{dN}\right) = -\frac{F_{NN}F_k - F_{kN}F_N}{F_k^2} = -\frac{(1-\alpha)}{\alpha}\frac{k}{N^2} < 0.$$

Exercise 29.2

1. From (29.31) $\hat{Y} = n$, independent of α, so $d\hat{Y}/d\alpha = 0$. Substitute (29.32) in (29.26) giving

$$s\bar{k}^\alpha - n\bar{k} = 0,$$

and so

$$\frac{d\bar{k}}{d\alpha} = -\frac{s\bar{k}^\alpha \log \bar{k}}{\alpha s \bar{k}^{\alpha-1} - n} > 0$$

(see text for why the denominator of this expression must be negative). Using (29.32),

$$\frac{d\bar{y}}{d\alpha} = \alpha \frac{d\bar{k}}{d\alpha}^{\alpha-1} + \bar{k}^\alpha \log \bar{k} > 0.$$

So an increase in capital's share of output increases the steady-state capital stock and output per head.

2. Aggregate consumption is

$$C = (1-s)Y$$

and steady-state consumption per head is

$$\bar{c} = (1-s)\bar{y}.$$

From which,

$$\frac{d\bar{c}}{ds} = -\bar{y} + (1-s)\frac{d\bar{y}}{ds} \gtrless 0$$

$$\frac{d\bar{c}}{dn} = (1-s)\frac{d\bar{y}}{dn} < 0 \quad \text{from (29.43).}$$

Index